Student Solutions Manual

for
Gustafson and Frisk's

Intermediate Algebra

Seventh Edition

Michael G. Welden
Rock Valley College

THOMSON

BROOKS/COLE

Australia • Canada • Mexico • Singapore • Spain • United Kingdom • United States

Printed in Canada
  2   3   4   5   6   7   08 07 06 05

Printer: Webcom Limited

ISBN: 0-534-46346-0

For more information about our products, contact us at:
Thomson Learning Academic Resource Center
1-800-423-0563

For permission to use material from this text or product, submit a request online at
http://www.thomsonrights.com.
Any additional questions about permissions can be submitted by email to **thomsonrights@thomson.com.**

Thomson Brooks/Cole
10 Davis Drive
Belmont, CA 94002-3098
USA

Asia
Thomson Learning
5 Shenton Way #01-01
UIC Building
Singapore 068808

Australia/New Zealand
Thomson Learning
102 Dodds Street
Southbank, Victoria 3006
Australia

Canada
Nelson
1120 Birchmount Road
Toronto, Ontario M1K 5G4
Canada

Europe/Middle East/South Africa
Thomson Learning
High Holborn House
50/51 Bedford Row
London WC1R 4LR
United Kingdom

Latin America
Thomson Learning
Seneca, 53
Colonia Polanco
11560 Mexico D.F.
Mexico

Spain/Portugal
Paraninfo
Calle/Magallanes, 25
28015 Madrid, Spain

Preface

This manual contains detailed solutions to all of the odd exercises of the text *Intermediate Algebra*, seventh edition, by R. David Gustafson and Peter D. Frisk. It also contains solutions to all chapter summary, chapter test, and cumulative review exercises found in the text.

Many of the exercises in the text may be solved using more than one method, but it is not feasible to list all possible solutions in this manual. Also, some of the exercises may have been solved in this manual using a method that differs slightly from that presented in the text. There are a few exercises in the text whose solutions may vary from person to person. Some of these solutions may not have been included in this manual. For the solution to an exercise like this, the notation "answers may vary" has been included.

Please remember that only reading a solution does not teach you how to solve a problem. To repeat a commonly used phrase, mathematics is not a spectator sport. You MUST make an honest attempt to solve each exercise in the text without using this manual first. This manual should be viewed more or less as a last resort. Above all, DO NOT simply copy the solution from this manual onto your own paper. Doing so will not help you learn how to do the exercise, nor will it help you to do better on quizzes or tests.

I would like to thank the members of the mathematics faculty at Rock Valley College and Rebecca Subity of Brooks/Cole Publishing Company for their help and support. This solutions manual was prepared using EXP 5.1.

This book is dedicated to my parents, Ed and Carol Welden, for their constant encouragement and support throughout my life.

May your study of this material be successful and rewarding.

Michael G. Welden

Contents

SECTION 1.1

Exercise 1.1 (page 11)

1. $\dfrac{6}{8} = \dfrac{\cancel{2} \cdot 3}{\cancel{2} \cdot 4} = \dfrac{3}{4}$

3. $\dfrac{32}{40} = \dfrac{\cancel{8} \cdot 4}{\cancel{8} \cdot 5} = \dfrac{4}{5}$

5. $\dfrac{1}{4} \cdot \dfrac{3}{5} = \dfrac{1 \cdot 3}{4 \cdot 5} = \dfrac{3}{20}$

7. $\dfrac{2}{3} \div \dfrac{3}{7} = \dfrac{2}{3} \cdot \dfrac{7}{3} = \dfrac{2 \cdot 7}{3 \cdot 3} = \dfrac{14}{9}$

9. $\dfrac{5}{9} + \dfrac{4}{9} = \dfrac{5+4}{9} = \dfrac{9}{9} = 1$

11. $\dfrac{2}{3} + \dfrac{4}{5} = \dfrac{2 \cdot 5}{3 \cdot 5} + \dfrac{4 \cdot 3}{5 \cdot 3} = \dfrac{10}{15} + \dfrac{12}{15} = \dfrac{22}{15}$

13. set **15.** even **17.** natural; 1; itself **19.** 0

21. rational **23.** $<$ **25.** $\approx$ **27.** natural: $1, 2, 9$

29. integer: $-3, 0, 1, 2, 9$ **31.** irrational number: $\sqrt{3}$ **33.** even natural number: 2

35. prime number: 2 **37.** odd composite number: 9

39.

41.

43. $\dfrac{7}{8} = 7 \div 8 = 0.875$; terminating

45. $-\dfrac{11}{15} = -11 \div 15 = -0.7\overline{3}$; repeating

47. $5 < 9$ **49.** $-5 > -10$ **51.** $-7 < 7$ **53.** $6 > -6$

55. $19 > 12 \Rightarrow 12 < 19$ **57.** $-6 \leq -5 \Rightarrow -5 \geq -6$

59. $5 \geq -3 \Rightarrow -3 \leq 5$ **61.** $-10 < 0 \Rightarrow 0 > -10$

63. $\{x | x > 3\} \Rightarrow$

65. $\{x | x \leq 7\} \Rightarrow$

67. $[-5, \infty) \Rightarrow$

69. $\{x | 2 < x < 5\} \Rightarrow$

71. $[-6, 9] \Rightarrow$

73. $\{x | x < -3 \text{ or } x > 3\} \Rightarrow$

75. $(-\infty, -6] \cup [5, \infty) \Rightarrow$

77. $|20| = 20$

1

79. $-|-6| = -(+6) = -6$ **81.** $|-5| + |-2| = 5 + 2 = 7$ **83.** $|-5| \cdot |4| = 5 \cdot 4 = 20$

85. $|x| = 3 \Rightarrow x = 3$ or $x = -3$ **87.** If $x = |x|$, then $x \geq 0$.

89.   **91.** Answers may vary.

93. Answers may vary.

95. If $|x| < 50$, then $x \in \{-49, -48, -47, ..., -1, 0, 1, ..., 47, 48, 49\}$. There are 99 integers.

97. Answers may vary.

Exercise 1.2 (page 24)

1. $\{x | x > 4\} \Rightarrow$

3. $(2, 10] \Rightarrow$

5. Cost of gasoline $= 32(1.29) = 41.28$; Cost of oil $= 3(1.35) = 4.05$
Tax on oil $= 0.05(4.05) = 0.2025 \approx 0.20$
Total cost $= 41.28 + 4.05 + 0.20 = 45.53 \Rightarrow$ The total cost is \$45.53.

7. absolute; common   **9.** change; add   **11.** negative

13. mean; median; mode   **15.** $(a \cdot b) \cdot c = a \cdot (b \cdot c)$ **17.** $a(b + c) = ab + ac$

19. 1   **21.** $-3 + (-5) = -(3 + 5) = -8$

23. $-7 + 2 = -(7 - 2) = -5$ **25.** $-3 - 4 = -3 + (-4) = -7$

27. $-33 - (-33) = -33 + (+33) = 0$

29. $-2(6) = -12$

31. $-3(-7) = 21$

33. $\dfrac{-8}{4} = -2$

35. $\dfrac{-16}{-4} = 4$

37. $\dfrac{1}{2} + \left(-\dfrac{1}{3}\right) = \dfrac{3}{6} + \left(-\dfrac{2}{6}\right) = \dfrac{1}{6}$

39. $\dfrac{1}{2} - \left(-\dfrac{3}{5}\right) = \dfrac{5}{10} + \dfrac{6}{10} = \dfrac{11}{10}$

41. $\dfrac{1}{3} - \dfrac{1}{2} = \dfrac{2}{6} - \dfrac{3}{6} = -\dfrac{1}{6}$

43. $\left(-\dfrac{3}{5}\right)\left(\dfrac{10}{7}\right) = -\dfrac{30}{35} = -\dfrac{6}{7}$

45. $\dfrac{3}{4} \div \left(-\dfrac{3}{8}\right) = \dfrac{3}{4}\left(-\dfrac{8}{3}\right) = -\dfrac{24}{12} = -2$

47. $-\dfrac{16}{5} \div \left(-\dfrac{10}{3}\right) = -\dfrac{16}{5}\left(-\dfrac{3}{10}\right) = \dfrac{48}{50} = \dfrac{24}{25}$

49. $3 + 4 \cdot 5 = 3 + 20 = 23$

51. $3 - 2 - 1 = 1 - 1 = 0$

53. $3 - (2 - 1) = 3 - 1 = 2$

55. $2 - 3 \cdot 5 = 2 - 15 = -13$

57. $8 \div 4 \div 2 = 2 \div 2 = 1$

59. $8 \div (4 \div 2) = 8 \div 2 = 4$

61. $2 + 6 \div 3 - 5 = 2 + 2 - 5 = 4 - 5 = -1$

63. $(2 + 6) \div (3 - 5) = 8 \div (-2) = -4$

65. $\dfrac{3(8 + 4)}{2 \cdot 3 - 9} = \dfrac{3(12)}{6 - 9} = \dfrac{36}{-3} = -12$

67. $\dfrac{100(2 - 4)}{1{,}000 \div 10 \div 10} = \dfrac{100(-2)}{100 \div 10} = \dfrac{-200}{10}$
$$= -20$$

69. $\text{mean} = \dfrac{7 + 5 + 9 + 10 + 8 + 6 + 6 + 7 + 9 + 12 + 9}{11} = \dfrac{88}{11} = 8$

71. In order: $5, 6, 6, 7, 7, 8, 9, 9, 9, 10, 12$. The mode is 9.

73. In order: $8, 8, 10, 12, 12, 12, 14, 16, 16, 23, 23, 26$. The mode is 12.

75. $ab + cd = (3)(-2) + (-1)(2)$
$= -6 + (-2) = -8$

77. $a(b + c) = 3[-2 + (-1)] = 3(-3) = -9$

79. $\dfrac{ad + c}{cd + b} = \dfrac{3(2) + (-1)}{-1(2) + (-2)} = \dfrac{6 + (-1)}{-2 + (-2)}$
$$= \dfrac{5}{-4} = -\dfrac{5}{4}$$

81. $\dfrac{ac - bd}{cd - ad} = \dfrac{3(-1) - (-2)(2)}{(-1)(2) - 3(2)} = \dfrac{-3 - (-4)}{-2 - 6}$
$$= \dfrac{1}{-8} = -\dfrac{1}{8}$$

83. $C = \dfrac{N(N - 1)}{2} = \dfrac{200(200 - 1)}{2} = \dfrac{200(199)}{2} = \dfrac{\cancel{2} \cdot 100(199)}{\cancel{2}} = 19{,}900 \text{ comparisons}$

85. $P = a + b + c = (23.5 + 37.2 + 39.7) \text{ ft} = 100.4 \text{ ft}$

87. commutative property of addition   **89.** distributive property

91. additive identity property   **93.** multiplicative inverse property

95. associative property of addition   **97.** commutative property of multiplication

Problems 99-101 are to be solved using a calculator. The keystrokes needed to solve each problem using a TI-83 graphing calculator appear in each solution. There may be other solutions. Keystrokes for other calculators may be slightly different.

99. $\boxed{(}\ \boxed{3}\ \boxed{7}\ \boxed{.}\ \boxed{9}\ \boxed{+}\ \boxed{2}\ \boxed{5}\ \boxed{.}\ \boxed{2}\ \boxed{)}\ \boxed{+}\ \boxed{1}\ \boxed{4}\ \boxed{.}\ \boxed{3}\ \boxed{\text{ENTER}}\ \{77.4\}$

$\boxed{3}\ \boxed{7}\ \boxed{.}\ \boxed{9}\ \boxed{+}\ \boxed{(}\ \boxed{2}\ \boxed{5}\ \boxed{.}\ \boxed{2}\ \boxed{+}\ \boxed{1}\ \boxed{4}\ \boxed{.}\ \boxed{3}\ \boxed{)}\ \boxed{\text{ENTER}}\ \{77.4\}$

associative property of addition

101. $\boxed{2}\ \boxed{.}\ \boxed{7}\ \boxed{3}\ \boxed{\times}\ \boxed{(}\ \boxed{4}\ \boxed{.}\ \boxed{5}\ \boxed{3}\ \boxed{4}\ \boxed{+}\ \boxed{5}\ \boxed{7}\ \boxed{.}\ \boxed{1}\ \boxed{2}\ \boxed{)}\ \boxed{\text{ENTER}}$

$\{168.31542\}$

$\boxed{2}\ \boxed{.}\ \boxed{7}\ \boxed{3}\ \boxed{\times}\ \boxed{4}\ \boxed{.}\ \boxed{5}\ \boxed{3}\ \boxed{4}\ \boxed{+}\ \boxed{2}\ \boxed{.}\ \boxed{7}\ \boxed{3}\ \boxed{\times}\ \boxed{5}\ \boxed{7}\ \boxed{.}\ \boxed{1}\ \boxed{2}$

$\boxed{\text{ENTER}}\ \{168.31542\}$ distributive property

103. $(+22.25) + (+39.75) = +62$
He earned \$62.

105. $(+17) + (-13) = +4$
The temperature rose $4°$.

107. $-3(-4) = +12$
It was $12°$ warmer 3 hours ago.

109. $-(5 \cdot 60)(+23) = -6900$
There were 6900 fewer gallons of water.

The keystrokes needed to solve problems 111-119 using a TI-83 graphing calculator appear in each solution. There may be other solutions. Keystrokes for other calculators may be slightly different.

111. $\boxed{(-)}\ \boxed{2}\ \boxed{3}\ \boxed{0}\ \boxed{0}\ \boxed{+}\ \boxed{1}\ \boxed{7}\ \boxed{5}\ \boxed{0}\ \boxed{+}\ \boxed{1}\ \boxed{8}\ \boxed{7}\ \boxed{5}\ \boxed{\text{ENTER}}\ \{1325.\}$
The army gained 1325 meters.

113. $\boxed{4}\ \boxed{3}\ \boxed{7}\ \boxed{.}\ \boxed{3}\ \boxed{7}\ \boxed{+}\ \boxed{1}\ \boxed{2}\ \boxed{5}\ \boxed{.}\ \boxed{1}\ \boxed{8}\ \boxed{+}\ \boxed{1}\ \boxed{3}\ \boxed{7}\ \boxed{.}\ \boxed{2}\ \boxed{6}\ \boxed{+}\ \boxed{1}$

$\boxed{4}\ \boxed{5}\ \boxed{.}\ \boxed{5}\ \boxed{6}\ \boxed{-}\ \boxed{1}\ \boxed{1}\ \boxed{7}\ \boxed{.}\ \boxed{1}\ \boxed{1}\ \boxed{-}\ \boxed{1}\ \boxed{8}\ \boxed{3}\ \boxed{.}\ \boxed{4}\ \boxed{9}\ \boxed{-}\ \boxed{1}\ \boxed{2}$

$\boxed{2}\ \boxed{.}\ \boxed{8}\ \boxed{9}\ \boxed{\text{ENTER}}\ \{421.88\}$ Her ending balance is \$421.88.

115. $\boxed{(}\ \boxed{1}\ \boxed{5}\ \boxed{2}\ \boxed{5}\ \boxed{+}\ \boxed{7}\ \boxed{8}\ \boxed{5}\ \boxed{+}\ \boxed{1}\ \boxed{6}\ \boxed{2}\ \boxed{8}\ \boxed{+}\ \boxed{1}\ \boxed{2}\ \boxed{1}\ \boxed{4}\ \boxed{+}\ \boxed{9}$

$\boxed{1}\ \boxed{7}\ \boxed{+}\ \boxed{1}\ \boxed{1}\ \boxed{9}\ \boxed{7}\ \boxed{)}\ \boxed{\div}\ \boxed{6}\ \boxed{\text{ENTER}}\ \{1211.\}$ The mean of daily sales is \$1211.

117. $\boxed{(}\ \boxed{7}\ \boxed{5}\ \boxed{+}\ \boxed{8}\ \boxed{2}\ \boxed{+}\ \boxed{8}\ \boxed{7}\ \boxed{+}\ \boxed{8}\ \boxed{0}\ \boxed{+}\ \boxed{7}\ \boxed{6}\ \boxed{)}\ \boxed{\div}\ \boxed{5}\ \boxed{\text{ENTER}}\ \{80.\}$
His mean score is 80.

119. $\boxed{(}\,\boxed{1}\,\boxed{0}\,\boxed{0}\,\boxed{0}\,\boxed{0}\,\boxed{0}\,\boxed{+}\,\boxed{4}\,\boxed{\times}\,\boxed{1}\,\boxed{0}\,\boxed{0}\,\boxed{0}\,\boxed{0}\,\boxed{0}\,\boxed{)}\,\boxed{\div}\,\boxed{5}\,\boxed{\text{ENTER}}$

$\{28000.\}$ The ad is misleading. Although the mean wage of all workers including the businessman is $28,000 as stated, a better measure of the average would be the median wage, which is $10,000.

121. $P = 4s = 4(7.5)\,\text{cm} = 30\,\text{cm}$ **123. Answers may vary.**

125. The mean increases by 7 as well. This property is always true.

127. The median is often the most appropriate average to use when one or two extremely high or low values occur. The following are situations when computing the median might be appropriate:
- finding an average salary of all employees when there are a few highly paid employees
- finding an average household income for a city
- finding an average score of all students on a test when there is one extremely high score

Exercise 1.3 (page 36)

1. $a + b + c = 4 + (-2) + 5 = 2 + 5 = 7$

3. $\dfrac{ab + 2c}{a + b} = \dfrac{4(-2) + 2(5)}{4 + (-2)} = \dfrac{-8 + 10}{2} = \dfrac{2}{2} = 1$

5. base; exponent   **7.** x^{m+n} **9.** $x^n y^n$ **11.** 1

13. x^{m-n} **15.** $A = s^2$ **17.** $A = \frac{1}{2}bh$ **19.** $A = \pi r^2$

21. $V = lwh$ **23.** $V = Bh$ **25.** $V = \frac{1}{3}Bh$

27. base $= 5$, exponent $= 3$ **29.** base $= x$, exponent $= 5$

31. base $= b$, exponent $= 6$ **33.** base $= -mn^2$, exponent $= 3$

35. $3^2 = 3 \cdot 3 = 9$ **37.** $-3^2 = -1 \cdot 3^2 = -1 \cdot 3 \cdot 3 = -9$

39. $(-3)^2 = (-3)(-3) = 9$ **41.** $5^{-2} = \dfrac{1}{5^2} = \dfrac{1}{25}$

43. $-5^{-2} = -1 \cdot 5^{-2} = -\dfrac{1}{5^2} = -\dfrac{1}{25}$ **45.** $(-5)^{-2} = \dfrac{1}{(-5)^2} = \dfrac{1}{25}$

47. $8^0 = 1$ **49.** $(-8)^0 = 1$

51. $(-2x)^5 = (-2)^5 x^5 = -32x^5$ **53.** $(-2x)^6 = (-2)^6 x^6 = 64x^6$

55. $x^2 x^3 = x^{2+3} = x^5$ **57.** $k^0 k^7 = 1 \cdot k^7 = k^7$

59. $x^2 x^3 x^5 = x^{2+3+5} = x^{10}$ **61.** $p^9 p p^0 = p^9 p^1 \cdot 1 = p^{9+1} = p^{10}$

5

63. $aba^3b^4 = a^{1+3}b^{1+4} = a^4b^5$

65. $(-x)^2y^4x^3 = (-1)^2x^2y^4x^3 = x^{2+3}y^4 = x^5y^4$

67. $\left(x^4\right)^7 = x^{4\cdot7} = x^{28}$

69. $\left(b^{-8}\right)^9 = b^{-8\cdot9} = b^{-72} = \dfrac{1}{b^{72}}$

71. $\left(x^3y^2\right)^4 = \left(x^3\right)^4\left(y^2\right)^4 = x^{3\cdot4}y^{2\cdot4} = x^{12}y^8$

73. $\left(r^{-3}s\right)^3 = \left(r^{-3}\right)^3s^3 = r^{-3\cdot3} = r^{-9}s^3 = \dfrac{s^3}{r^9}$

75. $\left(a^2a^3\right)^4 = \left(a^{2+3}\right)^4 = \left(a^5\right)^4 = a^{5\cdot4} = a^{20}$

77. $\left(-d^2\right)^3\left(d^{-3}\right)^3 = (-1)^3\left(d^2\right)^3d^{-3\cdot3} = -1\cdot d^{2\cdot3}d^{-9} = -1\cdot d^6d^{-9} = -1\cdot d^{6+(-9)} = -d^{-3} = -\dfrac{1}{d^3}$

79. $\left(3x^3y^4\right)^3 = 3^3\left(x^3\right)^3\left(y^4\right)^3 = 27x^{3\cdot3}y^{4\cdot3} = 27x^9y^{12}$

81. $\left(-\tfrac{1}{3}mn^2\right)^6 = \left(-\tfrac{1}{3}\right)^6m^6\left(n^2\right)^6 = \tfrac{1}{729}m^6n^{2\cdot6} = \tfrac{1}{729}m^6n^{12}$

83. $\left(\dfrac{a^3}{b^2}\right)^5 = \dfrac{\left(a^3\right)^5}{\left(b^2\right)^5} = \dfrac{a^{15}}{b^{10}}$

85. $\left(\dfrac{a^{-3}}{b^{-2}}\right)^{-2} = \dfrac{\left(a^{-3}\right)^{-2}}{\left(b^{-2}\right)^{-2}} = \dfrac{a^6}{b^4}$

87. $\dfrac{a^8}{a^3} = a^{8-3} = a^5$

89. $\dfrac{c^{12}c^5}{c^{10}} = \dfrac{c^{17}}{c^{10}} = c^{17-10} = c^7$

91. $\dfrac{m^9m^{-2}}{\left(m^2\right)^3} = \dfrac{m^7}{m^6} = m^{7-6} = m^1 = m$

93. $\dfrac{1}{a^{-4}} = a^4$

95. $\dfrac{3m^5m^{-7}}{m^2m^{-5}} = \dfrac{3m^{-2}}{m^{-3}} = 3m^{-2-(-3)} = 3m$

97. $\left(\dfrac{4a^{-2}b}{3ab^{-3}}\right)^3 = \left(\dfrac{4bb^3}{3aa^2}\right)^3 = \left(\dfrac{4b^4}{3a^3}\right)^3 = \dfrac{4^3\left(b^4\right)^3}{3^3\left(a^3\right)^3} = \dfrac{64b^{12}}{27a^9}$

99. $\left(\dfrac{3a^{-2}b^2}{17a^2b^3}\right)^0 = 1$

101. $\left(\dfrac{-2a^4b}{a^{-3}b^2}\right)^{-3} = \left(\dfrac{a^{-3}b^2}{-2a^4b}\right)^3 = \dfrac{\left(a^{-3}\right)^3\left(b^2\right)^3}{(-2)^3\left(a^4\right)^3b^3} = \dfrac{a^{-9}b^6}{-8a^{12}b^3} = \dfrac{b^3}{-8a^{21}} = -\dfrac{b^3}{8a^{21}}$

103. $\left(\dfrac{2a^3b^2}{3a^{-3}b^2}\right)^{-3} = \left(\dfrac{3a^{-3}b^2}{2a^3b^2}\right)^3 = \left(\dfrac{3}{2a^6}\right)^3 = \dfrac{3^3}{2^3\left(a^6\right)^3} = \dfrac{27}{8a^{18}}$

105. $\dfrac{\left(3x^2\right)^{-2}}{x^3x^{-4}x^0} = \dfrac{3^{-2}\left(x^2\right)^{-2}}{x^{-1}} = \dfrac{3^{-2}x^{-4}}{x^{-1}} = 3^{-2}x^{-4-(-1)} = 3^{-2}x^{-3} = \dfrac{1}{3^2x^3} = \dfrac{1}{9x^3}$

SECTION 1.3

107. $\dfrac{a^n a^3}{a^4} = \dfrac{a^{n+3}}{a^4} = a^{n+3-4} = a^{n-1}$

109. $\left(\dfrac{b^n}{b^3}\right)^3 = \dfrac{(b^n)^3}{(b^3)^3} = \dfrac{b^{3n}}{b^9} = b^{3n-9}$

111. $\dfrac{a^{-n} a^2}{a^3} = \dfrac{a^{-n+2}}{a^3} = a^{-n+2-3} = a^{-n-1}$

$\left(\text{an equivalent form is } \dfrac{1}{a^{n+1}}\right)$

113. $\dfrac{a^{-n} a^{-2}}{a^{-4}} = \dfrac{a^{-n-2}}{a^{-4}} = a^{-n-2-(-4)} = a^{2-n}$

Problems 115-125 are to be solved using a calculator. The keystrokes needed to solve each problem using a TI-83 graphing calculator appear in each solution. There may be other solutions. Keystrokes for other calculators may be slightly different.

115. $\boxed{1}\ \boxed{.}\ \boxed{2}\ \boxed{3}\ \boxed{\char`\^}\ \boxed{6}\ \boxed{\text{ENTER}}$
{3.462825992}

117. $\boxed{(-)}\ \boxed{6}\ \boxed{.}\ \boxed{2}\ \boxed{5}\ \boxed{\char`\^}\ \boxed{3}\ \boxed{\text{ENTER}}$
{−244.140625}

119. $\boxed{3}\ \boxed{.}\ \boxed{6}\ \boxed{8}\ \boxed{\char`\^}\ \boxed{0}\ \boxed{\text{ENTER}}$ {1.}

121. $\boxed{7}\ \boxed{.}\ \boxed{2}\ \boxed{x^2}\ \boxed{\times}\ \boxed{2}\ \boxed{.}\ \boxed{7}\ \boxed{x^2}\ \boxed{\text{ENTER}}$ {377.9136}
$\boxed{(}\ \boxed{7}\ \boxed{.}\ \boxed{2}\ \boxed{\times}\ \boxed{2}\ \boxed{.}\ \boxed{7}\ \boxed{)}\ \boxed{x^2}\ \boxed{\text{ENTER}}$ {377.9136}

123. $\boxed{3}\ \boxed{.}\ \boxed{2}\ \boxed{x^2}\ \boxed{\times}\ \boxed{3}\ \boxed{.}\ \boxed{2}\ \boxed{\char`\^}\ \boxed{(-)}\ \boxed{2}\ \boxed{\text{ENTER}}$ {1}

125. $\boxed{7}\ \boxed{.}\ \boxed{2}\ \boxed{3}\ \boxed{\char`\^}\ \boxed{(-)}\ \boxed{3}\ \boxed{\text{ENTER}}$ {0.002645971171398}
$\boxed{1}\ \boxed{\div}\ \boxed{(}\ \boxed{7}\ \boxed{.}\ \boxed{2}\ \boxed{3}\ \boxed{\char`\^}\ \boxed{3}\ \boxed{)}\ \boxed{\text{ENTER}}$ {0.002645971171398}

127. $x^2 y^3 = (-2)^2(3)^3 = 4(27) = 108$

129. $\dfrac{x^{-3}}{y^3} = \dfrac{1}{x^3 y^3} = \dfrac{1}{(-2)^3(3)^3} = \dfrac{1}{-8(27)}$
$= -\dfrac{1}{216}$

131. $(xy^2)^{-2} = x^{-2}y^{-4} = \dfrac{1}{x^2 y^4} = \dfrac{1}{(-2)^2(3)^4} = \dfrac{1}{4(81)} = \dfrac{1}{324}$

133. $(-yx^{-1})^3 = (-1)^3 y^3 (x^{-1})^3 = -y^3 x^{-3} = -\dfrac{y^3}{x^3} = -\dfrac{3^3}{(-2)^3} = -\dfrac{27}{-8} = \dfrac{27}{8}$

135. $A = lw = (3\,\text{m})(5\,\text{m}) = 15\,\text{m}^2$

137. $A = \pi r^2 = \pi(6\,\text{cm})^2 = \pi(36\,\text{cm}^2) \approx 113\,\text{cm}^2$

139. $A = \tfrac{1}{2}h(b_1 + b_2) = \tfrac{1}{2}(5\,\text{cm})(6\,\text{cm} + 12\,\text{cm}) = \tfrac{1}{2}(5\,\text{cm})(18\,\text{cm}) = \tfrac{1}{2}(90\,\text{cm}^2) = 45\,\text{cm}^2$

141. $A = s^2 + \tfrac{1}{2}bh = (15\,\text{cm})^2 + \tfrac{1}{2}(15\,\text{cm})(10\,\text{cm}) = (225 + 75)\,\text{cm}^2 = 300\,\text{cm}^2$

143. $V = s^3 = (7\,\text{m})^3 = 343\,\text{m}^3$

145. $V = lwh = (6\,\text{ft})(6\,\text{ft})(10\,\text{ft}) = 360\,\text{ft}^3$

147. $V = \tfrac{1}{3}Bh = \tfrac{1}{3}\pi r^2 h = \tfrac{1}{3}\pi(4\,\text{ft})^2(10\,\text{ft}) = \tfrac{1}{3}(160\,\text{ft}^3)\pi \approx 168\,\text{ft}^3$

149. $V = Bh + \frac{1}{2} \cdot \frac{4}{3}\pi r^3 = \pi r^2 h + \frac{2}{3}\pi r^3 = \pi(6\,\mathrm{m})^2(20\,\mathrm{m}) + \frac{2}{3}\pi(6\,\mathrm{m})^3 \approx 2714\,\mathrm{m}^3$

151. $\boxed{5}\,\boxed{0}\,\boxed{0}\,\boxed{0}\,\boxed{(}\,\boxed{1}\,\boxed{+}\,\boxed{.}\,\boxed{1}\,\boxed{1}\,\boxed{)}\,\boxed{\wedge}\,\boxed{5}\,\boxed{0}\,\boxed{\text{ENTER}}$
{922824.1337} The balance will be \$922,824.13.

153. Answers may vary.   **155.** Answers may vary.

157. $2^{-1} + 3^{-1} - 4^{-1} = \frac{1}{2} + \frac{1}{3} - \frac{1}{4} = \frac{6}{12} + \frac{4}{12} - \frac{3}{12} = \frac{7}{12}$

159. Let $x = 2$, $m = 3$ and $n = 4$: $x^m + x^n = 2^3 + 2^4 = 24$ while $x^{m+n} = 2^{3+4} = 2^7 = 128$.

Exercise 1.4 (page 44)

1. $\frac{3}{4} = 3 \div 4 = 0.75$ **3.** $\frac{13}{9} = 13 \div 9 = 1.\overline{4}$ **5.** $3^2 + 4^3 + 2^4 = 9 + 64 + 16$
$= 89$

7. 10^n **9.** left   **11.** $3900 = 3.9 \times 10^3$

13. $0.0078 = 7.8 \times 10^{-3}$ **15.** $-45{,}000 = -4.5 \times 10^4$ **17.** $-0.00021 = -2.1 \times 10^{-4}$

19. $17{,}600{,}000 = 1.76 \times 10^7$ **21.** $0.0000096 = 9.6 \times 10^{-6}$

23. $323 \times 10^5 = 3.23 \times 10^2 \times 10^5$ **25.** $6000 \times 10^{-7} = 6.0 \times 10^3 \times 10^{-7}$
$= 3.23 \times 10^7$ $= 6.0 \times 10^{-4}$

27. $0.0527 \times 10^5 = 5.27 \times 10^{-2} \times 10^5$ **29.** $0.0317 \times 10^{-2} = 3.17 \times 10^{-2} \times 10^{-2}$
$= 5.27 \times 10^3$ $= 3.17 \times 10^{-4}$

31. $2.7 \times 10^2 = 270$ **33.** $3.23 \times 10^{-3} = 0.00323$

35. $7.96 \times 10^5 = 796{,}000$ **37.** $3.7 \times 10^{-4} = 0.00037$

39. $5.23 \times 10^0 = 5.23$ **41.** $23.65 \times 10^6 = 23{,}650{,}000$

43. $\dfrac{(4000)(30{,}000)}{0.0006} = \dfrac{(4 \times 10^3)(3 \times 10^4)}{6 \times 10^{-4}} = \dfrac{(4)(3)}{6} \cdot \dfrac{10^3 10^4}{10^{-4}} = 2 \times 10^{11}$

45. $\dfrac{(640{,}000)(2{,}700{,}000)}{120{,}000} = \dfrac{(6.4 \times 10^5)(2.7 \times 10^6)}{1.2 \times 10^5} = \dfrac{(6.4)(2.7)}{1.2} \cdot \dfrac{10^5 10^6}{10^5} = 14.4 \times 10^6$
$= 1.44 \times 10^7$

47. $\dfrac{(0.006)(0.008)}{0.0012} = \dfrac{(6 \times 10^{-3})(8 \times 10^{-3})}{1.2 \times 10^{-3}} = \dfrac{(6)(8)}{1.2} \cdot \dfrac{10^{-3} 10^{-3}}{10^{-3}} = 40 \times 10^{-3} = 0.04$

49. $\dfrac{(220{,}000)(0.000009)}{0.00033} = \dfrac{(2.2 \times 10^5)(9 \times 10^{-6})}{3.3 \times 10^{-4}} = \dfrac{(2.2)(9)}{3.3} \cdot \dfrac{10^5 10^{-6}}{10^{-4}} = 6 \times 10^3 = 6000$

51. $\dfrac{(320{,}000)^2(0.0009)}{12{,}000^2} = \dfrac{(3.2 \times 10^5)^2(9 \times 10^{-4})}{(1.2 \times 10^4)^2} = \dfrac{[(3.2)^2 \times 10^{10}](9 \times 10^{-4})}{(1.2)^2 \times 10^8}$

$$= \dfrac{(3.2)^2(9)}{(1.2)^2} \cdot \dfrac{10^{10} 10^{-4}}{10^8}$$

$$= 64 \times 10^{-2} = 0.64$$

53. $23{,}437^3 \approx 1.2874 \times 10^{13}$ **55.** $(63{,}480)(893{,}322) \approx 5.671 \times 10^{10}$

57. $\dfrac{(69.4)^8(73.1)^2}{(0.0043)^3} \approx 3.6 \times 10^{25}$ **59.** gamma ray, x-ray, visible light, infrared, radio wave

61. $3.31 \times 10^4 \, \text{cm/sec} = \dfrac{3.31 \times 10^4 \, \text{cm}}{1 \, \text{sec}} \cdot \dfrac{60 \, \text{sec}}{1 \, \text{min}} \cdot \dfrac{60 \, \text{min}}{1 \, \text{hr}} = \dfrac{(3.31 \times 10^4)(6 \times 10^1)(6 \times 10^1) \, \text{cm}}{1 \, \text{hr}}$

$$= (3.31)(6)(6) \cdot (10^4 10^1 10^1) \, \text{cm/hour}$$

$$= 119.16 \times 10^6 \, \text{cm/hour}$$

$$= 1.1916 \times 10^8 \, \text{cm/hour}$$

$$\approx 1.19 \times 10^8 \, \text{cm/hour}$$

63. Mass of one million protons = Mass of one proton $\cdot \, 1{,}000{,}000 = 1.67248 \times 10^{-24} \cdot 1 \times 10^6 \, \text{g}$

$$= 1.67248 \times 10^{-18} \, \text{g}$$

65. $235{,}000 \, \text{miles} = \dfrac{2.35 \times 10^5 \, \text{mile}}{1} \cdot \dfrac{5280 \, \text{ft}}{1 \, \text{mile}} \cdot \dfrac{12 \, \text{in}}{1 \, \text{ft}} = \dfrac{(2.35 \times 10^5)(5.28 \times 10^3)(1.2 \times 10^1) \, \text{in}}{1}$

$$= (2.35)(5.28)(1.2) \cdot 10^5 10^3 10^1 \, \text{in.}$$

$$= 14.8896 \times 10^9 \, \text{in.}$$

$$= 1.48896 \times 10^{10} \, \text{in.} \approx 1.49 \times 10^{10} \, \text{in.}$$

67. $95{,}000 \, \text{km} = \dfrac{9.5 \times 10^4 \, \text{km}}{1} \cdot \dfrac{0.6214 \, \text{mile}}{1 \, \text{km}} = (9.5 \times 10^4)(6.214 \times 10^{-1}) \, \text{miles}$

$$= (9.5)(6.214) \cdot 10^4 10^{-1} \, \text{miles}$$

$$= 59.033 \times 10^3 \, \text{miles} \approx 5.9 \times 10^4 \, \text{miles}$$

69. $1 \, \text{in.} = \dfrac{1 \, \text{in.}}{1} \cdot \dfrac{25.4 \, \text{mm}}{1 \, \text{in.}} \cdot \dfrac{1 \, \text{angstrom}}{0.0000001 \, \text{mm}} = \dfrac{2.54 \times 10^1}{1 \times 10^{-7}} = 2.54 \times 10^8 \approx 3 \times 10^8 \, \text{angstroms}$

71. $\dfrac{3{,}574{,}000{,}000}{18{,}000}\,\text{hr} = \dfrac{3.574 \times 10^9}{1.8 \times 10^4}\,\text{hr} = \dfrac{3.574}{1.8} \cdot \dfrac{10^9}{10^4}\,\text{hr} = 1.98555555556 \times 10^5\,\text{hr}$

$1.98555555556 \times 10^5\,\text{hr} = \dfrac{1.98555555556 \times 10^5\,\text{hr}}{1} \cdot \dfrac{1\,\text{day}}{24\,\text{hr}} \cdot \dfrac{1\,\text{yr}}{365\,\text{day}}$

$\qquad = \dfrac{(1.98555555556 \times 10^5)}{(2.4 \times 10^1)(3.65 \times 10^2)}\,\text{years}$

$\qquad = \dfrac{1.98555555556}{(2.4)(3.65)} \cdot \dfrac{10^5}{10^1 10^2}\,\text{years}$

$\qquad = 0.2266615930999 \times 10^2\,\text{years} \approx 23\,\text{years}$

73. $186{,}000\,\text{mi/sec} = \dfrac{1.86 \times 10^5\,\text{mi}}{1\,\text{sec}} \cdot \dfrac{60\,\text{sec}}{1\,\text{min}} \cdot \dfrac{60\,\text{min}}{1\,\text{hr}} \cdot \dfrac{24\,\text{hr}}{1\,\text{day}} \cdot \dfrac{365\,\text{day}}{1\,\text{yr}}$

$\qquad = (1.86 \times 10^5)(6 \times 10^1)(6 \times 10^1)(2.4 \times 10^1)(3.65 \times 10^2)\,\text{mi/yr}$

$\qquad = (1.86)(6)(6)(2.4)(3.65) \cdot 10^5 10^1 10^1 10^1 10^2\,\text{mi/yr} = 586.5696 \times 10^{10}\,\text{mi/yr}$

$1.3\,\text{p} = \dfrac{1.3 \times 10^0\,\text{p}}{1} \cdot \dfrac{3.26\,\text{ly}}{1\,\text{p}} = 1.3(3.26 \cdot 586.5696 \times 10^{10})\,\text{mi} \approx 2.5 \times 10^{13}\,\text{miles}$

75. **Answers may vary.**

77. Answer will depend on calculator.

Exercise 1.5 (page 54)

1. $(-4)^3 = (-4)(-4)(-4) = -64$

3. $\left(\dfrac{x+y}{x-y}\right)^0 = 1$

5. $\left(\dfrac{x^2 x^5}{x^3}\right)^2 = \left(\dfrac{x^7}{x^3}\right)^2 = \left(x^4\right)^2 = x^8$

7. $(2x)^{-3} = \dfrac{1}{(2x)^3} = \dfrac{1}{8x^3}$

9. equation

11. equivalent

13. $c;\ \dfrac{b}{c}$

15. Like

17. identity

19.
$3x + 2 = 17$
$3(5) + 2 \overset{?}{=} 17$
$15 + 2 \overset{?}{=} 17$
$17 = 17$
5 is a solution.

21.
$\dfrac{3}{5}x - 5 = -2$
$\dfrac{3}{5}(5) - 5 \overset{?}{=} -2$
$3 - 5 \overset{?}{=} -2$
$-2 = -2$
5 is a solution.

23.
$x + 6 = 8$
$x + 6 - 6 = 8 - 6$
$x = 2$

25.
$a - 5 = 20$
$a - 5 + 5 = 20 + 5$
$a = 25$

27. $2u = 6$
$$\frac{2u}{2} = \frac{6}{2}$$
$$u = 3$$

29. $\dfrac{x}{4} = 7$
$$4 \cdot \frac{x}{4} = 4 \cdot 7$$
$$x = 28$$

31. $3x + 1 = 3$
$$3x + 1 - 1 = 3 - 1$$
$$3x = 2$$
$$\frac{3x}{3} = \frac{2}{3}$$
$$x = \frac{2}{3}$$

33. $2x + 1 = 13$
$$2x + 1 - 1 = 13 - 1$$
$$2x = 12$$
$$\frac{2x}{2} = \frac{12}{2}$$
$$x = 6$$

35. $3(x - 4) = -36$
$$3x - 12 = -36$$
$$3x - 12 + 12 = -36 + 12$$
$$3x = -24$$
$$\frac{3x}{3} = \frac{-24}{3}$$
$$x = -8$$

37. $3(r - 4) = -4$
$$3r - 12 = -4$$
$$3r - 12 + 12 = -4 + 12$$
$$3r = 8$$
$$\frac{3r}{3} = \frac{8}{3}$$
$$r = \frac{8}{3}$$

39. like terms; $2x + 6x = 8x$

41. not like terms

43. like terms; $3x^2 + (-5x^2) = -2x^2$

45. not like terms

47. $3a - 22 = -2a - 7$
$$3a + 2a - 22 = -2a + 2a - 7$$
$$5a - 22 = -7$$
$$5a - 22 + 22 = -7 + 22$$
$$5a = 15$$
$$\frac{5a}{5} = \frac{15}{5}$$
$$a = 3$$

49. $2(2x + 1) = 15 + 3x$
$$4x + 2 = 15 + 3x$$
$$4x - 3x + 2 = 15 + 3x - 3x$$
$$x + 2 = 15$$
$$x + 2 - 2 = 15 - 2$$
$$x = 13$$

51. $3(y - 4) - 6 = y$
$$3y - 12 - 6 = y$$
$$3y - 18 = y$$
$$3y - y - 18 = y - y$$
$$2y - 18 = 0$$
$$2y - 18 + 18 = 0 + 18$$
$$2y = 18$$
$$\frac{2y}{2} = \frac{18}{2}$$
$$y = 9$$

53. $5(5 - a) = 37 - 2a$
$$25 - 5a = 37 - 2a$$
$$25 - 5a + 2a = 37 - 2a + 2a$$
$$25 - 3a = 37$$
$$25 - 25 - 3a = 37 - 25$$
$$-3a = 12$$
$$\frac{-3a}{-3} = \frac{12}{-3}$$
$$a = -4$$

55.
$$4(y+1) = -2(4-y)$$
$$4y+4 = -8+2y$$
$$4y-2y+4 = -8+2y-2y$$
$$2y+4 = -8$$
$$2y+4-4 = -8-4$$
$$2y = -12$$
$$\frac{2y}{2} = \frac{-12}{2}$$
$$y = -6$$

57.
$$2(a-5)-(3a+1) = 0$$
$$2a-10-3a-1 = 0$$
$$-a-11 = 0$$
$$-a-11+11 = 0+11$$
$$-a = 11$$
$$a = -11$$

59.
$$3(y-5)+10 = 2(y+4)$$
$$3y-15+10 = 2y+8$$
$$3y-5 = 2y+8$$
$$3y-2y-5 = 2y-2y+8$$
$$y-5 = 8$$
$$y-5+5 = 8+5$$
$$y = 13$$

61.
$$9(x+2) = -6(4-x)+18$$
$$9x+18 = -24+6x+18$$
$$9x+18 = 6x-6$$
$$9x-6x+18 = 6x-6x-6$$
$$3x+18 = -6$$
$$3x+18-18 = -6-18$$
$$3x = -24$$
$$\frac{3x}{3} = \frac{-24}{3}$$
$$x = -8$$

63.
$$-4p-2(3p+5) = -6p+2(p+2)$$
$$-4p-6p-10 = -6p+2p+4$$
$$-10p-10 = -4p+4$$
$$-10p+4p-10 = -4p+4p+4$$
$$-6p-10 = 4$$
$$-6p-10+10 = 4+10$$
$$-6p = 14$$
$$\frac{-6p}{-6} = \frac{14}{-6}$$
$$p = -\frac{7}{3}$$

65.
$$4+4(n+2) = 3n-2(n-5)$$
$$4+4n+8 = 3n-2n+10$$
$$4n+12 = n+10$$
$$4n-n+12 = n-n+10$$
$$3n+12 = 10$$
$$3n+12-12 = 10-12$$
$$3n = -2$$
$$n = -\frac{2}{3}$$

67.
$$\frac{1}{2}x - 4 = -1 + 2x$$
$$2\left(\frac{1}{2}x - 4\right) = 2(-1 + 2x)$$
$$2\left(\frac{1}{2}x\right) + 2(-4) = 2(-1) + 2(2x)$$
$$x - 8 = -2 + 4x$$
$$x - 4x - 8 = -2 + 4x - 4x$$
$$-3x - 8 = -2$$
$$-3x - 8 + 8 = -2 + 8$$
$$-3x = 6$$
$$\frac{-3x}{-3} = \frac{6}{-3}$$
$$x = -2$$

69.
$$\frac{x}{2} - \frac{x}{3} = 4$$
$$6\left(\frac{x}{2} - \frac{x}{3}\right) = 6(4)$$
$$6\left(\frac{x}{2}\right) - 6\left(\frac{x}{3}\right) = 24$$
$$3x - 2x = 24$$
$$x = 24$$

71.
$$\frac{x}{6} + 1 = \frac{x}{3}$$
$$6\left(\frac{x}{6} + 1\right) = 6\left(\frac{x}{3}\right)$$
$$6\left(\frac{x}{6}\right) + 6(1) = 2x$$
$$x + 6 = 2x$$
$$x - 2x + 6 = 2x - 2x$$
$$-x + 6 = 0$$
$$-x + 6 - 6 = 0 - 6$$
$$-x = -6$$
$$x = 6$$

73.
$$5 - \frac{x+2}{3} = 7 - x$$
$$3\left(5 - \frac{x+2}{3}\right) = 3(7 - x)$$
$$3(5) - 3\left(\frac{x+2}{3}\right) = 3(7) + 3(-x)$$
$$15 - (x + 2) = 21 - 3x$$
$$15 - x - 2 = 21 - 3x$$
$$13 - x = 21 - 3x$$
$$13 - x + 3x = 21 - 3x + 3x$$
$$13 + 2x = 21$$
$$13 - 13 + 2x = 21 - 13$$
$$2x = 8$$
$$\frac{2x}{2} = \frac{8}{2}$$
$$x = 4$$

75.
$$\frac{4x-2}{2} = \frac{3x+6}{3}$$
$$6\left(\frac{4x-2}{2}\right) = 6\left(\frac{3x+6}{3}\right)$$
$$3(4x-2) = 2(3x+6)$$
$$12x - 6 = 6x + 12$$
$$12x - 6x - 6 = 6x - 6x + 12$$
$$6x - 6 = 12$$
$$6x - 6 + 6 = 12 + 6$$
$$6x = 18$$
$$\frac{6x}{6} = \frac{18}{6}$$
$$x = 3$$

77.
$$\frac{a+1}{3} + \frac{a-1}{5} = \frac{2}{15}$$
$$15\left(\frac{a+1}{3} + \frac{a-1}{5}\right) = 15\left(\frac{2}{15}\right)$$
$$15\left(\frac{a+1}{3}\right) + 15\left(\frac{a-1}{5}\right) = 2$$
$$5(a+1) + 3(a-1) = 2$$
$$5a + 5 + 3a - 3 = 2$$
$$8a + 2 = 2$$
$$8a + 2 - 2 = 2 - 2$$
$$8a = 0$$
$$\frac{8a}{8} = \frac{0}{8}$$
$$a = 0$$

79.
$$\frac{5a}{2} - 12 = \frac{a}{3} + 1$$
$$6\left(\frac{5a}{2} - 12\right) = 6\left(\frac{a}{3} + 1\right)$$
$$6\left(\frac{5a}{2}\right) - 6(12) = 6\left(\frac{a}{3}\right) + 6(1)$$
$$3(5a) - 72 = 2a + 6$$
$$15a - 72 = 2a + 6$$
$$15a - 2a - 72 = 2a - 2a + 6$$
$$13a - 72 = 6$$
$$13a - 72 + 72 = 6 + 72$$
$$13a = 78$$
$$\frac{13a}{13} = \frac{78}{13}$$
$$a = 6$$

81.
$$4(2 - 3t) + 6t = -6t + 8$$
$$8 - 12t + 6t = -6t + 8$$
$$8 - 6t = -6t + 8$$
$$-6t + 8 = -6t + 8$$
Identity

83.
$$\frac{a+1}{4} + \frac{2a-3}{4} = \frac{a}{2} - 2$$
$$4\left(\frac{a+1}{4} + \frac{2a-3}{4}\right) = 4\left(\frac{a}{2} - 2\right)$$
$$4\left(\frac{a+1}{4}\right) + 4\left(\frac{2a-3}{4}\right) = 4\left(\frac{a}{2}\right) + 4(-2)$$
$$a + 1 + 2a - 3 = 2a - 8$$
$$3a - 2 = 2a - 8$$
$$3a - 2a - 2 = 2a - 2a - 8$$
$$a - 2 = -8$$
$$a - 2 + 2 = -8 + 2$$
$$a = -6$$

85.
$$3(x - 4) + 6 = -2(x + 4) + 5x$$
$$3x - 12 + 6 = -2x - 8 + 5x$$
$$3x - 6 = 3x - 8$$
$$3x - 3x - 6 = 3x - 3x - 8$$
$$-6 = -8$$
Contradiction

87. $y(y+2)+1 = y^2 + 2y + 1$

$y^2 + 2y + 1 = y^2 + 2y + 1$

Identity

89. $A = lw$

$\dfrac{A}{l} = \dfrac{lw}{l}$

$\dfrac{A}{l} = w$, or $w = \dfrac{A}{l}$

91. $V = \dfrac{1}{3}Bh$

$3V = 3 \cdot \dfrac{1}{3}Bh$

$3V = Bh$

$\dfrac{3V}{h} = \dfrac{Bh}{h}$

$\dfrac{3V}{h} = B$, or $B = \dfrac{3V}{h}$

93. $I = prt$

$\dfrac{I}{pr} = \dfrac{prt}{pr}$

$\dfrac{I}{pr} = t$, or $t = \dfrac{I}{pr}$

95. $p = 2l + 2w$

$p - 2l = 2l - 2l + 2w$

$p - 2l = 2w$

$\dfrac{p - 2l}{2} = \dfrac{2w}{2}$

$\dfrac{p - 2l}{2} = w$, or $w = \dfrac{p - 2l}{2}$

97. $A = \dfrac{1}{2}h(B + b)$

$2A = 2 \cdot \dfrac{1}{2}h(B + b)$

$2A = h(B + b)$

$\dfrac{2A}{h} = \dfrac{h(B + b)}{h}$

$\dfrac{2A}{h} = B + b$

$\dfrac{2A}{h} - b = B + b - b$

$\dfrac{2A}{h} - b = B$, or $B = \dfrac{2A}{h} - b$

99. $y = mx + b$

$y - b = mx + b - b$

$y - b = mx$

$\dfrac{y - b}{m} = \dfrac{mx}{m}$

$\dfrac{y - b}{m} = x$, or $x = \dfrac{y - b}{m}$

101. $l = a + (n - 1)d$

$l = a + nd - d$

$l - a + d = a - a + nd - d + d$

$l - a + d = nd$

$\dfrac{l - a + d}{d} = \dfrac{nd}{d}$

$\dfrac{l - a + d}{d} = n$, or $n = \dfrac{l - a + d}{d}$

103.
$$S = \frac{a - lr}{1 - r}$$
$$(1 - r)S = (1 - r) \cdot \frac{a - lr}{1 - r}$$
$$S - rS = a - lr$$
$$S - Sr - a = a - a - lr$$
$$S - Sr - a = -lr$$
$$\frac{S - Sr - a}{-r} = \frac{-lr}{-r}$$
$$\frac{-S + Sr + a}{r} = l, \text{ or } l = \frac{a - S + Sr}{r}$$

105.
$$S = \frac{n(a + l)}{2}$$
$$2S = 2 \cdot \frac{n(a + l)}{2}$$
$$2S = n(a + l)$$
$$2S = na + nl$$
$$2S - na = na - na + nl$$
$$2S - na = nl$$
$$\frac{2S - na}{n} = \frac{nl}{n}$$
$$\frac{2S - na}{n} = l, \text{ or } l = \frac{2S - na}{n}$$

107.
$$F = \frac{GmM}{d^2}$$
$$d^2 F = d^2 \cdot \frac{GmM}{d^2}$$
$$Fd^2 = GmM$$
$$\frac{Fd^2}{GM} = \frac{GmM}{GM}$$
$$\frac{Fd^2}{GM} = m, \text{ or } m = \frac{Fd^2}{GM}$$

109.
$$F = \frac{9}{5}C + 32$$
$$F - 32 = \frac{9}{5}C + 32 - 32$$
$$F - 32 = \frac{9}{5}C$$
$$\frac{5}{9}(F - 32) = \frac{5}{9} \cdot \frac{9}{5}C$$
$$\frac{5}{9}(F - 32) = C, \text{ or } C = \frac{5}{9}(F - 32)$$

$$F = 32°: \ C = \frac{5}{9}(32 - 32) = \frac{5}{9}(0) = 0°$$
$$F = 70°: \ C = \frac{5}{9}(70 - 32) = \frac{5}{9}(38) \approx 21.1°$$
$$F = 212°: \ C = \frac{5}{9}(212 - 32) = \frac{5}{9}(180) = 100°$$

111.
$$C = 0.07n + 6.50$$
$$C - 6.50 = 0.07n + 6.50 - 6.50$$
$$C - 6.50 = 0.07n$$
$$\frac{C - 6.50}{0.07} = \frac{0.07n}{0.07}$$
$$\frac{C - 6.50}{0.07} = n, \text{ or } n = \frac{C - 6.50}{0.07}$$

$$C = 49.97: \ n = \frac{49.97 - 6.50}{0.07} = \frac{43.47}{0.07} = 621 \text{ kwh}$$
$$C = 76.50: \ n = \frac{76.50 - 6.50}{0.07} = \frac{70}{0.07} = 1000 \text{ kwh}$$
$$C = 125: \ n = \frac{125 - 6.50}{0.07} = \frac{118.50}{0.07} \approx 1692.9 \text{ kwh}$$

113. $E = IR$

$$\frac{E}{I} = \frac{IR}{I}$$

$$\frac{E}{I} = R, \text{ or } R = \frac{E}{I}$$

$$R = \frac{56}{7} = 8 \text{ ohms}$$

115.

$$a = 180\left(1 - \frac{2}{n}\right)$$

$$a = 180 - \frac{360}{n}$$

$$a - 180 = 180 - 180 - \frac{360}{n}$$

$$a - 180 = -\frac{360}{n}$$

$$n(a - 180) = n\left(-\frac{360}{n}\right)$$

$$n(a - 180) = -360$$

$$\frac{n(a - 180)}{a - 180} = \frac{-360}{a - 180}$$

$$n = \frac{360}{180 - a}$$

$$n = \frac{360}{180 - a} = \frac{360}{180 - 135} = 8 \text{ sides}$$

117. Answers may vary.

119. The distributive property was not used correctly.

Exercise 1.6 (page 63)

1. $\left(\dfrac{3x^{-3}}{4x^2}\right)^{-4} = \left(\dfrac{4x^2}{3x^{-3}}\right)^{4} = \left(\dfrac{4x^5}{3}\right)^{4} = \dfrac{256x^{20}}{81}$

3. $\dfrac{a^m a^3}{a^2} = \dfrac{a^{m+3}}{a^2} = a^{m+3-2} = a^{m+1}$

5. $5x + 4$ **7.** $40x$ **9.** $180°$

11. complementary   **13.** right   **15.** vertex

17. Let x = length of 1st piece.
Then $2x$, $4x$ and $8x$ = the other lengths.

$$\boxed{\text{Sum of lengths}} = \boxed{\text{total length}}$$

$$x + 2x + 4x + 8x = 60$$

$$15x = 60$$

$$x = 4$$

The longest piece has a length of $8x$,
or $8(4) = 32$ feet.

19. Let x = the length of the shorter piece.
Then $2x + 1$ = the length of the other piece.

$$\boxed{\begin{array}{c}\text{shorter}\\\text{length}\end{array}} + \boxed{\begin{array}{c}\text{other}\\\text{length}\end{array}} = \boxed{\text{total length}}$$

$$x + 2x + 1 = 22$$

$$3x + 1 = 22$$

$$3x = 21$$

$$x = 7$$

The pieces have lengths of 7 feet and 15 feet.

21. Let $x =$ the cost of the VCR.

Then $x + 55 =$ the cost of the TV.

$$\boxed{\begin{smallmatrix} \text{VCR} \\ \text{cost} \end{smallmatrix}} + \boxed{\begin{smallmatrix} \text{TV} \\ \text{cost} \end{smallmatrix}} = \boxed{\text{total cost}}$$

$$x + x + 55 = 655$$
$$2x + 55 = 655$$
$$2x = 600$$
$$x = 300$$

The TV costs $x + 55 = 300 + 55 = \$355$.

23. Percent markdown $= \dfrac{\text{amount of markdown}}{\text{regular price}}$

$$= \dfrac{726 - 580.80}{726}$$
$$= \dfrac{145.20}{726}$$
$$= 0.20 = 20\%$$

25. Percent markup $= \dfrac{\text{amount of markup}}{\text{regular price}} = \dfrac{40 - 12}{12} = \dfrac{28}{12} \approx 2.33 = 233\%.$

27. Let $B = \#$ of shares in Big Bank. Then $500 - B = \#$ of shares in Safe Savings.

$$\boxed{\begin{smallmatrix} \text{Big Bank} \\ \text{shares} \end{smallmatrix}} + \boxed{\begin{smallmatrix} \text{Safe Savings} \\ \text{shares} \end{smallmatrix}} = \boxed{\begin{smallmatrix} \text{Total} \\ \text{value} \end{smallmatrix}}$$

$$115B + 97(500 - B) = 53900$$
$$115B + 48500 - 97B = 53900$$
$$18B = 5400$$
$$B = 300$$

The student owns 300 shares of Big Bank and 200 shares of Safe Savings.

29. Let $B = \#$ of scientific model sold. Then $85 - B = \#$ of graphing model sold.

$$\boxed{\begin{smallmatrix} \text{Value of} \\ \text{scientific} \end{smallmatrix}} + \boxed{\begin{smallmatrix} \text{Value of} \\ \text{graphing} \end{smallmatrix}} = \boxed{\begin{smallmatrix} \text{Total} \\ \text{value} \end{smallmatrix}}$$

$$15B + 67(85 - B) = 3875$$
$$15B + 5695 - 67B = 3875$$
$$-52B = -1820$$
$$B = 35$$

The store sold 35 scientific calculators and 50 graphing calculators.

31. Let $r =$ the number of roses he buys.

$$\boxed{\begin{smallmatrix} \text{Cost of} \\ \text{roses} \end{smallmatrix}} + \boxed{\begin{smallmatrix} \text{Delivery} \\ \text{charge} \end{smallmatrix}} = \boxed{\text{Total cost}}$$

$$1.25r + 5 = 21.25$$
$$1.25r = 16.25$$
$$r = 13$$

The man can buy 13 roses.

33. Let $m =$ the number of miles he drives.

$$\boxed{\begin{smallmatrix} \text{Mileage} \\ \text{cost} \end{smallmatrix}} + \boxed{\begin{smallmatrix} \text{Daily} \\ \text{cost} \end{smallmatrix}} = \boxed{\text{Total cost}}$$

$$0.10m + 2 \cdot 12 = 30$$
$$0.10m = 6$$
$$m = 60$$

$$\boxed{\begin{smallmatrix} \text{Mileage} \\ \text{cost} \end{smallmatrix}} + \boxed{\begin{smallmatrix} \text{Daily} \\ \text{cost} \end{smallmatrix}} = \boxed{\text{Total cost}}$$

$$0.10m + 2 \cdot 12 = 36$$
$$0.10m = 12$$
$$m = 120$$

He can drive 60 miles for \$30 and 120 miles for \$36.

35. Let $w =$ the width. Then $2w =$ the length.

$$\boxed{\text{Perimeter of garden}} = 72$$

$$w + 2w + w + 2w = 72$$
$$6w = 72$$
$$w = 12$$

The dimension are 12 m by 24 m.

37. Let $w =$ the width. Then $2w =$ the length.

$$\boxed{\text{Total amount of fence}} = 624$$

$$w + 2w + w = 624$$
$$4w = 624$$
$$w = 156$$

The dimension are 156 feet by 312 feet.

39.

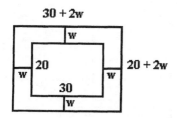

$$\boxed{\text{Total fencing}} = 180$$
$$2(30 + 2w) + 2(20 + 2w) = 180$$
$$60 + 4w + 40 + 4w = 180$$
$$8w + 100 = 180$$
$$8w = 80$$
$$w = 10$$

The walkway will be 10 feet wide.

41. Let x = the measure of the smaller angle.
Then $x + 35$ = the other measure.

$$\boxed{\begin{array}{c}\text{The sum of}\\ \text{the angles}\end{array}} = 180°$$
$$x + x + 35 = 180$$
$$2x + 35 = 180$$
$$2x = 145$$
$$x = 72.5$$

The smaller angle has a measure of 72.5°.

43. Let x = the measure of the smaller angle.
Then $x + 22$ = the other measure.

$$\boxed{\begin{array}{c}\text{sum of}\\ \text{angles}\end{array}} = 90°$$
$$x + x + 22 = 90$$
$$2x + 22 = 90$$
$$2x = 68$$
$$x = 34$$

The larger angle has a measure of 56°.

45.

$$\boxed{\begin{array}{c}\text{The sum of the}\\ \text{measures of } \angle 1 \text{ and } \angle 2\end{array}} = 180°$$
$$x + 50 + 2x - 20 = 180°$$
$$3x + 30 = 180°$$
$$3x = 150°$$
$$x = 50°$$

47. Let x = the measure of each angle.

$$\boxed{\text{Sum of measures of all angles}} = 180°$$
$$3x = 180°$$
$$x = 60°$$

Each angle has a measure of 60°.

49.

$$\boxed{\begin{array}{c}\text{Measure}\\ \text{of } \angle 2\end{array}} = \boxed{\begin{array}{c}\text{Measure}\\ \text{of } \angle 4\end{array}}$$
$$6x + 20 = 8x - 20$$
$$-2x = -40$$
$$x = 20$$
$$m(\angle 2) = 6x + 20 = 6(20) + 20 = 140°$$
$$m(\angle 1) = 180 - 140 = 40°$$

51. Let x = the original height. Then $3x$ = the new height.

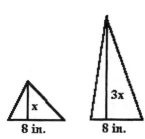

$$\boxed{\text{Old Area}} + 96 = \boxed{\text{New Area}}$$
$$\frac{1}{2}(8)(x) + 96 = \frac{1}{2}(8)(3x)$$
$$4x + 96 = 12x$$
$$96 = 8x$$
$$12 = x$$

The original height is 12 inches.

53.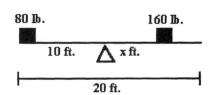

$$\boxed{\begin{array}{c}\text{Boy's force}\\\text{times distance}\end{array}} = \boxed{\begin{array}{c}\text{Father's force}\\\text{times distance}\end{array}}$$

$$80(10) = 160x$$
$$800 = 160x$$
$$5 = x$$

The father should sit 5 feet from the fulcrum.

55.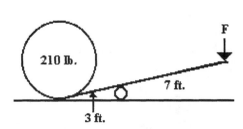

$$\boxed{\begin{array}{c}\text{Stone's force}\\\text{times distance}\end{array}} = \boxed{\begin{array}{c}\text{Woman's force}\\\text{times distance}\end{array}}$$

$$210(3) = F(7)$$
$$630 = 7F$$
$$90 = F$$

The woman needs to exert 90 pounds of force.

57. $\boxed{\text{Sum of forces times distances on left}} = \boxed{\text{Sum of forces times distances on right}}$

$$100(12) + 70(8) = 40(x) + 200(8)$$
$$1200 + 560 = 40x + 1600$$
$$160 = 40x$$
$$4 = x \quad \text{The distance should be 4 feet.}$$

59. $F = \frac{5}{9}(F - 32)$
$9F = 5(F - 32)$
$9F = 5F - 160$
$4F = -160$
$F = -40$

The temperatures are the same at $-40°$.

61. Answers may vary.

63. $40(6) + 10x = 50(4)$
$240 + 10x = 200$
$10x = -40$
$x = -4$

Exercise 1.7 (page 72)

1. $9x - 3 = 6x$
$9x - 6x - 3 = 6x - 6x$
$3x - 3 + 3 = 0 + 3$
$3x = 3$
$x = 1$

3. $\dfrac{8(y - 5)}{3} = 2(y - 4)$

$3 \cdot \dfrac{8(y - 5)}{3} = 3 \cdot 2(y - 4)$

$8(y - 5) = 6(y - 4)$
$8y - 40 = 6y - 24$
$2y = 16$
$y = 8$

SECTION 1.7

5. principal; rate; time

7. value; price; number

9. Let x = amount invested at 8%. Then $12,000 - x$ = amount invested at 9%.

$$\boxed{\begin{array}{c}\text{Interest}\\\text{at 8\%}\end{array}} + \boxed{\begin{array}{c}\text{Interest}\\\text{at 9\%}\end{array}} = \boxed{\begin{array}{c}\text{Total}\\\text{interest}\end{array}}$$

$$0.08x + 0.09(12000 - x) = 1060$$
$$0.08x + 1080 - 0.09x = 1060$$
$$-0.01x = -20$$
$$x = \frac{-20}{-0.01}$$
$$x = 2000$$

She invested $2000 in the money market account (8%) and $10,000 in the CD (9%).

11. Let r = the needed interest rate for the remainder of the money.

$$\boxed{\begin{array}{c}\text{Interest}\\\text{at 7\%}\end{array}} + \boxed{\begin{array}{c}\text{Interest}\\\text{at } r\%\end{array}} = \boxed{\begin{array}{c}\text{Total}\\\text{interest}\end{array}}$$

$$0.07(6000) + r(10000) = 1500$$
$$420 + 10000r = 1500$$
$$10000r = 1080$$
$$r = \frac{1080}{10000}$$
$$r = 0.1080$$

She needs to earn 10.8% interest on the remainder of the money.

13. Let x = the amount she has to invest. She needs $x + 3000$ to earn 11% interest.

$$\boxed{\begin{array}{c}\text{Interest}\\\text{at 11\%}\end{array}} = 2 \cdot \boxed{\begin{array}{c}\text{Interest}\\\text{at 7.5\%}\end{array}}$$

$$0.11(x + 3000) = 2(0.075)(x)$$
$$0.11x + 330 = 0.15x$$
$$-0.04x = -330$$
$$x = \frac{-330}{-0.04}$$
$$x = 8250$$

She has $8,250 on hand to invest.

15. Let s = the # of student tickets sold. Then $200 - s$ = the # of adult tickets sold.

$$\boxed{\begin{array}{c}\text{Student}\\\$\end{array}} + \boxed{\begin{array}{c}\text{Adult}\\\$\end{array}} = \boxed{\begin{array}{c}\text{Total}\\\$\end{array}}$$

$$2s + 4(200 - s) = 750$$
$$2s + 800 - 4s = 750$$
$$-2s + 800 = 750$$
$$-2s = -50$$
$$s = \frac{-50}{-2}$$
$$r = 25$$

25 student tickets were sold.

17. Let t = time for the cars to meet.

	Rate	Time	Dist.
First car	50	t	$50t$
Second car	48	t	$48t$

$$\boxed{\begin{array}{c}\text{1st}\\\text{dist.}\end{array}} + \boxed{\begin{array}{c}\text{2nd}\\\text{dist.}\end{array}} = \boxed{\begin{array}{c}\text{Total}\\\text{dist.}\end{array}}$$

$$50t + 48t = 343$$
$$98t = 343$$
$$t = \frac{343}{98}$$
$$t = \frac{7}{2} = 3\frac{1}{2}$$

It will take the cars $3\frac{1}{2}$ hours to meet.

19. Let t = time car travels. Then $t + 1$ = time cyclist travels.

	Rate	Time	Dist.
Cyclist	18	$t + 1$	$18(t + 1)$
Car	45	t	$45t$

$$\boxed{\text{Cyclist distance}} = \boxed{\text{Car distance}}$$

$$18(t + 1) = 45t$$
$$18t + 18 = 45t$$
$$-27t = -18$$
$$t = \frac{-18}{-27}$$
$$t = \frac{2}{3}$$

It will take $\frac{2}{3}$ of an hour to pass the cyclist.

21. Let $t =$ time for them to run.

	Rate	Time	Dist.
1st Runner	12	t	$12t$
2nd Runner	10	t	$10t$

1st dist.	$-$	2nd dist.	$=$	Dist. between them

$$12t - 10t = 0.25$$
$$2t = \tfrac{1}{4}$$
$$t = \tfrac{1}{8}$$

It will take $\tfrac{1}{8}$ hour to be $\tfrac{1}{4}$ mile apart.

23. Let $t =$ time to return.

	Rate	Time	Dist.
Upstream	8	3	24
Downstream	16	t	$16t$

Downstream distance	$=$	Upstream distance

$$16t = 24$$
$$t = \tfrac{24}{16} = \tfrac{3}{2} = 1\tfrac{1}{2}$$

It will take the rider $1\tfrac{1}{2}$ hours to return.

25. Let $t =$ the time at the slower rate.

	Rate	Time	Dist.
Slower	45	t	$45t$
Faster	55	$8-t$	$55(8-t)$

Slower dist.	$+$	Faster dist.	$=$	Total dist.

$$45t + 55(8 - t) = 400$$
$$45t + 440 - 55t = 400$$
$$-10t = -40$$
$$t = \frac{-40}{-10} = 4$$

He traveled 4 hours at 45 mph, and 4 hours at 55 mph.

27. Let $x =$ pounds of cheaper candy. Then $30 - x =$ pounds of other candy.

	Price per pound	# pounds	Value
Cheaper	0.95	x	$0.95x$
Other	1.10	$30 - x$	$1.10(30 - x)$
Mixture	1.00	30	$1.00(30)$

Value of cheaper	$+$	Value of other	$=$	Value of mixture

$$0.95x + 1.1(30 - x) = 30$$
$$0.95x + 33 - 1.1x = 30$$
$$-0.15x = -3$$
$$x = \frac{-3}{-0.15} = 20 \Rightarrow 30 - x = 30 - 20 = 10$$

The owner should use 20 pounds of the cheaper candy and 10 pounds of the other candy.

29. Let $x =$ ounces of water added (0% alcohol).

$$\boxed{\begin{array}{c}\text{Alcohol}\\\text{at start}\end{array}} + \boxed{\begin{array}{c}\text{Alcohol}\\\text{added}\end{array}} = \boxed{\begin{array}{c}\text{Alcohol}\\\text{at end}\end{array}}$$

$$0.15(20) + 0(x) = 0.10(20 + x)$$
$$3 + 0 = 2 + 0.10x$$
$$1 = 0.10x$$
$$x = \frac{1}{0.10} = 10$$

10 ounces of water should be added.

31. Let $x =$ gallons of cream used.

Then $20 - x =$ gallons of 2% milk used.

$$\boxed{\begin{array}{c}\text{Butterfat}\\\text{in cream}\end{array}} + \boxed{\begin{array}{c}\text{Butterfat}\\\text{in 2\% milk}\end{array}} = \boxed{\begin{array}{c}\text{Butterfat}\\\text{in 4\% milk}\end{array}}$$

$$0.22x + 0.02(20 - x) = 0.04(20)$$
$$0.22x + 0.4 - 0.02x = 0.8$$
$$0.2x = 0.4$$
$$x = \frac{0.4}{0.2} = 2$$

2 gallons of cream should be used.

33. Let $x =$ the number of additional problems the student gets correct.

$$\boxed{\begin{array}{c}\text{Original}\\\text{\# correct}\end{array}} + \boxed{\begin{array}{c}\text{Additional}\\\text{\# correct}\end{array}} = \boxed{\begin{array}{c}\text{Total}\\\text{\# correct}\end{array}}$$

$$0.70(30) + x = 0.80(45)$$
$$21 + x = 36$$
$$x = 15$$

The student must get all 15 of the additional problems correct to receive a grade of 80%.

35. Let $x =$ number of points earned on the final.

$$\boxed{\begin{array}{c}\text{Original}\\\text{points}\end{array}} + \boxed{\begin{array}{c}\text{Points}\\\text{on final}\end{array}} = \boxed{\begin{array}{c}\text{Points}\\\text{at end}\end{array}}$$

$$375 + x = 0.90(450)$$
$$375 + x = 405$$
$$x = 30$$

She must earn 30 points on the final to receive a grade of 90% (A).

37. Let $w =$ the wholesale price of the textbook.

$$\boxed{\begin{array}{c}\text{Wholesale}\\\text{price}\end{array}} + \boxed{\text{Markup}} = \boxed{\begin{array}{c}\text{Retail}\\\text{price}\end{array}}$$

$$w + 0.30w = 65$$
$$1.3w = 65$$
$$w = \frac{65}{1.3} = 50$$

The bookstore pays $50 for the book.

39. Let $x =$ space in front of first partition. Then $x + 3$ and $x + 6$ are the other spaces.

$$\boxed{\text{Lengths of spaces}} + \boxed{\text{Lengths of partitions}} = \boxed{\text{Total length}}$$

$$x + x + 3 + x + 6 + 2(0.5) = 28$$
$$3x + 10 = 28$$
$$3x = 18$$
$$x = 6 \quad \text{The first space should be 6 inches.}$$

41. **Answers may vary.**

43. We need to know how far or how long he drove while at the other rate in order to solve the problem.

Chapter 1 Summary   (page 76)

1. whole numbers: $0, 1, 2, 4$

2. natural numbers: $1, 2, 4$

3.   rational numbers: $-4, -\frac{2}{3}, 0, 1, 2, 4$

4.   integers: $-4, 0, 1, 2, 4$

5.   irrational numbers: π

6.   real numbers: $-4, -\frac{2}{3}, 0, 1, 2, \pi, 4$

7.   negative numbers: $-4, -\frac{2}{3}$ **8.**   positive numbers: $1, 2, \pi, 4$ **9.**   prime numbers: 2

10.   composite numbers: 4 **11.**   even integers: $-4, 0, 2, 4$ **12.**   odd integers: 1

13.

14.

15. $\{x | x \geq -4\} \Rightarrow$

16. $\{x | -2 < x \leq 6\} \Rightarrow$

17. $(-2, 3) \Rightarrow$

18. $[2, 6] \Rightarrow$

19. $\{x | x > 2\} \Rightarrow$

20. $(-\infty, -1) \Rightarrow$

21. $(-\infty, 0] \cup (2, \infty) \Rightarrow$

22. $|0| = 0$ **23.** $|-1| = 1$ **24.** $|8| = 8$ **25.** $-|8| = -(8) = -8$

26. $3 + (+5) = +8$

27. $-6 + (-3) = -(6 + 3) = -9$

28. $-15 + (-13) = -(15 + 13) = -28$

29. $25 + 32 = +(25 + 32) = +57$

30. $-2 + 5 = +(5 - 2) = 3$

31. $3 + (-12) = -(12 - 3) = -9$

32. $8 + (-3) = +(8 - 3) = +5$

33. $7 + (-9) = -(9 - 7) = -2$

34. $-25 + 12 = -(25 - 12) = -13$

35. $-30 + 35 = +(35 - 30) = +5$

36. $-3 - 10 = -3 + (-10) = -13$

37. $-8 - (-3) = -8 + 3 = -5$

38. $27 - (-12) = 27 + 12 = 39$

39. $38 - (-15) = 38 + 15 = 53$

40. $(+5)(+7) = +35$ **41.** $(-6)(-7) = +42$ **42.** $\dfrac{-16}{-4} = 4$ **43.** $\dfrac{-25}{-5} = 5$

44. $4(-3) = -12$ **45.** $-3(8) = -24$ **46.** $\dfrac{-8}{2} = -4$ **47.** $\dfrac{8}{-4} = -2$

48. $-4(3-6) = -4(-3) = 12$ **49.** $3[8 - (-1)] = 3(9) = 27$

50. $-[4 - 2(6-4)] = -[4 - 2(2)] = -[4-4] = -[0] = 0$

51. $3[-5 + 3(2-7)] = 3[-5 + 3(-5)] = 3[-5 + (-15)] = 3[-20] = -60$

52. $\dfrac{3-8}{10-5} = \dfrac{-5}{5} = -1$ **53.** $\dfrac{-32-8}{6-16} = \dfrac{-40}{-10} = 4$

54. mean $= \dfrac{12 + 13 + 14 + 14 + 15 + 15 + 15 + 17 + 19 + 20}{10} = \dfrac{154}{10} = 15.4$

55. median $= \dfrac{15 + 15}{2} = \dfrac{30}{2} = 15$ **56.** mode $= 15$

57. Yes. The mean, median and mode **could** be the same value for a group of numbers.

58. $\dfrac{3a - 2b}{cd} = \dfrac{3(5) - 2(-2)}{(-3)(2)} = \dfrac{15 - (-4)}{-6}$
$= \dfrac{19}{-6} = -\dfrac{19}{6}$

59. $\dfrac{3b + 2d}{ac} = \dfrac{3(-2) + 2(2)}{5(-3)} = \dfrac{-6 + 4}{-15}$
$= \dfrac{-2}{-15} = \dfrac{2}{15}$

60. $\dfrac{ab + cd}{c(b-d)} = \dfrac{5(-2) + (-3)(2)}{-3(-2-2)} = \dfrac{-10 + (-6)}{-3(-4)} = \dfrac{-16}{12} = -\dfrac{4}{3}$

61. $\dfrac{ac - bd}{a(d+c)} = \dfrac{5(-3) - (-2)(2)}{5[2 + (-3)]} = \dfrac{-15 - (-4)}{5(-1)} = \dfrac{-11}{-5} = \dfrac{11}{5}$

62. distributive property   **63.** commutative property of addition

64. associative property of addition   **65.** additive identity property

66. additive inverse property   **67.** commutative property of multiplication

68. associative property of multiplication   **69.** multiplicative identity property

70. multiplicative inverse property   **71.** double negative rule

72. $3^6 = 3 \cdot 3 \cdot 3 \cdot 3 \cdot 3 \cdot 3 = 729$

73. $-2^6 = -1 \cdot 2^6 = -1 \cdot 2 \cdot 2 \cdot 2 \cdot 2 \cdot 2 \cdot 2$
$= -1 \cdot 64 = -64$

74. $(-4)^3 = (-4)(-4)(-4) = -64$ **75.** $-5^{-4} = -1 \cdot 5^{-4} = -1 \cdot \dfrac{1}{5^4} = -\dfrac{1}{625}$

76. $(3x^4)(-2x^2) = 3(-2)x^4x^2 = -6x^{4+2}$
$= -6x^6$

77. $(-x^5)(3x^3) = -1(3)x^5x^3 = -3x^{5+3}$
$= -3x^8$

78. $x^{-4}x^3 = x^{-4+3} = x^{-1} = \dfrac{1}{x}$

79. $x^{-10}x^{12} = x^{-10+12} = x^2$

80. $(3x^2)^3 = 3^3(x^2)^3 = 27x^6$

81. $(4x^4)^4 = 4^4(x^4)^4 = 256x^{16}$

82. $(-2x^2)^5 = (-2)^5(x^2)^5 = -32x^{10}$

83. $-(-3x^3)^5 = -(-3)^5(x^3)^5 = -(-243)x^{15}$
$= 243x^{15}$

84. $(x^2)^{-5} = x^{-10} = \dfrac{1}{x^{10}}$

85. $(x^{-4})^{-5} = x^{20}$

86. $(3x^{-3})^{-2} = 3^{-2}x^6 = \dfrac{x^6}{3^2} = \dfrac{x^6}{9}$

87. $(2x^{-4})^4 = 2^4x^{-16} = \dfrac{16}{x^{16}}$

88. $\dfrac{x^6}{x^4} = x^{6-4} = x^2$

89. $\dfrac{x^{12}}{x^7} = x^{12-7} = x^5$

90. $\dfrac{a^7}{a^{12}} = a^{7-12} = a^{-5} = \dfrac{1}{a^5}$

91. $\dfrac{a^4}{a^7} = a^{4-7} = a^{-3} = \dfrac{1}{a^3}$

92. $\dfrac{y^{-3}}{y^4} = y^{-3-4} = y^{-7} = \dfrac{1}{y^7}$

93. $\dfrac{y^5}{y^{-4}} = y^{5-(-4)} = y^9$

94. $\dfrac{x^{-5}}{x^{-4}} = x^{-5-(-4)} = x^{-1} = \dfrac{1}{x}$

95. $\dfrac{x^{-6}}{x^{-9}} = x^{-6-(-9)} = x^3$

96. $(3x^2y^3)^2 = 3^2(x^2)^2(y^3)^2 = 9x^4y^6$

97. $(-3a^3b^2)^{-4} = \dfrac{1}{(-3a^3b^2)^4} = \dfrac{1}{(-3)^4(a^3)^4(b^2)^4} = \dfrac{1}{81a^{12}b^8}$

98. $\left(\dfrac{3x^2}{4y^3}\right)^{-3} = \left(\dfrac{4y^3}{3x^2}\right)^3 = \dfrac{(4y^3)^3}{(3x^2)^3} = \dfrac{4^3(y^3)^3}{3^3(x^2)^3} = \dfrac{64y^9}{27x^6}$

99. $\left(\dfrac{4y^{-2}}{5y^{-3}}\right)^3 = \dfrac{4^3(y^{-2})^3}{5^3(y^{-3})^3} = \dfrac{64y^{-6}}{125y^{-9}} = \dfrac{64y^3}{125}$

100. $19{,}300{,}000{,}000 = 1.93 \times 10^{10}$

101. $0.0000000273 = 2.73 \times 10^{-8}$

102. $7.2 \times 10^7 = 72{,}000{,}000$

103. $8.3 \times 10^{-9} = 0.0000000083$

104. $\dfrac{270{,}000{,}000 \text{ persons}}{1} \cdot \dfrac{1640 \text{ gallons}}{1 \text{ person}} = (2.7 \times 10^8)(1.64 \times 10^3) \text{ gallons} = 4.428 \times 10^{11} \text{ gallons}$

105.
$$5x + 12 = 37$$
$$5x + 12 - 12 = 37 - 12$$
$$5x = 25$$
$$\frac{5x}{5} = \frac{25}{5}$$
$$x = 5$$

106.
$$-3x - 7 = 20$$
$$-3x - 7 + 7 = 20 + 7$$
$$-3x = 27$$
$$\frac{-3x}{-3} = \frac{27}{-3}$$
$$x = -9$$

107.
$$4(y - 1) = 28$$
$$4y - 4 = 28$$
$$4y - 4 + 4 = 28 + 4$$
$$4y = 32$$
$$\frac{4y}{4} = \frac{32}{4}$$
$$y = 8$$

108.
$$3(x + 7) = 42$$
$$3x + 21 = 42$$
$$3x + 21 - 21 = 42 - 21$$
$$3x = 21$$
$$\frac{3x}{3} = \frac{21}{3}$$
$$x = 7$$

109.
$$13(x - 9) - 2 = 7x - 5$$
$$13x - 117 - 2 = 7x - 5$$
$$13x - 119 = 7x - 5$$
$$13x - 7x - 119 = 7x - 7x - 5$$
$$6x - 119 + 119 = -5 + 119$$
$$6x = 114$$
$$\frac{6x}{6} = \frac{114}{6}$$
$$x = 19$$

110.
$$\frac{8(x - 5)}{3} = 2(x - 4)$$
$$3 \cdot \frac{8(x - 5)}{3} = 3 \cdot 2(x - 4)$$
$$8(x - 5) = 6(x - 4)$$
$$8x - 40 = 6x - 24$$
$$8x - 6x - 40 = 6x - 6x - 24$$
$$2x - 40 + 40 = -24 + 40$$
$$2x = 16$$
$$\frac{2x}{2} = \frac{16}{2}$$
$$x = 8$$

111.
$$\frac{3y}{4} - 13 = -\frac{y}{3}$$
$$12\left(\frac{3y}{4} - 13\right) = 12\left(-\frac{y}{3}\right)$$
$$12\left(\frac{3y}{4}\right) - 12(13) = -4y$$
$$3(3y) - 156 = -4y$$
$$9y - 156 = -4y$$
$$9y + 4y - 156 = -4y + 4y$$
$$13y - 156 + 156 = 0 + 156$$
$$13y = 156$$
$$\frac{13y}{13} = \frac{156}{13}$$
$$y = 12$$

112.
$$\frac{2y}{5} + 5 = \frac{14y}{10}$$
$$10\left(\frac{2y}{5} + 5\right) = 10\left(\frac{14y}{10}\right)$$
$$10\left(\frac{2y}{5}\right) + 10(5) = 14y$$
$$2(2y) + 50 = 14y$$
$$4y + 50 = 14y$$
$$4y - 14y + 50 = 14y - 14y$$
$$-10y + 50 - 50 = 0 - 50$$
$$-10y = -50$$
$$\frac{-10y}{-10} = \frac{-50}{-10}$$
$$y = 5$$

113. $V = \dfrac{4}{3}\pi r^3$

$3V = 3\left(\dfrac{4}{3}\pi r^3\right)$

$3V = 4\pi r^3$

$\dfrac{3V}{4\pi} = \dfrac{4\pi r^3}{4\pi}$

$\dfrac{3V}{4\pi} = r^3$, or $r^3 = \dfrac{3V}{4\pi}$

114. $V = \dfrac{1}{3}\pi r^2 h$

$3V = 3\left(\dfrac{1}{3}\pi r^2 h\right)$

$3V = \pi r^2 h$

$\dfrac{3V}{\pi r^2} = \dfrac{\pi r^2 h}{\pi r^2}$

$\dfrac{3V}{\pi r^2} = h$, or $h = \dfrac{3V}{\pi r^2}$

115. $v = \dfrac{1}{6}ab(x+y)$

$6v = 6 \cdot \dfrac{1}{6}ab(x+y)$

$6v = ab(x+y)$

$\dfrac{6v}{ab} = \dfrac{ab(x+y)}{ab}$

$\dfrac{6v}{ab} = x+y$

$\dfrac{6v}{ab} - y = x+y-y$

$\dfrac{6v}{ab} - y = x$, or $x = \dfrac{6v}{ab} - y$

116. $V = \pi h^2\left(r - \dfrac{h}{3}\right)$

$\dfrac{V}{\pi h^2} = \dfrac{\pi h^2\left(r - \frac{h}{3}\right)}{\pi h^2}$

$\dfrac{V}{\pi h^2} = r - \dfrac{h}{3}$

$\dfrac{V}{\pi h^2} + \dfrac{h}{3} = r - \dfrac{h}{3} + \dfrac{h}{3}$

$\dfrac{V}{\pi h^2} + \dfrac{h}{3} = r$, or $r = \dfrac{V}{\pi h^2} + \dfrac{h}{3}$

117. Let x = the length of the first piece.
Then $3x$ = the length of the other piece.

$\boxed{\text{Sum of lengths}} = \boxed{\text{Total length}}$

$x + 3x = 20$

$4x = 20$

$x = 5$

He should cut the board 5 feet from one end.

118. Let w = the width of the rectangle. Then $w + 4$ = the length of the rectangle.

$\boxed{\text{Perimeter}} = 28$

$w + w + w + 4 + w + 4 = 28$

$4w + 8 = 28$

$4w = 20$

$w = 5$

The dimensions are 5 meters by 9 meters, for an area of $45\,\text{m}^2$.

119.

48 lb.   180 lb.

10 ft. △ x ft.

20 ft.

$\boxed{\begin{array}{c}\text{Sue's force}\\\text{times distance}\end{array}} = \boxed{\begin{array}{c}\text{Father's force}\\\text{times distance}\end{array}}$

$48(10) = 180x$

$480 = 180x$

$\dfrac{480}{180} = x$

$\dfrac{8}{3} = x$

$2\dfrac{2}{3} = x$

The father should sit $2\dfrac{2}{3}$ feet from the fulcrum.

CHAPTER 1 SUMMARY

120. Let x = the amount she invests at 10%.
Then $25{,}000 - x$ = amount invested at 9%.

$$\boxed{\begin{array}{c}\text{Interest}\\\text{at 10\%}\end{array}} + \boxed{\begin{array}{c}\text{Interest}\\\text{at 9\%}\end{array}} = \boxed{\begin{array}{c}\text{Total}\\\text{interest}\end{array}}$$

$$0.10x + 0.09(25000 - x) = 2430$$
$$0.10x + 2250 - 0.09x = 2430$$
$$0.01x = 180$$
$$x = 18000$$

She invested $18,000 at 10% interest and $7000 at 9% interest.

121. Let x = liters of water added.

$$\boxed{\begin{array}{c}\text{Alcohol}\\\text{at start}\end{array}} + \boxed{\begin{array}{c}\text{Alcohol}\\\text{added}\end{array}} = \boxed{\begin{array}{c}\text{Alcohol}\\\text{at end}\end{array}}$$

$$0.12(20) + 0(x) = 0.08(20 + x)$$
$$2.4 = 1.6 + 0.08x$$
$$0.8 = 0.08x$$
$$10 = x$$

10 liters of water should be added.

122. Let t = time for them to travel.

	Rate	Time	Dist.
Car	55	t	$55t$
Motorcycle	40	t	$40t$

$$\boxed{\text{Car distance}} - \boxed{\text{Motorcycle distance}} = \boxed{\text{Distance between them}}$$

$$55t - 40t = 5$$
$$15t = 5$$
$$t = \tfrac{5}{15} = \tfrac{1}{3}$$

It will take $\tfrac{1}{3}$ of an hour for the vehicles to be 5 miles apart.

Chapter 1 Test (page 81)

1. 1, 2 and 5 are natural numbers.

2. $\sqrt{7}$ is an irrational number.

3.

4.

5. $\{x | x > 4\} \Rightarrow$

6. $[-3, \infty) \Rightarrow$

7. $\{x | -2 \le x < 4\} \Rightarrow$

8. $(-\infty, -1] \cup [2, \infty) \Rightarrow$

9. $-|8| = -8$

10. $|-5| = 5$

11. $7 + (-5) = 2$

12. $-5(-4) = 20$

13. $\dfrac{12}{-3} = -4$

14. $-4 - \dfrac{-15}{3} = -4 - (-5) = -4 + 5 = 1$

15. mean $= \dfrac{-2 + 0 + 2 + (-2) + 3 + (-1) + (-1) + 1 + 1 + 2}{10} = \dfrac{3}{10} = 0.3$

16. $-2, -2, -1, -1, 0, 1, 1, 2, 2, 3$: median $= \dfrac{0 + 1}{2} = \dfrac{1}{2} = 0.5$

17. $ab = (2)(-3) = -6$

18. $a + bc = 2 + (-3)(4) = 2 + (-12) = -10$

19. $\begin{aligned} ab - bc &= 2(-3) - (-3)(4) = -6 - (-12) \\ &= -6 + 12 = 6 \end{aligned}$

20. $\dfrac{-3b + a}{ac - b} = \dfrac{-3(-3) + 2}{2(4) - (-3)} = \dfrac{9 + 2}{8 + 3} = \dfrac{11}{11}$
$= 1$

21. commutative property of addition

22. distributive property

23. $x^3 x^5 = x^{3+5} = x^8$

24. $\left(2x^2 y^3\right)^3 = 2^3 \left(x^2\right)^3 \left(y^3\right)^3 = 8x^6 y^9$

25. $\left(m^{-4}\right)^2 = m^{-8} = \dfrac{1}{m^8}$

26. $\left(\dfrac{m^2 n^3}{m^4 n^{-2}}\right)^{-2} = \left(\dfrac{m^4 n^{-2}}{m^2 n^3}\right)^2 = \left(\dfrac{m^2}{n^5}\right)^2$
$= \dfrac{m^4}{n^{10}}$

27. $4{,}700{,}000 = 4.7 \times 10^6$

28. $0.00000023 = 2.3 \times 10^{-7}$

29. $6.53 \times 10^5 = 653{,}000$

30. $24.5 \times 10^{-3} = 0.0245$

31.
$$\begin{aligned} 9(x + 4) + 4 &= 4(x - 5) \\ 9x + 36 + 4 &= 4x - 20 \\ 9x + 40 &= 4x - 20 \\ 9x - 4x + 40 &= 4x - 4x - 20 \\ 5x + 40 &= -20 \\ 5x + 40 - 40 &= -20 - 40 \\ 5x &= -60 \\ \tfrac{5x}{5} &= \tfrac{-60}{5} \\ x &= -12 \end{aligned}$$

32.
$$\begin{aligned} \dfrac{y - 1}{5} + 2 &= \dfrac{2y - 3}{3} \\ 15\left(\dfrac{y - 1}{5} + 2\right) &= 15\left(\dfrac{2y - 3}{3}\right) \\ 15\left(\dfrac{y - 1}{5}\right) + 15(2) &= 5(2y - 3) \\ 3(y - 1) + 30 &= 10y - 15 \\ 3y - 3 + 30 &= 10y - 15 \\ 3y + 27 &= 10y - 15 \\ 3y - 10y + 27 &= 10y - 10y - 15 \\ -7y + 27 - 27 &= -15 - 27 \\ -7y &= -42 \\ \tfrac{-7y}{-7} &= \tfrac{-42}{-7} \\ y &= 6 \end{aligned}$$

33.

$$p = L + \frac{s}{f}i$$

$$p - L = L - L + \frac{s}{f}i$$

$$p - L = \frac{s}{f}i$$

$$f(p - L) = f \cdot \frac{s}{f}i$$

$$f(p - L) = si$$

$$\frac{f(p - L)}{s} = \frac{si}{s}$$

$$\frac{f(p - L)}{s} = i, \text{ or } i = \frac{f(p - L)}{s}$$

34. Let w = the width of the rectangle. Then $w + 5$ = the length of the rectangle.

$$\boxed{\text{Perimeter}} = 26$$

$$w + w + w + 5 + w + 5 = 26$$

$$4w + 10 = 26$$

$$4w = 16$$

$$w = 4$$

The dimensions are 4 cm by 9 cm, for an area of 36 cm^2.

35. Let x = the amount invested at 9%. Then $10000 - x$ = the amount at 8%.

$$\boxed{\begin{array}{c}\text{Interest} \\ \text{at 9\%}\end{array}} + \boxed{\begin{array}{c}\text{Interest} \\ \text{at 8\%}\end{array}} = \boxed{\begin{array}{c}\text{Total} \\ \text{interest}\end{array}}$$

$$0.09x + 0.08(10000 - x) = 860$$

$$0.09x + 800 - 0.08x = 860$$

$$0.01x = 60$$

$$x = 6000$$

He has $6000 invested at 9% interest, and **$4000 invested at 8% interest.**

36. Let x = liters of water added.

$$\boxed{\begin{array}{c}\text{Salt} \\ \text{at start}\end{array}} + \boxed{\begin{array}{c}\text{Salt} \\ \text{added}\end{array}} = \boxed{\begin{array}{c}\text{Salt} \\ \text{at end}\end{array}}$$

$$0.05(20) + 0(x) = 0.01(20 + x)$$

$$1 + 0 = 0.2 + 0.01x$$

$$0.8 = 0.01x$$

$$80 = x$$

80 liters of water should be added.

Exercise 2.1 (page 95)

1. $-3 - 3(-5) = -3 + 15$
$$= 12$$

3. $\dfrac{-3 + 5(2)}{9 + 5} = \dfrac{-3 + 10}{14}$
$$= \dfrac{7}{14} = \dfrac{1}{2}$$

5.
$$-4x + 7 = -21$$
$$-4x + 7 - 7 = -21 - 7$$
$$-4x = -28$$
$$\frac{-4x}{-4} = \frac{-28}{-4}$$
$$x = 7$$

7. ordered pair

9. origin

11. rectangular coordinate

13. the y-intercept

15. vertical

17. sub 1

19-25.

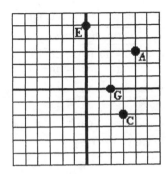

27. $(2, 4)$

29. $(-2, -1)$

31. $(4, 0)$

33. $(0, 0)$

35. $y = -x + 4$
   $y = -(-1) + 4 = 1 + 4 = 5$
   $y = -(0) + 4 = 0 + 4 = 4$
   $y = -2 + 4 = 2$

x	y
-1	5
0	4
2	2

37. $y = 2x - 3$
   $y = 2(-1) - 3 = -2 - 3 = -5$
   $y = 2(0) - 3 = 0 - 3 = -3$
   $y = 2(3) - 3 = 6 - 3 = 3$

x	y
-1	-5
0	-3
3	3

39. From Exercise #35:

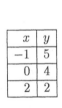

x	y
-1	5
0	4
2	2

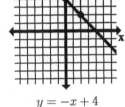

$$y = -x + 4$$

41. From Exercise #37:

x	y
-1	-5
0	-3
3	3

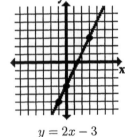

$$y = 2x - 3$$

43. $3x + 4y = 12$ $\qquad$ $3x + 4y = 12$
   $3x + 4(0) = 12$ $\qquad$ $3(0) + 4y = 12$
   $\qquad 3x = 12$ $\qquad\qquad$ $4y = 12$
   $\qquad\quad x = 4$ $\qquad\qquad\quad$ $y = 3$
   x-intercept: $(4, 0)$ $\quad$ y-intercept: $(0, 3)$

$3x + 4y = 12$

45.

$$y = -3x + 2$$
$$0 = -3x + 2$$
$$3x = 2$$
$$x = \frac{2}{3}$$

x-intercept: $\left(\frac{2}{3}, 0\right)$

$$y = -3x + 2$$
$$y = -3(0) + 2$$
$$y = 2$$

y-intercept: $(0, 2)$

$$y = -3x + 2$$

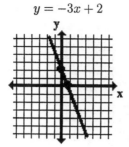

47.

$$y = \frac{3}{2}x$$
$$0 = \frac{3}{2}x$$
$$2(0) = 2 \cdot \frac{3}{2}x$$
$$0 = 3x$$
$$x = \frac{0}{3} = 0$$

x-intercept: $(0, 0)$

Pick a value for x
such as $x = 4$.
$$y = \frac{3}{2}x$$
$$y = \frac{3}{2}(4)$$
$$y = 6$$

point: $(4, 6)$

$$y = \frac{3}{2}x$$

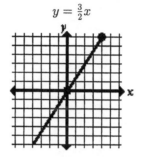

49.

$$3y = 6x - 9$$
$$3(0) = 6x - 9$$
$$0 = 6x - 9$$
$$-6x = -9$$
$$x = \frac{-9}{-6} = \frac{3}{2}$$

x-intercept: $\left(\frac{3}{2}, 0\right)$

$$3y = 6x - 9$$
$$3y = 6(0) - 9$$
$$3y = 0 - 9$$
$$3y = -9$$
$$y = \frac{-9}{3} = -3$$

y-intercept: $(0, -3)$

$$3y = 6x - 9$$

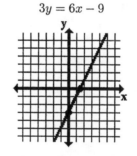

51.

$$3x + 4y - 8 = 0$$
$$3x + 4(0) - 8 = 0$$
$$3x - 8 = 0$$
$$3x = 8$$
$$x = \frac{8}{3}$$

x-intercept: $\left(\frac{8}{3}, 0\right)$

$$3x + 4y - 8 = 0$$
$$3(0) + 4y - 8 = 0$$
$$4y - 8 = 0$$
$$4y = 8$$
$$y = 2$$

y-intercept: $(0, 2)$

$$3x + 4y - 8 = 0$$

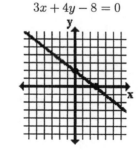

53. $x = 3$
vertical line with
x-coordinate of 3

$x = 3$

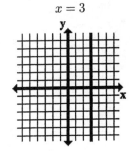

55. $-3y + 2 = 5$
$ -3y = 3$
$ y = -1$
horizontal line
with y-coordinate
of -1

$y = -1$

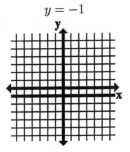

57. $x = \frac{x_1 + x_2}{2} = \frac{0+6}{2} = \frac{6}{2} = 3$
$y = \frac{y_1 + y_2}{2} = \frac{0+8}{2} = \frac{8}{2} = 4$
midpoint: $(3, 4)$

59. $x = \frac{x_1 + x_2}{2} = \frac{6+12}{2} = \frac{18}{2} = 9$
$y = \frac{y_1 + y_2}{2} = \frac{8+16}{2} = \frac{24}{2} = 12$
midpoint: $(9, 12)$

61. $x = \frac{x_1 + x_2}{2} = \frac{2+5}{2} = \frac{7}{2}$
$y = \frac{y_1 + y_2}{2} = \frac{4+8}{2} = \frac{12}{2} = 6$
midpoint: $\left(\frac{7}{2}, 6\right)$

63. $x = \frac{x_1 + x_2}{2} = \frac{-2+3}{2} = \frac{1}{2}$
$y = \frac{y_1 + y_2}{2} = \frac{-8+4}{2} = \frac{-4}{2} = -2$
midpoint: $\left(\frac{1}{2}, -2\right)$

65. $x = \frac{x_1 + x_2}{2} = \frac{-3+(-5)}{2} = \frac{-8}{2} = -4$
$y = \frac{y_1 + y_2}{2} = \frac{5+(-5)}{2} = \frac{0}{2} = 0$
midpoint: $(-4, 0)$

67. $x = \frac{x_1 + x_2}{2} \qquad y = \frac{y_1 + y_2}{2}$
$-2 = \frac{-8+x_2}{2} \qquad 3 = \frac{5+y_2}{2}$
$-4 = -8 + x_2 \qquad 6 = 5 + y_2$
$4 = x_2 \qquad\quad\ 1 = y_2$
Q has coordinates $(4, 1)$.

69. $y = 3.7x - 4.5$; x-intercept: $(1.22, 0)$

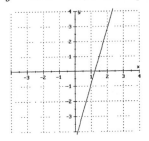

71. $1.5x - 3y = 7$; x-intercept: $(4.67, 0)$

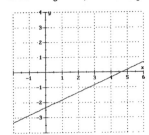

73.

x	2	4	5	6
y	12	24	30	36

Graph the ordered pairs
and connect with a line:

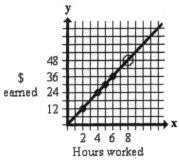

The point $(8, 48)$ is on the graph. The
student will earn $48 for 8 hours of work.

75.

x	0	1	3
y	15,000	12,000	6000

Graph the ordered pairs
and connect with a line:

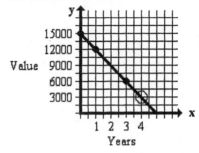

The point $(4, 3000)$ is on the graph.
The car will be worth $3000 in 4 years.

77. To find the value of the house after 5 years,
let $x = 5$:

$$y = 7500x + 125,000$$
$$= 7500(5) + 125,000$$
$$= 37,500 + 125,000$$
$$= 162,500$$

It will be worth $162,500 after 5 years.

79. To find the number of TVs sold,
let $p = 150$:

$$p = -\frac{1}{10}q + 170$$
$$150 = -\frac{1}{10}q + 170$$
$$\tfrac{1}{10}q = 20$$
$$q = 200$$

200 TVs will be sold at a price of $150.

81. The shuttle is launched from the point $(3, 0)$ along the line $x = 3$. Let $x = 3$ in the equation:
$y = 2x + 6 = 2(3) + 6 = 6 + 6 = 12 \Rightarrow$ The shuttle will stay in range for 12 miles.

83. $y = 0.25x + 5$

x	y
4	6
8	7
12	8
16	9

It will cost $10 to make 20 calls.

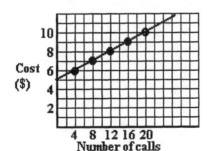

85. Answers may vary.

87. If the line $y = ax + b$ passes through only quadrants I and II, then $a = 0$ and $b > 0$.

Exercise 2.2 (page 106)

1. $(x^3y^2)^3 = x^9y^6$

3. $(x^{-3}y^2)^{-4} = x^{12}y^{-8} = \dfrac{x^{12}}{y^8}$

5. $\left(\dfrac{3x^2y^3}{8}\right)^0 = 1$

7. $y;\ x$

9. $\dfrac{y_2 - y_1}{x_2 - x_1}$

11. run

13. vertical

15. Parallel

17. $m = \dfrac{\Delta y}{\Delta x} = \dfrac{5 - (-3)}{2 - (-2)} = \dfrac{8}{4} = 2$

19. $m = \dfrac{\Delta y}{\Delta x} = \dfrac{9 - 0}{3 - 0} = \dfrac{9}{3} = 3$

21. $m = \dfrac{\Delta y}{\Delta x} = \dfrac{1 - 8}{6 - (-1)} = \dfrac{-7}{7} = -1$

23. $m = \dfrac{\Delta y}{\Delta x} = \dfrac{2 - (-1)}{-6 - 3} = \dfrac{3}{-9} = -\dfrac{1}{3}$

25. $m = \dfrac{\Delta y}{\Delta x} = \dfrac{5 - 5}{-9 - 7} = \dfrac{0}{-16} = 0$

27. $m = \dfrac{\Delta y}{\Delta x} = \dfrac{-2 - (-5)}{-7 - (-7)} = \dfrac{3}{0}$: undefined

29. $m = \dfrac{\Delta y}{\Delta x} = \dfrac{2.5 - 3.7}{3.7 - 2.5} = \dfrac{-1.2}{1.2} = -1$

31. Find two points on the line:

Let $x = 0$:   Let $y = 0$:

$3x + 2y = 12$ $3x + 2y = 12$

$3(0) + 2y = 12$ $3x + 2(0) = 12$

$2y = 12$ $3x = 12$

$y = 6$ $x = 4$

$(0, 6)$ $(4, 0)$

$m = \dfrac{\Delta y}{\Delta x} = \dfrac{6 - 0}{0 - 4} = \dfrac{6}{-4} = -\dfrac{3}{2}$

33. Find two points on the line:

Let $x = 0$:   Let $y = 0$:

$3x = 4y - 2$ $3x = 4y - 2$

$3(0) = 4y - 2$ $3x = 4(0) - 2$

$0 = 4y - 2$ $3x = -2$

$2 = 4y$ $x = -\dfrac{2}{3}$

$\dfrac{1}{2} = y$ $\left(-\dfrac{2}{3}, 0\right)$

$\left(0, \dfrac{1}{2}\right)$

$m = \dfrac{\Delta y}{\Delta x} = \dfrac{\frac{1}{2} - 0}{0 - \left(-\frac{2}{3}\right)} = \dfrac{\frac{1}{2}}{\frac{2}{3}} = \dfrac{3}{4}$

35. Find two points on the line:

Let $x = 0$:   Let $y = 0$:

$y = \dfrac{x - 4}{2}$ $y = \dfrac{x - 4}{2}$

$y = \dfrac{0 - 4}{2}$ $0 = \dfrac{x - 4}{2}$

$y = \dfrac{-4}{2}$ $0 = x - 4$

$y = -2$ $4 = x$

$(0, -2)$ $(4, 0)$

$m = \dfrac{\Delta y}{\Delta x} = \dfrac{-2 - 0}{0 - 4} = \dfrac{-2}{-4} = \dfrac{1}{2}$

37. $4y = 3(y+2)$
$4y = 3y + 6$
$y = 6$
horizontal line $\Rightarrow m = 0$

39. negative

41. positive

43. undefined

45. $m_1 \neq m_2 \Rightarrow$ not parallel
$m_1 \cdot m_2 = 3\left(-\dfrac{1}{3}\right) = -1 \Rightarrow$ perpendicular

47. $m_1 \neq m_2 \Rightarrow$ not parallel
$m_1 \cdot m_2 = 4(0.25) = 1 \Rightarrow$ not perpendicular

49. $m_1 = m_2 \Rightarrow$ parallel

51. $m_{\overline{PQ}} = \dfrac{\Delta y}{\Delta x} = \dfrac{2-4}{4-3} = \dfrac{-2}{1} = -2$
same slope $\Rightarrow$ parallel

53. $m_{\overline{PQ}} = \dfrac{\Delta y}{\Delta x} = \dfrac{5-1}{6-(-2)} = \dfrac{4}{8} = \dfrac{1}{2}$
opposite reciprocal slope $\Rightarrow$ perpendicular

55. $m_{\overline{PQ}} = \dfrac{\Delta y}{\Delta x} = \dfrac{6-4}{6-5} = \dfrac{2}{1} = 2$
neither parallel nor perpendicular

57. $m_{\overline{PQ}} = \dfrac{\Delta y}{\Delta x} = \dfrac{8-4}{4-(-2)} = \dfrac{4}{6} = \dfrac{2}{3}$
$m_{\overline{PR}} = \dfrac{\Delta y}{\Delta x} = \dfrac{12-4}{8-(-2)} = \dfrac{8}{10} = \dfrac{4}{5}$
Since they do not have the same slope,
the three points are not on the same line.

59. $m_{\overline{PQ}} = \dfrac{\Delta y}{\Delta x} = \dfrac{0-10}{-6-(-4)} = \dfrac{-10}{-2} = 5$
$m_{\overline{PR}} = \dfrac{\Delta y}{\Delta x} = \dfrac{5-10}{-1-(-4)} = \dfrac{-5}{3} = -\dfrac{5}{3}$
Since they do not have the same slope,
the three points are not on the same line.

61. $m_{\overline{PQ}} = \dfrac{\Delta y}{\Delta x} = \dfrac{8-4}{0-(-2)} = \dfrac{4}{2} = 2$
$m_{\overline{PR}} = \dfrac{\Delta y}{\Delta x} = \dfrac{12-4}{2-(-2)} = \dfrac{8}{4} = 2$
Since they have the same slope and a common
point, the three points are on the same line.

63. On the x-axis, all y-coordinates are 0.
The equation is $y = 0$, and the slope is 0.

65. Call the points $A(-3,4)$, $B(4,1)$ and $C(-1,-1)$. Compute these slopes:
$m_{\overline{AB}} = \dfrac{\Delta y}{\Delta x} = \dfrac{1-4}{4-(-3)} = \dfrac{-3}{7} = -\dfrac{3}{7}$
$m_{\overline{AC}} = \dfrac{\Delta y}{\Delta x} = \dfrac{-1-4}{-1-(-3)} = \dfrac{-5}{2} = -\dfrac{5}{2}$
$m_{\overline{BC}} = \dfrac{\Delta y}{\Delta x} = \dfrac{-1-1}{-1-4} = \dfrac{-2}{-5} = \dfrac{2}{5}$
Since $\overline{AC}$ and $\overline{BC}$ are perpendicular, it is a right triangle.

67. Call the points $A(a,0)$, $B(0,a)$, $C(-a,0)$ and $D(0,-a)$. Compute these slopes:

$$m_{\overline{AB}} = \frac{\Delta y}{\Delta x} = \frac{a-0}{0-a} = \frac{a}{-a} = -1$$

$$m_{\overline{BC}} = \frac{\Delta y}{\Delta x} = \frac{0-a}{-a-0} = \frac{-a}{-a} = 1$$

$$m_{\overline{CD}} = \frac{\Delta y}{\Delta x} = \frac{-a-0}{0-(-a)} = \frac{-a}{a} = -1$$

$$m_{\overline{DA}} = \frac{\Delta y}{\Delta x} = \frac{0-(-a)}{a-0} = \frac{a}{a} = 1$$

Thus, $\overline{AB} \perp \overline{BC}$, $\overline{BC} \perp \overline{CD}$, $\overline{CD} \perp \overline{DA}$, and $\overline{DA} \perp \overline{AB}$.

69. Call the points $A(0,0)$, $B(0,a)$, $C(b,c)$ and $D(b,a+c)$. Compute these slopes:

$$m_{\overline{AB}} = \frac{\Delta y}{\Delta x} = \frac{0-a}{0-0} = \frac{-a}{0} \Rightarrow \text{undefined}$$

$$m_{\overline{BD}} = \frac{\Delta y}{\Delta x} = \frac{a-(a+c)}{0-b} = \frac{a-a-c}{-b} = \frac{-c}{-b} = \frac{c}{b}$$

$$m_{\overline{DC}} = \frac{\Delta y}{\Delta x} = \frac{c-(a+c)}{b-b} = \frac{c-a-c}{0} = \frac{-a}{0} \Rightarrow \text{undefined}$$

$$m_{\overline{CA}} = \frac{\Delta y}{\Delta x} = \frac{c-0}{b-0} = \frac{c}{b}$$

Thus, $\overline{AB} \parallel \overline{DC}$ and $\overline{BD} \parallel \overline{CA}$ and the figure is a parallelogram.

71. $\text{slope} = \dfrac{\text{rise}}{\text{run}} = \dfrac{32 \text{ ft}}{1 \text{ mi}} = \dfrac{32 \text{ ft}}{5280 \text{ ft}} = \dfrac{1}{165}$

73. $m = \dfrac{\Delta y}{\Delta x} = \dfrac{2}{50} = 0.04$; $m = \dfrac{\Delta y}{\Delta x} = \dfrac{5}{50} = 0.1$; $m = \dfrac{\Delta y}{\Delta x} = \dfrac{8}{50} = 0.16$

75. $\text{slope} = \dfrac{\text{rise}}{\text{run}} = \dfrac{18 \text{ ft}}{5 \text{ ft}} = \dfrac{18}{5}$

77. Let x represent the number of years, and let y represent the sales. Then we know two points on the line: $(1, 85000)$ and $(3, 125000)$. Find the slope:

$$m = \frac{\Delta y}{\Delta x} = \frac{125000-85000}{3-1} = \frac{40000}{2} = \$20{,}000 \text{ per year.}$$

79. Answers may vary.

81. Find two points on the line:

Let $x = 0$:

$$Ax + By = C$$
$$A(0) + By = C$$
$$By = C$$
$$y = \frac{C}{B}$$

$\left(0, \frac{C}{B}\right)$ is on the line.

Let $y = 0$:

$$Ax + By = C$$
$$Ax + B(0) = C$$
$$Ax = C$$
$$x = \frac{C}{A}$$

$\left(\frac{C}{A}, 0\right)$ is on the line.

$$m = \frac{\Delta y}{\Delta x} = \frac{\frac{C}{B}-0}{0-\frac{C}{A}} = \frac{\frac{C}{B}}{-\frac{C}{A}} = -\frac{A}{B}$$

83. If the three points are on a line, then all slopes must be equal:

$$m = \frac{\Delta y}{\Delta x} = \frac{7-10}{5-7} = \frac{-3}{-2} = \frac{3}{2} \qquad m = \frac{\Delta y}{\Delta x} = \frac{a-7}{3-5} = \frac{a-7}{-2}$$

$$\frac{a-7}{-2} = \frac{3}{2}$$

$$-2 \cdot \frac{a-7}{-2} = -2 \cdot \frac{3}{2}$$

$$a - 7 = -3$$

$$a = 4$$

Exercise 2.3 (page 119)

1. $3(x+2) + x = 5x$

$3x + 6 + x = 5x$

$4x + 6 = 5x$

$6 = x$

3. $\dfrac{5(2-x)}{3} - 1 = x + 5$

$\dfrac{10 - 5x}{3} = x + 6$

$3 \cdot \dfrac{10 - 5x}{3} = 3(x+6)$

$10 - 5x = 3x + 18$

$10 = 8x + 18$

$-8 = 8x$

$-1 = x$

5. We want the alloy to contain 25% gold. Let $x =$ the amount of copper added to the alloy.

$$\boxed{\text{Gold at start}} + \boxed{\text{Gold added}} = \boxed{\text{Gold at end}}$$

$20 + 0 = 0.25(60 + x)$

$20 = 15 + 0.25x$

$5 = 0.25x$

$20 = x$

20 oz of copper should be added to the alloy.

7. $y - y_1 = m(x - x_1)$

9. $Ax + By = C$

11. perpendicular

13. $y - y_1 = m(x - x_1)$

$y - 7 = 5(x - 0)$

$y - 7 = 5x$

$-5x + y = 7$

$5x - y = -7$

15. $y - y_1 = m(x - x_1)$

$y - 0 = -3(x - 2)$

$y = -3x + 6$

$3x + y = 6$

17. Notice that the line goes through $P(2,5)$ and the point $(-1, 3)$. Find the slope:

$$m = \frac{\Delta y}{\Delta x} = \frac{5 - 3}{2 - (-1)} = \frac{2}{3}$$

Use point-slope form to find the equation:

$y - y_1 = m(x - x_1)$

$y - 5 = \dfrac{2}{3}(x - 2)$

$y - 5 = \dfrac{2}{3}(x - 2)$

$3(y - 5) = 3 \cdot \dfrac{2}{3}(x - 2)$

$3y - 15 = 2x - 4$

$-2x + 3y = 11$

$2x - 3y = -11$

19. Find the slope of the line:
$$m = \frac{\Delta y}{\Delta x} = \frac{4 - 0}{4 - 0} = \frac{4}{4} = 1$$
Use point-slope form to find the equation:
$$y - y_1 = m(x - x_1)$$
$$y - 0 = 1(x - 0)$$
$$y = x$$

21. Find the slope of the line:
$$m = \frac{\Delta y}{\Delta x} = \frac{4 - (-3)}{3 - 0} = \frac{7}{3}$$
Use point-slope form to find the equation:
$$y - y_1 = m(x - x_1)$$
$$y - 4 = \tfrac{7}{3}(x - 3)$$
$$y - 4 = \tfrac{7}{3}x - 7$$
$$y = \tfrac{7}{3}x - 3$$

23. Find the slope of the line:
$$m = \frac{\Delta y}{\Delta x} = \frac{4 - (-5)}{-2 - 3} = \frac{9}{-5} = -\frac{9}{5}$$
Use point-slope form to find the equation:
$$y - y_1 = m(x - x_1)$$
$$y - 4 = -\tfrac{9}{5}(x - (-2))$$
$$y - 4 = -\tfrac{9}{5}(x + 2)$$
$$y - 4 = -\tfrac{9}{5}x - \tfrac{18}{5}$$
$$y = -\tfrac{9}{5}x - \tfrac{18}{5} + 4$$
$$y = -\tfrac{9}{5}x + \tfrac{2}{5}$$

25.
$$y = mx + b$$
$$y = 3x + 17$$

27.
$$y = mx + b$$
$$5 = -7(7) + b$$
$$5 = -49 + b$$
$$54 = b$$
$$y = -7x + 54$$

29.
$$y = mx + b$$
$$-4 = 0(2) + b$$
$$-4 = b$$
$$y = 0x + (-4)$$
$$y = -4$$

31. Find the slope of the line:
$$m = \frac{\Delta y}{\Delta x} = \frac{8 - 10}{6 - 2} = \frac{-2}{4} = -\frac{1}{2}$$
$$y = mx + b$$
$$8 = -\frac{1}{2}(6) + b$$
$$8 = -3 + b$$
$$11 = b$$
$$y = -\frac{1}{2}x + 11$$

33.
$$y + 1 = x$$
$$y = x - 1$$
$$m = 1; b = -1 \Rightarrow (0, -1)$$

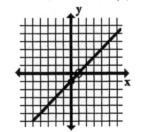

35.
$$x = \frac{3}{2}y - 3$$
$$2x = 3y - 6$$
$$-3y = -2x + 6$$
$$y = \frac{2}{3}x + 2$$
$$m = \frac{2}{3}; b = 2 \Rightarrow (0, 2)$$

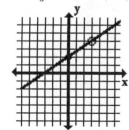

37.
$$3(y - 4) = -2(x - 3)$$
$$3y - 12 = -2x + 6$$
$$3y = -2x + 18$$
$$y = -\frac{2}{3}x + 6$$
$$m = -\frac{2}{3}; b = 6 \Rightarrow (0, 6)$$

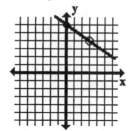

39.
$$3x - 2y = 8$$
$$-2y = -3x + 8$$
$$y = \frac{3}{2}x - 4$$
$$m = \frac{3}{2}; b = -4 \Rightarrow (0, -4)$$

41.
$$-2(x + 3y) = 5$$
$$-2x - 6y = 5$$
$$-6y = 2x + 5$$
$$y = -\frac{1}{3}x - \frac{5}{6}$$
$$m = -\frac{1}{3}; b = -\frac{5}{6} \Rightarrow \left(0, -\frac{5}{6}\right)$$

43.
$$x = \frac{2y - 4}{7}$$
$$7x = 2y - 4$$
$$-2y = -7x - 4$$
$$y = \frac{7}{2}x + 2$$
$$m = \frac{7}{2}; b = 2 \Rightarrow (0, 2)$$

45.
$$y = 3x + 4 \qquad y = 3x - 7$$
$$m = 3 \qquad\qquad m = 3$$
$$\text{parallel}$$

47.
$$x + y = 2 \qquad y = x + 5$$
$$y = -x + 2 \qquad m = 1$$
$$m = -1$$
$$\text{perpendicular}$$

49.
$$y = 3x + 7 \qquad 2y = 6x - 9$$
$$m = 3 \qquad\qquad y = 3x - \frac{9}{2}$$
$$m = 3$$
$$\text{parallel}$$

51.
$$x = 3y + 4 \qquad y = -3x + 7$$
$$-3y = -x + 4 \qquad m = -3$$
$$y = \frac{1}{3}x - \frac{4}{3}$$
$$m = \frac{1}{3}$$
$$\text{perpendicular}$$

53.
$$y = 3 \qquad\qquad x = 4$$
$$\text{horizontal line} \quad \text{vertical line}$$
$$\text{perpendicular}$$

55.
$$x = \frac{y - 2}{3} \qquad 3(y - 3) + x = 0$$
$$3x = y - 2 \qquad 3y - 9 + x = 0$$
$$-y = -3x - 2 \qquad\qquad 3y = -x + 9$$
$$y = 3x + 2 \qquad\qquad y = -\frac{1}{3}x + 3$$
$$m = 3 \qquad\qquad\qquad m = -\frac{1}{3}$$
$$\text{perpendicular}$$

57. Find the slope of the given line:

$y = 4x - 7 \Rightarrow m = 4$

Use the parallel slope.

$y - y_1 = m(x - x_1)$

$y - 0 = 4(x - 0)$

$y = 4x$

59. Find the slope of the given line:

$4x - y = 7$

$-y = -4x + 7$

$y = 4x - 7 \Rightarrow m = 4$

Use the parallel slope.

$y - y_1 = m(x - x_1)$

$y - 5 = 4(x - 2)$

$y - 5 = 4x - 8$

$y = 4x - 3$

61. Find the slope of the given line:

$x = \frac{5}{4}y - 2$

$4x = 5y - 8$

$-5y = -4x - 8$

$y = \frac{4}{5}x + \frac{8}{5} \Rightarrow m = \frac{4}{5}$

Use the parallel slope.

$y - y_1 = m(x - x_1)$

$y - (-2) = \frac{4}{5}(x - 4)$

$y + 2 = \frac{4}{5}x - \frac{16}{5}$

$y = \frac{4}{5}x - \frac{16}{5} - 2$

$y = \frac{4}{5}x - \frac{26}{5}$

63. Find the slope of the given line:

$y = 4x - 7 \Rightarrow m = 4$

Use the perpendicular slope.

$y - y_1 = m(x - x_1)$

$y - 0 = -\frac{1}{4}(x - 0)$

$y = -\frac{1}{4}x$

65. Find the slope of the given line:

$4x - y = 7$

$-y = -4x + 7$

$y = 4x - 7 \Rightarrow m = 4$

Use the perpendicular slope.

$y - y_1 = m(x - x_1)$

$y - 5 = -\frac{1}{4}(x - 2)$

$y - 5 = -\frac{1}{4}x + \frac{1}{2}$

$y = -\frac{1}{4}x + \frac{1}{2} + 5$

$y = -\frac{1}{4}x + \frac{11}{2}$

67. Find the slope of the given line:

$x = \frac{5}{4}y - 2$

$4x = 5y - 8$

$-5y = -4x - 8$

$y = \frac{4}{5}x + \frac{8}{5} \Rightarrow m = \frac{4}{5}$

Use the perpendicular slope.

$y - y_1 = m(x - x_1)$

$y - (-2) = -\frac{5}{4}(x - 4)$

$y + 2 = -\frac{5}{4}x + 5$

$y = -\frac{5}{4}x + 3$

69. Find the slope of the 1st line.   Find the slope of the 2nd line.   perpendicular

$m = -\frac{A}{B} = -\frac{4}{5}$ $m = -\frac{A}{B} = -\frac{5}{-4} = \frac{5}{4}$

71. Find the slope of the 1st line.   Find the slope of the 2nd line.   parallel

$m = -\frac{A}{B} = -\frac{2}{3}$ $m = -\frac{A}{B} = -\frac{6}{9} = -\frac{2}{3}$

73. The line $y = 3$ is horizontal, so any perpendicular line will be vertical.
Find the midpoint of the described segment:
$$x = \frac{x_1 + x_2}{2} = \frac{2 + (-6)}{2} = \frac{-4}{2} = -2; \quad y = \frac{y_1 + y_2}{2} = \frac{4 + 10}{2} = \frac{14}{2} = 7$$
The vertical line through the point $(-2, 7)$ is $x = -2$.

75. The line $x = 3$ is vertical, so any parallel line will be vertical.
Find the midpoint of the described segment:
$$x = \frac{x_1 + x_2}{2} = \frac{2 + 8}{2} = \frac{10}{2} = 5; \quad y = \frac{y_1 + y_2}{2} = \frac{-4 + 12}{2} = \frac{8}{2} = 4$$
The vertical line through the point $(5, 4)$ is $x = 5$.

77.
$$Ax + By = C$$
$$By = -Ax + C$$
$$\frac{By}{B} = \frac{-Ax + C}{B}$$
$$y = -\frac{A}{B}x + \frac{C}{B}$$
$$m = -\frac{A}{B}; \quad b = \frac{C}{B}$$

79. Let x represent the number of years since the truck was purchased, and let y represent the value of the truck. Then we know two points on the depreciation line: $(0, 19984)$ and $(8, 1600)$. Find the slope and then use point-slope form to find the depreciation equation.
$$m = \frac{\Delta y}{\Delta x} = \frac{19984 - 1600}{0 - 8} = \frac{18384}{-8} = -2298 \qquad y - y_1 = m(x - x_1)$$
$$y - 19984 = -2298(x - 0)$$
$$y = -2298x + 19984$$

81. Let x represent the number of years since the painting was purchased, and let y represent the value of the painting. Then we know two points on the appreciation line: $(0, 250000)$ and $(5, 500000)$. Find the slope and then use point-slope form to find the appreciation equation.
$$m = \frac{\Delta y}{\Delta x} = \frac{500000 - 250000}{5 - 0} = \frac{250000}{5} = 50000 \qquad y - y_1 = m(x - x_1)$$
$$y - 250000 = 50000(x - 0)$$
$$y = 50,000x + 250,000$$

83. Let x represent the number of years since the TV was purchased, and let y represent the value of the TV. Then we know two points on the depreciation line: $(0, 1750)$ and $(3, 800)$. Find the slope and then use point-slope form to find the depreciation equation.
$$m = \frac{\Delta y}{\Delta x} = \frac{1750 - 800}{0 - 3} = \frac{950}{-3} = -\frac{950}{3} \qquad y - y_1 = m(x - x_1)$$
$$y - 1750 = -\frac{950}{3}(x - 0)$$
$$y = -\frac{950}{3}x + 1750$$

85. Let x represent the number of years since the copy machine was purchased, and let y represent its value. Since the value decreases \$180 per year, the slope of the depreciation line is -180. Since the copier is worth \$1750 new, the y-intercept is 1750. Thus the depreciation equation can be found:

$y = mx + b$ Now let $x = 7$: $y = -180x + 1750 = -180(7) + 1750$

$y = -180x + 1750$ $= -1260 + 1750 = \$490$

87. Let x represent the number of years since the house was purchased, and let y represent its value. Since the value increases \$4000 per year, the slope of the depreciation line is 4000. We also know a point on the line: $(2, 122000)$. Use point-slope form to find the appreciation equation.

$y - y_1 = m(x - x_1)$ Substitute $x = 10$ and find its value:

$y - 122000 = 4000(x - 2)$ $y = 4000x + 114000 = 4000(10) + 114000$

$y - 122000 = 4000x - 8000$ $= 40000 + 114000 = \$154,000$

$y = 4000x + 114000$

89. The relationship between number of copies and the total cost is linear. Let x represent the number of copies (in hundreds), and let y represent the total cost. The slope of the line will be the charge per one hundred copies, or 15. We also know a point on the line: $(3, 75)$. Use point-slope form to find the equation of the line:

$y - y_1 = m(x - x_1)$ Let $x = 10$ and find the cost:

$y - 75 = 15(x - 3)$ $y = 15x + 30 = 15(10) + 30 = 150 + 30 = \$180$

$y - 75 = 15x - 45$

$y = 15x + 30$

91. Answers may vary.   **93.** Answers may vary.

95. Answers may vary.   **97.** Answers may vary.

99. To pass through II and IV, the slope must be negative. To pass through II and not III, the y-intercept must be positive. Thus $a < 0$ and $b > 0$.

Exercise 2.4 (page 129)

1. $\dfrac{y + 2}{2} = 4(y + 2)$ **3.** $\dfrac{2a}{3} + \dfrac{1}{2} = \dfrac{6a - 1}{6}$

$2 \cdot \dfrac{y + 2}{2} = 2 \cdot 4(y + 2)$ $6\left(\dfrac{2a}{3} + \dfrac{1}{2}\right) = 6 \cdot \dfrac{6a - 1}{6}$

$y + 2 = 8(y + 2)$ $2(2a) + 3 = 6a - 1$

$y + 2 = 8y + 16$ $4a + 3 = 6a - 1$

$-7y = 14$ $-2a = -4$

$y = -2$ $a = 2$

5.   input   **7.** x **9.**   function; input; output

11.   range   **13.**   0   **15.** $mx + b$

SECTION 2.4

17. $y = 2x + 3$ **is a function**, since each value of x corresponds to exactly one value of y.

19. $y = 2x^2$ **is a function**, since each value of x corresponds to exactly one value of y.

21. $y = 3 + 7x^2$ **is a function**, since each value of x corresponds to exactly one value of y.

23. $x = |y|$ **is not a function**, since $x = 2$ corresponds to both $y = 2$ and $y = -2$.

25. $f(x) = 3x$
$f(3) = 3(3) = 9$
$f(-1) = 3(-1) = -3$

27. $f(x) = 2x - 3$
$f(3) = 2(3) - 3 = 3$
$f(-1) = 2(-1) - 3 = -5$

29. $f(x) = 7 + 5x$
$f(3) = 7 + 5(3) = 22$
$f(-1) = 7 + 5(-1) = 2$

31. $f(x) = 9 - 2x$
$f(3) = 9 - 2(3) = 3$
$f(-1) = 9 - 2(-1) = 11$

33. $f(x) = x^2$
$f(2) = 2^2 = 4$
$f(3) = 3^2 = 9$

35. $f(x) = x^3 - 1$
$f(2) = 2^3 - 1 = 8 - 1$
$\quad = 7$
$f(3) = 3^3 - 1 = 27 - 1$
$\quad = 26$

37. $f(x) = (x + 1)^2$
$f(2) = (2 + 1)^2 = 3^2 = 9$
$f(3) = (3 + 1)^2 = 4^2 = 16$

39. $f(x) = 2x^2 - x$
$f(2) = 2(2)^2 - 2$
$\quad = 2(4) - 2$
$\quad = 8 - 2 = 6$
$f(3) = 2(3)^2 - 3$
$\quad = 2(9) - 3$
$\quad = 18 - 3 = 15$

41. $f(x) = |x| + 2$
$f(2) = |2| + 2$
$\quad = 2 + 2 = 4$
$f(-2) = |-2| + 2$
$\quad = 2 + 2 = 4$

43. $f(x) = x^2 - 2$
$f(2) = 2^2 - 2$
$\quad = 4 - 2 = 2$
$f(-2) = (-2)^2 - 2$
$\quad = 4 - 2 = 2$

45. $f(x) = \dfrac{1}{x + 3}$
$f(2) = \dfrac{1}{2 + 3} = \dfrac{1}{5}$
$f(-2) = \dfrac{1}{-2 + 3} = \dfrac{1}{1} = 1$

47. $f(x) = \dfrac{x}{x - 3}$
$f(2) = \dfrac{2}{2 - 3} = \dfrac{2}{-1}$
$\quad = -2$
$f(-2) = \dfrac{-2}{-2 - 3} = \dfrac{-2}{-5}$
$\quad = \dfrac{2}{5}$

49. $g(x) = 2x$
$g(w) = 2w$
$g(w + 1) = 2(w + 1) = 2w + 2$

51. $g(x) = 3x - 5$
$g(w) = 3w - 5$
$g(w + 1) = 3(w + 1) - 5 = 3w + 3 - 5$
$\quad = 3w - 2$

53. $f(3) + f(2) = [2(3) + 1] + [2(2) + 1]$
$\quad = 6 + 1 + 4 + 1 = 12$

55. $f(b) - f(a) = [2b + 1] - [2a + 1]$
$\quad = 2b + 1 - 2a - 1 = 2b - 2a$

57. $f(b) - 1 = 2b + 1 - 1 = 2b$

59. $f(0) + f\left(-\frac{1}{2}\right) = [2(0) + 1] + \left[2\left(-\frac{1}{2}\right) + 1\right] = 0 + 1 - 1 + 1 = 1$

61. domain $= \{-2, 4, 6\}$; range $= \{3, 5, 7\}$

63. To ensure that the denominator is not 0, x cannot equal 4. The domain is $(-\infty, 4) \cup (4, \infty)$. Because the numerator will never be 0, the range is $(-\infty, 0) \cup (0, \infty)$.

65. not a function

67. The domain is the set of all x-coordinates on the graph. The domain is $(-\infty, \infty)$.
The range is the set of all y-coordinates on the graph. The range is $(-\infty, \infty)$.
Since each vertical line passes through the graph at most once, it is the graph of a function.

69. $f(x) = 2x - 1$; D $= (-\infty, \infty)$
R $= (-\infty, \infty)$

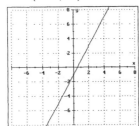

71. $2x - 3y = 6$; D $= (-\infty, \infty)0$
R $= (-\infty, \infty)$

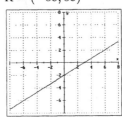

73. $y = 3x^2 + 2 \Rightarrow$ This is not a linear function because the exponent on x is 2.

75. $x = 3y - 4 \Rightarrow y = \frac{1}{3}x + \frac{4}{3} \Rightarrow$ This is a linear function $\left(m = \frac{1}{3}, b = \frac{4}{3}\right)$.

77. Let $t = 3$ and find s.
$$s = f(t) = -16t^2 + 256t$$
$$= -16(3)^2 + 256(3)$$
$$= -16(9) + 768$$
$$= -144 + 768$$
$$= 624$$
The bullet will have a height of 624 ft.

79. Let $h = 16$ and find t.
$$h = -16t^2 + 32t$$
$$16 = -16t^2 + 32t$$
$$16t^2 - 32t + 16 = 0$$
$$16(t^2 - 2t + 1) = 0$$
$$16(t - 1)(t - 1) = 0$$
$$t - 1 = 0 \quad \textbf{OR} \quad t - 1 = 0$$
$$t = 1 \qquad\qquad t = 1$$
It will take 1 second to reach the hand.

81. Let $C = 25$ and find F.

$$F(C) = \frac{9}{5}C + 32$$
$$= \frac{9}{5}(25) + 32$$
$$= 45 + 32$$
$$= 77$$

The temperature is $77°F$.

83. Set $R(x)$ equal to $C(x)$ and solve for x.

$$R(x) = C(x)$$
$$120x = 57.50x + 12000$$
$$62.50x = 12000$$
$$x = \frac{12000}{62.50}$$
$$x = 192$$

The company must sell 192 players.

85. **Answers may vary.**

87. $f(x) + g(x) = 2x + 1 + x^2 = x^2 + 2x + 1 = g(x) + f(x)$

Exercise 2.5 (page 141)

1. 41, 43, 47   **3.** $a \cdot b = b \cdot a$ **5.** 1   **7.** squaring

9. absolute value   **11.** horizontal   **13.** 2; down   **15.** 4; to the left

17. $f(x) = x^2 - 3$; Shift $y = x^2$ down 3.

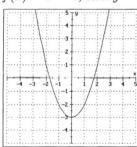

19. $f(x) = (x - 1)^3$; Shift $y = x^3$ right 1.

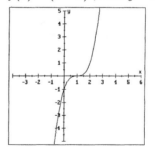

21. $f(x) = |x| - 2$; Shift $y = |x|$ down 2.

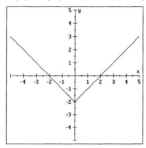

23. $f(x) = |x - 1|$; Shift $y = |x|$ right 1.

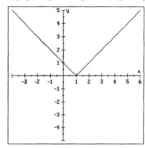

25. $f(x) = x^2 + 8$
x: $[-4, 4]$, y: $[-4, 4]$

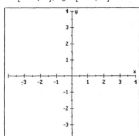

27. $f(x) = |x + 5|$
x: $[-4, 4]$, y: $[-4, 4]$

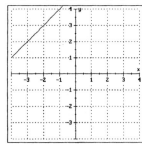

29. $f(x) = (x - 6)^2$
x: $[-4, 4]$, y: $[-4, 4]$

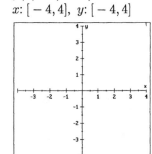

x: $[-7, 7]$, y: $[-2, 12]$

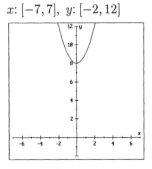

x: $[-12, 2]$, y: $[-4, 10]$

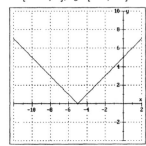

x: $[-2, 12]$, y: $[-2, 12]$

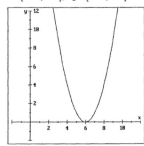

31. $f(x) = x^3 + 8$
x: $[-4, 4]$, y: $[-4, 4]$

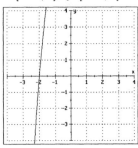

33. $f(x) = x^2 - 5$
Shift $f(x) = x^2$ D 5.

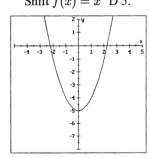

35. $f(x) = (x - 1)^3$
Shift $f(x) = x^3$ R 1.

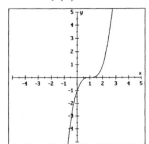

x: $[-10, 10]$, y: $[-4, 16]$

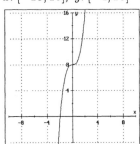

SECTION 2.5

37. $f(x) = |x - 2| - 1$
Shift $f(x) = |x|$ D 1, R 2.

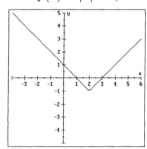

39. $f(x) = (x + 1)^3 - 2$
Shift $f(x) = x^3$ D 2, L 1.

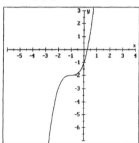

41. $f(x) = -|x| + 1$
Shift $f(x) = -|x|$ U 1.

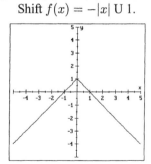

43. $f(x) = -(x - 1)^2$
Shift $f(x) = -x^2$ R 1.

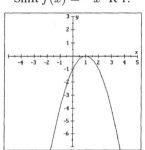

45. Graph $y = 3x + 6$ and $y = 0$, and find the x-coordinates of any points of intersection.

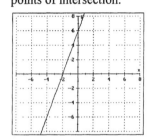

solution: $x = -2$

47. Graph $y = 4(x - 1)$ and $y = 3x$, and find the x-coordinates of any points of intersection.

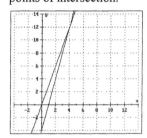

solution: $x = 4$

49. Graph $y = 11x + 6(3 - x)$ and $y = 3$, and find the x-coordinates of any points of intersection.

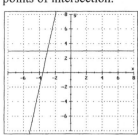

solution: $x = -3$

51. **Answers may vary.**

53. **Answers may vary.**

55. **Answers may vary.**

57. **Answers may vary.**

59. **Answers may vary.**

Chapter 2 Summary (page 144)

1.

x	y
-9	$\boxed{-10}$
$\boxed{-6}$	-8
-3	$\boxed{-6}$
$\boxed{0}$	-4
3	$\boxed{-2}$
$\boxed{6}$	0
$\boxed{9}$	2

$$2x - 3y = 12$$
$$2(-9) - 3y = 12$$
$$-18 - 3y = 12$$
$$-3y = 30$$
$$y = -10$$

$$2x - 3y = 12$$
$$2x - 3(-8) = 12$$
$$2x + 24 = 12$$
$$2x = -12$$
$$x = -6$$

$$2x - 3y = 12$$
$$2(-3) - 3y = 12$$
$$-6 - 3y = 12$$
$$-3y = 18$$
$$y = -6$$

$$2x - 3y = 12$$
$$2x - 3(-4) = 12$$
$$2x + 12 = 12$$
$$2x = 0$$
$$x = 0$$

$$2x - 3y = 12$$
$$2(3) - 3y = 12$$
$$6 - 3y = 12$$
$$-3y = 6$$
$$y = -2$$

$$2x - 3y = 12$$
$$2x - 3(0) = 12$$
$$2x = 12$$
$$x = 6$$

$$2x - 3y = 12$$
$$2x - 3(2) = 12$$
$$2x - 6 = 12$$
$$2x = 18$$
$$x = 9$$

2.

$$x + y = 4 \qquad x + y = 4$$
$$0 + y = 4 \qquad x + 0 = 4$$
$$y = 4 \qquad\quad x = 4$$
$$(0, 4) \qquad\quad (4, 0)$$

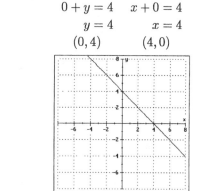

3.

$$2x - y = 8 \qquad 2x - y = 8$$
$$2(0) - y = 8 \qquad 2x - 0 = 8$$
$$-y = 8 \qquad\quad 2x = 8$$
$$y = -8 \qquad\quad x = 4$$
$$(0, -8) \qquad\quad (4, 0)$$

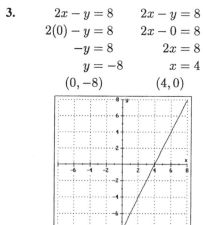

4.

$$y = 3x + 4 \qquad y = 3x + 4$$
$$y = 3(0) + 4 \qquad y = 3(1) + 4$$
$$y = 0 + 4 \qquad\quad y = 3 + 4$$
$$y = 4 \qquad\qquad y = 7$$
$$(0, 4) \qquad\qquad (1, 7)$$

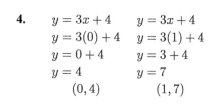

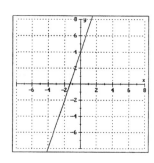

5.
$$x = 4 - 2y \quad x = 4 - 2y$$
$$0 = 4 - 2y \quad x = 4 - 2(0)$$
$$2y = 4 \quad\quad x = 4$$
$$y = 2 \quad\quad\quad (4, 0)$$
$$(0, 2)$$

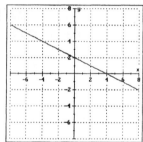

6. $\quad y = 4$ (horizontal)

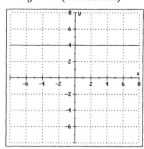

7. $\quad x = -2$ (vertical)

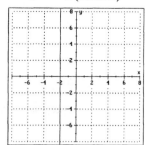

8.
$$2(x + 3) = x + 2$$
$$2x + 6 = x + 2$$
$$x = -4 \text{ (vertical)}$$

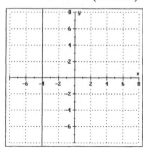

9.
$$3y = 2(y - 1)$$
$$3y = 2y - 2$$
$$y = -2 \text{ (horiozntal)}$$

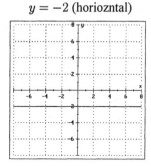

10. $x = \dfrac{x_1 + x_2}{2} = \dfrac{-3 + 6}{2} = \dfrac{3}{2}; \; y = \dfrac{y_1 + y_2}{2} = \dfrac{5 + 11}{2} = \dfrac{16}{2} = 8$; midpoint: $\left(\frac{3}{2}, 8\right)$

11. $m = \dfrac{\Delta y}{\Delta x} = \dfrac{8 - 5}{5 - 2} = \dfrac{3}{3} = 1$

12. $m = \dfrac{\Delta y}{\Delta x} = \dfrac{12 - (-2)}{6 - (-3)} = \dfrac{14}{9}$

13. $m = \dfrac{\Delta y}{\Delta x} = \dfrac{-6 - 4}{-5 - (-3)} = \dfrac{-10}{-2} = 5$

14. $m = \dfrac{\Delta y}{\Delta x} = \dfrac{-9 - (-4)}{-6 - 5} = \dfrac{-5}{-11} = \dfrac{5}{11}$

15. $m = \dfrac{\Delta y}{\Delta x} = \dfrac{4-4}{8-(-2)} = \dfrac{0}{10} = 0$

16. $m = \dfrac{\Delta y}{\Delta x} = \dfrac{8-(-4)}{-5-(-5)} = \dfrac{12}{0}$

undefined slope

17. $2x - 3y = 18$

$\qquad -3y = -2x + 18$

$\qquad \dfrac{-3y}{-3} = \dfrac{-2x}{-3} + \dfrac{18}{-3}$

$\qquad y = \tfrac{2}{3}x - 6 \Rightarrow m = \tfrac{2}{3}$

18. $2x + y = 8$

$\qquad y = -2x + 8 \Rightarrow m = -2$

19. $-2(x - 3) = 10$

$\qquad -2x + 6 = 10$

$\qquad -2x = 4$

$\qquad x = -2$

The slope is undefined (vertical).

20. $3y + 1 = 7$

$\qquad 3y = 6$

$\qquad y = 2$

$\qquad m = 0 \text{ (horizontal)}$

21. $m_1 \neq m_2 \Rightarrow$ not parallel

$\quad m_1 \cdot m_2 = 4\left(-\tfrac{1}{4}\right) = -1 \Rightarrow$ perpendicular

22. $m_1 = m_2 \Rightarrow$ parallel

23. $m_1 \neq m_2 \Rightarrow$ not parallel

$\quad m_1 \cdot m_2 = 0.5\left(-\tfrac{1}{2}\right) = -0.25$

$\qquad \Rightarrow$ not perpendicular

24. $m_1 \neq m_2 \Rightarrow$ not parallel

$\quad m_1 \cdot m_2 = 5(-0.2) = -1$

$\qquad \Rightarrow$ perpendicular

25. Let $y =$ sales and let $x =$ year of business.

$m = \dfrac{\Delta y}{\Delta x} = \dfrac{y_2 - y_1}{x_1 - x_1} = \dfrac{130{,}000 - 65{,}000}{4 - 1} = \dfrac{65{,}000}{3} \approx \$21{,}666.67 \text{ per year}$

26. $y - y_1 = m(x - x_1)$

$\qquad y - 5 = 3(x + 8)$

$\qquad y - 5 = 3x + 24$

$\quad -3x + y = 29$

$\qquad 3x - y = -29$

27. $m = \dfrac{\Delta y}{\Delta x} = \dfrac{4-(-9)}{-2-6} = \dfrac{13}{-8} = -\dfrac{13}{8}$

$\qquad y - y_1 = m(x - x_1)$

$\qquad y - 4 = -\dfrac{13}{8}(x + 2)$

$\qquad 8(y - 4) = 8\left(-\dfrac{13}{8}\right)(x + 2)$

$\qquad 8y - 32 = -13(x + 2)$

$\qquad 8y - 32 = -13x - 26$

$\qquad 13x + 8y = 6$

28. Find the slope of the given line:
$$3x - 2y = 7$$
$$-2y = -3x + 7$$
$$y = \tfrac{3}{2}x - \tfrac{7}{2} \Rightarrow m = \tfrac{3}{2}$$
Use the parallel slope:
$$y - y_1 = m(x - x_1)$$
$$y + 5 = \tfrac{3}{2}(x + 3)$$
$$2(y + 5) = 2 \cdot \tfrac{3}{2}(x + 3)$$
$$2y + 10 = 3(x + 3)$$
$$2y + 10 = 3x + 9$$
$$-3x + 2y = -1 \Rightarrow 3x - 2y = 1$$

29. Find the slope of the given line:
$$3x - 2y = 7$$
$$-2y = -3x + 7$$
$$y = \tfrac{3}{2}x - \tfrac{7}{2} \Rightarrow m = \tfrac{3}{2}$$
Use the perpendicular slope:
$$y - y_1 = m(x - x_1)$$
$$y + 5 = -\tfrac{2}{3}(x + 3)$$
$$3(y + 5) = 3\left(-\tfrac{2}{3}\right)(x + 3)$$
$$3y + 15 = -2(x + 3)$$
$$3y + 15 = -2x - 6$$
$$2x + 3y = -21$$

30. Let x represent the number of years since the copy machine was purchased, and let y represent its value. We know two points on the line: $(0, 8700)$ and $(5, 100)$.

Find the slope:
$$m = \frac{\Delta y}{\Delta x} = \frac{8700 - 100}{0 - 5} = \frac{8600}{-5} = -1720$$

Find the equation of the line:
$$y - y_1 = m(x - x_1)$$
$$y - 8700 = -1720(x - 0)$$
$$y = -1720x + 8700$$

31. $y = 6x - 4$ **is a function**, since each value of x corresponds to exactly one value of y.

32. $y = 4 - x$ **is a function**, since each value of x corresponds to exactly one value of y.

33. $y^2 = x$ **is not a function**, since $x = 9$ corresponds to both $y = 3$ and $y = -3$.

34. $|y| = x^2$ **is not a function**, since $x = 2$ corresponds to both $y = 4$ and $y = -4$.

35. $f(x) = 3x + 2$
$$f(-3) = 3(-3) + 2 = -9 + 2 = -7$$

36. $g(x) = x^2 - 4$
$$g(8) = 8^2 - 4 = 64 - 4 = 60$$

37. $g(x) = x^2 - 4$
$$g(-2) = (-2)^2 - 4 = 4 - 4 = 0$$

38. $f(x) = 3x + 2$
$$f(5) = 3(5) + 2 = 15 + 2 = 17$$

39. Since any value can be substituted for x, the domain is $(-\infty, \infty)$. The range is also $(-\infty, \infty)$.

40. Since any value can be substituted for x, the domain is $(-\infty, \infty)$. The range is also $(-\infty, \infty)$.

41. Since any value can be substituted for x, the domain is $(-\infty, \infty)$. Since $x^2 \geq 0$, $x^2 + 1$ must be greater than or equal to 1. Thus the range is $[1, \infty)$.

42. Since $x = 2$ makes the denominator equal to 0, the domain is $(-\infty, 2) \cup (2, \infty)$. Since the numerator will never equal 0, the fraction will never equal 0. The range is $(-\infty, 0) \cup (0, \infty)$.

43. Since $x = 3$ makes the denominator equal to 0, the domain is $(-\infty, 3) \cup (3, \infty)$. Since the numerator will never equal 0, the fraction will never equal 0. The range is $(-\infty, 0) \cup (0, \infty)$.

44. x can be any real number, so the domain is $(-\infty, \infty)$. y always has a value of 7, so the range is $\{7\}$.

45. Since each vertical line passes through the graph at most once, it is a function.

46. Since a vertical line can pass through the graph more than once, it is not a function.

47. Since a vertical line can pass through the graph more than once, it is not a function.

48. Since each vertical line passes through the graph at most once, it is a function.

49. $f(x) = x^2 - 3$
          Shift $y = x^2$ down 3.

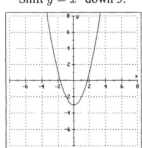

50. $f(x) = |x| - 4$
          Shift $y = |x|$ down 4.

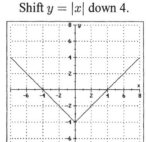

51. $f(x) = (x - 2)^3$
          Shift $y = x^3$ right 2.

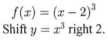
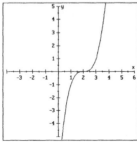

52. $f(x) = (x + 4)^2 - 3$
          Shift $y = x^2$ down 3, left 4.

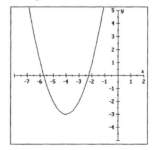

53-56. Compare your graphs to the graphs in numbers **49-52.**

57. $y = 3x + 2 \Rightarrow$ linear $(m = 3, b = 2)$

58. $y = \dfrac{x + 5}{4} \Rightarrow y = \dfrac{1}{4}x + \dfrac{5}{4} \Rightarrow$ linear $\left(m = \frac{1}{4}, b = \frac{5}{4}\right)$

59. $4x - 3y = 12 \Rightarrow -3y = -4x + 12 \Rightarrow y = \dfrac{4}{3}x - 4 \Rightarrow$ linear $\left(m = \dfrac{4}{3}, b = -4\right)$

60. $y = x^2 - 25 \Rightarrow$ not linear (The exponent on x is 2.)

61. $f(x) = -|x - 3|$

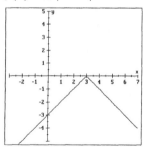

Chapter 2 Test (page 148)

1.
$$2x - 5y = 10 \qquad 2x - 5y = 10$$
$$2(0) - 5y = 10 \qquad 2x - 5(0) = 10$$
$$-5y = 10 \qquad 2x = 10$$
$$y = -2 \qquad x = 5$$
$$(0, -2) \qquad (5, 0)$$

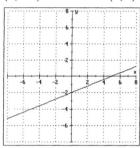

2.
$$y = \frac{x - 3}{5} \qquad y = \frac{x - 3}{5}$$
$$0 = \frac{x - 3}{5} \qquad y = \frac{0 - 3}{5}$$
$$5(0) = 5 \cdot \frac{x - 3}{5} \qquad y = -\frac{3}{5}$$
$$0 = x - 3 \qquad \text{y-intercept: } \left(0, -\tfrac{3}{5}\right)$$
$$3 = x$$
$$x\text{-intercept: } (3, 0)$$

3. $x = \dfrac{x_1 + x_2}{2} = \dfrac{-3 + 4}{2} = \dfrac{1}{2}; \ y = \dfrac{y_1 + y_2}{2} = \dfrac{3 + (-2)}{2} = \dfrac{1}{2}; \text{midpoint: } \left(\tfrac{1}{2}, \tfrac{1}{2}\right)$

4. $m = \dfrac{\Delta y}{\Delta x} = \dfrac{4 - 8}{-2 - 6} = \dfrac{-4}{-8} = \dfrac{1}{2}$

5.
$$2x - 3y = 8$$
$$-3y = -2x + 8$$
$$y = \frac{2}{3}x - \frac{8}{3} \Rightarrow m = \frac{2}{3}$$

6. The graph of $x = 12$ is a vertical line, so the slope is undefined.

7. The graph of $y = 12$ is a horizontal line, so the slope is 0.

8. $y - y_1 = m(x - x_1)$

$y + 5 = \dfrac{2}{3}(x - 4)$

$y + 5 = \dfrac{2}{3}x - \dfrac{8}{3}$

$y = \dfrac{2}{3}x - \dfrac{8}{3} - 5$

$y = \dfrac{2}{3}x - \dfrac{23}{3}$

9. $m = \dfrac{\Delta y}{\Delta x} = \dfrac{6 - (-10)}{-2 - (-4)} = \dfrac{16}{2} = 8$

$y - y_1 = m(x - x_1)$

$y - 6 = 8(x + 2)$

$y - 6 = 8x + 16$

$-8x + y = 22$

$8x - y = -22$

10. $-2(x - 3) = 3(2y + 5)$

$-2x + 6 = 6y + 15$

$6y + 15 = -2x + 6$

$6y = -2x - 9$

$y = \dfrac{-2}{6}x - \dfrac{9}{6}$

$y = -\dfrac{1}{3}x - \dfrac{3}{2}$

$m = -\dfrac{1}{3}, b = -\dfrac{3}{2} \Rightarrow \left(0, -\dfrac{3}{2}\right)$

11. $4x - y = 12$

$-y = -4x + 12$

$y = 4x - 12 \Rightarrow m = 4$

$y = \frac{1}{4}x + 3 \Rightarrow m = \frac{1}{4}$

neither parallel nor perpendicular

12. $y = -\frac{2}{3}x + 4 \Rightarrow m = -\frac{2}{3}$

$2y = 3x - 3$

$y = \frac{3}{2}x - \frac{3}{2} \Rightarrow m = \frac{3}{2}$

perpendicular

13. $y = \frac{3}{2}x - 7 \Rightarrow m = \frac{3}{2}$

Use the parallel slope:

$y - y_1 = m(x - x_1)$

$y - 0 = \frac{3}{2}(x - 0)$

$y = \frac{3}{2}x$

14. $y = -\frac{2}{3}x - 7 \Rightarrow m = -\frac{2}{3}$

Use the perpendicular slope:

$y - y_1 = m(x - x_1)$

$y - 6 = \frac{3}{2}(x + 3)$

$y - 6 = \frac{3}{2}x + \frac{9}{2}$

$y = \frac{3}{2}x + \frac{21}{2}$

15. $|y| = x$ **is not a function**, since $x = 2$ corresponds to both $y = 2$ and $y = -2$.

16. Domain: $(-\infty, \infty)$; Range: $[0, \infty)$

17. Domain: $(-\infty, \infty)$; Range: $(-\infty, \infty)$

18. $f(x) = 3x + 1$

$f(3) = 3(3) + 1 = 9 + 1 = 10$

19. $g(x) = x^2 - 2$

$g(0) = 0^2 - 2 = 0 - 2 = -2$

20. $f(x) = 3x + 1$

$f(a) = 3a + 1$

21. $g(x) = x^2 - 2$

$g(-x) = (-x)^2 - 2 = x^2 - 2$

22. function

23. not a function

24.
$$f(x) = x^2 - 1$$
Shift $y = x^2$ down 1.

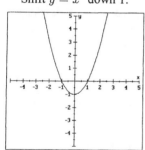

25.
$$f(x) = -|x + 2|$$
Shift $y = -|x|$ left 2.

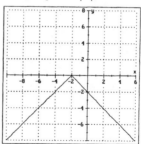

Cumulative Review Exercises (page 149)

1. natural numbers: $1, 2, 6, 7$

2. whole numbers: $0, 1, 2, 6, 7$

3. rational numbers: $-2, 0, 1, 2, \frac{13}{12}, 6, 7$

4. irrational numbers: $\sqrt{5}, \pi$

5. negative numbers: -2

6. real numbers: $-2, 0, 1, 2, \frac{13}{12}, 6, 7, \sqrt{5}, \pi$

7. prime numbers: $2, 7$

8. composite numbers: 6

9. even numbers: $-2, 0, 2, 6$

10. odd numbers: $1, 7$

11. $\{x \mid -2 < x \le 5\} \Rightarrow$

12. $[-5, 0) \cup [3, 6] \Rightarrow$

13. $-|5| + |-3| = -5 + 3 = -2$

14. $\dfrac{|-5| + |-3|}{-|4|} = \dfrac{5+3}{-4} = \dfrac{8}{-4} = -2$

15. $2 + 4 \cdot 5 = 2 + 20 = 22$

16. $\dfrac{8-4}{2-4} = \dfrac{4}{-2} = -2$

17. $20 \div (-10 \div 2) = 20 \div (-5) = -4$

18. $\dfrac{6 + 3(6+4)}{2(3-9)} = \dfrac{6 + 3(10)}{2(-6)} = \dfrac{6 + 30}{-12}$
$$= \dfrac{36}{-12} = -3$$

19. $-x - 2y = -2 - 2(-3) = -2 + 6 = 4$

20. $\dfrac{x^2 - y^2}{2x + y} = \dfrac{2^2 - (-3)^2}{2(2) + (-3)} = \dfrac{4 - 9}{4 - 3} = \dfrac{-5}{1}$
$$= -5$$

21. associative property of addition

22. distributive property

23. commutative property of addition

24. associative property of multiplication

25. $\left(x^2y^3\right)^4 = \left(x^2\right)^4\left(y^3\right)^4 = x^8y^{12}$

26. $\dfrac{c^4c^8}{\left(c^5\right)^2} = \dfrac{c^{12}}{c^{10}} = c^2$

27. $\left(-\dfrac{a^3b^{-2}}{ab}\right)^{-1} = \left(-\dfrac{a^2}{b^3}\right)^{-1} = -\dfrac{b^3}{a^2}$

28. $\left(\dfrac{-3a^3b^{-2}}{6a^{-2}b^3}\right)^0 = 1$

29. $0.00000497 = 4.97 \times 10^{-6}$

30. $9.32 \times 10^8 = 932{,}000{,}000$

31. $2x - 5 = 11$
$\quad\quad\ 2x = 16$
$\quad\quad\ \ x = 8$

32. $\dfrac{2x - 6}{3} = x + 7$

$3 \cdot \dfrac{2x - 6}{3} = 3(x + 7)$

$\quad\ 2x - 6 = 3x + 21$

$\quad\quad\ -x = 27$

$\quad\quad\quad x = -27$

33. $4(y - 3) + 4 = -3(y + 5)$
$\quad 4y - 12 + 4 = -3y - 15$
$\quad\quad\ 4y - 8 = -3y - 15$
$\quad\quad\quad\ 7y = -7$
$\quad\quad\quad\ \ y = -1$

34. $2x - \dfrac{3(x - 2)}{2} = 7 - \dfrac{x - 3}{3}$

$6(2x) - 6 \cdot \dfrac{3x - 6}{2} = 6(7) - 6 \cdot \dfrac{x - 3}{3}$

$12x - 3(3x - 6) = 42 - 2(x - 3)$

$12x - 9x + 18 = 42 - 2x + 6$

$\quad\quad 3x + 18 = 48 - 2x$

$\quad\quad\quad\ \ 5x = 30$

$\quad\quad\quad\quad\ x = 6$

35. $S = \dfrac{n(a + l)}{2}$

$2S = 2 \cdot \dfrac{n(a + l)}{2}$

$2S = n(a + l)$

$\dfrac{2S}{n} = \dfrac{n(a + l)}{n}$

$\dfrac{2S}{n} = a + l$

$\dfrac{2S}{n} - l = a,\ \text{or}\ a = \dfrac{2S}{n} - l$

36. $A = \dfrac{1}{2}h(b_1 + b_2)$

$2A = 2 \cdot \dfrac{1}{2}h(b_1 + b_2)$

$2A = h(b_1 + b_2)$

$\dfrac{2A}{b_1 + b_2} = \dfrac{h(b_1 + b_2)}{b_1 + b_2}$

$\dfrac{2A}{b_1 + b_2} = h,\ \text{or}\ h = \dfrac{2A}{b_1 + b_2}$

37. Let x represent the first even integer.
Then $x + 2$ and $x + 4$ represent the others.

$$\boxed{\text{1st}} + \boxed{\text{2nd}} + \boxed{\text{3rd}} = \boxed{\begin{array}{c}\text{Sum of the}\\\text{integers}\end{array}}$$

$$x + x + 2 + x + 4 = 90$$
$$3x + 6 = 90$$
$$3x = 84$$
$$x = 28$$

The integers are 28, 30 and 32.

38. Let w represent the width of the rectangle.
Then $3w$ represents the length.

$$2 \cdot \boxed{\text{Length}} + 2 \cdot \boxed{\text{Width}} = \boxed{\text{Perimeter}}$$

$$2(3w) + 2w = 112$$
$$6w + 2w = 112$$
$$8w = 112$$
$$w = 14$$

The dimensions are 14 cm by 42 cm.

39. $2x - 3y = 6$

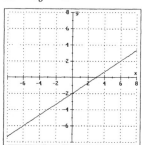

The equation defines a function.

40. $m = \dfrac{\Delta y}{\Delta x} = \dfrac{5 - (-9)}{-2 - 8} = \dfrac{14}{-10} = -\dfrac{7}{5}$

41. Find the slope and use slope-intercept form:

$$m = \frac{\Delta y}{\Delta x} = \frac{5 - (-9)}{-2 - 8} = \frac{14}{-10} = -\frac{7}{5}$$
$$y - y_1 = m(x - x_1)$$
$$y - 5 = -\frac{7}{5}(x + 2)$$
$$y - 5 = -\frac{7}{5}x - \frac{14}{5}$$
$$y = -\frac{7}{5}x + \frac{11}{5}$$

42. Find the slope of the given line, and use that
slope to find the equation of the desired line:
$$3x + y = 8$$
$$y = -3x + 8 \Rightarrow m = -3$$
$$y - y_1 = m(x - x_1)$$
$$y - 3 = -3(x + 2)$$
$$y - 3 = -3x - 6$$
$$y = -3x - 3$$

43. $f(x) = 3x^2 + 2$
$$f(-1) = 3(-1)^2 + 2$$
$$f(-1) = 3(1) + 2 = 3 + 2 = 5$$

44. $g(x) = 2x - 1$
$$g(0) = 2(0) - 1 = 0 - 1 = -1$$

45. $g(x) = 2x - 1$
$$g(t) = 2t - 1$$

46. $f(x) = 3x^2 + 2$
$$f(-r) = 3(-r)^2 + 2 = 3r^2 + 2$$

47. $y = -x^2 + 1$

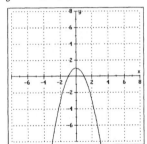

The equation describes a function.
domain: $(-\infty, \infty)$; range: $(-\infty, 1]$

48. $y = \left|\frac{1}{2}x - 3\right|$

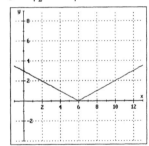

The equation describes a function.
domain: $(-\infty, \infty)$; range: $[0, \infty)$

Exercise 3.1 (page 157)

1. $93{,}000{,}000 = 9.3 \times 10^7$

3. $345 \times 10^2 = 3.45 \times 10^2 \times 10^2 = 3.45 \times 10^4$

5. system

7. inconsistent

9. dependent

11.
$y = 2x \qquad y = \frac{1}{2}x + \frac{3}{2}$
$2 \stackrel{?}{=} 2(1) \quad 2 \stackrel{?}{=} \frac{1}{2}(1) + \frac{3}{2}$
$2 = 2 \qquad 2 \stackrel{?}{=} \frac{1}{2} + \frac{3}{2}$
$\qquad\qquad 2 = \frac{4}{2}$
$(1, 2)$ is a solution.

13.
$y = \frac{1}{2}x - 2$
$-3 \stackrel{?}{=} \frac{1}{2}(2) - 2$
$-3 \stackrel{?}{=} 1 - 2$
$-3 \neq -1$
$(2, -3)$ is not a solution.

15.
$\begin{cases} x + y = 6 \\ x - y = 2 \end{cases}$

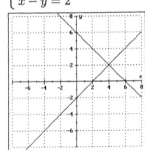

$(4, 2)$ is the solution.

17.
$\begin{cases} 2x + y = 1 \\ x - 2y = -7 \end{cases}$

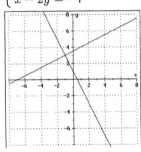

$(-1, 3)$ is the solution.

19.
$\begin{cases} 2x + 3y = 0 \\ 2x + y = 4 \end{cases}$

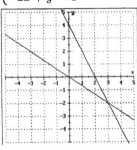

$(3, -2)$ is the solution.

21.
$\begin{cases} x = 13 - 4y \\ 3x = 4 + 2y \end{cases}$

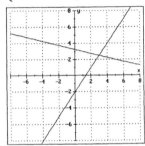

$\left(3, \frac{5}{2}\right)$ is the solution.

23. $\begin{cases} x = 3 - 2y \\ 2x + 4y = 6 \end{cases}$

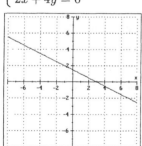

dependent system

25. $\begin{cases} x = 2 \\ y = \dfrac{4 - x}{2} \end{cases}$

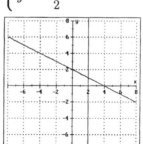

$(2, 1)$ is the solution.

27. $\begin{cases} y = 3 \\ x = 2 \end{cases}$

$(2, 3)$ is the solution.

29. $\begin{cases} x = \dfrac{11 - 2y}{3} \\ y = \dfrac{11 - 6x}{4} \end{cases}$

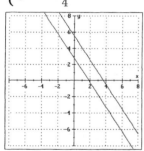

inconsistent system

31. $\begin{cases} \dfrac{5}{2}x + y = \dfrac{1}{2} \\ 2x - \dfrac{3}{2}y = 5 \end{cases}$

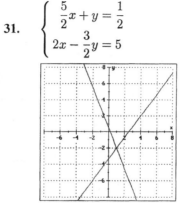

$(1, -2)$ is the solution.

33. $\begin{cases} x = \dfrac{5y - 4}{2} \\ x - \dfrac{5}{3}y + \dfrac{1}{3} = 0 \end{cases}$

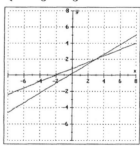

$(3, 2)$ is the solution.

35. $\begin{cases} x = -\dfrac{3}{2}y \\ x = \dfrac{3}{2}y - 2 \end{cases}$

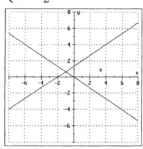

$\left(-1, \frac{2}{3}\right)$ is the solution.

37. $\begin{cases} y = 3.2x - 1.5 \\ y = -2.7x - 3.7 \end{cases}$

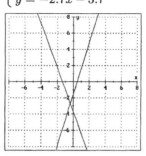

solution: $(-0.37, -2.69)$

39. $\begin{cases} 1.7x + 2.3y = 3.2 \\ y = 0.25x + 8.95 \end{cases}$

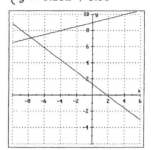

solution: $(-7.64, 7.04)$

41. **a.** The point $(15, 2.0)$ is on the graph of the cost function, so it costs $2 million to manufacture 15,000 cameras.

  b. The point $(20, 3.0)$ is on the graph of the revenue function, so there is a revenue of $3 million when 20,000 cameras are sold.

  c. The graphs of the cost function and the revenue function meet at the point $(10, 1.5)$, so the revenue and cost functions are equal for 10,000 cameras.

43. $\begin{cases} 2x + 3y = 6 \\ 2x - 3y = 9 \end{cases}$

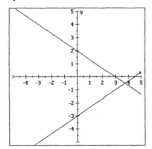

  a. There is a possibility of a collision.
  b. The danger point is at $(3.75, -0.5)$.
  c. The collision is not certain, since the ships could be there at different times.

45. Let $x =$ the number of hours spent and let $y =$ the total cost.

$\begin{cases} y = 50x \\ y = 40x + 30 \end{cases}$

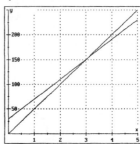

The lines meet at $(3, 150)$. Thus, the repair takes 3 hours.

47. **Answers may vary.**

49. One possible answer: $\begin{cases} x + y = -3 \\ x - y = -7 \end{cases}$

Exercise 3.2 (page 171)

1. $\left(a^2 a^3\right)^2 \left(a^4 a^2\right)^2 = \left(a^5\right)^2 \left(a^6\right)^2$
$$= a^{10} a^{12} = a^{22}$$

3. $\left(\dfrac{-3x^3 y^4}{x^{-5} y^3}\right)^{-4} = \left(\dfrac{x^{-5} y^3}{-3x^3 y^4}\right)^4$
$$= \left(\dfrac{1}{-3x^8 y}\right)^4 = \dfrac{1}{81x^{32} y^4}$$

5. setup; unit

7. parallelogram

9. Opposite

11. $\begin{cases} (1) \quad y = x \\ (2) \quad x + y = 4 \end{cases}$

Substitute $y = x$ from (1) into (2):

$x + \boldsymbol{y} = 4$

$x + \boldsymbol{x} = 4$

$2x = 4$

$x = 2$

Substitute this and solve for y:

$y = x = 2$

Solution: $(2, 2)$

13. $\begin{cases} (1) \quad x - y = 2 \\ (2) \quad 2x + y = 13 \end{cases}$

Substitute $x = y + 2$ from (1) into (2):

$2x + y = 13$

$2(\boldsymbol{y + 2}) + y = 13$

$2y + 4 + y = 13$

$3y = 9$

$y = 3$

Substitute this and solve for x:

$x = y + 2 = 3 + 2 = 5$

Solution: $(5, 3)$

15. $\begin{cases} (1) \quad x + 2y = 6 \\ (2) \quad 3x - y = -10 \end{cases}$

Substitute $x = -2y + 6$ from (1) into (2):

$3x - y = -10$

$3(\boldsymbol{-2y + 6}) - y = -10$

$-6y + 18 - y = -10$

$-7y = -28$

$y = 4$

Substitute this and solve for x:

$x = -2y + 6 = -2(4) + 6 = -2$

Solution: $(-2, 4)$

17. $\begin{cases} (1) \quad 3x = 2y - 4 \\ (2) \quad 6x - 4y = -4 \end{cases}$

Substitute $x = \dfrac{2y - 4}{3}$ from (1) into (2):

$6\boldsymbol{x} - 4y = -4$

$6\left(\dfrac{\boldsymbol{2y - 4}}{\boldsymbol{3}}\right) - 4y = -4$

$2(2y - 4) - 4y = -4$

$4y - 8 - 4y = -4$

$-8 = -4$

Impossible $\Rightarrow$ no solution

19. $\begin{cases} (1) \quad 3x - 4y = 9 \\ (2) \quad x + 2y = 8 \end{cases}$

Substitute $x = -2y + 8$ from (2) into (1):

$3\boldsymbol{x} - 4y = 9$

$3(\boldsymbol{-2y + 8}) - 4y = 9$

$-6y + 24 - 4y = 9$

$-10y = -15$

$y = \frac{3}{2}$

Substitute this and solve for x:

$x = -2y + 8 = -2\left(\dfrac{3}{2}\right) + 8 = 5$

Solution: $\left(5, \frac{3}{2}\right)$

21. $\begin{cases} (1) \quad 2x + 2y = -1 \\ (2) \quad 3x + 4y = 0 \end{cases}$

Substitute $y = \dfrac{-2x - 1}{2}$ from (1) into (2):

$3x + 4\boldsymbol{y} = 0$

$3x + 4\left(\dfrac{\boldsymbol{-2x - 1}}{\boldsymbol{2}}\right) = 0$

$3x + 2(-2x - 1) = 0$

$3x - 4x - 2 = 0$

$-x = 2$

$x = -2$

Substitute this and solve for y:

$y = \dfrac{-2x - 1}{2} = \dfrac{-2(-2) - 1}{2} = \dfrac{3}{2}$

Solution: $\left(-2, \frac{3}{2}\right)$

23.
$$x - y = 3$$
$$\underline{x + y = 7}$$
$$2x \quad = 10$$
$$x \quad = 5$$
Substitute and solve for y:
$$x + y = 7$$
$$5 + y = 7$$
$$y = 2$$
The solution is $(5, 2)$.

25.
$$2x + y = -10$$
$$\underline{2x - y = -6}$$
$$4x \quad = -16$$
$$x \quad = -4$$
Substitute and solve for y:
$$2x + y = -10$$
$$2(-4) + y = -10$$
$$-8 + y = -10$$
$$y = -2$$
The solution is $(-4, -2)$.

27.
$$2x + 3y = 8 \Rightarrow \times (2) \quad 4x + 6y = 16$$
$$3x - 2y = -1 \Rightarrow \times (3) \quad \underline{9x - 6y = -3}$$
$$13x \quad = 13$$
$$x \quad = 1$$

$2x + 3y = 8$ Solution:
$2(1) + 3y = 8$ $\boxed{(1, 2)}$
$3y = 6$
$y = 2$

29.
$$4x + 9y = 8$$
$$2x - 6y = -3 \Rightarrow \times (-2) \quad \underline{-4x + 12y = 6}$$
$$4x + 9y = 8$$
$$21y = 14$$
$$y = \frac{14}{21} = \frac{2}{3}$$

$2x - 6y = -3$ Solution:
$2x - 6\left(\frac{2}{3}\right) = -3$ $\boxed{\left(\frac{1}{2}, \frac{2}{3}\right)}$
$2x - 4 = -3$
$2x = 1$
$x = \frac{1}{2}$

31.
$$8x - 4y = 16 \Rightarrow 8x - 4y = 16 \Rightarrow \quad 8x - 4y = 16$$
$$2x - 4 = y \Rightarrow 2x - y = 4 \Rightarrow \times (-4) \quad \underline{-8x + 4y = -16}$$
$$0 = 0 \Rightarrow \boxed{\text{Dependent equations}}$$

33.
$$x = \frac{3}{2}y + 5 \Rightarrow \times (2) \quad 2x = 3y + 10 \Rightarrow 2x - 3y = 10 \Rightarrow \quad 2x - 3y = 10$$
$$2x - 3y = 8 \Rightarrow \quad 2x - 3y = 8 \Rightarrow 2x - 3y = 8 \Rightarrow \times (-1) \quad \underline{-2x + 3y = -8}$$
$$0 \neq 2 \Rightarrow \boxed{\text{No solution}}$$

35.
$$\frac{x}{2} + \frac{y}{2} = 6 \Rightarrow \times 2 \quad x + y = 12$$
$$\frac{x}{2} - \frac{y}{2} = -2 \Rightarrow \times 2 \quad \underline{x - y = -4}$$
$$2x \quad = 8$$
$$x \quad = 4$$

$x + y = 12$ Solution:
$4 + y = 12$ $\boxed{(4, 8)}$
$y = 8$

37.
$$\frac{3}{4}x + \frac{2}{3}y = 7 \Rightarrow \times 12 \quad 9x + 8y = 84 \Rightarrow \times 2 \quad 18x + 16y = 168$$
$$\frac{3}{5}x - \frac{1}{2}y = 18 \Rightarrow \times 10 \quad 6x - 5y = 180 \Rightarrow \times (-3) \quad \underline{-18x + 15y = -540}$$
$$31y = -372$$
$$y = -12$$

$6x - 5y = 180$
$6x - 5(-12) = 180$
$6x + 60 = 180$
$6x = 120$
$x = 20$

Solution: $\boxed{(20, -12)}$

39. $\dfrac{3x}{2} - \dfrac{2y}{3} = 0 \Rightarrow \times 6 \quad 9x - 4y = 0 \Rightarrow \qquad 9x - 4y = 0 \qquad 9x - 4y = 0$

$\dfrac{3x}{4} + \dfrac{4y}{3} = \dfrac{5}{2} \Rightarrow \times 12 \quad 9x + 16y = 30 \Rightarrow \times(-1) \quad -9x - 16y = -30 \qquad 9x - 4\left(\tfrac{3}{2}\right) = 0$

$$\begin{array}{r} 9x - 6 = 0 \\ -20y = -30 \qquad\qquad 9x = 6 \\ y = \tfrac{-30}{-20} \\ y = \tfrac{3}{2} \qquad x = \tfrac{6}{9} = \tfrac{2}{3} \end{array}$$

Solution: $\boxed{\left(\tfrac{2}{3}, \tfrac{3}{2}\right)}$

41. $\dfrac{2}{5}x - \dfrac{1}{6}y = \dfrac{7}{10} \Rightarrow \times 30 \quad 12x - 5y = 21 \Rightarrow \times 3 \qquad 36x - 15y = 63$

$\dfrac{3}{4}x - \dfrac{2}{3}y = \dfrac{19}{8} \Rightarrow \times 24 \quad 18x - 16y = 57 \Rightarrow \times(-2) \quad -36x + 32y = -114$

$$\begin{array}{r} 17y = -51 \\ y = -3 \end{array}$$

$$12x - 5y = 21$$
$$12x - 5(-3) = 21$$
$$12x + 15 = 21$$
$$12x = 6$$
$$x = \tfrac{6}{12} = \tfrac{1}{2} \Rightarrow \text{Solution: } \boxed{\left(\tfrac{1}{2}, -3\right)}$$

43. Let $x = 0.33333\overline{3}$. Then $10x = 3.33333\overline{3}$.

$$\begin{array}{r} 10x = 3.33333\overline{3} \\ x = 0.33333\overline{3} \\ \hline 9x = 3 \\ \dfrac{9x}{9} = \dfrac{3}{9} \\ x = \dfrac{1}{3} \end{array}$$

45. Let $x = -0.34898989\overline{89}$.

Then $100x = -34.89898989\overline{89}$.

$$\begin{array}{r} 100x = -34.89898989\overline{89} \\ x = -0.34898989\overline{89} \\ \hline 99x = -34.55 \\ \dfrac{99x}{99} = -\dfrac{34.55}{99} \\ x = -\dfrac{34.55(100)}{99(100)} \\ x = -\dfrac{3455}{9900} = -\dfrac{691}{1980} \end{array}$$

The fraction is $-\dfrac{691}{1980}$.

For problems #47-49, begin each problem by letting $m = \dfrac{1}{x}$ and $n = \dfrac{1}{y}$. Solve for m and n, and then solve for x and y.

47. $\dfrac{1}{x} + \dfrac{1}{y} = \dfrac{5}{6} \Rightarrow \quad m + n = \dfrac{5}{6}$

Solve for n: $\qquad$ Solve for x: $\quad$ Solve for y: $\quad$ Solution:

$\dfrac{1}{x} - \dfrac{1}{y} = \dfrac{1}{6} \Rightarrow \quad m - n = \dfrac{1}{6}$

$$\begin{array}{r} 2m = \dfrac{6}{6} \\ 2m = 1 \\ m = \dfrac{1}{2} \end{array}$$

Solve for n:
$$m + n = \dfrac{5}{6}$$
$$\dfrac{1}{2} + n = \dfrac{5}{6}$$
$$n = \dfrac{5}{6} - \dfrac{1}{2}$$
$$n = \dfrac{1}{3}$$

Solve for x:
$$m = \dfrac{1}{x}$$
$$\dfrac{1}{2} = \dfrac{1}{x}$$
$$2 = x$$

Solve for y:
$$n = \dfrac{1}{y}$$
$$\dfrac{1}{3} = \dfrac{1}{y}$$
$$3 = y$$

Solution: $\boxed{(2, 3)}$

49.

$$\frac{1}{x} + \frac{2}{y} = -1 \Rightarrow m + 2n = -1 \Rightarrow \qquad m + 2n = -1$$

$$\frac{2}{x} - \frac{1}{y} = -7 \Rightarrow 2m - n = -7 \Rightarrow \times 2 \quad 4m - 2n = -14$$

$$\overline{\qquad\qquad\qquad} \qquad \overline{\qquad\qquad\qquad} \qquad \begin{array}{rl} 5m &= -15 \\ m &= -3 \end{array}$$

Solve for n: Solve for x: Solve for y: Solution: $\boxed{\left(-\frac{1}{3}, 1\right)}$

$$\begin{array}{l} m + 2n = -1 \\ -3 + 2n = -1 \\ \quad 2n = 2 \\ \quad n = 1 \end{array} \qquad \begin{array}{l} m = \dfrac{1}{x} \\ -3 = \dfrac{1}{x} \\ -\dfrac{1}{3} = x \end{array} \qquad \begin{array}{l} n = \dfrac{1}{y} \\ 1 = \dfrac{1}{y} \\ 1 = y \end{array}$$

51. Let $x =$ the cost of the pair of shoes and $y =$ the cost of the sweater.

(1) $x + y = 98$ $x + y = 98$ $y = x + 16$ The sweater cost \$57.
(2) $y = x + 16$ $x + x + 16 = 98$ $y = 41 + 16$
                      $2x = 82$ $y = 57$
                      $x = 41$

53.

(1) $R_1 + R_2 = 1375$ $R_1 + R_2 = 1375$ $R_1 = R_2 + 125$ The resistances
(2) $R_1 = R_2 + 125$ $R_2 + 125 + R_2 = 1375$ $R_1 = 625 + 125$ are $R_1 = 750$
                             $2R_2 = 1250$ $R_1 = 750$ ohms and
                              $R_2 = 625$ $R_2 = 625$ ohms.

55. Let $l =$ the length of the field and $w =$ the width of the field.

$$\begin{array}{l} 2w + 2l = 72 \Rightarrow \times(-1) \\ 3w + 2l = 88 \Rightarrow \end{array} \qquad \begin{array}{rl} -2w - 2l &= -72 \\ 3w + 2l &= 88 \\ \hline w &= 16 \end{array} \qquad \begin{array}{l} 2w + 2l = 72 \\ 2(16) + 2l = 72 \\ 32 + 2l = 72 \\ 2l = 40 \\ l = 20 \end{array}$$

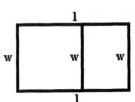

The dimensions of the field are
20 meters by 16 meters.

57. Let $x =$ the amount invested at 10% and $y =$ the amount invested at 12%.

$$\begin{array}{l} x + y = 8000 \Rightarrow \\ 0.10x + 0.12y = 900 \Rightarrow \times 100 \end{array} \quad \begin{array}{l} x + y = 8000 \Rightarrow \times(-10) \\ 10x + 12y = 90000 \Rightarrow \end{array} \quad \begin{array}{rl} -10x - 10y &= -80000 \\ 10x + 12y &= 90000 \\ \hline 2y &= 10000 \\ y &= 5000 \end{array}$$

$$\begin{array}{l} x + y = 8000 \\ x + 5000 = 8000 \\ \quad x = 3000 \Rightarrow \$3000 \text{ was invested at } 10\%, \text{ while } \$5000 \text{ was invested at } 12\%. \end{array}$$

SECTION 3.2

59. Let x = the # of ounces of the 8% solution and y = the # of ounces of the 15% solution.

$$
\begin{array}{llll}
x + \quad y = 100 \Rightarrow & x + \quad y = 100 \Rightarrow \times(-8) & -8x - \quad 8y = -800 \\
0.08x + 0.15y = 12.2 \Rightarrow \times 100 & 8x + 15y = 1220 \Rightarrow & \underline{8x + 15y = 1220} \\
& & 7y = \quad 420 \\
& & y = \quad 60
\end{array}
$$

$x + y = 100$
$x + 60 = 100$
$\quad x = 40 \Rightarrow$ 40 oz of the 8% and 60 oz of the 15% solution should be used.

61. Let c = the speed of the car and p = the speed of the plane. Remember: distance = rate · time, so time = $\dfrac{\text{distance}}{\text{rate}}$. Form one equation from the fact that the car travels 50 miles in the same time that the plane travels 180 miles:

Car time = Plane time $\Rightarrow \dfrac{50}{c} = \dfrac{180}{p} \Rightarrow 50p = 180c \Rightarrow 50p - 180c = 0.$

Form a second equation from the relationship given between the rates: $p = c + 143.$

$$
\begin{array}{lll}
(1) \quad 50p - 180c = 0 & 50p - 180c = 0 & \text{The car's speed is 55 mph.} \\
(2) \quad\quad\quad p = c + 143 & 50(c + 143) - 180c = 0 \\
& 50c + 7150 - 180c = 0 \\
& -130c = -7150 \\
& c = 55
\end{array}
$$

63. Let r = the number of racing bikes and m = the number of mountain bikes.

$$
\begin{array}{llll}
60r + 90m = 15900 \Rightarrow \times(-7) & -420r - 630m = -111300 & 60r + 90m = 15900 \\
55r + 70m = 13075 \Rightarrow \times 9 & \underline{495r + 630m = \quad 117675} & 60(85) + 90m = 15900 \\
& 75r \qquad\quad = \quad 6375 & 5100 + 90m = 15900 \\
& r \qquad\quad = \quad\quad 85 & 90m = 10800 \\
& & m = 120
\end{array}
$$

85 racing bikes and 120 mountain bikes can be built.

65. Let x = the number of plates produced at the break-even point. Let C_1 = the cost of the first machine: $C_1 = 300 + 2x$. Let C_2 = the cost of the second machine: $C_2 = 500 + x$. To find the break point, find x such that $C_1 = C_2$:

$$
\begin{array}{l}
\quad\quad C_1 = C_2 \\
300 + 2x = 500 + x \\
\quad\quad\quad x = 200 \Rightarrow \text{The break point is 200 plates.}
\end{array}
$$

67. Let x = the number of computers sold at the break-even point. Let C = the costs of the store: $C = 8925 + 850x$. Let R = the revenue of the store: $R = 1275x$. To find the break-even point, find x such that $C = R$:

$$
\begin{array}{l}
\quad\quad C = R \\
8925 + 850x = 1275x \\
\quad\quad 8925 = 425x \\
\quad\quad\quad 21 = x \Rightarrow \text{The break-even point is 21 computers.}
\end{array}
$$

69. Let $x =$ the number of pieces of software sold at the break-even point. Let $C =$ the costs of the business: $C = 18375 + 5.45x$. Let $R =$ the revenue of the business: $R = 29.95x$. To find the break-even point, find x such that $C = R$:

$$C = R$$
$$18375 + 5.45x = 29.95x$$
$$18375 = 24.50x$$
$$750 = x \Rightarrow \text{The break-even point is 750 pieces of software.}$$

71. Let $x =$ the number of gallons of paint A sold at the break-even point. Let $C =$ the costs of the company: $C = 32500 + 13x$. Let $R =$ the revenue of the business: $R = 18x$. To find the break-even point, find x such that $C = R$:

$$C = R$$
$$32500 + 13x = 18x$$
$$32500 = 5x$$
$$6500 = x \Rightarrow \text{The break-even point is 6500 gallons of paint A.}$$

73. Calculate the profit made using each process (profit $=$ revenue $-$ cost)
A: revenue $-$ cost $= 18(6000) - (32500 + 13 \cdot 6000) = 108000 - 110500 = -2500$
B: revenue $-$ cost $= 18(6000) - (80600 + 5 \cdot 6000) = 108000 - 110600 = -2600$
Since the loss is less with process A, process A should be used.

75. Let $x =$ the number of pumps using process A sold at the break-even point. Let $C =$ the costs of the company: $C = 12390 + 29x$. Let $R =$ the revenue of the business: $R = 50x$. To find the break-even point, find x such that $C = R$:

$$C = R$$
$$12390 + 29x = 50x$$
$$12390 = 21x$$
$$590 = x \Rightarrow \text{The break-even point is 590 pumps using process A.}$$

77. Calculate the profit made using each process (profit $=$ revenue $-$ cost)
A: revenue $-$ cost $= 550(50) - (12390 + 550 \cdot 29) = 27500 - 28340 = -840$
B: revenue $-$ cost $= 550(50) - (20460 + 550 \cdot 17) = 27500 - 29810 = -2310$
Since the loss is less with process A, process A should be used.

79. Calculate the profit made using each process (profit $=$ revenue $-$ cost)
A: revenue $-$ cost $= 650(50) - (12390 + 650 \cdot 29) = 32500 - 31240 = 1260$
B: revenue $-$ cost $= 650(50) - (20460 + 650 \cdot 17) = 32500 - 31510 = 990$
Since the profit is more with process A, process A should be used.

81. Let $x =$ the measure of the first angle and $y =$ the measure of the second angle.

$$
\begin{array}{ll}
x + y = 180 & \quad x + y = 180 \quad \text{The angles have measures of } 35° \text{ and } 145°. \\
\underline{x - y = 110} & \quad 145 + y = 180 \\
2x \quad\;\;\, = 290 & \qquad y = 35 \\
x \quad\;\;\, = 145 &
\end{array}
$$

83. $\angle A$ and $\angle B$ are supplementary $\Rightarrow 2x + y + 3x = 180 \Rightarrow 5x + y = 180$.
$\angle A$ and $\angle D$ are supplementary $\Rightarrow 2x + y + y = 180 \Rightarrow 2x + 2y = 180$.

$5x + y = 180 \Rightarrow \times(-2) \quad -10x - 2y = -360$ Solve for y:
$2x + 2y = 180 \Rightarrow$

$$\begin{array}{rl} 2x + 2y = & 180 \\ \hline -8x \quad = & -180 \\ x \quad = & 22.5 \end{array}$$

$$2x + 2y = 180$$
$$2(22.5) + 2y = 180$$
$$45 + 2y = 180$$
$$2y = 135$$
$$y = 67.5$$

85. Let $x =$ the measure of the angle of the range of motion and $y =$ the measure of the second angle.

(1) $x + y = 90$ $x + y = 90$ $x = 4y$ The range of motion is $72°$.
(2) $x = 4y$ $4y + y = 90$ $x = 4(18)$
                    $5y = 90$ $x = 72$
                    $y = 18$

87. Set $X_L = X_C$:

$$X_L = X_C$$
$$2\pi f L = \frac{1}{2\pi f C}$$
$$2\pi f L \cdot 2\pi f C = \frac{1}{2\pi f C} \cdot 2\pi f C$$
$$4\pi^2 f^2 L C = 1$$
$$\frac{4\pi^2 f^2 L C}{4\pi^2 L C} = \frac{1}{4\pi^2 L C}$$
$$f^2 = \frac{1}{4\pi^2 L C}$$

89. Answers may vary.

91. Answers may vary.

Exercise 3.3 (page 182)

1. $m = \dfrac{\Delta y}{\Delta x} = \dfrac{-4 - 5}{-2 - 3} = \dfrac{-9}{-5} = \dfrac{9}{5}$

3. $f(0) = 2(0)^2 + 1 = 2(0) + 1 = 0 + 1 = 1$

5. $f(s) = 2s^2 + 1$

7. plane

9. infinitely

11. $x - y + z = 2$ $2x + y - z = 4$ $2x - 3y + z = 2$
$2 - 1 + 1 \overset{?}{=} 2$ $2(2) + 1 - 1 \overset{?}{=} 4$ $2(2) - 3(1) + 1 \overset{?}{=} 2$
$2 = 2$ $4 + 1 - 1 \overset{?}{=} 4$ $4 - 3 + 1 \overset{?}{=} 2$
              $4 = 4$ $2 = 2$

$(2, 1, 1)$ is a solution to the system.

13.

(1)	$x+y+z=4$	(1)	$x+y+z=4$	(2)	$2x+y-z=1$
(2)	$2x+y-z=1$	(2)	$2x+y-z=1$	(3)	$2x-3y+z=1$
(3)	$2x-3y+z=1$	(4)	$\overline{3x+2y\quad=5}$	(5)	$\overline{4x-2y\quad=2}$

$$
\begin{aligned}
&\text{(4)}\ 3x+2y=5 & 3x+2y&=5 & x+y+z&=4 \\
&\text{(5)}\ \underline{4x-2y=2} & 3(1)+2y&=5 & 1+1+z&=4 \\
&\qquad\ \ \overline{7x\quad\ \ =7} & 3+2y&=5 & 2+z&=4 \\
&\qquad\quad\ x\quad\ =1 & 2y&=2 & z&=2 \quad \boxed{\text{The solution is } (1,1,2).} \\
& & y&=1
\end{aligned}
$$

15.

(1)	$2x+2y+3z=10$	(1)	$2x+2y+3z=10$	(3)	$x+y+2z=6$
(2)	$3x+y-z=0$	$3\cdot(2)$	$\underline{9x+3y-3z=0}$	$2\cdot(2)$	$\underline{6x+2y-2z=0}$
(3)	$x+y+2z=6$	(4)	$\overline{11x+5y\quad=10}$	(5)	$\overline{7x+3y\quad=6}$

$$
\begin{aligned}
11x+5y=10 &\Rightarrow \times 3 & 33x+15y&=30 & 11x+5y&=10 & x+y+2z&=6 \\
\underline{7x+3y=6} &\Rightarrow \times(-5) & \underline{-35x-15y}&=\underline{-30} & 11(0)+5y&=10 & 0+2+2z&=6 \\
& & -2x\quad\ &=0 & 0+5y&=10 & 2+2z&=6 \\
& & x\quad\ &=0 & 5y&=10 & 2z&=4 \\
& & & & y&=2 & z&=2
\end{aligned}
$$

Solution: $\boxed{(0,2,2)}$

17.

(1)	$a+b+2c=7$	(1)	$a+b+2c=7$	(1)	$a+b+2c=7$
(2)	$a+2b+c=8$	$-2\cdot(2)$	$\underline{-2a-4b-2c=-16}$	$-2\cdot(3)$	$\underline{-4a-2b-2c=-18}$
(3)	$2a+b+c=9$	(4)	$\overline{-a-3b\quad=-9}$	(5)	$\overline{-3a-b\quad=-11}$

$$
\begin{aligned}
-a-3b=-9 &\Rightarrow \times(-3) & 3a+9b&=27 & -a-3b&=-9 & 2a+b+c&=9 \\
\underline{-3a-b=-11} &\Rightarrow & \underline{-3a-b}&=\underline{-11} & -a-3(2)&=-9 & 2(3)+2+c&=9 \\
& & 8b&=16 & -a-6&=-9 & 6+2+c&=9 \\
& & b&=2 & -a&=-3 & 8+c&=9 \\
& & & & a&=3 & c&=1
\end{aligned}
$$

Solution: $\boxed{(3,2,1)}$

19.

(1)	$2x+y-z=1$	(2)	$x+2y+2z=2$	(3)	$4x+5y+3z=3$
(2)	$x+2y+2z=2$	$2\cdot(1)$	$\underline{4x+2y-2z=2}$	$3\cdot(1)$	$\underline{6x+3y-3z=3}$
(3)	$4x+5y+3z=3$	(4)	$\overline{5x+4y\quad=4}$	(5)	$\overline{10x+8y\quad=6}$

$$
\begin{aligned}
5x+4y=4 &\Rightarrow \times(-2) & -10x-8y&=-8 \\
\underline{10x+8y=6} &\Rightarrow & \underline{10x+8y}&=\underline{6} \\
& & 0&=-2
\end{aligned}
$$

Since this equation is always false, there is no solution. The system is inconsistent.

21.

(1) $\quad 4x + 3z = 4$ (2) $\quad 2y - 6z = -1$ (2) $\quad\quad 2y - 6z = -1$
(2) $\quad 2y - 6z = -1$ $2 \cdot (1)$ $8x \quad\quad + 6z = 8$ $2 \cdot (3)$ $16x + 8y + 6z = 18$
(3) $\quad 8x + 4y + 3z = 9$ (4) $\overline{8x + 2y \quad\quad = 7}$ (5) $\overline{16x + 10y \quad\quad = 17}$

$$
\begin{aligned}
8x + 2y &= 7 \Rightarrow \times(-2) & -16x - 4y &= -14 & 8x + 2y &= 7 & 4x + 3z &= 4 \\
16x + 10y &= 17 \Rightarrow & 16x + 10y &= 17 & 8x + 2(\tfrac{1}{2}) &= 7 & 4(\tfrac{3}{4}) + 3z &= 4 \\
& & 6y &= 3 & 8x + 1 &= 7 & 3 + 3z &= 4 \\
& & y &= \tfrac{1}{2} & 8x &= 6 & 3z &= 1 \\
& & & & x &= \tfrac{3}{4} & z &= \tfrac{1}{3}
\end{aligned}
$$

Solution: $\boxed{\left(\tfrac{3}{4}, \tfrac{1}{2}, \tfrac{1}{3}\right)}$

23.

(1) $\quad 2x + 3y + 4z = 6$ (3) $\quad 4x + 6y + 8z = 12$
(2) $\quad 2x - 3y - 4z = -4$ $-2 \cdot (1)$ $\underline{-4x - 6y - 8z = -12}$
(3) $\quad 4x + 6y + 8z = 12$ (4) $ 0 = 0$

Since equation (4) is always true, the equations are dependent.

25.

$$
\begin{aligned}
x + \tfrac{1}{3}y + z &= 13 & \Rightarrow \times 3 & \quad (1)\ 3x + y + 3z = 39 \\
\tfrac{1}{2}x - y + \tfrac{1}{3}z &= -2 & \Rightarrow \times 6 & \quad (2)\ 3x - 6y + 2z = -12 \\
x + \tfrac{1}{2}y - \tfrac{1}{3}z &= 2 & \Rightarrow \times 6 & \quad (3)\ 6x + 3y - 2z = 12
\end{aligned}
$$

(2) $\quad 3x - 6y + 2z = -12$ (2) $\quad 3x - 6y + 2z = -12$
$6 \cdot (1)$ $\underline{18x + 6y + 18z = 234}$ $2 \cdot (3)$ $\underline{12x + 6y - 4z = 24}$
(4) $\quad 21x + 20z = 222$ (5) $\quad 15x - 2z = 12$

$$
\begin{aligned}
21x + 20z &= 222 \Rightarrow & 21x + 20z &= 222 & 15x - 2z &= 12 & 3x + y + 3z &= 39 \\
15x - 2z &= 12 \Rightarrow \times 10 & \underline{150x - 20z} &= \underline{120} & 15(2) - 2z &= 12 & 3(2) + y + 3(9) &= 39 \\
& & 171x &= 342 & 30 - 2z &= 12 & 6 + y + 27 &= 39 \\
& & x &= 2 & -2z &= -18 & y &= 6 \\
& & & & z &= 9
\end{aligned}
$$

Solution: $\boxed{(2, 6, 9)}$

27. Let $x =$ the first integer, $y =$ the second integer and $z =$ the third integer.

$$
\begin{aligned}
x + y + z &= 18 & \Rightarrow & \quad (1)\ x + y + z = 18 & (1) & \quad x + y + z = 18 \\
z &= 4y & \Rightarrow & \quad (2)\ -4y + z = 0 & (3) & \quad \underline{-x + y = 6} \\
y &= x + 6 & \Rightarrow & \quad (3)\ -x + y = 6 & (4) & \quad 2y + z = 24
\end{aligned}
$$

$$
\begin{aligned}
2y + z &= 24 \Rightarrow \times 2 & 4y + 2z &= 48 & 2y + z &= 24 & x + y + z &= 18 & &\text{The integers} \\
-4y + z &= 0 \Rightarrow & \underline{-4y + z} &= \underline{0} & 2y + 16 &= 24 & x + 4 + 16 &= 18 & &\text{are } -2, 4 \text{ and } 16. \\
& & 3z &= 48 & 2y &= 8 & x &= -2 & & \\
& & z &= 16 & y &= 4 & & &
\end{aligned}
$$

29. Let A, B and C represent the measures of the three angles.

$$A + B + C = 180 \qquad \Rightarrow \quad (1) \; A + B + C = 180 \qquad (1) \quad A + B + C = 180$$
$$A = B + C - 100 \quad \Rightarrow \quad (2) \; A - B - C = -100 \qquad (2) \quad A - B - C = -100$$
$$C = 2B - 40 \qquad \Rightarrow \quad (3) \; {-2B} + C = -40 \qquad (4) \; \underline{2A \qquad\qquad = 80}$$
$$A \qquad\qquad\quad = 40$$

$$
\begin{array}{llllll}
A + B + C = 180 & (-1)\cdot 3 & 2B - C = 40 & B + C = 140 & \text{The angles have measures} \\
40 + B + C = 180 & \underline{(5)} & \underline{B + C = 140} & 60 + C = 140 & \text{of } 40^\circ, 60^\circ \text{ and } 80^\circ. \\
(5) \quad B + C = 140 & & 3B \qquad\;\; = 180 & C = 80 & \\
& & B \qquad\;\; = 60 & &
\end{array}
$$

31. Let $A = $ the units of food A, $B = $ the units of food B and $C = $ the units of food C.

$$
\begin{array}{lll}
(1) \quad A + 2B + 2C = 11 \quad \text{(fat)} & \qquad (1) \qquad A + 2B + 2C = 11 \\
(2) \qquad A + B + C = 6 \quad \text{(carbohydrate)} & -2 \cdot (2) \quad {-2A} - 2B - 2C = -12 \\
(3) \quad 2A + B + 2C = 10 \quad \text{(protein)} & \qquad (4) \quad \underline{{-A} \qquad\qquad\;\; = -1} \\
& \qquad\qquad\qquad\qquad A \qquad\qquad\quad = 1
\end{array}
$$

$$
\begin{array}{llll}
 (1) & A + 2B + 2C = 11 & -3A - 2C = -9 & A + B + C = 6 \\
-2 \cdot (3) & \underline{{-4A} - 2B - 4C = -20} & -3(1) - 2C = -9 & 1 + B + 3 = 6 \\
 (5) & {-3A} \qquad\quad\;\; - 2C = {-9} & -3 - 2C = -9 & B + 4 = 6 \\
& & -2C = -6 & B = 2 \\
& & C = 3 &
\end{array}
$$

1 unit of food A, 2 units of food B and 3 units of food C should be used.

33. Let $x = $ the number of \$5 statues, $y = $ the number of \$4 statues and $z = $ the number of \$3 statues.

$$
\begin{array}{lll}
(1) \qquad\quad x + y + z = 180 \quad \text{(total number made)} & -3 \cdot (1) \quad {-3x} - 3y - 3z = -540 \\
(2) \qquad 5x + 4y + 3z = 650 \quad \text{(total cost)} & \qquad (2) \quad \underline{5x + 4y + 3z = 650} \\
(3) \quad 20x + 12y + 9z = 2100 \quad \text{(total revenue)} & \qquad (4) \quad 2x + y \qquad\quad = 110
\end{array}
$$

$$
\begin{array}{lll}
-9 \cdot (1) & -9x - 9y - 9z = -1620 & 2x + y = 110 \Rightarrow \times(-3) \quad -6x - 3y = -330 \\
 (3) & \underline{20x + 12y + 9z = 2100} & 11x + 3y = 480 \Rightarrow \qquad\qquad\;\; \underline{11x + 3y = 480} \\
 (5) & 11x + 3y \qquad\quad = 480 & 5x \qquad\;\; = 150 \\
& & x \qquad\;\; = 30
\end{array}
$$

$$
\begin{array}{lll}
2x + y = 110 & x + y + z = 180 & \text{30 of the \$5, 50 of the \$4 and 100 of the \$3 statues} \\
2(30) + y = 110 & 30 + 50 + z = 180 & \text{should be made.} \\
60 + y = 110 & z = 100 & \\
y = 50 & &
\end{array}
$$

35. Let x = the number of $5 tickets, y = the number of $3 tickets and z = the number of $2 tickets.

(1) $\quad x + y + z = 750$ (total sold) $\qquad 2 \cdot (1)\ 2x + 2y + 2z = 1500$
(2) $\qquad\qquad x = 2z \Rightarrow x - 2z = 0$ (twice as many) $\qquad$ (2) $\ x \qquad\quad - 2z = \qquad 0$
(3) $\ 5x + 3y + 2z = 2625$ (total revenue) $\qquad$ (4) $\overline{\ 3x + 2y \qquad\ = 1500}$

(2) $\quad x \qquad - 2z = \quad 0 \qquad 3x + 2y = 1500 \Rightarrow \times(-2)\ \ -6x - 4y = -3000$
(3) $\ \underline{5x + 3y + 2z = 2625}\qquad \underline{6x + 3y = 2625} \Rightarrow \qquad\quad \underline{6x + 3y = \ \ 2625}$
(5) $\ 6x + 3y \qquad\ = 2625 \qquad\qquad\qquad\qquad\qquad\qquad -y = \ -375$
$\qquad\qquad\qquad\qquad\qquad\qquad\qquad\qquad\qquad\qquad\qquad\quad\ y = \qquad 375$

$\qquad 3x + 2y = 1500 \qquad\qquad x + y + z = 750 \qquad$ 250 of the $5, 375 of the $3 and 125 of the $2
$\quad 3x + 2(375) = 1500 \qquad 250 + 375 + z = 750 \qquad$ tickets were sold.
$\qquad 3x + 750 = 1500 \qquad\qquad 625 + z = 750$
$\qquad\qquad 3x = 750 \qquad\qquad\qquad\quad z = 125$
$\qquad\qquad\quad x = 250$

37. Let x = the number of totem poles, y = the number of bears and z = the number of deer.

(1) $\quad 2x + 2y + z = 14$ (carving) $\qquad -2 \cdot (1)\ \ -4x - 4y - 2z = -28$
(2) $\quad\ x + 2y + 2z = 15$ (sanding) $\qquad\qquad$ (2) $\quad\ x + 2y + 2z = \quad 15$
(3) $\ 3x + 2y + 2z = 21$ (painting) $\qquad\qquad$ (4) $\overline{-3x - 2y \qquad\ = -13}$

$-2 \cdot (1)\ \ -4x - 4y - 2z = -28 \qquad -3x - 2y = -13 \Rightarrow \times(-1)\ \ 3x + 2y = \ 13$
$\quad\ (3)\ \underline{\ \ 3x + 2y + 2z = \quad 21}\qquad \underline{-x - 2y = \ -7} \Rightarrow \qquad\qquad \underline{-x - 2y = -7}$
$\quad\ (5)\ \ -x - 2y \qquad\ = \quad -7 \qquad\qquad\qquad\qquad\qquad\qquad\quad 2x \qquad\ = \quad 6$
$\qquad\qquad\qquad\qquad\qquad\qquad\qquad\qquad\qquad\qquad\qquad\qquad\quad x \qquad\ = \quad 3$

$\quad 3x + 2y = 13 \qquad\qquad 2x + 2y + z = 14$
$\ 3(3) + 2y = 13 \qquad\ 2(3) + 2(2) + z = 14$
$\quad 9 + 2y = 13 \qquad\qquad\ 6 + 4 + z = 14$
$\qquad\ 2y = 4 \qquad\qquad\qquad\qquad z = 4 \qquad$ | 3 totem poles, 2 bears and 4 deer should be made. |
$\qquad\ \ y = 2$

39. Substitute the coordinates of each point for x and y in the equation $y = ax^2 + bx + c$.

$y = ax^2 + bx + c \qquad\qquad y = ax^2 + bx + c \qquad\quad y = ax^2 + bx + c$
$0 = a(0)^2 + b(0) + c \quad\ -4 = a(2)^2 + b(2) + c \quad\ 0 = a(4)^2 + b(4) + c$
$0 = c \qquad\qquad\qquad\quad -4 = 4a + 2b + c \qquad\quad\ 0 = 16a + 4b + c$

Solve the system of equations formed from the three equations:

(1) $\qquad\qquad c = 0 \qquad 4a + 2b = -4 \Rightarrow \times(-2)\ \ -8a - 4b = 8 \qquad 16a + 4b = 0$
(2) $\quad 4a + 2b + c = -4 \qquad \underline{16a + 4b = \quad 0} \Rightarrow \qquad\ \underline{16a + 4b = 0} \qquad 16(1) + 4b = 0$
(3) $\ \ 16a + 4b + c = 0 \qquad\qquad\qquad\qquad\qquad\qquad\quad 8a \qquad\ = 8 \qquad\ 16 + 4b = 0$
$\qquad\qquad\qquad\qquad\qquad\qquad\qquad\qquad\qquad\qquad\qquad\quad\ a \qquad\ = 1 \qquad\qquad 4b = -16$
$\qquad\qquad\qquad\qquad\qquad\qquad\qquad\qquad\qquad\qquad\qquad\qquad\qquad\qquad\qquad\qquad\quad b = -4$

The equation is $y = x^2 - 4x$.

41. Substitute the coordinates of each point for x and y in the equation $x^2 + y^2 + cx + dy + e = 0$.

$$x^2 + y^2 + cx + dy + e = 0 \qquad\qquad x^2 + y^2 + cx + dy + e = 0$$
$$(1)^2 + (3)^2 + c(1) + d(3) + e = 0 \qquad (3)^2 + (1)^2 + c(3) + d(1) + e = 0$$
$$1 + 9 + c + 3d + e = 0 \qquad\qquad 9 + 1 + 3c + d + e = 0$$
$$c + 3d + e = -10 \qquad\qquad\qquad 3c + d + e = -10$$

$$x^2 + y^2 + cx + dy + e = 0 \qquad (1) \quad c + 3d + e = -10$$
$$(1)^2 + (-1)^2 + c(1) + d(-1) + e = 0 \qquad (2) \quad 3c + d + e = -10$$
$$1 + 1 + c - d + e = 0 \qquad\qquad (3) \quad c - d + e = -2$$
$$c - d + e = -2$$

$$\begin{array}{lrr} (1) & c + 3d + e = -10 \\ -1 \cdot (2) & -3c - d - e = 10 \\ (4) & -2c + 2d = 0 \end{array} \qquad \begin{array}{lrr} (1) & c + 3d + e = -10 \\ -1 \cdot (3) & -c + d - e = 2 \\ (5) & 4d = -8 \\ & d = -2 \end{array} \qquad \begin{array}{l} -2c + 2d = 0 \\ -2c + 2(-2) = 0 \\ -2c - 4 = 0 \\ -2c = 4 \\ c = -2 \end{array}$$

$$\begin{array}{l} c + 3d + e = -10 \\ -2 + 3(-2) + e = -10 \\ -2 - 6 + e = -10 \\ -8 + e = -10 \\ e = -2 \end{array} \qquad \text{The equation is } x^2 + y^2 - 2x - 2y - 2 = 0.$$

43. **Answers may vary.**

45.
$$\begin{array}{ll} (1) & x + y + z + w = 3 \\ (2) & x - y - z - w = -1 \\ (3) & x + y - z - w = 1 \\ (4) & x + y - z + w = 3 \end{array} \qquad \begin{array}{lr} (1) & x + y + z + w = 3 \\ (2) & x - y - z - w = -1 \\ & 2x = 2 \\ & x = 1 \end{array} \qquad \begin{array}{lr} (1) & x + y + z + w = 3 \\ (3) & x + y - z - w = 1 \\ & 2x + 2y = 4 \end{array}$$

$$\begin{array}{lr} (1) & x + y + z + w = 3 \\ (4) & x + y - z + w = 3 \\ & 2x + 2y + 2w = 6 \end{array} \qquad \begin{array}{l} 2x + 2y = 4 \\ 2(1) + 2y = 4 \\ 2y = 2 \\ y = 1 \end{array} \qquad \begin{array}{l} 2x + 2y + 2w = 6 \\ 2(1) + 2(1) + 2w = 6 \\ 4 + 2w = 6 \\ 2w = 2 \\ w = 1 \end{array} \qquad \begin{array}{l} x + y + z + w = 3 \\ 1 + 1 + z + 1 = 3 \\ 3 + z = 3 \\ z = 0 \end{array}$$

$$\boxed{\text{The solution is } (1, 1, 0, 1).}$$

Exercise 3.4 (page 192)

Note: The notation $3R_1 + R_3 \Rightarrow R_2$ means to multiply Row #1 of the previous matrix, add that result to Row #3 of the previous matrix, and write the final result in Row #2 of the current matrix.

1. $93{,}000{,}000 = 9.3 \times 10^7$

3. $63 \times 10^3 = 6.3 \times 10^1 \times 10^3 = 6.3 \times 10^4$

5. matrix

7. 3; columns

9. augmented

11. type 1

SECTION 3.4

13. nonzero

15.
$$\begin{bmatrix} 2 & 1 & 1 \\ 5 & 4 & 1 \end{bmatrix} \xRightarrow{R_2 + (-R_1) \Rightarrow R_2} \begin{bmatrix} 2 & 1 & 1 \\ 3 & 3 & \boxed{0} \end{bmatrix}$$

17.
$$\begin{bmatrix} 3 & -2 & 1 \\ -1 & 2 & 4 \end{bmatrix} \xRightarrow{2R_2 \Rightarrow R_2} \begin{bmatrix} 3 & -2 & 1 \\ -2 & 4 & \boxed{8} \end{bmatrix}$$

19.
$$\begin{bmatrix} 1 & 1 & | & 2 \\ 1 & -1 & | & 0 \end{bmatrix} \xRightarrow{R_1 + (-R_2) \Rightarrow R_2} \begin{bmatrix} 1 & 1 & | & 2 \\ 0 & 2 & | & 2 \end{bmatrix} \xRightarrow{\frac{1}{2}R_2 \Rightarrow R_2} \begin{bmatrix} 1 & 1 & | & 2 \\ 0 & 1 & | & 1 \end{bmatrix}$$

From R_2, $y = 1$.　From R_1:　　The solution is $(1,1)$.
$$x + y = 2$$
$$x + 1 = 2 \Rightarrow x = 1$$

21.
$$\begin{bmatrix} 1 & 2 & | & -4 \\ 2 & 1 & | & 1 \end{bmatrix} \xRightarrow{-2R_1 + R_2 \Rightarrow R_2} \begin{bmatrix} 1 & 2 & | & -4 \\ 0 & -3 & | & 9 \end{bmatrix} \xRightarrow{-\frac{1}{3}R_2 \Rightarrow R_2} \begin{bmatrix} 1 & 2 & | & -4 \\ 0 & 1 & | & -3 \end{bmatrix}$$

From R_2, $y = -3$.　From R_1:　　The solution is $(2,-3)$.
$$x + 2y = -4$$
$$x + 2(-3) = -4$$
$$x - 6 = -4 \Rightarrow x = 2$$

23.
$$\begin{bmatrix} 3 & 4 & | & -12 \\ 9 & -2 & | & 6 \end{bmatrix} \xRightarrow{-3R_1 + R_2 \Rightarrow R_2} \begin{bmatrix} 3 & 4 & | & -12 \\ 0 & -14 & | & 42 \end{bmatrix} \xRightarrow{-\frac{1}{14}R_2 \Rightarrow R_2} \begin{bmatrix} 3 & 4 & | & -12 \\ 0 & 1 & | & -3 \end{bmatrix}$$

From R_2, $y = -3$.　From R_1:　　The solution is $(0,-3)$.
$$3x + 4y = -12$$
$$3x + 4(-3) = -12$$
$$3x - 12 = -12$$
$$3x = 0 \Rightarrow x = 0$$

25.
$$\begin{cases} 5a = 24 + 2b \\ 5b = 3a + 16 \end{cases} \Rightarrow \begin{cases} 5a - 2b = 24 \\ -3a + 5b = 16 \end{cases}$$

$$\begin{bmatrix} 5 & -2 & | & 24 \\ -3 & 5 & | & 16 \end{bmatrix} \xRightarrow{2R_2 + R_1 \Rightarrow R_1} \begin{bmatrix} -1 & 8 & | & 56 \\ -3 & 5 & | & 16 \end{bmatrix} \xRightarrow{-R_1 \Rightarrow R_1} \begin{bmatrix} 1 & -8 & | & -56 \\ -3 & 5 & | & 16 \end{bmatrix} \xRightarrow{3R_1 + R_2 \Rightarrow R_2} \begin{bmatrix} 1 & -8 & | & -56 \\ 0 & -19 & | & -152 \end{bmatrix}$$

$$\xRightarrow{-\frac{1}{19}R_2 \Rightarrow R_2} \begin{bmatrix} 1 & -8 & | & -56 \\ 0 & 1 & | & 8 \end{bmatrix}$$

From R_2, $b = 8$.　From R_1:　　The solution is $(8,8)$.
$$a - 8b = -56$$
$$a - 8(8) = -56$$
$$a - 64 = -56 \Rightarrow a = 8$$

27.

$$-R_1 + R_2 \Rightarrow R_2$$
$$-R_1 + R_3 \Rightarrow R_3$$

$$\begin{bmatrix} 1 & 1 & 1 & | & 6 \\ 1 & 2 & 1 & | & 8 \\ 1 & 1 & 2 & | & 9 \end{bmatrix} \Rightarrow \begin{bmatrix} 1 & 1 & 1 & | & 6 \\ 0 & 1 & 0 & | & 2 \\ 0 & 0 & 1 & | & 3 \end{bmatrix}$$

From R_3, $z = 3$. From R_2, $y = 2$. From R_1:   The solution is $(1, 2, 3)$.

$$x + y + z = 6$$
$$x + 2 + 3 = 6$$
$$x + 5 = 6$$
$$x = 1$$

29.

$$R_1 + R_2 \Rightarrow R_2 \qquad\qquad -\tfrac{1}{4} R_3 \Rightarrow R_2$$
$$-2R_1 + R_3 \Rightarrow R_3 \qquad\qquad \tfrac{1}{4} R_2 \Rightarrow R_3$$

$$\begin{bmatrix} 2 & 1 & 3 & | & 3 \\ -2 & -1 & 1 & | & 5 \\ 4 & -2 & 2 & | & 2 \end{bmatrix} \Rightarrow \begin{bmatrix} 2 & 1 & 3 & | & 3 \\ 0 & 0 & 4 & | & 8 \\ 0 & -4 & -4 & | & -4 \end{bmatrix} \Rightarrow \begin{bmatrix} 2 & 1 & 3 & | & 3 \\ 0 & 1 & 1 & | & 1 \\ 0 & 0 & 1 & | & 2 \end{bmatrix}$$

From R_3, $z = 2$. From R_2:   From R_1:   The solution is $(-1, -1, 2)$.

$$y + z = 1 \qquad 2x + y + 3z = 3$$
$$y + 2 = 1 \qquad 2x + (-1) + 3(2) = 3$$
$$y = -1 \qquad 2x - 1 + 6 = 3$$
$$2x + 5 = 3$$
$$2x = -2$$
$$x = -1$$

31.

$$-3R_2 + R_1 \Rightarrow R_2$$
$$-2R_1 + R_3 \Rightarrow R_3 \qquad\qquad \tfrac{2}{5} R_2 + R_3 \Rightarrow R_3$$

$$\begin{bmatrix} 3 & -2 & 4 & | & 4 \\ 1 & 1 & 1 & | & 3 \\ 6 & -2 & -3 & | & 10 \end{bmatrix} \Rightarrow \begin{bmatrix} 3 & -2 & 4 & | & 4 \\ 0 & -5 & 1 & | & -5 \\ 0 & 2 & -11 & | & 2 \end{bmatrix} \Rightarrow \begin{bmatrix} 3 & -2 & 4 & | & 4 \\ 0 & -5 & 1 & | & -5 \\ 0 & 0 & -\tfrac{53}{5} & | & 0 \end{bmatrix}$$

From R_3, $z = 0$. From R_2:   From R_1:   The solution is $(2, 1, 0)$.

$$-5y + z = -5 \qquad 3x - 2y + 4z = 4$$
$$-5y + 0 = -5 \qquad 3x - 2(1) + 4(0) = 4$$
$$-5y = -5 \qquad 3x - 2 = 4$$
$$y = 1 \qquad 3x = 6$$
$$x = 2$$

33.

$$-3R_1 + R_2 \Rightarrow R_2 \qquad -\tfrac{1}{4} R_2 \Rightarrow R_2$$
$$-2R_1 + R_3 \Rightarrow R_3 \qquad -4R_3 + R_2 \Rightarrow R_3$$

$$\begin{bmatrix} 1 & 1 & | & 3 \\ 3 & -1 & | & 1 \\ 2 & 1 & | & 4 \end{bmatrix} \Rightarrow \begin{bmatrix} 1 & 1 & | & 3 \\ 0 & -4 & | & -8 \\ 0 & -1 & | & -2 \end{bmatrix} \Rightarrow \begin{bmatrix} 1 & 1 & | & 3 \\ 0 & 1 & | & 2 \\ 0 & 0 & | & 0 \end{bmatrix}$$

continued on next page...

33. continued

From R_2, $y = 2$.   From R_1:   The solution is $(1, 2)$.

$$x + y = 3$$
$$x + 2 = 3$$
$$x = 1$$

35.

$$-2R_2 + R_1 \Rightarrow R_2 \qquad -\tfrac{1}{7}R_2 \Rightarrow R_2$$
$$2R_3 + R_1 \Rightarrow R_3 \qquad -\tfrac{7}{9}R_2 + R_3 \Rightarrow R_3$$

$$\begin{bmatrix} 2 & -1 & 4 \\ 1 & 3 & 2 \\ -1 & -4 & -2 \end{bmatrix} \Rightarrow \begin{bmatrix} 2 & -1 & 4 \\ 0 & -7 & 0 \\ 0 & -9 & 0 \end{bmatrix} \Rightarrow \begin{bmatrix} 2 & -1 & 4 \\ 0 & 1 & 0 \\ 0 & 0 & 0 \end{bmatrix}$$

From R_2, $y = 0$.   From R_1:   The solution is $(2, 0)$.

$$2x - y = 4$$
$$2x - 0 = 4$$
$$2x = 4$$
$$x = 2$$

37.

$$-2R_2 + R_1 \Rightarrow R_2 \qquad \tfrac{1}{3}R_2 \Rightarrow R_2$$
$$2R_3 + R_1 \Rightarrow R_3 \qquad -\tfrac{7}{3}R_2 + R_3 \Rightarrow R_3$$

$$\begin{bmatrix} 2 & 1 & 7 \\ 1 & -1 & 2 \\ -1 & 3 & -2 \end{bmatrix} \Rightarrow \begin{bmatrix} 2 & 1 & 7 \\ 0 & 3 & 3 \\ 0 & 7 & 3 \end{bmatrix} \Rightarrow \begin{bmatrix} 2 & 1 & 7 \\ 0 & 1 & 1 \\ 0 & 0 & -4 \end{bmatrix}$$

R_3: $0x + 0y = -4$, or $0 = -4$, which is an impossible equation.   NO SOLUTION

39.

$$-R_2 + R_1 \Rightarrow R_2 \qquad \tfrac{1}{2}R_2 \Rightarrow R_2$$
$$-3R_1 + R_3 \Rightarrow R_3 \qquad 4R_2 + R_3 \Rightarrow R_3$$

$$\begin{bmatrix} 1 & 3 & 7 \\ 1 & 1 & 3 \\ 3 & 1 & 5 \end{bmatrix} \Rightarrow \begin{bmatrix} 1 & 3 & 7 \\ 0 & 2 & 4 \\ 0 & -8 & -16 \end{bmatrix} \Rightarrow \begin{bmatrix} 1 & 3 & 7 \\ 0 & 1 & 2 \\ 0 & 0 & 0 \end{bmatrix}$$

From R_2, $y = 2$.   From R_1:   The solution is $(1, 2)$.

$$x + 3y = 7$$
$$x + 3(2) = 7$$
$$x + 6 = 7 \Rightarrow x = 1$$

41.

$$R_1 + R_2 \Rightarrow R_2$$

$$\begin{bmatrix} 1 & 2 & 3 & -2 \\ -1 & -1 & -2 & 4 \end{bmatrix} \Rightarrow \begin{bmatrix} 1 & 2 & 3 & -2 \\ 0 & 1 & 1 & 2 \end{bmatrix}$$

From R_2:   From R_1:   The solution is

$$y + z = 2 \qquad x + 2y + 3z = -2 \qquad (-6 - z, 2 - z, z).$$
$$y = 2 - z \qquad x + 2(2 - z) + 3z = -2$$
$$x + 4 - 2z + 3z = -2$$
$$x + z = -6$$
$$x = -6 - z$$

43.

$$\begin{array}{c} \\ \begin{bmatrix} 1 & -1 & 0 & | & 1 \\ 0 & 1 & 1 & | & 1 \\ 1 & 0 & 1 & | & 2 \end{bmatrix} \end{array} \xRightarrow{-R_1 + R_3 \Rightarrow R_3} \begin{bmatrix} 1 & -1 & 0 & | & 1 \\ 0 & 1 & 1 & | & 1 \\ 0 & 1 & 1 & | & 1 \end{bmatrix} \xRightarrow{-R_2 + R_3 \Rightarrow R_3} \begin{bmatrix} 1 & -1 & 0 & | & 1 \\ 0 & 1 & 1 & | & 1 \\ 0 & 0 & 0 & | & 0 \end{bmatrix}$$

From R_2:   From R_1:   The solution is

$y + z = 1$ $x - y = 1$ $(2 - z, 1 - z, z)$.

$\quad y = 1 - z$ $x - (1 - z) = 1$

$\qquad\qquad\quad x - 1 + z = 1$

$\qquad\qquad\qquad\quad x + z = 2$

$\qquad\qquad\qquad\qquad x = 2 - z$

45. Let $x = $ the measure of the first angle and $y = $ the measure of the second angle. Form and solve this

system of equations: $\begin{cases} x + y = 90 \\ \quad\; y = x + 46 \end{cases} \Rightarrow \begin{cases} \quad x + y = 90 \\ -x + y = 46 \end{cases}$

$$\begin{bmatrix} 1 & 1 & | & 90 \\ -1 & 1 & | & 46 \end{bmatrix} \xRightarrow{R_1 + R_2 \Rightarrow R_2} \begin{bmatrix} 1 & 1 & | & 90 \\ 0 & 2 & | & 136 \end{bmatrix} \xRightarrow{\frac{1}{2}R_2 \Rightarrow R_2} \begin{bmatrix} 1 & 1 & | & 90 \\ 0 & 1 & | & 68 \end{bmatrix}$$

From R_2, $y = 68$.   From R_1:   The angles have measures

$\qquad\qquad\qquad\qquad\qquad x + y = 90$ of $22°$ and $68°$.

$\qquad\qquad\qquad\qquad\qquad x + 68 = 90$

$\qquad\qquad\qquad\qquad\qquad\quad x = 22$

47. Let A, B and C represent the measures of the three angles.

$$\begin{cases} A + B + C = 180 \\ \qquad\quad B = A + 25 \\ \qquad\quad C = 2A - 5 \end{cases} \Rightarrow \begin{cases} A + B + C = 180 \\ \quad -A + B = 25 \\ \quad -2A + C = -5 \end{cases}$$

$$\begin{bmatrix} 1 & 1 & 1 & | & 180 \\ -1 & 1 & 0 & | & 25 \\ -2 & 0 & 1 & | & -5 \end{bmatrix} \xRightarrow[2R_1 + R_3 \Rightarrow R_3]{R_1 + R_2 \Rightarrow R_2} \begin{bmatrix} 1 & 1 & 1 & | & 180 \\ 0 & 2 & 1 & | & 205 \\ 0 & 2 & 3 & | & 355 \end{bmatrix} \xRightarrow{-R_2 + R_3 \Rightarrow R_3} \begin{bmatrix} 1 & 1 & 1 & | & 180 \\ 0 & 2 & 1 & | & 205 \\ 0 & 0 & 2 & | & 150 \end{bmatrix} \xRightarrow{\frac{1}{2}R_3 \Rightarrow R_3} \begin{bmatrix} 1 & 1 & 1 & | & 180 \\ 0 & 2 & 1 & | & 205 \\ 0 & 0 & 1 & | & 75 \end{bmatrix}$$

From R_3, $C = 75$.   From R_2:   From R_1:

$\qquad\qquad\qquad\qquad\qquad 2B + C = 205$ $A + B + C = 180$

$\qquad\qquad\qquad\qquad\qquad 2B + 75 = 205$ $A + 65 + 75 = 180$

$\qquad\qquad\qquad\qquad\qquad\quad 2B = 130$ $A + 140 = 180$

$\qquad\qquad\qquad\qquad\qquad\qquad B = 65$ $\qquad\quad A = 40$

The angles have measures of $40°$, $65°$ and $75°$.

49. Plug the coordinates of the points into the general equation to form and solve a system of equations.

$y = ax^2 + bx + c$ $y = ax^2 + bx + c$ $y = ax^2 + bx + c$

$1 = a(0)^2 + b(0) + c$ $2 = a(1)^2 + b(1) + c$ $4 = a(-1)^2 + b(-1) + c$

$1 = c$ $2 = a + b + c$ $4 = a - b + c$

$$\begin{cases} a + b + c = 2 \\ a - b + c = 4 \\ c = 1 \end{cases} \Rightarrow \begin{bmatrix} 1 & 1 & 1 & 2 \\ 1 & -1 & 1 & 4 \\ 0 & 0 & 1 & 1 \end{bmatrix} \Rightarrow \begin{matrix} -R_2 + R_1 \Rightarrow R_2 \\ \begin{bmatrix} 1 & 1 & 1 & 2 \\ 0 & 2 & 0 & -2 \\ 0 & 0 & 1 & 1 \end{bmatrix} \end{matrix} \Rightarrow \begin{matrix} \frac{1}{2}R_2 \Rightarrow R_2 \\ \begin{bmatrix} 1 & 1 & 1 & 2 \\ 0 & 1 & 0 & -1 \\ 0 & 0 & 1 & 1 \end{bmatrix} \end{matrix}$$

From R_3, $c = 1$.   From R_2, $b = -1$. From R_1:

$$a + b + c = 2$$
$$a + (-1) + 1 = 2$$
$$a = 2$$

The equation is $y = 2x^2 - x + 1$.

51. Answers may vary.

53. The last equation represents the equation $0x + 0y + 0z = k$, or $0 = k$. If $k = 0$, then the system can be solved. However, if $k \neq 0$, the system will have no solution.

Exercise 3.5 (page 201)

1.
$$3(x + 2) - (2 - x) = x - 5$$
$$3x + 6 - 2 + x = x - 5$$
$$4x + 4 = x - 5$$
$$3x = -9$$
$$x = -3$$

3.
$$\frac{5}{3}(5x + 6) - 10 = 0$$
$$3 \cdot \frac{5}{3}(5x + 6) - 3 \cdot 10 = 3 \cdot 0$$
$$5(5x + 6) - 30 = 0$$
$$25x + 30 - 30 = 0$$
$$25x = 0$$
$$x = 0$$

5. number

7. $\begin{vmatrix} a_2 & c_2 \\ a_3 & c_3 \end{vmatrix}$

9. $\begin{vmatrix} 3 & 4 \\ 2 & -3 \end{vmatrix}$

11. $\begin{vmatrix} 2 & 3 \\ -2 & 1 \end{vmatrix} = 2(1) - 3(-2)$
$$= 2 + 6 = 8$$

13. $\begin{vmatrix} -1 & 2 \\ 3 & -4 \end{vmatrix} = -1(-4) - 2(3)$
$$= 4 - 6 = -2$$

15. $\begin{vmatrix} x & y \\ y & x \end{vmatrix} = x(x) - y(y) = x^2 - y^2$

17. $\begin{vmatrix} 1 & 0 & 1 \\ 0 & 1 & 0 \\ 1 & 1 & 1 \end{vmatrix} = 1\begin{vmatrix} 1 & 0 \\ 1 & 1 \end{vmatrix} - 0\begin{vmatrix} 0 & 0 \\ 1 & 1 \end{vmatrix} + 1\begin{vmatrix} 0 & 1 \\ 1 & 1 \end{vmatrix} = 1(1) - 0(0) + 1(-1) = 1 - 0 - 1 = 0$

19. $\begin{vmatrix} -1 & 2 & 1 \\ 2 & 1 & -3 \\ 1 & 1 & 1 \end{vmatrix} = -1\begin{vmatrix} 1 & -3 \\ 1 & 1 \end{vmatrix} - 2\begin{vmatrix} 2 & -3 \\ 1 & 1 \end{vmatrix} + 1\begin{vmatrix} 2 & 1 \\ 1 & 1 \end{vmatrix} = -1(4) - 2(5) + 1(1) = -4 - 10 + 1$

$$= -13$$

21. $\begin{vmatrix} 1 & -2 & 3 \\ -2 & 1 & 1 \\ -3 & -2 & 1 \end{vmatrix} = 1\begin{vmatrix} 1 & 1 \\ -2 & 1 \end{vmatrix} - (-2)\begin{vmatrix} -2 & 1 \\ -3 & 1 \end{vmatrix} + 3\begin{vmatrix} -2 & 1 \\ -3 & -2 \end{vmatrix} = 1(3) + 2(1) + 3(7) = 3 + 2 + 21$

$$= 26$$

23. $\begin{vmatrix} 1 & 2 & 3 \\ 4 & 5 & 6 \\ 7 & 8 & 9 \end{vmatrix} = 1\begin{vmatrix} 5 & 6 \\ 8 & 9 \end{vmatrix} - 2\begin{vmatrix} 4 & 6 \\ 7 & 9 \end{vmatrix} + 3\begin{vmatrix} 4 & 5 \\ 7 & 8 \end{vmatrix} = 1(-3) - 2(-6) + 3(-3) = -3 + 12 - 9 = 0$

25. $\begin{vmatrix} a & 2a & -a \\ 2 & -1 & 3 \\ 1 & 2 & -3 \end{vmatrix} = a\begin{vmatrix} -1 & 3 \\ 2 & -3 \end{vmatrix} - 2a\begin{vmatrix} 2 & 3 \\ 1 & -3 \end{vmatrix} + (-a)\begin{vmatrix} 2 & -1 \\ 1 & 2 \end{vmatrix} = a(-3) - 2a(-9) - a(5)$

$$= -3a + 18a - 5a = 10a$$

27. $\begin{vmatrix} 1 & a & b \\ 1 & 2a & 2b \\ 1 & 3a & 3b \end{vmatrix} = 1\begin{vmatrix} 2a & 2b \\ 3a & 3b \end{vmatrix} - a\begin{vmatrix} 1 & 2b \\ 1 & 3b \end{vmatrix} + b\begin{vmatrix} 1 & 2a \\ 1 & 3a \end{vmatrix} = 1(0) - a(b) + b(a) = 0 - ab + ab = 0$

29. $x = \dfrac{\begin{vmatrix} 6 & 1 \\ 2 & -1 \end{vmatrix}}{\begin{vmatrix} 1 & 1 \\ 1 & -1 \end{vmatrix}} = \dfrac{-6-2}{-1-1} = \dfrac{-8}{-2} = 4;\ y = \dfrac{\begin{vmatrix} 1 & 6 \\ 1 & 2 \end{vmatrix}}{\begin{vmatrix} 1 & 1 \\ 1 & -1 \end{vmatrix}} = \dfrac{2-6}{-2} = \dfrac{-4}{-2} = 2;\ \text{solution: } (4, 2)$

31. $x = \dfrac{\begin{vmatrix} 1 & 1 \\ -7 & -2 \end{vmatrix}}{\begin{vmatrix} 2 & 1 \\ 1 & -2 \end{vmatrix}} = \dfrac{-2-(-7)}{-4-1} = \dfrac{5}{-5} = -1;\ y = \dfrac{\begin{vmatrix} 2 & 1 \\ 1 & -7 \end{vmatrix}}{\begin{vmatrix} 2 & 1 \\ 1 & -2 \end{vmatrix}} = \dfrac{-14-1}{-5} = \dfrac{-15}{-5} = 3$

solution: $(-1, 3)$

33. $x = \dfrac{\begin{vmatrix} 0 & 3 \\ -4 & -6 \end{vmatrix}}{\begin{vmatrix} 2 & 3 \\ 4 & -6 \end{vmatrix}} = \dfrac{0-(-12)}{-12-12} = \dfrac{12}{-24} = -\dfrac{1}{2};\ y = \dfrac{\begin{vmatrix} 2 & 0 \\ 4 & -4 \end{vmatrix}}{\begin{vmatrix} 2 & 3 \\ 4 & -6 \end{vmatrix}} = \dfrac{-8-0}{-24} = \dfrac{-8}{-24} = \dfrac{1}{3}$

solution: $\left(-\dfrac{1}{2}, \dfrac{1}{3}\right)$

35. $\begin{cases} y = \dfrac{-2x+1}{3} \\ 3x - 2y = 8 \end{cases} \Rightarrow \begin{cases} 3y = -2x+1 \\ 3x - 2y = 8 \end{cases} \Rightarrow \begin{cases} 2x + 3y = 1 \\ 3x - 2y = 8 \end{cases}$

$x = \dfrac{\begin{vmatrix} 1 & 3 \\ 8 & -2 \end{vmatrix}}{\begin{vmatrix} 2 & 3 \\ 3 & -2 \end{vmatrix}} = \dfrac{-2 - 24}{-4 - 9} = \dfrac{-26}{-13} = 2; \; y = \dfrac{\begin{vmatrix} 2 & 1 \\ 3 & 8 \end{vmatrix}}{\begin{vmatrix} 2 & 3 \\ 3 & -2 \end{vmatrix}} = \dfrac{16 - 3}{-13} = \dfrac{13}{-13} = -1; \text{ solution: } (2, -1)$

37. $\begin{cases} y = \dfrac{11 - 3x}{2} \\ x = \dfrac{11 - 4y}{6} \end{cases} \Rightarrow \begin{cases} 2y = 11 - 3x \\ 6x = 11 - 4y \end{cases} \Rightarrow \begin{cases} 3x + 2y = 11 \\ 6x + 4y = 11 \end{cases}$

$x = \dfrac{\begin{vmatrix} 11 & 2 \\ 11 & 4 \end{vmatrix}}{\begin{vmatrix} 3 & 2 \\ 6 & 4 \end{vmatrix}} = \dfrac{44 - 22}{12 - 12} = \dfrac{22}{0} \Rightarrow \text{ denominator } = 0, \text{ numerator } \neq 0 \Rightarrow \text{ no solution}$

39. $\begin{cases} x = \dfrac{5y - 4}{2} \\ y = \dfrac{3x - 1}{5} \end{cases} \Rightarrow \begin{cases} 2x = 5y - 4 \\ 5y = 3x - 1 \end{cases} \Rightarrow \begin{cases} 2x - 5y = -4 \\ -3x + 5y = -1 \end{cases}$

$x = \dfrac{\begin{vmatrix} -4 & -5 \\ -1 & 5 \end{vmatrix}}{\begin{vmatrix} 2 & -5 \\ -3 & 5 \end{vmatrix}} = \dfrac{-20 - 5}{10 - 15} = \dfrac{-25}{-5} = 5; \; y = \dfrac{\begin{vmatrix} 2 & -4 \\ -3 & -1 \end{vmatrix}}{\begin{vmatrix} 2 & -5 \\ -3 & 5 \end{vmatrix}} = \dfrac{-2 - 12}{-5} = \dfrac{-14}{-5} = \dfrac{14}{5}$

solution: $\left(5, \dfrac{14}{5}\right)$

Note: In the following problems, D stands for the denominator determinant, while N_x, N_y and N_z stand for the numerator determinants for x, y and z, respectively.

41. $D = \begin{vmatrix} 1 & 1 & 1 \\ 1 & 1 & -1 \\ 1 & -1 & 1 \end{vmatrix} = 1\begin{vmatrix} 1 & -1 \\ -1 & 1 \end{vmatrix} - 1\begin{vmatrix} 1 & -1 \\ 1 & 1 \end{vmatrix} + 1\begin{vmatrix} 1 & 1 \\ 1 & -1 \end{vmatrix} = 1(0) - 1(2) + 1(-2) = -4$

$N_x = \begin{vmatrix} 4 & 1 & 1 \\ 0 & 1 & -1 \\ 2 & -1 & 1 \end{vmatrix} = 4\begin{vmatrix} 1 & -1 \\ -1 & 1 \end{vmatrix} - 1\begin{vmatrix} 0 & -1 \\ 2 & 1 \end{vmatrix} + 1\begin{vmatrix} 0 & 1 \\ 2 & -1 \end{vmatrix} = 4(0) - 1(2) + 1(-2) = -4$

$N_y = \begin{vmatrix} 1 & 4 & 1 \\ 1 & 0 & -1 \\ 1 & 2 & 1 \end{vmatrix} = 1\begin{vmatrix} 0 & -1 \\ 2 & 1 \end{vmatrix} - 4\begin{vmatrix} 1 & -1 \\ 1 & 1 \end{vmatrix} + 1\begin{vmatrix} 1 & 0 \\ 1 & 2 \end{vmatrix} = 1(2) - 4(2) + 1(2) = -4$

$N_z = \begin{vmatrix} 1 & 1 & 4 \\ 1 & 1 & 0 \\ 1 & -1 & 2 \end{vmatrix} = 1\begin{vmatrix} 1 & 0 \\ -1 & 2 \end{vmatrix} - 1\begin{vmatrix} 1 & 0 \\ 1 & 2 \end{vmatrix} + 4\begin{vmatrix} 1 & 1 \\ 1 & -1 \end{vmatrix} = 1(2) - 1(2) + 4(-2) = -8$

$x = \dfrac{N_x}{D} = \dfrac{-4}{-4} = 1; \; y = \dfrac{N_y}{D} = \dfrac{-4}{-4} = 1; \; z = \dfrac{N_z}{D} = \dfrac{-8}{-4} = 2 \Rightarrow \text{ solution: } (1, 1, 2)$

43. $D = \begin{vmatrix} 1 & 1 & 2 \\ 1 & 2 & 1 \\ 2 & 1 & 1 \end{vmatrix} = 1\begin{vmatrix} 2 & 1 \\ 1 & 1 \end{vmatrix} - 1\begin{vmatrix} 1 & 1 \\ 2 & 1 \end{vmatrix} + 2\begin{vmatrix} 1 & 2 \\ 2 & 1 \end{vmatrix} = 1(1) - 1(-1) + 2(-3) = -4$

$N_x = \begin{vmatrix} 7 & 1 & 2 \\ 8 & 2 & 1 \\ 9 & 1 & 1 \end{vmatrix} = 7\begin{vmatrix} 2 & 1 \\ 1 & 1 \end{vmatrix} - 1\begin{vmatrix} 8 & 1 \\ 9 & 1 \end{vmatrix} + 2\begin{vmatrix} 8 & 2 \\ 9 & 1 \end{vmatrix} = 7(1) - 1(-1) + 2(-10) = -12$

$N_y = \begin{vmatrix} 1 & 7 & 2 \\ 1 & 8 & 1 \\ 2 & 9 & 1 \end{vmatrix} = 1\begin{vmatrix} 8 & 1 \\ 9 & 1 \end{vmatrix} - 7\begin{vmatrix} 1 & 1 \\ 2 & 1 \end{vmatrix} + 2\begin{vmatrix} 1 & 8 \\ 2 & 9 \end{vmatrix} = 1(-1) - 7(-1) + 2(-7) = -8$

$N_z = \begin{vmatrix} 1 & 1 & 7 \\ 1 & 2 & 8 \\ 2 & 1 & 9 \end{vmatrix} = 1\begin{vmatrix} 2 & 8 \\ 1 & 9 \end{vmatrix} - 1\begin{vmatrix} 1 & 8 \\ 2 & 9 \end{vmatrix} + 7\begin{vmatrix} 1 & 2 \\ 2 & 1 \end{vmatrix} = 1(10) - 1(-7) + 7(-3) = -4$

$x = \dfrac{N_x}{D} = \dfrac{-12}{-4} = 3$; $y = \dfrac{N_y}{D} = \dfrac{-8}{-4} = 2$; $z = \dfrac{N_z}{D} = \dfrac{-4}{-4} = 1 \Rightarrow$ solution: $(3, 2, 1)$

45. $D = \begin{vmatrix} 2 & 1 & -1 \\ 1 & 2 & 2 \\ 4 & 5 & 3 \end{vmatrix} = 2\begin{vmatrix} 2 & 2 \\ 5 & 3 \end{vmatrix} - 1\begin{vmatrix} 1 & 2 \\ 4 & 3 \end{vmatrix} + (-1)\begin{vmatrix} 1 & 2 \\ 4 & 5 \end{vmatrix} = 2(-4) - 1(-5) - 1(-3) = 0$

$N_x = \begin{vmatrix} 1 & 1 & -1 \\ 2 & 2 & 2 \\ 3 & 5 & 3 \end{vmatrix} = 1\begin{vmatrix} 2 & 2 \\ 5 & 3 \end{vmatrix} - 1\begin{vmatrix} 2 & 2 \\ 3 & 3 \end{vmatrix} + (-1)\begin{vmatrix} 2 & 2 \\ 3 & 5 \end{vmatrix} = 1(-4) - 1(0) - 1(4) = -8$

$x = \dfrac{N_x}{D} = \dfrac{-8}{0} \Rightarrow$ denominator $= 0$, numerator $\neq 0 \Rightarrow$ **no solution.**

47. $D = \begin{vmatrix} 2 & 1 & 1 \\ 1 & -2 & 3 \\ 1 & 1 & -4 \end{vmatrix} = 2\begin{vmatrix} -2 & 3 \\ 1 & -4 \end{vmatrix} - 1\begin{vmatrix} 1 & 3 \\ 1 & -4 \end{vmatrix} + 1\begin{vmatrix} 1 & -2 \\ 1 & 1 \end{vmatrix} = 2(5) - 1(-7) + 1(3) = 20$

$N_x = \begin{vmatrix} 5 & 1 & 1 \\ 10 & -2 & 3 \\ -3 & 1 & -4 \end{vmatrix} = 5\begin{vmatrix} -2 & 3 \\ 1 & -4 \end{vmatrix} - 1\begin{vmatrix} 10 & 3 \\ -3 & -4 \end{vmatrix} + 1\begin{vmatrix} 10 & -2 \\ -3 & 1 \end{vmatrix}$

$\qquad = 5(5) - 1(-31) + 1(4) = 60$

$N_y = \begin{vmatrix} 2 & 5 & 1 \\ 1 & 10 & 3 \\ 1 & -3 & -4 \end{vmatrix} = 2\begin{vmatrix} 10 & 3 \\ -3 & -4 \end{vmatrix} - 5\begin{vmatrix} 1 & 3 \\ 1 & -4 \end{vmatrix} + 1\begin{vmatrix} 1 & 10 \\ 1 & -3 \end{vmatrix}$

$\qquad = 2(-31) - 5(-7) + 1(-13) = -40$

$N_z = \begin{vmatrix} 2 & 1 & 5 \\ 1 & -2 & 10 \\ 1 & 1 & -3 \end{vmatrix} = 2\begin{vmatrix} -2 & 10 \\ 1 & -3 \end{vmatrix} - 1\begin{vmatrix} 1 & 10 \\ 1 & -3 \end{vmatrix} + 5\begin{vmatrix} 1 & -2 \\ 1 & 1 \end{vmatrix} = 2(-4) - 1(-13) + 5(3) = 20$

$x = \dfrac{N_x}{D} = \dfrac{60}{20} = 3$; $y = \dfrac{N_y}{D} = \dfrac{-40}{20} = -2$; $z = \dfrac{N_z}{D} = \dfrac{20}{20} = 1 \Rightarrow$ solution: $(3, -2, 1)$

49. $D = \begin{vmatrix} 2 & 3 & 4 \\ 2 & -3 & -4 \\ 4 & 6 & 8 \end{vmatrix} = 2\begin{vmatrix} -3 & -4 \\ 6 & 8 \end{vmatrix} - 3\begin{vmatrix} 2 & -4 \\ 4 & 8 \end{vmatrix} + 4\begin{vmatrix} 2 & -3 \\ 4 & 6 \end{vmatrix} = 2(0) - 3(32) + 4(24) = 0$

continued on next page...

49. continued

$$N_x = \begin{vmatrix} 6 & 3 & 4 \\ -4 & -3 & -4 \\ 12 & 6 & 8 \end{vmatrix} = 6\begin{vmatrix} -3 & -4 \\ 6 & 8 \end{vmatrix} - 3\begin{vmatrix} -4 & -4 \\ 12 & 8 \end{vmatrix} + 4\begin{vmatrix} -4 & -3 \\ 12 & 6 \end{vmatrix} = 6(0) - 3(16) + 4(12) = 0$$

$$x = \frac{N_x}{D} = \frac{0}{0} \Rightarrow \text{denominator} = 0, \text{numerator} = 0 \Rightarrow \textbf{dependent} \text{ equations.}$$

51. $\begin{cases} x + y = 1 \Rightarrow \\ \frac{1}{2}y + z = \frac{5}{2} \Rightarrow \\ x - z = -3 \Rightarrow \end{cases} \begin{aligned} x + y &= 1 \\ y + 2z &= 5 \\ x - z &= -3 \end{aligned}$

$$D = \begin{vmatrix} 1 & 1 & 0 \\ 0 & 1 & 2 \\ 1 & 0 & -1 \end{vmatrix} = 1\begin{vmatrix} 1 & 2 \\ 0 & -1 \end{vmatrix} - 1\begin{vmatrix} 0 & 2 \\ 1 & -1 \end{vmatrix} + 0\begin{vmatrix} 0 & 1 \\ 1 & 0 \end{vmatrix} = 1(-1) - 1(-2) + 0(-1) = 1$$

$$N_x = \begin{vmatrix} 1 & 1 & 0 \\ 5 & 1 & 2 \\ -3 & 0 & -1 \end{vmatrix} = 1\begin{vmatrix} 1 & 2 \\ 0 & -1 \end{vmatrix} - 1\begin{vmatrix} 5 & 2 \\ -3 & -1 \end{vmatrix} + 0\begin{vmatrix} 5 & 1 \\ -3 & 0 \end{vmatrix} = 1(-1) - 1(1) + 0(3) = -2$$

$$N_y = \begin{vmatrix} 1 & 1 & 0 \\ 0 & 5 & 2 \\ 1 & -3 & -1 \end{vmatrix} = 1\begin{vmatrix} 5 & 2 \\ -3 & -1 \end{vmatrix} - 1\begin{vmatrix} 0 & 2 \\ 1 & -1 \end{vmatrix} + 0\begin{vmatrix} 0 & 5 \\ 1 & -3 \end{vmatrix} = 1(1) - 1(-2) + 0(-5) = 3$$

$$N_z = \begin{vmatrix} 1 & 1 & 1 \\ 0 & 1 & 5 \\ 1 & 0 & -3 \end{vmatrix} = 1\begin{vmatrix} 1 & 5 \\ 0 & -3 \end{vmatrix} - 1\begin{vmatrix} 0 & 5 \\ 1 & -3 \end{vmatrix} + 1\begin{vmatrix} 0 & 1 \\ 1 & 0 \end{vmatrix} = 1(-3) - 1(-5) + 1(-1) = 1$$

$$x = \frac{N_x}{D} = \frac{-2}{1} = -2; \quad y = \frac{N_y}{D} = \frac{3}{1} = 3; \quad z = \frac{N_z}{D} = \frac{1}{1} = 1 \Rightarrow \text{solution: } (-2, 3, 1)$$

53. $\begin{cases} 2x - y + 4z + 2 = 0 \Rightarrow \\ 5x + 8y + 7z = -8 \Rightarrow \\ x + 3y + z + 3 = 0 \Rightarrow \end{cases} \begin{aligned} 2x - y + 4z &= -2 \\ 5x + 8y + 7z &= -8 \\ x + 3y + z &= -3 \end{aligned}$

$$D = \begin{vmatrix} 2 & -1 & 4 \\ 5 & 8 & 7 \\ 1 & 3 & 1 \end{vmatrix} = 2\begin{vmatrix} 8 & 7 \\ 3 & 1 \end{vmatrix} - (-1)\begin{vmatrix} 5 & 7 \\ 1 & 1 \end{vmatrix} + 4\begin{vmatrix} 5 & 8 \\ 1 & 3 \end{vmatrix} = 2(-13) + 1(-2) + 4(7) = 0$$

$$N_x = \begin{vmatrix} -2 & -1 & 4 \\ -8 & 8 & 7 \\ -3 & 3 & 1 \end{vmatrix} = -2\begin{vmatrix} 8 & 7 \\ 3 & 1 \end{vmatrix} - (-1)\begin{vmatrix} -8 & 7 \\ -3 & 1 \end{vmatrix} + 4\begin{vmatrix} -8 & 8 \\ -3 & 3 \end{vmatrix} = -2(-13) + 1(13) + 4(0)$$
$$= 39$$

$$x = \frac{N_x}{D} = \frac{39}{0} \Rightarrow \text{denominator} = 0, \text{numerator} \neq 0 \Rightarrow \text{no solution.}$$

55. $\begin{vmatrix} x & 1 \\ 3 & 2 \end{vmatrix} = 1$

$$2x - 3 = 1$$
$$2x = 4$$
$$x = 2$$

57. $\begin{vmatrix} x & -2 \\ 3 & 1 \end{vmatrix} = \begin{vmatrix} 4 & 2 \\ x & 3 \end{vmatrix}$

$$x - (-6) = 12 - 2x$$
$$x + 6 = 12 - 2x$$
$$3x = 6$$
$$x = 2$$

59. Let $x =$ the amount invested in HiTech, $y =$ the amount invested in SaveTel and $z =$ the amount invested in HiGas. Form and solve the following system of equations:

$$\begin{cases} x + y + z = 20000 \Rightarrow & x + y + z = 20000 \\ 0.10x + 0.05y + 0.06z = 0.066(20000) \Rightarrow & 10x + 5y + 6z = 132000 \\ y + z = 3x \Rightarrow & -3x + y + z = 0 \end{cases}$$

$$D = \begin{vmatrix} 1 & 1 & 1 \\ 10 & 5 & 6 \\ -3 & 1 & 1 \end{vmatrix} = 1\begin{vmatrix} 5 & 6 \\ 1 & 1 \end{vmatrix} - 1\begin{vmatrix} 10 & 6 \\ -3 & 1 \end{vmatrix} + 1\begin{vmatrix} 10 & 5 \\ -3 & 1 \end{vmatrix} = 1(-1) - 1(28) + 1(25) = -4$$

$$N_x = \begin{vmatrix} 20000 & 1 & 1 \\ 132000 & 5 & 6 \\ 0 & 1 & 1 \end{vmatrix} = 20000\begin{vmatrix} 5 & 6 \\ 1 & 1 \end{vmatrix} - 1\begin{vmatrix} 132000 & 6 \\ 0 & 1 \end{vmatrix} + 1\begin{vmatrix} 132000 & 5 \\ 0 & 1 \end{vmatrix}$$

$$= 20000(-1) - 1(132000) + 1(132000) = -20000$$

$$N_y = \begin{vmatrix} 1 & 20000 & 1 \\ 10 & 132000 & 6 \\ -3 & 0 & 1 \end{vmatrix} = 1\begin{vmatrix} 132000 & 6 \\ 0 & 1 \end{vmatrix} - 20000\begin{vmatrix} 10 & 6 \\ -3 & 1 \end{vmatrix} + 1\begin{vmatrix} 10 & 132000 \\ -3 & 0 \end{vmatrix}$$

$$= 1(132000) - 20000(28) + 1(396000) = -32000$$

$$N_z = \begin{vmatrix} 1 & 1 & 20000 \\ 10 & 5 & 132000 \\ -3 & 1 & 0 \end{vmatrix} = 1\begin{vmatrix} 5 & 132000 \\ 1 & 0 \end{vmatrix} - 1\begin{vmatrix} 10 & 132000 \\ -3 & 0 \end{vmatrix} + 20000\begin{vmatrix} 10 & 5 \\ -3 & 1 \end{vmatrix}$$

$$= 1(-132000) - 1(396000) + 20000(25) = -28000$$

$$x = \frac{N_x}{D} = \frac{-20000}{-4} = 5000; \quad y = \frac{N_y}{D} = \frac{-32000}{-4} = 8000; \quad z = \frac{N_z}{D} = \frac{-28000}{-4} = 7000$$

He should invest \$5000 in HiTech, \$8000 in SaveTel and \$7000 in HiGas.

61. $\begin{vmatrix} 2 & -3 & 4 \\ -1 & 2 & 4 \\ 3 & -3 & 1 \end{vmatrix} = -23$ **63.** $\begin{vmatrix} 2 & 1 & -3 \\ -2 & 2 & 4 \\ 1 & -2 & 2 \end{vmatrix} = 26$ **65.** **Answers may vary.**

67.

$$\begin{vmatrix} x & y & 1 \\ -2 & 3 & 1 \\ 3 & 5 & 1 \end{vmatrix} = 0$$

$$x\begin{vmatrix} 3 & 1 \\ 5 & 1 \end{vmatrix} - y\begin{vmatrix} -2 & 1 \\ 3 & 1 \end{vmatrix} + 1\begin{vmatrix} -2 & 3 \\ 3 & 5 \end{vmatrix} = 0$$

$$x(3-5) - y(-2-3) + 1(-10-9) = 0$$

$$-2x + 5y - 19 = 0$$

$$2x - 5y = -19 \quad \text{Verify that both points satisfy this equation.}$$

69.

$$\begin{vmatrix} 1 & 0 & 2 & 1 \\ 2 & 1 & 1 & 3 \\ 1 & 1 & 1 & 1 \\ 2 & 1 & 1 & 1 \end{vmatrix} = 1\begin{vmatrix} 1 & 1 & 3 \\ 1 & 1 & 1 \\ 1 & 1 & 1 \end{vmatrix} - 0\begin{vmatrix} 2 & 1 & 3 \\ 1 & 1 & 1 \\ 2 & 1 & 1 \end{vmatrix} + 2\begin{vmatrix} 2 & 1 & 3 \\ 1 & 1 & 1 \\ 2 & 1 & 1 \end{vmatrix} - 1\begin{vmatrix} 2 & 1 & 1 \\ 1 & 1 & 1 \\ 2 & 1 & 1 \end{vmatrix}$$

$$= 1(0) - 0(???) + 2(-2) - 1(0) = -4$$

Chapter 3 Summary (page 205)

1. $\begin{cases} 2x + y = 11 \\ -x + 2y = 7 \end{cases}$

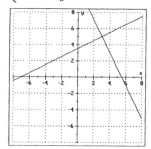

$(3, 5)$ is the solution.

2. $\begin{cases} 3x + 2y = 0 \\ 2x - 3y = -13 \end{cases}$

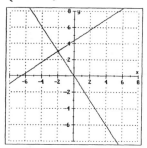

$(-2, 3)$ is the solution.

3. $\begin{cases} \frac{1}{2}x + \frac{1}{3}y = 2 \\ y = 6 - \frac{3}{2}x \end{cases}$

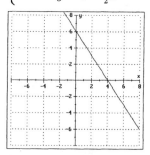

dependent equations

4. $\begin{cases} \frac{1}{3}x - \frac{1}{2}y = 1 \\ 6x - 9y = 2 \end{cases}$

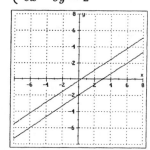

inconsistent system

5. $\begin{cases} (1) \quad y = x + 4 \\ (2) \quad 2x + 3y = 7 \end{cases}$

Substitute $y = x + 4$ from (1) into (2):
$$2x + 3\boldsymbol{y} = 7$$
$$2x + 3(\boldsymbol{x + 4}) = 7$$
$$2x + 3x + 12 = 7$$
$$5x = -5$$
$$x = -1$$

Substitute this and solve for y:
$$y = x + 4 = -1 + 4 = 3$$
Solution: $(-1, 3)$

6. $\begin{cases} (1) \quad y = 2x + 5 \\ (2) \quad 3x - 5y = -4 \end{cases}$

Substitute $y = 2x + 5$ from (1) into (2):
$$3x - 5\boldsymbol{y} = -4$$
$$3x - 5(\boldsymbol{2x + 5}) = -4$$
$$3x - 10x - 25 = -4$$
$$-7x = 21$$
$$x = -3$$

Substitute this and solve for y:
$$y = 2x + 5 = 2(-3) + 5 = -1$$
Solution: $(-3, -1)$

7. $\begin{cases} (1) & x + 2y = 11 \\ (2) & 2x - y = 2 \end{cases}$

Substitute $x = -2y + 11$ from (1) into (2):
$$2x - y = 2$$
$$2(-2y + 11) - y = 2$$
$$-4y + 22 - y = 2$$
$$-5y = -20$$
$$y = 4$$
Substitute this and solve for x:
$$x = -2y + 11 = -2(4) + 11 = 3$$
Solution: $(3, 4)$

8. $\begin{cases} (1) & 2x + 3y = -2 \\ (2) & 3x + 5y = -2 \end{cases}$

Substitute $x = \frac{-3y-2}{2}$ from (1) into (2):
$$3x + 5y = -2$$
$$3 \cdot \frac{-3y-2}{2} + 5y = -2$$
$$3(-3y - 2) + 10y = -4$$
$$-9y - 6 + 10y = -4$$
$$y = 2$$
Substitute this and solve for y:
$$x = \frac{-3y-2}{2} = \frac{-3(2)-2}{2} = \frac{-8}{2} = -4$$
Solution: $(-4, 2)$

9.
$$\begin{array}{l} x + y = -2 \Rightarrow \times(-2) \\ 2x + 3y = -3 \Rightarrow \end{array} \quad \begin{array}{r} -2x - 2y = 4 \\ 2x + 3y = -3 \\ \hline y = 1 \end{array}$$

Substitute and solve for x:
$$x + y = -2$$
$$x + 1 = -2$$
$$x = -3 \quad \text{Solution: } \boxed{(-3, 1)}$$

10.
$$\begin{array}{l} 3x + 2y = 1 \Rightarrow \times 3 \\ 2x - 3y = 5 \Rightarrow \times 2 \end{array} \quad \begin{array}{r} 9x + 6y = 3 \\ 4x - 6y = 10 \\ \hline 13x = 13 \\ x = 1 \end{array}$$

Substitute and solve for y:
$$3x + 2y = 1$$
$$3(1) + 2y = 1$$
$$3 + 2y = 1$$
$$2y = -2$$
$$y = -1 \quad \text{Solution: } \boxed{(1, -1)}$$

11.
$$\begin{array}{l} x + \frac{1}{2}y = 7 \quad \Rightarrow \times 2 \\ -2x = 3y - 6 \Rightarrow \end{array} \quad \begin{array}{r} 2x + y = 14 \Rightarrow \\ -2x - 3y = -6 \Rightarrow \\ \hline -2y = 8 \\ y = -4 \end{array}$$

Solve for x:
$$2x + y = 14$$
$$2x + (-4) = 14$$
$$2x = 18$$
$$x = 9 \quad \boxed{\text{Solution: } (9, -4)}$$

12.
$$\begin{array}{l} y = \dfrac{x-3}{2} \Rightarrow \times 2 \\ x = \dfrac{2y+7}{2} \Rightarrow \times 2 \end{array} \quad \begin{array}{l} 2y = x - 3 \Rightarrow -x + 2y = -3 \\ 2x = 2y + 7 \Rightarrow 2x - 2y = 7 \\ \hline x = 4 \end{array}$$

Solve for y:
$$y = \frac{x-3}{2}$$
$$y = \frac{4-3}{2} = \frac{1}{2} \quad \text{Solution: } \boxed{\left(4, \tfrac{1}{2}\right)}$$

13.
$$\begin{array}{ll} (1) & x + y + z = 6 \\ (2) & x - y - z = -4 \\ (3) & -x + y - z = -2 \end{array} \quad \begin{array}{ll} (1) & x + y + z = 6 \\ (2) & x - y - z = -4 \\ \hline (4) & 2x = 2 \\ & x = 1 \end{array} \quad \begin{array}{ll} (1) & x + y + z = 6 \\ (3) & -x + y - z = -2 \\ \hline (5) & 2y = 4 \\ & y = 2 \end{array}$$

$$x + y + z = 6$$
$$1 + 2 + z = 6$$
$$3 + z = 6$$
$$z = 3 \quad \text{Solution: } \boxed{(1, 2, 3)}$$

14. (1) $2x + 3y + z = -5$ (1) $2x + 3y + z = -5$ (3) $3x + y + 2z = 4$
    (2) $-x + 2y - z = -6$ (2) $-x + 2y - z = -6$ $2 \cdot (2)$ $-2x + 4y - 2z = -12$
    (3) $3x + y + 2z = 4$ (4) $\overline{x + 5y\ \ \ \ \ = -11}$ (5) $\overline{x + 5y\ \ \ \ \ = -8}$

$$\begin{aligned} x + 5y = -11 &\Rightarrow \times(-1) & -x - 5y &= 11 \\ \underline{x + 5y = \ -8} &\Rightarrow & \underline{x + 5y} &= -8 \\ & & 0 &= 3 \end{aligned}$$

Since this is an impossible equation, there is no solution. It is an inconsistent system.

15.
$$\begin{bmatrix} 1 & 2 & | & 4 \\ 2 & -1 & | & 3 \end{bmatrix} \xRightarrow[]{-2R_1 + R_2 \Rightarrow R_2} \begin{bmatrix} 1 & 2 & | & 4 \\ 0 & -5 & | & -5 \end{bmatrix} \xRightarrow[]{-\frac{1}{5}R_2 \Rightarrow R_2} \begin{bmatrix} 1 & 2 & | & 4 \\ 0 & 1 & | & 1 \end{bmatrix}$$

From R_2, $y = 1$. From R_1: Solution: $\boxed{(2,1)}$
$$\begin{aligned} x + 2y &= 4 \\ x + 2(1) &= 4 \\ x &= 2 \end{aligned}$$

16.
$$\begin{bmatrix} 1 & 1 & 1 & | & 6 \\ 2 & -1 & 1 & | & 1 \\ 4 & 1 & -1 & | & 5 \end{bmatrix} \xRightarrow[-4R_1 + R_3 \Rightarrow R_3]{-2R_1 + R_2 \Rightarrow R_2} \begin{bmatrix} 1 & 1 & 1 & | & 6 \\ 0 & -3 & -1 & | & -11 \\ 0 & -3 & -5 & | & -19 \end{bmatrix} \xRightarrow[-R_3 + R_2 \Rightarrow R_3]{-R_2 \Rightarrow R_2} \begin{bmatrix} 1 & 1 & 1 & | & 6 \\ 0 & 3 & 1 & | & 11 \\ 0 & 0 & 4 & | & 8 \end{bmatrix} \xRightarrow[]{\frac{1}{4}R_3 \Rightarrow R_3} \begin{bmatrix} 1 & 1 & 1 & | & 6 \\ 0 & 3 & 1 & | & 11 \\ 0 & 0 & 1 & | & 2 \end{bmatrix}$$

From R_3, $z = 2$. From R_2: From R_1: Solution: $\boxed{(1,3,2)}$
$$\begin{aligned} 3y + z &= 11 & x + y + z &= 6 \\ 3y + 2 &= 11 & x + 3 + 2 &= 6 \\ 3y &= 9 & x &= 1 \\ y &= 3 \end{aligned}$$

17.
$$\begin{bmatrix} 1 & 1 & | & 3 \\ 1 & -2 & | & -3 \\ 2 & 1 & | & 4 \end{bmatrix} \xRightarrow[-2R_1 + R_3 \Rightarrow R_3]{-R_1 + R_2 \Rightarrow R_2} \begin{bmatrix} 1 & 1 & | & 3 \\ 0 & -3 & | & -6 \\ 0 & -1 & | & -2 \end{bmatrix} \xRightarrow[-3R_3 + R_2 \Rightarrow R_3]{-\frac{1}{3}R_2 \Rightarrow R_2} \begin{bmatrix} 1 & 1 & | & 3 \\ 0 & 1 & | & 2 \\ 0 & 0 & | & 0 \end{bmatrix}$$

From R_2, $y = 2$. From R_1: Solution: $\boxed{(1,2)}$
$$\begin{aligned} x + y &= 3 \\ x + 2 &= 3 \\ x &= 1 \end{aligned}$$

CHAPTER 3 SUMMARY

18.
$$\begin{bmatrix} 1 & 2 & 1 & | & 2 \\ 2 & 5 & 4 & | & 5 \end{bmatrix} \xRightarrow{-2R_1 + R_2 \Rightarrow R_2} \begin{bmatrix} 1 & 2 & 1 & | & 2 \\ 0 & 1 & 2 & | & 1 \end{bmatrix}$$

From R_2:
$y + 2z = 1$
$y = 1 - 2z$

From R_1:
$x + 2y + z = 2$
$x + 2(1 - 2z) + z = 2$
$x + 2 - 4z + z = 2$
$x - 3z = 0$
$x = 3z$

Solution: $\boxed{(3z, 1 - 2z, z)}$

19. $\begin{vmatrix} 2 & 3 \\ -4 & 3 \end{vmatrix} = 2(3) - 3(-4) = 6 + 12 = 18$

20. $\begin{vmatrix} -3 & -4 \\ 5 & -6 \end{vmatrix} = -3(-6) - (-4)(5) = 18 - (-20) = 18 + 20 = 38$

21. $\begin{vmatrix} -1 & 2 & -1 \\ 2 & -1 & 3 \\ 1 & -2 & 2 \end{vmatrix} = -1\begin{vmatrix} -1 & 3 \\ -2 & 2 \end{vmatrix} - 2\begin{vmatrix} 2 & 3 \\ 1 & 2 \end{vmatrix} + (-1)\begin{vmatrix} 2 & -1 \\ 1 & -2 \end{vmatrix} = -1(4) - 2(1) - 1(-3) = -3$

22. $\begin{vmatrix} 3 & -2 & 2 \\ 1 & -2 & -2 \\ 2 & 1 & -1 \end{vmatrix} = 3\begin{vmatrix} -2 & -2 \\ 1 & -1 \end{vmatrix} - (-2)\begin{vmatrix} 1 & -2 \\ 2 & -1 \end{vmatrix} + 2\begin{vmatrix} 1 & -2 \\ 2 & 1 \end{vmatrix} = 3(4) + 2(3) + 2(5) = 28$

23. $x = \dfrac{\begin{vmatrix} 10 & 4 \\ 1 & -3 \end{vmatrix}}{\begin{vmatrix} 3 & 4 \\ 2 & -3 \end{vmatrix}} = \dfrac{-30 - 4}{-9 - 8} = \dfrac{-34}{-17} = 2; \; y = \dfrac{\begin{vmatrix} 3 & 10 \\ 2 & 1 \end{vmatrix}}{\begin{vmatrix} 3 & 4 \\ 2 & -3 \end{vmatrix}} = \dfrac{3 - 20}{-17} = \dfrac{-17}{-17} = 1$

Solution: $\boxed{(2, 1)}$

24. $x = \dfrac{\begin{vmatrix} -17 & -5 \\ 3 & 2 \end{vmatrix}}{\begin{vmatrix} 2 & -5 \\ 3 & 2 \end{vmatrix}} = \dfrac{-34 + 15}{4 + 15} = \dfrac{-19}{19} = -1; \; y = \dfrac{\begin{vmatrix} 2 & -17 \\ 3 & 3 \end{vmatrix}}{\begin{vmatrix} 2 & -5 \\ 3 & 2 \end{vmatrix}} = \dfrac{6 + 51}{19} = \dfrac{57}{19} = 3$

Solution: $\boxed{(-1, 3)}$

25. $D = \begin{vmatrix} 1 & 2 & 1 \\ 2 & 1 & 1 \\ 1 & 1 & 2 \end{vmatrix} = 1\begin{vmatrix} 1 & 1 \\ 1 & 2 \end{vmatrix} - 2\begin{vmatrix} 2 & 1 \\ 1 & 2 \end{vmatrix} + 1\begin{vmatrix} 2 & 1 \\ 1 & 1 \end{vmatrix} = 1(1) - 2(3) + 1(1) = -4$

$N_x = \begin{vmatrix} 0 & 2 & 1 \\ 3 & 1 & 1 \\ 5 & 1 & 2 \end{vmatrix} = 0\begin{vmatrix} 1 & 1 \\ 1 & 2 \end{vmatrix} - 2\begin{vmatrix} 3 & 1 \\ 5 & 2 \end{vmatrix} + 1\begin{vmatrix} 3 & 1 \\ 5 & 1 \end{vmatrix} = 0(1) - 2(1) + 1(-2) = -4$

continued on next page

25. continued

$$N_y = \begin{vmatrix} 1 & 0 & 1 \\ 2 & 3 & 1 \\ 1 & 5 & 2 \end{vmatrix} = 1\begin{vmatrix} 3 & 1 \\ 5 & 2 \end{vmatrix} - 0\begin{vmatrix} 2 & 1 \\ 1 & 2 \end{vmatrix} + 1\begin{vmatrix} 2 & 3 \\ 1 & 5 \end{vmatrix} = 1(1) - 0(3) + 1(7) = 8$$

$$N_z = \begin{vmatrix} 1 & 2 & 0 \\ 2 & 1 & 3 \\ 1 & 1 & 5 \end{vmatrix} = 1\begin{vmatrix} 1 & 3 \\ 1 & 5 \end{vmatrix} - 2\begin{vmatrix} 2 & 3 \\ 1 & 5 \end{vmatrix} + 0\begin{vmatrix} 2 & 1 \\ 1 & 1 \end{vmatrix} = 1(2) - 2(7) + 0(1) = -12$$

$$x = \frac{N_x}{D} = \frac{-4}{-4} = 1; \; y = \frac{N_y}{D} = \frac{8}{-4} = -2; \; z = \frac{N_z}{D} = \frac{-12}{-4} = 3 \Rightarrow \text{solution: } (1, -2, 3)$$

26. $D = \begin{vmatrix} 2 & 3 & 1 \\ 1 & 3 & 2 \\ 1 & -1 & -1 \end{vmatrix} = 2\begin{vmatrix} 3 & 2 \\ -1 & -1 \end{vmatrix} - 3\begin{vmatrix} 1 & 2 \\ 1 & -1 \end{vmatrix} + 1\begin{vmatrix} 1 & 3 \\ 1 & -1 \end{vmatrix} = 2(-1) - 3(-3) + 1(-4)$

$$= 3$$

$$N_x = \begin{vmatrix} 2 & 3 & 1 \\ 7 & 3 & 2 \\ -7 & -1 & -1 \end{vmatrix} = 2\begin{vmatrix} 3 & 2 \\ -1 & -1 \end{vmatrix} - 3\begin{vmatrix} 7 & 2 \\ -7 & -1 \end{vmatrix} + 1\begin{vmatrix} 7 & 3 \\ -7 & -1 \end{vmatrix}$$

$$= 2(-1) - 3(7) + 1(14) = -9$$

$$N_y = \begin{vmatrix} 2 & 2 & 1 \\ 1 & 7 & 2 \\ 1 & -7 & -1 \end{vmatrix} = 2\begin{vmatrix} 7 & 2 \\ -7 & -1 \end{vmatrix} - 2\begin{vmatrix} 1 & 2 \\ 1 & -1 \end{vmatrix} + 1\begin{vmatrix} 1 & 7 \\ 1 & -7 \end{vmatrix}$$

$$= 2(7) - 2(-3) + 1(-14) = 6$$

$$N_z = \begin{vmatrix} 2 & 3 & 2 \\ 1 & 3 & 7 \\ 1 & -1 & -7 \end{vmatrix} = 2\begin{vmatrix} 3 & 7 \\ -1 & -7 \end{vmatrix} - 3\begin{vmatrix} 1 & 7 \\ 1 & -7 \end{vmatrix} + 2\begin{vmatrix} 1 & 3 \\ 1 & -1 \end{vmatrix}$$

$$= 2(-14) - 3(-14) + 2(-4) = 6$$

$$x = \frac{N_x}{D} = \frac{-9}{3} = -3; \; y = \frac{N_y}{D} = \frac{6}{3} = 2; \; z = \frac{N_z}{D} = \frac{6}{3} = 2 \Rightarrow \text{solution: } (-3, 2, 2)$$

Chapter 3 Test (page 207)

1. $\begin{cases} 2x + y = 5 \\ y = 2x - 3 \end{cases}$

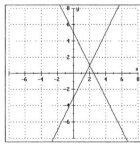

$(2, 1)$ is the solution.

2. $\begin{cases} (1) \quad 2x - 4y = 14 \\ (2) \quad x = -2y + 7 \end{cases}$

Substitute $x = -2y + 7$ from (2) into (1):
$$2x - 4y = 14$$
$$2(-2y + 7) - 4y = 14$$
$$-4y + 14 - 4y = 14$$
$$-8y = 0$$
$$y = 0$$

Substitute this and solve for x:
$$x = -2y + 7$$
$$x = -2(0) + 7 = 7$$

Solution: $(7, 0)$

89

3. $\quad 2x + 3y = -5 \Rightarrow \times 2 \quad 4x + 6y = -10 \qquad 2x + 3y = -5 \qquad$ Solution:

$\quad \underline{3x - 2y = 12 \Rightarrow \times 3 \quad 9x - 6y = 36} \qquad 2(2) + 3y = -5 \qquad \boxed{(2, -3)}$

$\qquad\qquad\qquad\qquad\qquad 13x \qquad = 26 \qquad 4 + 3y = -5$

$\qquad\qquad\qquad\qquad\qquad\quad x \qquad = \ 2 \qquad\qquad 3y = -9$

$\qquad\qquad\qquad\qquad\qquad\qquad\qquad\qquad\qquad\qquad y = -3$

4. $\quad \dfrac{x}{2} - \dfrac{y}{4} = -4 \Rightarrow \times 4 \ \ 2x - y = -16 \qquad x + y = -2 \qquad$ Solution:

$\quad \underline{x + y = -2 \Rightarrow \qquad\qquad\qquad} \quad \underline{x + y = \ -2} \qquad -6 + y = -2 \qquad \boxed{(-6, 4)}$

$\qquad\qquad\qquad\qquad\qquad\qquad\quad 3x \quad\ = -18 \qquad\qquad y = 4$

$\qquad\qquad\qquad\qquad\qquad\qquad\qquad x \quad\ = \ -6$

5. $\quad 3(x + y) = x - 3 \Rightarrow \quad 2x + 3y = -3 \qquad\qquad 2x + 3y = -3 \Rightarrow \qquad\qquad\qquad 2x + 3y = -3$

$\qquad -y = \dfrac{2x + 3}{3} \quad \Rightarrow \qquad 2x + 3y = -3 \qquad \underline{2x + 3y = -3 \Rightarrow \times(-1)} \underline{-2x - 3y \ = 3}$

$\qquad\qquad\qquad\qquad\qquad\qquad\qquad\qquad\qquad\qquad\qquad\qquad\qquad\qquad\qquad\qquad 0 = 0$

The equation is an identity, so the system has infinitely many solutions. $\Rightarrow$ dependent equations

6. See **#5**. Since the system has at least one solution, it is a consistent system.

7. $\begin{bmatrix} 1 & 2 & -1 \\ 2 & -2 & 3 \end{bmatrix} \Rightarrow \overset{\displaystyle -3R_1 + R_2 \Rightarrow R_2}{\begin{bmatrix} 1 & 2 & -1 \\ -1 & -8 & \boxed{6} \end{bmatrix}}$

8. $\begin{bmatrix} -1 & 3 & 6 \\ 3 & -2 & 4 \end{bmatrix} \Rightarrow \overset{\displaystyle -2R_1 + R_2 \Rightarrow R_2}{\begin{bmatrix} -1 & 3 & 6 \\ 5 & -8 & \boxed{-8} \end{bmatrix}}$

9. $\left[\begin{array}{ccc|c} 1 & 1 & 1 & 4 \\ 1 & 1 & -1 & 6 \\ 2 & -3 & 1 & -1 \end{array}\right]$

10. $\begin{bmatrix} 1 & 1 & 1 \\ 1 & 1 & -1 \\ 2 & -3 & 1 \end{bmatrix}$

11. $\left[\begin{array}{cc|c} 1 & 1 & 4 \\ 2 & -1 & 2 \end{array}\right] \Rightarrow \overset{\displaystyle -2R_1 + R_2 \Rightarrow R_2}{\left[\begin{array}{cc|c} 1 & 1 & 4 \\ 0 & -3 & -6 \end{array}\right]} \Rightarrow \overset{\displaystyle -\frac{1}{3}R_2 \Rightarrow R_2}{\left[\begin{array}{cc|c} 1 & 1 & 4 \\ 0 & 1 & 2 \end{array}\right]}$

From R_2, $y = 2$. From R_1: Solution: $\boxed{(2, 2)}$

$\qquad\qquad\qquad\qquad\qquad\qquad x + y = 4$

$\qquad\qquad\qquad\qquad\qquad\qquad x + 2 = 4$

$\qquad\qquad\qquad\qquad\qquad\qquad\quad x = 2$

12. $\left[\begin{array}{cc|c} 1 & 1 & 2 \\ 1 & -1 & -4 \\ 2 & 1 & 1 \end{array}\right] \Rightarrow \overset{\substack{\displaystyle -R_2 + R_1 \Rightarrow R_2 \\ \displaystyle -2R_1 + R_3 \Rightarrow R_3}}{\left[\begin{array}{cc|c} 1 & 1 & 2 \\ 0 & 2 & 6 \\ 0 & -1 & -3 \end{array}\right]} \Rightarrow \overset{\substack{\displaystyle \frac{1}{2}R_2 \Rightarrow R_2 \\ \displaystyle 2R_3 + R_2 \Rightarrow R_3}}{\left[\begin{array}{cc|c} 1 & 1 & 2 \\ 0 & 1 & 3 \\ 0 & 0 & 0 \end{array}\right]}$

From R_2, $y = 3$. From R_1: Solution: $\boxed{(-1, 3)}$

$\qquad\qquad\qquad\qquad\qquad\qquad x + y = 2$

$\qquad\qquad\qquad\qquad\qquad\qquad x + 3 = 2$

$\qquad\qquad\qquad\qquad\qquad\qquad\quad x = -1$

13. $\begin{vmatrix} 2 & -3 \\ 4 & 5 \end{vmatrix} = 2(5) - (-3)(4) = 10 - (-12)$ **14.** $\begin{vmatrix} -3 & -4 \\ -2 & 3 \end{vmatrix} = -3(3) - (-4)(-2)$
$$= 22 \qquad\qquad\qquad = -9 - 8 = -17$$

15. $\begin{vmatrix} 1 & 2 & 0 \\ 2 & 0 & 3 \\ 1 & -2 & 2 \end{vmatrix} = 1\begin{vmatrix} 0 & 3 \\ -2 & 2 \end{vmatrix} - 2\begin{vmatrix} 2 & 3 \\ 1 & 2 \end{vmatrix} + 0\begin{vmatrix} 2 & 0 \\ 1 & -2 \end{vmatrix} = 1(6) - 2(1) + 0(-4) = 4$

16. $\begin{vmatrix} 2 & -1 & 1 \\ 3 & 1 & 0 \\ 0 & 1 & 2 \end{vmatrix} = 2\begin{vmatrix} 1 & 0 \\ 1 & 2 \end{vmatrix} - (-1)\begin{vmatrix} 3 & 0 \\ 0 & 2 \end{vmatrix} + 1\begin{vmatrix} 3 & 1 \\ 0 & 1 \end{vmatrix} = 2(2) + 1(6) + 1(3) = 13$

17. $\begin{vmatrix} -6 & -1 \\ -6 & 1 \end{vmatrix}$ **18.** $\begin{vmatrix} 1 & -1 \\ 3 & 1 \end{vmatrix}$

19. $x = \dfrac{\begin{vmatrix} -6 & -1 \\ -6 & 1 \end{vmatrix}}{\begin{vmatrix} 1 & -1 \\ 3 & 1 \end{vmatrix}} = \dfrac{-6 - 6}{1 - (-3)} = \dfrac{-12}{4} = -3$ **20.** $y = \dfrac{\begin{vmatrix} 1 & -6 \\ 3 & -6 \end{vmatrix}}{\begin{vmatrix} 1 & -1 \\ 3 & 1 \end{vmatrix}} = \dfrac{-6 - (-18)}{1 - (-3)} = \dfrac{12}{4} = 3$

21. $D = \begin{vmatrix} 1 & 1 & 1 \\ 1 & 1 & -1 \\ 2 & -3 & 1 \end{vmatrix} = 1\begin{vmatrix} 1 & -1 \\ -3 & 1 \end{vmatrix} - 1\begin{vmatrix} 1 & -1 \\ 2 & 1 \end{vmatrix} + 1\begin{vmatrix} 1 & 1 \\ 2 & -3 \end{vmatrix} = 1(-2) - 1(3) + 1(-5) = -10$

$N_x = \begin{vmatrix} 4 & 1 & 1 \\ 6 & 1 & -1 \\ -1 & -3 & 1 \end{vmatrix} = 4\begin{vmatrix} 1 & -1 \\ -3 & 1 \end{vmatrix} - 1\begin{vmatrix} 6 & -1 \\ -1 & 1 \end{vmatrix} + 1\begin{vmatrix} 6 & 1 \\ -1 & -3 \end{vmatrix}$
$$= 4(-2) - 1(5) + 1(-17) = -30$$

$x = \dfrac{N_x}{D} = \dfrac{-30}{-10} = 3$

22. See #21. $D = -10$.

$N_z = \begin{vmatrix} 1 & 1 & 4 \\ 1 & 1 & 6 \\ 2 & -3 & -1 \end{vmatrix} = 1\begin{vmatrix} 1 & 6 \\ -3 & -1 \end{vmatrix} - 1\begin{vmatrix} 1 & 6 \\ 2 & -1 \end{vmatrix} + 4\begin{vmatrix} 1 & 1 \\ 2 & -3 \end{vmatrix} = 1(17) - 1(-13) + 4(-5) = 10$

$z = \dfrac{N_z}{D} = \dfrac{10}{-10} = -1$

SECTION 4.1

Exercise 4.1 (page 217)

1. $\left(\dfrac{t^3t^5t^{-6}}{t^2t^{-4}}\right)^{-3} = \left(\dfrac{t^2t^{-4}}{t^3t^5t^{-6}}\right)^3 = \left(\dfrac{t^{-2}}{t^2}\right)^3 = \left(\dfrac{1}{t^4}\right)^3 = \dfrac{1}{t^{12}}$

3. Let x = the number of pies made.

$\boxed{\text{Expenses}} = \boxed{\text{Income}}$

$1200 + 3.40x = 5.95x$

$\qquad\quad 1200 = 2.55x$

$\qquad 470.59 = x$

He must sell at least 471 pies.

5. $\neq$

7. $<$

9. $\geq$

11. $a < c$ **13.** reversed **15.** $c < x;\ x < d$ **17.** open

19.
$x + 4 < 5$
$x + 4 - 4 < 5 - 4$
$x < 1$
solution set: $\{x | x < 1\}$

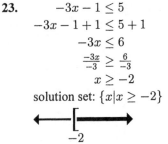

21.
$2x + 3 < 9$
$2x + 3 - 3 < 9 - 3$
$2x < 6$
$\dfrac{2x}{2} < \dfrac{6}{2}$
$x < 3$
solution set: $\{x | x < 3\}$

23.
$-3x - 1 \leq 5$
$-3x - 1 + 1 \leq 5 + 1$
$-3x \leq 6$
$\dfrac{-3x}{-3} \geq \dfrac{6}{-3}$
$x \geq -2$
solution set: $\{x | x \geq -2\}$

25.
$5x - 3 > 7$
$5x - 3 + 3 > 7 + 3$
$5x > 10$
$\dfrac{5x}{5} > \dfrac{10}{5}$
$x > 2$
solution set: $\{x | x > 2\}$

27.
$-4 < 2x < 8$
$\dfrac{-4}{2} < \dfrac{2x}{2} < \dfrac{8}{2}$
$-2 < x < 4$
sol'n set: $\{x | -2 < x < 4\}$

29.
$8x + 30 > -2x$
$8x + 2x + 30 > -2x + 2x$
$10x + 30 - 30 > 0 - 30$
$10x > -30$
$\dfrac{10x}{10} > \dfrac{-30}{10}$
$x > -3$
solution set: $(-3, \infty)$

31.
$-3x + 14 \geq 20$
$-3x + 14 - 14 \geq 20 - 14$
$-3x \geq 6$
$\dfrac{-3x}{-3} \leq \dfrac{6}{-3}$
$x \leq -2$
solution set: $(-\infty, -2]$

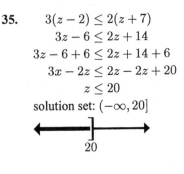

33.
$4(x + 5) < 12$
$4x + 20 < 12$
$4x + 20 - 20 < 12 - 20$
$4x < -8$
$\dfrac{4x}{4} < \dfrac{-8}{4}$
$x < -2$
solution set: $(-\infty, -2)$

35.
$3(z - 2) \leq 2(z + 7)$
$3z - 6 \leq 2z + 14$
$3z - 6 + 6 \leq 2z + 14 + 6$
$3z - 2z \leq 2z - 2z + 20$
$z \leq 20$
solution set: $(-\infty, 20]$

37.
$$-11(2-b) < 4(2b+2)$$
$$-22 + 11b < 8b + 8$$
$$-22 + 22 + 11b < 8b + 8 + 22$$
$$11b - 8b < 8b - 8b + 30$$
$$3b < 30$$
$$\frac{3b}{3} < \frac{30}{3}$$
$$b < 10$$
solution set: $(-\infty, 10)$

39.
$$\frac{1}{2}y + 2 \geq \frac{1}{3}y - 4$$
$$6\left(\frac{1}{2}y + 2\right) \geq 6\left(\frac{1}{3}y - 4\right)$$
$$3y + 12 \geq 2y - 24$$
$$3y + 12 - 12 \geq 2y - 24 - 12$$
$$3y - 2y \geq 2y - 2y - 36$$
$$y \geq -36$$
solution set: $[-36, \infty)$

41.
$$\frac{2}{3}x + \frac{3}{2}(x-5) \leq x$$
$$6\left[\frac{2}{3}x + \frac{3}{2}(x-5)\right] \leq 6x$$
$$4x + 9(x-5) \leq 6x$$
$$4x + 9x - 45 \leq 6x$$
$$13x - 45 + 45 \leq 6x + 45$$
$$13x - 6x \leq 6x - 6x + 45$$
$$7x \leq 45$$
$$\frac{7x}{7} \leq \frac{45}{7}$$
$$x \leq \frac{45}{7}$$
solution set: $\left(-\infty, \frac{45}{7}\right]$

43.
$$-2 < -b + 3 < 5$$
$$-2 - 3 < -b + 3 - 3 < 5 - 3$$
$$-5 < -b < 2$$
$$\frac{-5}{-1} > \frac{-b}{-1} > \frac{2}{-1}$$
$$5 > b > -2, \text{ or } -2 < b < 5$$
solution set: $(-2, 5)$

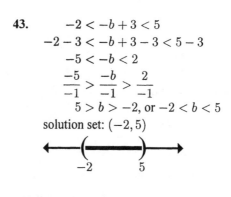

45.
$$15 > 2x - 7 > 9$$
$$15 + 7 > 2x - 7 + 7 > 9 + 7$$
$$22 > 2x > 16$$
$$\frac{22}{2} > \frac{2x}{2} > \frac{16}{2}$$
$$11 > x > 8, \text{ or } 8 < x < 11$$
solution set: $(8, 11)$

47.
$$-6 < -3(x-4) \leq 24$$
$$\frac{-6}{-3} > \frac{-3(x-4)}{-3} \geq \frac{24}{-3}$$
$$2 > x - 4 \geq -8$$
$$2 + 4 > x - 4 + 4 \geq -8 + 4$$
$$6 > x \geq -4, \text{ or } -4 \leq x < 6$$
solution set: $[-4, 6)$

SECTION 4.1

49. $0 \geq \frac{1}{2}x - 4 > 6$

This inequality indicates that $0 \geq 6$ (by the transitive property). Since this is not possible, there is no solution to the inequality.

51.
$$0 \leq \frac{4-x}{3} \leq 2$$
$$3(0) \leq 3\left(\frac{4-x}{3}\right) \leq 3(2)$$
$$0 \leq 4 - x \leq 6$$
$$0 - 4 \leq 4 - 4 - x \leq 6 - 4$$
$$-4 \leq -x \leq 2$$
$$\frac{-4}{-1} \geq \frac{-x}{-1} \geq \frac{2}{-1}$$
$$4 \geq x \geq -2, \text{ or } -2 \leq x \leq 4$$
solution set: $[-2, 4]$

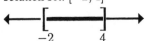

53.
$$x + 3 < 3x - 1 < 2x + 2$$

$x + 3 < 3x - 1$	and	$3x - 1 < 2x + 2$

$$x + 3 < 3x - 1 \qquad\qquad\qquad 3x - 1 < 2x + 2$$
$$x + 3 - 3 < 3x - 1 - 3 \qquad\qquad 3x - 1 + 1 < 2x + 2 + 1$$
$$x - 3x < 3x - 3x - 4 \qquad\qquad 3x - 2x < 2x - 2x + 3$$
$$-2x < -4 \qquad\qquad\qquad\qquad x < 3$$
$$\frac{-2x}{-2} > \frac{-4}{-2}$$
$$x > 2$$

If $x > 2$ **and** $x < 3$, then $2 < x < 3$. The solution set is $(2, 3)$.

55.
$$4x \geq -x + 5 \geq 3x - 4$$

$4x \geq -x + 5$	and	$-x + 5 \geq 3x - 4$

$$4x \geq -x + 5 \qquad\qquad\qquad -x + 5 \geq 3x - 4$$
$$4x + x \geq -x + x + 5 \qquad\qquad -x + 5 - 5 \geq 3x - 4 - 5$$
$$5x \geq 5 \qquad\qquad\qquad\qquad -x - 3x \geq 3x - 3x - 9$$
$$\frac{5x}{5} \geq \frac{5}{5} \qquad\qquad\qquad\qquad -4x \geq -9$$
$$x \geq 1 \qquad\qquad\qquad\qquad \frac{-4x}{-4} \leq \frac{-9}{-4}$$
$$x \leq \frac{9}{4}$$

If $x \geq 1$ **and** $x \leq \frac{9}{4}$, then $1 \leq x \leq \frac{9}{4}$. The solution set is $\left[1, \frac{9}{4}\right]$.

SECTION 4.1

57.
$$5(x+1) \leq 4(x+3) < 3(x-1)$$

$$5(x+1) \leq 4(x+3) \qquad \text{and} \qquad 4(x+3) < 3(x-1)$$
$$5x+5-5 \leq 4x+12-5 \qquad\qquad 4x+12-12 < 3x-3-12$$
$$5x-4x \leq 4x-4x+7 \qquad\qquad 4x-3x < 3x-3x-15$$
$$x \leq 7 \qquad\qquad\qquad\qquad x < -15$$

If $x \leq 7$ **and** $x < -15$, then $x < -15$. The solution set is $(-\infty, -15)$.

$$-15$$

59.
$$3x+2 < 8 \qquad \text{or} \qquad 2x-3 > 11$$
$$3x+2-2 < 8-2 \qquad\qquad 2x-3+3 > 11+3$$
$$3x < 6 \qquad\qquad\qquad 2x > 14$$
$$x < 2 \qquad\qquad\qquad x > 7$$

If $x < 2$ **or** $x > 7$, then the solution set is $(-\infty, 2) \cup (7, \infty)$.

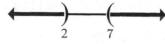
$$2 \qquad 7$$

61.
$$-4(x+2) \geq 12 \qquad \text{or} \qquad 3x+8 < 11$$
$$-4x-8+8 \geq 12+8 \qquad\qquad 3x+8-8 < 11-8$$
$$-4x \geq 20 \qquad\qquad\qquad 3x < 3$$
$$\frac{-4x}{-4} \leq \frac{20}{-4} \qquad\qquad\qquad x < 1$$
$$x \leq -5$$

If $x \leq -5$ **or** $x < 1$, then $x < 1$. The solution set is $(-\infty, 1)$.

$$1$$

63.
$$x < -3 \qquad \text{and} \qquad x > 3$$

It is impossible for x to be both less than -3 and greater than 3. There is no solution.

65. Let $x =$ the number of additional hours the person rents the rototiller.
$$15.50 + 7.95x < 50$$
$$7.95x < 34.50$$
$$x < 4.34$$
The person may rent the rototiller for 4 additional hours, or for 5 total hours.

67. Let $x =$ the number of children.
$$205 + 175 + 90x \leq 750$$
$$90x \leq 370$$
$$x \leq 4.11$$
At most 4 children can ride along.

69. Let $x =$ the amount invested at 9%.
$$0.08(10000) + 0.09x > 1250$$
$$800 + 0.09x > 1250$$
$$0.09x > 450$$
$$x > 5000$$
She must invest more than $5000.

71. Let $x =$ the # of compact discs bought.
$$175 + 8.50x \leq 330$$
$$8.50x \leq 155$$
$$x \leq 18.24$$
He can buy at most 18 discs.

73. Let $x =$ the student's score on 4th exam.
$$\frac{70 + 77 + 85 + x}{4} \geq 80$$
$$232 + x \geq 320$$
$$x \geq 88$$
The student needs a score of at least **88**.

75. Let $x =$ the # of hours he works at the library. Then $20 - x =$ the # of hours he works construction.
$$5x + 9(20 - x) > 125$$
$$5x + 180 - 9x > 125$$
$$-4x > -55$$
$$x < 13.75$$
He can work up to 13 full hours at the library.

77. Let $x =$ the total bill. Under Plan 1, the employee will pay $100, plus 30% of the rest (the amount over $100). This amount could be represented by $100 + 0.30(x - 100)$. Similarly, under Plan 2 the employee would pay $200 + 0.20(x - 200)$.

$$\boxed{\text{Amount paid in Plan 1}} > \boxed{\text{Amount paid in Plan 2}}$$
$$100 + 0.30(x - 100) > 200 + 0.20(x - 200)$$
$$100 + 0.30x - 30 > 200 + 0.20x - 40$$
$$70 + 0.30x > 160 + 0.20x$$
$$0.10x > 90$$
$$x > 900$$
If the total bill is over $900, then Plan 2 is better than Plan 1.

79. $2x + 3 < 5$; Graph $y = 2x + 3$ and $y = 5$ and find the x-coordinates when the first graph is below the second:

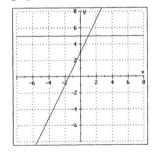

Solution: $x < 1$

81. $5x + 2 \geq -18$; Graph $y = 5x + 2$ and $y = -18$ and find the x-coordinates when the first graph is on or above the second:

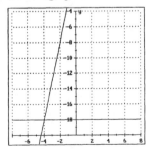

Solution: $x \geq -4$

83. We have values of $\sigma = 120$ and $E = 20$. Substitute into the formula:

$$\frac{3.84\sigma^2}{N} < E^2$$

$$\frac{3.84(120)^2}{N} < 20^2$$

$$\frac{55296}{N} < 400$$

$$55296 < 400N$$

$$138.24 < N$$

The sample size must be at least 139.

85. **Answers may vary.**

87. Let $x = 2$.

Then $x > -3$, but $x^2 = 2^2 = 4 < 9$.

89. Transitive: $=$, $\leq$ and $\not\geq$

Exercise 4.2 (page 231)

1.
$$3(2a - 1) = 2a$$
$$6a - 3 = 2a$$
$$4a = 3$$
$$a = \frac{3}{4}$$

3.
$$\frac{5x}{2} - 1 = \frac{x}{3} + 12$$
$$6\left(\frac{5x}{2} - 1\right) = 6\left(\frac{x}{3} + 12\right)$$
$$15x - 6 = 2x + 72$$
$$13x = 78$$
$$x = 6$$

5.
$$A = p + prt$$
$$A - p = prt$$
$$\frac{A - p}{pr} = t, \text{ or } t = \frac{A - p}{pr}$$

7. x

9. 0

11. reflected

13. $a = b$ or $a = -b$

15. $x \leq -k$ or $x \geq k$

17. $|8| = 8$

19. $-|2| = -2$

21. $-|-30| = -30$

23. $|\pi - 4| = -(\pi - 4) = 4 - \pi; (\pi - 4 \text{ is less than zero.})$

25. $f(x) = |x| - 2$

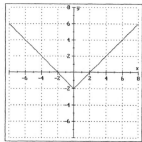

27. $f(x) = -|x + 4|$

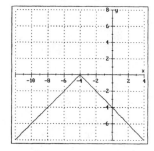

29.
$$|x| = 4$$
$$x = 4 \quad \text{or} \quad x = -4$$

31.
$$|x - 3| = 6$$
$$x - 3 = 6 \quad \text{or} \quad x - 3 = -6$$
$$x = 9 \qquad\qquad x = -3$$

33.
$$|2x - 3| = 5$$
$$2x - 3 = 5 \quad \text{or} \quad 2x - 3 = -5$$
$$2x = 8 \qquad\qquad 2x = -2$$
$$x = 4 \qquad\qquad x = -1$$

35.
$$|3x + 2| = 16$$
$$3x + 2 = 16 \quad \text{or} \quad 3x + 2 = -16$$
$$3x = 14 \qquad\qquad 3x = -18$$
$$x = \tfrac{14}{3} \qquad\qquad x = -6$$

37.
$$\left|\tfrac{7}{2}x + 3\right| = -5$$
Since an absolute value cannot be negative, there is no solution.

39.
$$\left|\tfrac{x}{2} - 1\right| = 3$$
$$\frac{x}{2} - 1 = 3 \quad \text{or} \quad \frac{x}{2} - 1 = -3$$
$$\frac{x}{2} = 4 \qquad\qquad \frac{x}{2} = -2$$
$$x = 8 \qquad\qquad x = -4$$

41.
$$|3 - 4x| = 5$$
$$3 - 4x = 5 \quad \text{or} \quad 3 - 4x = -5$$
$$-4x = 2 \qquad\qquad -4x = -8$$
$$x = -\tfrac{1}{2} \qquad\qquad x = 2$$

43.
$$|3x + 24| = 0$$
$$3x + 24 = 0 \quad \text{or} \quad 3x + 24 = -0$$
$$3x = -24 \qquad\qquad 3x = -24$$
$$x = -8 \qquad\qquad x = -8$$

45.
$$\left|\frac{3x + 48}{3}\right| = 12$$
$$\frac{3x + 48}{3} = 12 \quad \text{or} \quad \frac{3x + 48}{3} = -12$$
$$3x + 48 = 36 \qquad\qquad 3x + 48 = -36$$
$$3x = -12 \qquad\qquad 3x = -84$$
$$x = -4 \qquad\qquad x = -28$$

47.
$$|x + 3| + 7 = 10$$
$$|x + 3| = 3$$
$$x + 3 = 3 \quad \text{or} \quad x + 3 = -3$$
$$x = 0 \qquad\qquad x = -6$$

49.
$$\left|\tfrac{3}{5}x - 4\right| - 2 = -2$$
$$\left|\tfrac{3}{5}x - 4\right| = 0$$
$$\tfrac{3}{5}x - 4 = 0 \quad \text{or} \quad \tfrac{3}{5}x - 4 = -0$$
$$\tfrac{3}{5}x = 4 \qquad\qquad \tfrac{3}{5}x = 4$$
$$x = \tfrac{20}{3} \qquad\qquad x = \tfrac{20}{3}$$

51.
$$|2x + 1| = |3x + 3|$$
$$2x + 1 = 3x + 3 \quad \text{or} \quad 2x + 1 = -(3x + 3)$$
$$-x = 2 \qquad\qquad 2x + 1 = -3x - 3$$
$$x = -2 \qquad\qquad 5x = -4$$
$$\qquad\qquad x = -\tfrac{4}{5}$$

53.
$$|3x - 1| = |x + 5|$$
$$3x - 1 = x + 5 \quad \text{or} \quad 3x - 1 = -(x + 5)$$
$$2x = 6 \qquad\qquad 3x - 1 = -x - 5$$
$$x = 3 \qquad\qquad 4x = -4$$
$$\qquad\qquad x = -1$$

55.
$$|2 - x| = |3x + 2|$$
$$2 - x = 3x + 2 \quad \text{or} \quad 2 - x = -(3x + 2)$$
$$-4x = 0 \qquad\qquad 2 - x = -3x - 2$$
$$x = 0 \qquad\qquad 2x = -4$$
$$\qquad\qquad x = -2$$

57.
$$\left|\frac{x}{2} + 2\right| = \left|\frac{x}{2} - 2\right|$$

$\frac{x}{2} + 2 = \frac{x}{2} - 2$ **or** $\frac{x}{2} + 2 = -\left(\frac{x}{2} - 2\right)$

     $0 = -4$ $\frac{x}{2} + 2 = -\frac{x}{2} + 2$

(no solution from this part) $x = 0$

59.
$$\left|x + \tfrac{1}{3}\right| = |x - 3|$$

$x + \tfrac{1}{3} = x - 3$ **or** $x + \tfrac{1}{3} = -(x - 3)$

$3x + 1 = 3x - 9$ $x + \tfrac{1}{3} = -x + 3$

    $0 = -10$ $3x + 1 = -3x + 9$

(no solution from this part) $6x = 8$

                           $x = \tfrac{4}{3}$

61. $|3x + 7| = -|8x - 2|$; Since an absolute value cannot be **negative**, there is no solution.

63. $|2x| < 8$
$-8 < 2x < 8$
$\frac{-8}{2} < \frac{2x}{2} < \frac{8}{2}$
$-4 < x < 4$
solution set: $(-4, 4)$

65. $|x + 9| \leq 12$
$-12 \leq x + 9 \leq 12$
$-12 - 9 \leq x + 9 - 9 \leq 12 - 9$
$-21 \leq x \leq 3$
solution set: $[-21, 3]$

67. $|3x + 2| \leq -3$; Since an absolute value can never be negative, this inequality has no solution.

69. $|4x - 1| \leq 7$
$-7 \leq 4x - 1 \leq 7$
$-7 + 1 \leq 4x - 1 + 1 \leq 7 + 1$
$-6 \leq 4x \leq 8$
$\frac{-6}{4} \leq \frac{4x}{4} \leq \frac{8}{4}$
$-\frac{3}{2} \leq x \leq 2$
solution set: $\left[-\frac{3}{2}, 2\right]$

71. $|3 - 2x| < 7$
$-7 < 3 - 2x < 7$
$-7 - 3 < 3 - 3 - 2x < 7 - 3$
$-10 < -2x < 4$
$\frac{-10}{-2} > \frac{-2x}{-2} > \frac{4}{-2}$
$5 > x > -2$, or $-2 < x < 5$
solution set: $(-2, 5)$

73. $|5x| > 5$
$5x < -5$ **or** $5x > 5$
$x < -1$ $x > 1$
solution set: $(-\infty, -1) \cup (1, \infty)$

75. $|x - 12| > 24$
$x - 12 < -24$ **or** $x - 12 > 24$
$x < -12$ $x > 36$
solution set: $(-\infty, -12) \cup (36, \infty)$

77.
$$|3x + 2| > 14$$
$$3x + 2 < -14 \quad \text{or} \quad 3x + 2 > 14$$
$$3x < -16 \qquad\qquad 3x > 12$$
$$x < -\frac{16}{3} \qquad\qquad x > 4$$
solution set: $\left(-\infty, -\frac{16}{3}\right) \cup (4, \infty)$

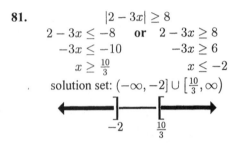

79.
$$|4x + 3| > -5$$
Since an absolute value is always at least 0, this inequality is true for all real numbers.
solution set: $(-\infty, \infty)$

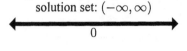

81.
$$|2 - 3x| \geq 8$$
$$2 - 3x \leq -8 \quad \text{or} \quad 2 - 3x \geq 8$$
$$-3x \leq -10 \qquad\qquad -3x \geq 6$$
$$x \geq \frac{10}{3} \qquad\qquad x \leq -2$$
solution set: $(-\infty, -2] \cup \left[\frac{10}{3}, \infty\right)$

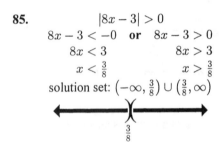

83.
$$-|2x - 3| < -7$$
$$|2x - 3| > 7$$
$$2x - 3 < -7 \quad \text{or} \quad 2x - 3 > 7$$
$$2x < -4 \qquad\qquad 2x > 10$$
$$x < -2 \qquad\qquad x > 5$$
solution set: $(-\infty, -2) \cup (5, \infty)$

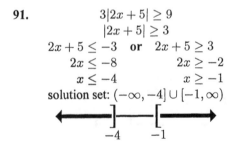

85.
$$|8x - 3| > 0$$
$$8x - 3 < -0 \quad \text{or} \quad 8x - 3 > 0$$
$$8x < 3 \qquad\qquad 8x > 3$$
$$x < \frac{3}{8} \qquad\qquad x > \frac{3}{8}$$
solution set: $\left(-\infty, \frac{3}{8}\right) \cup \left(\frac{3}{8}, \infty\right)$

87.
$$\left|\frac{x - 2}{3}\right| \leq 4$$
$$-4 \leq \frac{x - 2}{3} \leq 4$$
$$-12 \leq x - 2 \leq 12$$
$$-10 \leq x \leq 14$$
solution set: $[-10, 14]$

89.
$$|3x + 1| + 2 < 6$$
$$|3x + 1| < 4$$
$$-4 < 3x + 1 < 4$$
$$-5 < 3x < 3$$
$$-\frac{5}{3} < x < 1$$
solution set: $\left(-\frac{5}{3}, 1\right)$

91.
$$3|2x + 5| \geq 9$$
$$|2x + 5| \geq 3$$
$$2x + 5 \leq -3 \quad \text{or} \quad 2x + 5 \geq 3$$
$$2x \leq -8 \qquad\qquad 2x \geq -2$$
$$x \leq -4 \qquad\qquad x \geq -1$$
solution set: $(-\infty, -4] \cup [-1, \infty)$

93. $|5x - 1| + 4 \le 0$
$|5x - 1| \le -4$
Since an absolute value can never be negative, this inequality has no solution.

95. $\left|\frac{1}{3}x + 7\right| + 5 > 6$
$\left|\frac{1}{3}x + 7\right| > 1$
$\frac{1}{3}x + 7 < -1$ or $\frac{1}{3}x + 7 > 1$
$x + 21 < -3$ $x + 21 > 3$
$x < -24$ $x > -18$
solution set: $(-\infty, -24) \cup (-18, \infty)$

97. $\left|\frac{1}{5}x - 5\right| + 4 > 4$
$\left|\frac{1}{5}x - 5\right| > 0$
$\frac{1}{5}x - 5 < -0$ or $\frac{1}{5}x - 5 > 0$
$\frac{1}{5}x < 5$ $\frac{1}{5}x > 5$
$x < 25$ $x > 25$
solution set: $(-\infty, 25) \cup (25, \infty)$

99. $\left|\frac{1}{7}x + 1\right| \le 0$
Since an absolute value can never be less than zero, the only solution for this inequality is when the absolute value is equal to 0.
$\left|\frac{1}{7}x + 1\right| = 0$
$\frac{1}{7}x + 1 = 0$ or $\frac{1}{7}x + 1 = -0$
$\frac{1}{7}x = -1$ $\frac{1}{7}x = -1$
$x = -7$ $x = -7$
solution set: $[-7, -7]$

101. $\left|\dfrac{x - 5}{10}\right| \le 0$

Since an absolute value can never be less than zero, the only solution for this inequality is when the absolute value is equal to 0.
$\left|\dfrac{x - 5}{10}\right| = 0$
$\dfrac{x - 5}{10} = 0$ or $\dfrac{x - 5}{10} = -0$
$x - 5 = 0$ $x - 5 = 0$
$x = 5$ $x = 5$
solution set: $[5, 5]$

103. $-4 < x < 4 \Rightarrow |x| < 4$

105. $x + 3 < -6$ or $x + 3 > 6 \Rightarrow |x + 3| > 6$

107. $|d - 5| \le 1$
$-1 \le d - 5 \le 1$
$-1 + 5 \le d - 5 + 5 \le 1 + 5$
$4 \text{ ft} \le d \le 6 \text{ ft}$

109. $|t - 78°| \le 8°$
$-8° \le t - 78° \le 8°$
$-8° + 78° \le t - 78° + 78° \le 8° + 78°$
$70° \le t \le 86°$

111. $0.6° - 0.5° \le c \le 0.6° + 0.5°$
$-0.5° \le c - 0.6° \le 0.5°$
$|c - 0.6°| \le 0.5°$

113. Answers may vary.

115. Answers may vary.

117. $|x| + k = 0$

$\qquad |x| = -k$

If this equation has exactly two solutions, then $-k > 0$, or $k < 0$.

119. $|x| + |y| > |x + y|$ when x and y have different signs.

Exercise 4.3 (page 239)

1.
$$x + y = 4$$
$$\underline{x - y = 2}$$
$$2x \quad\;\; = 6$$
$$x \quad\;\; = 3$$
Substitute and solve for y:
$$x + y = 4$$
$$3 + y = 4$$
$$y = 1$$
The solution is $(3, 1)$.

3.
$$3x + y = 3 \Rightarrow \times 3 \qquad 9x + 3y = 9$$
$$2x - 3y = 13 \qquad\qquad\;\; \underline{2x - 3y = 13}$$
$$\qquad\qquad\qquad\qquad\qquad 11x \qquad = 22$$
$$\qquad\qquad\qquad\qquad\qquad x \qquad = 2$$
Substitute and solve for y:
$$3x + y = 3$$
$$3(2) + y = 3$$
$$6 + y = 3$$
$$y = -3$$
The solution is $(2, -3)$.

5. linear

7. edge

9. $y > x + 1$

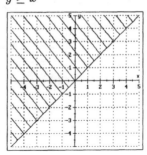

11. $y \geq x$

13. $2x + y \leq 6$

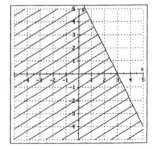

15. $3x \geq -y + 3$

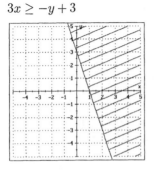

17. $y \geq 1 - \dfrac{3}{2}x$

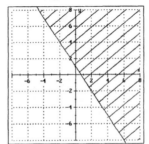

19. $x < 4$

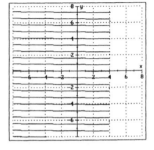

21. $-2 \leq x < 0$

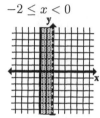

23. $y < -2$ or $y > 3$

25. The boundary line goes through $(2,0)$ and $(0,3)$. Use point-slope form to find the equation:

$$m = \frac{\Delta y}{\Delta x} = \frac{3-0}{0-2} = \frac{3}{-2} = -\frac{3}{2}$$

$$y - y_1 = m(x - x_1)$$

$$y - 0 = -\frac{3}{2}(x - 2)$$

$$y = -\frac{3}{2}x + 3$$

$$3x + 2y = 6$$

Pick a point which is shaded, and substitute the coordinates for x and y:

$$3x + 2y \; \boxed{} \; 6$$
$$3(2) + 2(1) \; \boxed{} \; 6$$
$$5 + 2 > 6$$

Since the line is dotted, do not include the line:

$$\boxed{3x + 2y > 6}$$

27. The boundary line is $x = 3$. Since points with x-coordinates less than 3 are shaded, and since the line is not dotted, the inequality is $\boxed{x \leq 3}$.

29. The boundary line goes through $(0,0)$ and $(1,1)$. Use point-slope form to find the equation:

$$m = \frac{\Delta y}{\Delta x} = \frac{1-0}{1-0} = \frac{1}{1} = 1$$

$$y - y_1 = m(x - x_1)$$

$$y - 0 = 1(x - 0)$$

$$y = x$$

Pick a point which is shaded, and substitute the coordinates for x and y:

$$y \; \boxed{\phantom{<}} \; x$$
$$0 < 1$$

Since the line is not dotted, include the line:

$$\boxed{y \leq x}$$

31. The points between the lines $x = -2$ and $x = 3$ are shaded. Both lines are not dotted, so they should be included. $\boxed{-2 \leq x \leq 3}$

33. The points above the line $y = -1$ (dotted) and below the line $y = -3$ (not dotted) are shaded. $\boxed{y > -1 \text{ or } y \leq -3}$

35. $y < 0.27x - 1$

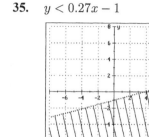

37. $y \geq -2.37x + 1.5$

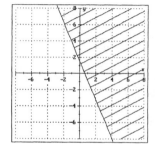

SECTION 4.3

39. Let $x =$ the number of simple returns completed, and let $y =$ the number of complicated returns completed. The inequality is $x + 3y \le 9$. Some ordered pairs are: $(1,1), (2,1), (2,2)$

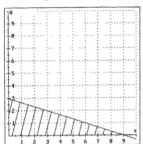

43. Let $x =$ the number of shares of Traffico, and let $y =$ the number of shares of Cleanco. The inequality is $50x + 60y \le 6000$. Some ordered pairs are: $(40,20), (60,40), (80,20)$

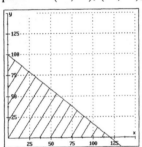

41. Let $x =$ the number of hours she uses the first, and let $y =$ the number of hours she uses the second. The inequality is $6x + 7y \le 42$. Some ordered pairs are: $(2,2), (3,3), (5,1)$

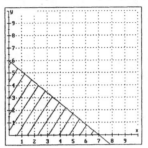

45. Answers may vary.

47. An inequality such as $x + (-x) < 10$ is an identity.

Exercise 4.4 (page 246)

1.
$$A = p + prt$$
$$A - p = prt$$
$$\frac{A-p}{pt} = r, \text{ or } r = \frac{A-p}{pt}$$

3.
$$z = \frac{x-\mu}{\sigma}$$
$$z\sigma = x - \mu$$
$$z\sigma + \mu = x, \text{ or } x = z\sigma + \mu$$

5.
$$l = a + (n-1)d$$
$$l - a = (n-1)d$$
$$\frac{l-a}{n-1} = d, \text{ or } d = \frac{l-a}{n-1}$$

7. intersect

104

9. $\begin{cases} y < 3x + 2 \\ y < -2x + 3 \end{cases}$

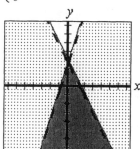

11. $\begin{cases} 3x + 2y > 6 \\ x + 3y \leq 2 \end{cases}$

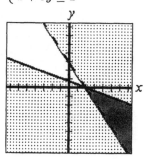

13. $\begin{cases} 3x + y \leq 1 \\ -x + 2y \geq 6 \end{cases}$

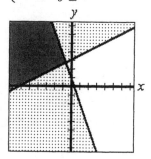

15. $\begin{cases} 2x - y > 4 \\ y < -x^2 + 2 \end{cases}$

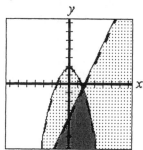

17. $\begin{cases} y > x^2 - 4 \\ y < -x^2 + 4 \end{cases}$

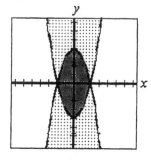

19. $\begin{cases} 2x + 3y \leq 6 \\ 3x + y \leq 1 \\ x \leq 0 \end{cases}$

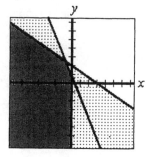

21. $\begin{cases} x - y < 4 \\ y \leq 0 \\ x \geq 0 \end{cases}$

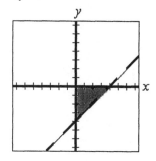

23. $\begin{cases} x \geq 0, \qquad y \geq 0 \\ 9x + 3y \leq 18 \\ 3x + 6y \leq 18 \end{cases}$

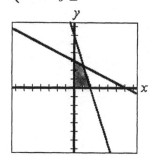

25. See #9.

27. Let $x =$ the number of \$10 discs and let $y =$ the number of \$15 discs.

$$\begin{cases} 10x + 15y \geq 30, & x \geq 0 \\ 10x + 15y \leq 60, & y \geq 0 \end{cases}$$

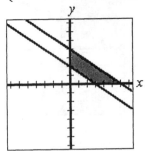

Two solutions are $(1, 2)$ and $(4, 1)$. A customer could buy one \$10 and two \$15 discs, or four \$10 and one \$15 disc.

29. Let $x =$ the # of desk chairs, and let $y =$ the # of side chairs.

$$\begin{cases} 150x + 100y \leq 900, & y > x \\ x \geq 0, & y \geq 0 \end{cases}$$

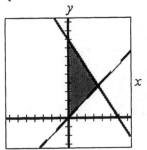

Two solutions are $(2, 4)$ and $(1, 5)$. Best could buy 2 desk chairs and 4 side chairs, or 1 desk chair and 5 side chairs.

31. **Answers may vary.**

33. No.

Exercise 4.5 (page 256)

1. $m = \dfrac{\Delta y}{\Delta x} = \dfrac{4 - 7}{-2 - 5} = \dfrac{-3}{-7} = \dfrac{3}{7}$

3. $y - y_1 = m(x - x_1)$
$y - 7 = \frac{3}{7}(x - 5)$
$y - 7 = \frac{3}{7}x - \frac{15}{7}$
$y = \frac{3}{7}x + \frac{34}{7}$

5. constraints

7. objective

9.

Vertex	$P = 2x + 3y$	Maximum?
$(0, 0)$	$= 0$	No
$(4, 0)$	$= 8$	No
$(0, 4)$	$= 12$	YES

P has a maximum value of 12 at $(0, 4)$.

11.

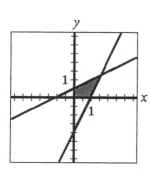

Vertex	$P = y + \frac{1}{2}x$	Maximum?
$(0,0)$	$= 0$	No
$(1,0)$	$= \frac{1}{2}$	No
$\left(\frac{5}{3}, \frac{4}{3}\right)$	$= \frac{13}{6}$	YES
$\left(0, \frac{1}{2}\right)$	$= \frac{1}{2}$	No

P has a maximum value of $\frac{13}{6}$ at $\left(\frac{5}{3}, \frac{4}{3}\right)$.

13.

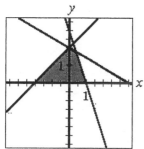

Vertex	$P = 2x + y$	Maximum?
$(1,0)$	$= 0$	No
$\left(\frac{3}{7}, \frac{12}{7}\right)$	$= \frac{18}{7}$	YES
$(0,2)$	$= 2$	No
$(-2,0)$	$= -4$	No

P has a maximum value of $\frac{18}{7}$ at $\left(\frac{3}{7}, \frac{12}{7}\right)$.

15.

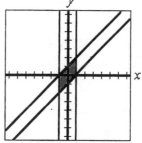

Vertex	$P = 3x - 2y$	Maximum?
$(1,0)$	$= 3$	YES
$(1,2)$	$= -1$	No
$(-1,0)$	$= -3$	No
$(-1,-2)$	$= 1$	No

P has a maximum value of 3 at $(1,0)$.

17.

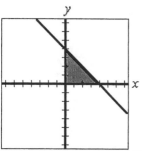

Vertex	$P = 5x + 12y$	Minimum?
$(4,0)$	$= 20$	No
$(0,4)$	$= 48$	No
$(0,0)$	$= 0$	YES

P has a minimum value of 0 at $(0,0)$.

19.

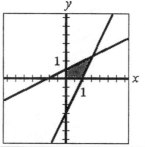

Vertex	$P = 3y + x$	Minimum?
$(0,0)$	$= 0$	YES
$(1,0)$	$= 1$	No
$\left(\frac{5}{3}, \frac{4}{3}\right)$	$= \frac{17}{3}$	No
$\left(0, \frac{1}{2}\right)$	$= \frac{3}{2}$	No

P has a minimum value of 0 at $(0,0)$.

21.

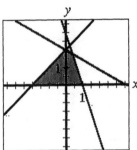

Vertex	$P = 6x + 2y$	Minimum?
$(1, 0)$	$= 6$	No
$\left(\frac{3}{7}, \frac{12}{7}\right)$	$= 6$	No
$(0, 2)$	$= 4$	No
$(-2, 0)$	$= -12$	YES

P has a minimum value of -12 at $(-2, 0)$.

23.

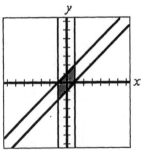

Vertex	$P = 2x - 2y$	Minimum?
$(1, 0)$	$= 2$	No
$(1, 2)$	$= -2$	YES
$(-1, 0)$	$= -2$	YES
$(-1, -2)$	$= 2$	No

P has a minimum value of -2 at $(1, 2)$ and at $(-1, 0)$.

25. Let x = the number of tables made, and let y = the number of chairs made.

$$\begin{cases} x \geq 0 & y \geq 0 \\ 2x + 3y \leq 42 & \text{(Tom's hours)} \\ 6x + 2y \leq 42 & \text{(Carlos' hours)} \end{cases}$$

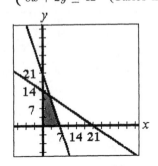

Profit $= P = 100x + 80y.$

Vertex	$P = 100x + 80y$
$(0, 0)$	$= 0$
$(7, 0)$	$= 700$
$(3, 12)$	$= 1260$
$(0, 14)$	$= 1120$

The maximum income of $1260 results when they make 3 tables and 12 chairs.

SECTION 4.5

27. Let x = the number of IBMs stocked, and let y = the number of Macintoshes stocked.

$$\begin{cases} 20 \le x \le 30 \\ 30 \le y \le 50 \\ x + y \le 60 \quad \text{(Total stock)} \end{cases}$$

Let the objective function be the total commissions:
$P = 50x + 40y$.

Vertex	$P = 50x + 40y$
$(20, 30)$	$= 2200$
$(30, 30)$	$= 2700$
$(20, 40)$	$= 2600$

The maximum amount of commissions is $2700, which results from stocking 30 of each type.

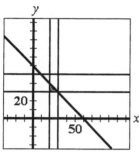

29. Let x = the number of VCRs made, and let y = the number of TVs made.

$$\begin{cases} x \ge 0 \qquad\quad y \ge 0 \\ 3x + 4y \le 180 \quad \text{(electronics hours)} \\ 2x + 3y \le 120 \quad \text{(assembly hours)} \\ 2x + y \le 60 \quad \text{(finishing hours)} \end{cases}$$

Let the objective function be the profit:
$P = 40x + 32y$.

Vertex	$P = 40x + 32y$
$(0, 0)$	$= 0$
$(0, 30)$	$= 960$
$(15, 30)$	$= 1560$
$(0, 40)$	$= 1280$

The maximum profit of $1560 results when they make 15 VCRs and 30 TVs.

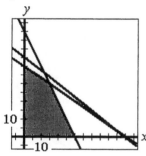

31. Let x = the amount in stocks, and let y = the amount in bonds.

$$\begin{cases} x \ge 100000 \\ y \ge 50000 \\ x + y \le 200000 \quad \text{(total amount)} \end{cases}$$

Let the objective function be the income:
$P = 0.09x + 0.07y$.

Vertex	$P = 0.09x + 0.07y$
$(100000, 50000)$	$= 12500$
$(150000, 50000)$	$= 17000$
$(100000, 100000)$	$= 16000$

The maximum income of $17,000 results when she invests $150,000 in stocks and $50,000 in bonds.

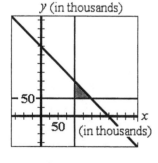

33. Answers may vary.

35. Answers may vary.

Chapter 4 Summary (page 259)

1.
$$5(x - 2) \le 5$$
$$5x - 10 + 10 \le 5 + 10$$
$$5x \le 15$$
$$x \le 3$$
solution set: $(-\infty, 3]$

2.
$$3x + 4 > 10$$
$$3x + 4 - 4 > 10 - 4$$
$$3x > 6$$
$$x > 2$$
solution set: $(2, \infty)$

3.
$$\frac{1}{3}x - 2 \ge \frac{1}{2}x + 2$$
$$6\left(\frac{1}{3}x - 2\right) \ge 6\left(\frac{1}{2}x + 2\right)$$
$$2x - 12 + 12 \ge 3x + 12 + 12$$
$$2x - 3x \ge 3x - 3xy + 24$$
$$-x \ge 24$$
$$x \le -24$$
solution set: $(-\infty, -24]$

4.
$$\frac{7}{4}(x + 3) < \frac{3}{8}(x - 3)$$
$$8 \cdot \frac{7}{4}(x + 3) < 8 \cdot \frac{3}{8}(x - 3)$$
$$14(x + 3) < 3(x - 3)$$
$$14x + 42 < 3x - 9$$
$$14x < 3x - 51$$
$$11x < -51$$
$$x < -\frac{51}{11}$$
solution set: $\left(-\infty, -\frac{51}{11}\right)$

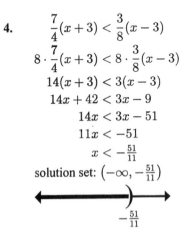

5.
$$3 < 3x + 4 < 10$$
$$3 - 4 < 3x + 4 - 4 < 10 - 4$$
$$-1 < 3x < 6$$
$$-\tfrac{1}{3} < \tfrac{3x}{3} < \tfrac{6}{3}$$
$$-\tfrac{1}{3} < x < 2 \Rightarrow \text{solution set: } \left(-\tfrac{1}{3}, 2\right) \Rightarrow$$

6.
$$4x > 3x + 2 > x - 3$$

$$4x > 3x + 2 \qquad \text{and} \qquad 3x + 2 > x - 3$$
$$4x - 3x > 3x - 3x + 2 \qquad 3x + 2 - 2 > x - 3 - 2$$
$$x > 2 \qquad 3x - x > x - x - 5$$
$$x > 2 \qquad 2x > -5$$
$$x > -\tfrac{5}{2}$$

If $x > 2$ **and** $x > -\frac{5}{2}$, $x > 2$. The solution set is $(2, \infty)$.

7.
$$-5 \le 2x - 3 < 5$$
$$-5 + 3 \le 2x - 3 + 2 < 5 + 3$$
$$-2 \le 2x < 8$$

$-1 \le x < 4 \Rightarrow$ solution set: $[-1, 4) \Rightarrow$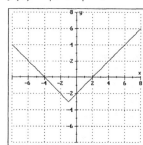

8. Let $x =$ the amount invested at 7%.
$$0.06(10000) + 0.07x \ge 2000$$
$$600 + 0.07x \ge 2000$$
$$0.07x \ge 1400$$
$$x \ge 20000 \Rightarrow \text{She must invest at least } \$20,000.$$

9. $|-7| = 7$ **10.** $|8| = 8$ **11.** $-|7| = -7$ **12.** $-|-12| = -12$

13. $f(x) = |x + 1| - 3$

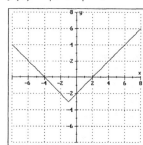

14. $f(x) = |x - 2| + 1$

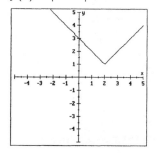

15.
$$|3x + 1| = 10$$
$$3x + 1 = 10 \quad \text{or} \quad 3x + 1 = -10$$
$$3x = 9 \qquad\qquad 3x = -11$$
$$x = 3 \qquad\qquad x = -\frac{11}{3}$$

16.
$$\left|\frac{3}{2}x - 4\right| = 9$$
$$\frac{3}{2}x - 4 = 9 \quad \text{or} \quad \frac{3}{2}x - 4 = -9$$
$$\frac{3}{2}x = 13 \qquad\qquad \frac{3}{2}x = -5$$
$$x = \frac{26}{3} \qquad\qquad x = -\frac{10}{3}$$

17.
$$\left|\frac{2 - x}{3}\right| = 4$$
$$\frac{2 - x}{3} = 4 \quad \text{or} \quad \frac{2 - x}{3} = -4$$
$$2 - x = 12 \qquad\qquad 2 - x = -12$$
$$-x = 10 \qquad\qquad -x = -14$$
$$x = -10 \qquad\qquad x = 14$$

18.
$$|3x + 2| = |2x - 3|$$
$$3x + 2 = 2x - 3 \quad \text{or} \quad 3x + 2 = -(2x - 3)$$
$$x = -5 \qquad\qquad 3x + 2 = -2x + 3$$
$$5x = 1$$
$$x = \frac{1}{5}$$

19.
$$|5x - 4| = |4x - 5|$$
$$5x - 4 = 4x - 5 \quad \text{or} \quad 5x - 4 = -(4x - 5)$$
$$x = -1 \qquad\qquad 5x - 4 = -4x + 5$$
$$9x = 9$$
$$x = 1$$

20.
$$\left|\frac{3-2x}{2}\right| = \left|\frac{3x-2}{3}\right|$$

$$\frac{3-2x}{2} = \frac{3x-2}{3} \quad \text{or} \quad \frac{3-2x}{2} = -\frac{3x-2}{3}$$
$$3(3-2x) = 2(3x-2) \qquad 3(3-2x) = -2(3x-2)$$
$$9 - 6x = 6x - 4 \qquad 9 - 6x = -6x + 4$$
$$-12x = -13 \qquad\qquad 0 = -5$$
$$x = \tfrac{13}{12} \qquad\qquad \text{(no solution from this part)}$$

21.
$$|2x + 7| < 3$$
$$-3 < 2x + 7 < 3$$
$$-3 - 7 < 2x + 7 - 7 < 3 - 7$$
$$-10 < 2x < -4$$
$$-5 < x < -2$$
solution set: $(-5, -2)$

22.
$$|5 - 3x| \le 14$$
$$-14 \le 5 - 3x \le 14$$
$$-14 - 5 \le 5 - 5 - 3x \le 14 - 5$$
$$-19 \le -3x \le 9$$
$$\frac{-19}{-3} \ge \frac{-3x}{-3} \ge \frac{9}{-3}$$
$$\frac{19}{3} \ge x \ge -3, \text{ or } -3 \le x \le \frac{19}{3}$$
solution set: $\left[-3, \tfrac{19}{3}\right]$

23. $\left|\tfrac{2}{3}x + 14\right| < 0$
Since an absolute value can never be negative, there is no solution.

24.
$$\left|\frac{1-5x}{3}\right| > 7$$
$$\frac{1-5x}{3} < -7 \quad \text{or} \quad \frac{1-5x}{3} > 7$$
$$1 - 5x < -21 \qquad 1 - 5x > 21$$
$$-5x < -22 \qquad\quad -5x > 20$$
$$x > \tfrac{22}{5} \qquad\qquad x < -4$$
solution set: $(-\infty, -4) \cup \left(\tfrac{22}{5}, \infty\right)$

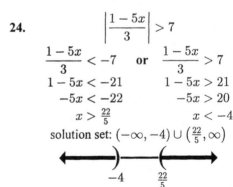

25.
$$|3x - 8| \ge 4$$
$$3x - 8 \le -4 \quad \text{or} \quad 3x - 8 \ge 4$$
$$3x \le 4 \qquad\qquad 3x \ge 12$$
$$x \le \tfrac{4}{3} \qquad\qquad x \ge 4$$
solution set: $\left(-\infty, \tfrac{4}{3}\right] \cup [4, \infty)$

26.
$$\left|\tfrac{3}{2}x - 14\right| \ge 0$$
Since an absolute value is always at least 0, this inequality is true for all real #s. Solution set: $(-\infty, \infty)$

27. $2x + 3y > 6$

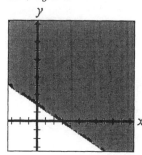

28. $y \leq 4 - x$

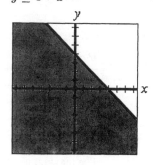

29. $-2 < x < 4$

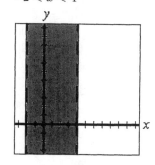

30. $y \leq -2$ or $y > 1$

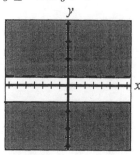

31. $\begin{cases} y \geq x + 1 \\ 3x + 2y < 6 \end{cases}$

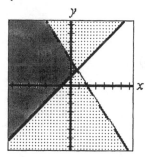

32. $\begin{cases} y \geq x^2 - 4 \\ y < x + 3 \end{cases}$

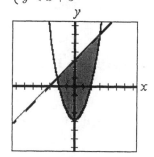

33.

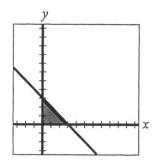

Vertex	$P = 2x + y$	Maximum?
$(0,0)$	$= 0$	No
$(3,0)$	$= 6$	YES
$(0,3)$	$= 3$	No

P has a maximum value of 6 at $(3,0)$.

34. Let $x =$ the number of bags of fertilizer X, and let $y =$ the number of bags of fertilizer Y.

$$\begin{cases} x \geq 0 \qquad\qquad y \geq 0 \\ 6x + 10y \leq 20000 \quad \text{(Nitrogen)} \\ 8x + 6y \leq 16400 \quad \text{(Phosphorus)} \\ 6x + 4y \leq 12000 \quad \text{(Potash)} \end{cases}$$

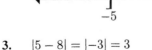

Let the objective function be the total profit:
$P = 6x + 5y.$

Vertex	$P = 6x + 5y$
$(2000, 0)$	$= 12000$
$(1600, 600)$	$= 12600$
$(1000, 1400)$	$= 13000$
$(0, 2000)$	$= 10000$

The maximum profit is $13,000, which results from using 1000 bags of fertilizer X and 1400 bags of fertilizer Y.

Chapter 4 Test   (page 262)

1. $-2(2x + 3) \geq 14$
$-4x - 6 + 6 \geq 14 + 6$
$\quad\quad -4x \geq 20$
$\quad\quad \dfrac{-4x}{-4} \leq \dfrac{20}{-4}$
$\quad\quad\quad x \leq -5$
solution set: $(-\infty, -5]$

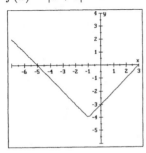

2. $-2 < \dfrac{x-4}{3} < 4$
$-6 < x - 4 < 12$
$-6 + 4 < x - 4 + 4 \leq 12 + 4$
$-2 < x < 16$
solution set: $(-2, 16)$

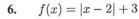

3. $|5 - 8| = |-3| = 3$

4. $|4\pi - 4| = 4\pi - 4 \ (4\pi - 4 > 0)$

5. $f(x) = |x + 1| - 4$

6. $f(x) = |x - 2| + 3$

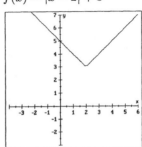

7.
$$|2x + 3| = 11$$
$$2x + 3 = 11 \quad \textbf{or} \quad 2x + 3 = -11$$
$$2x = 8 \qquad\qquad 2x = -14$$
$$x = 4 \qquad\qquad x = -7$$

8.
$$|4 - 3x| = 19$$
$$4 - 3x = 19 \quad \textbf{or} \quad 4 - 3x = -19$$
$$-3x = 15 \qquad\qquad -3x = -23$$
$$x = -5 \qquad\qquad x = \frac{23}{3}$$

9.
$$|3x + 4| = |x + 12|$$
$$3x + 4 = x + 12 \quad \textbf{or} \quad 3x + 4 = -(x + 12)$$
$$2x = 8 \qquad\qquad 3x + 4 = -x - 12$$
$$x = 4 \qquad\qquad 4x = -16$$
$$x = -4$$

10.
$$|3 - 2x| = |2x + 3|$$
$$3 - 2x = 2x + 3 \quad \textbf{or} \quad 3 - 2x = -(2x + 3)$$
$$-4x = 0 \qquad\qquad 3 - 2x = -2x - 3$$
$$x = 0 \qquad\qquad 0 = -6$$
(no solution from this part)

11. $|x + 3| \le 4$
$$-4 \le x + 3 \le 4$$
$$-4 - 3 \le x + 3 - 3 \le 4 - 3$$
$$-7 \le x \le 1$$
solution set: $[-7, 1]$

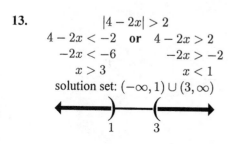

12.
$$|2x - 4| > 22$$
$$2x - 4 < -22 \quad \textbf{or} \quad 2x - 4 > 22$$
$$2x < -18 \qquad\qquad 2x > 26$$
$$x < -9 \qquad\qquad x > 13$$
solution set: $(-\infty, -9) \cup (13, \infty)$

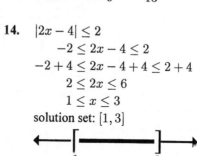

13.
$$|4 - 2x| > 2$$
$$4 - 2x < -2 \quad \textbf{or} \quad 4 - 2x > 2$$
$$-2x < -6 \qquad\qquad -2x > -2$$
$$x > 3 \qquad\qquad x < 1$$
solution set: $(-\infty, 1) \cup (3, \infty)$

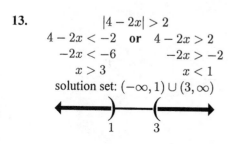

14. $|2x - 4| \le 2$
$$-2 \le 2x - 4 \le 2$$
$$-2 + 4 \le 2x - 4 + 4 \le 2 + 4$$
$$2 \le 2x \le 6$$
$$1 \le x \le 3$$
solution set: $[1, 3]$

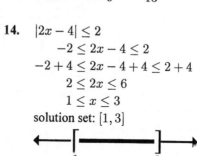

15. $3x + 2y \geq 6$

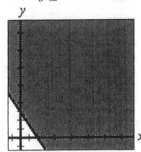

16. $-2 \leq y < 5$

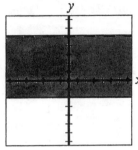

17. $\begin{cases} 2x - 3y \geq 6 \\ y \leq -x + 1 \end{cases}$

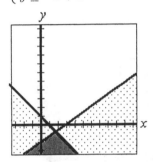

18. $\begin{cases} y \geq x^2 \\ y < x + 3 \end{cases}$

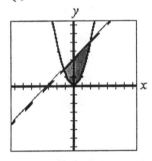

19.

Vertex	$P = 3x - y$	Maximum?
$(0, 1)$	$= -1$	No
$(1, 1)$	$= 2$	YES
$(1, 2)$	$= 1$	No
$(\frac{1}{3}, 2)$	$= -1$	No

P has a maximum value of 2 at $(1, 1)$.

Cumulative Review Exercises (page 263)

1.

2. The additive inverse of -5 is 5.

3. $x - xy = 2 - 2(-4) = 2 + 8 = 10$

4. $\dfrac{x^2 - y^2}{3x + y} = \dfrac{2^2 - (-4)^2}{3(2) + (-4)} = \dfrac{4 - 16}{6 - 4}$

$= \dfrac{-12}{2} = -6$

CUMULATIVE REVIEW EXERCISES

5. $\left(x^2x^3\right)^2 = \left(x^5\right)^2 = x^{10}$

6. $\left(x^2\right)^3\left(x^4\right)^2 = x^6x^8 = x^{14}$

7. $\left(\dfrac{x^3}{x^5}\right)^{-2} = \left(\dfrac{x^5}{x^3}\right)^2 = \left(x^2\right)^2 = x^4$

8. $\dfrac{a^2b^n}{a^nb^2} = a^{2-n}b^{n-2}$

9. $32{,}600{,}000 = 3.26 \times 10^7$

10. $0.000012 = 1.2 \times 10^{-5}$

11. $3x - 6 = 20$
$3x = 26$
$x = \dfrac{26}{3}$

12. $6(x-1) = 2(x+3)$
$6x - 6 = 2x + 6$
$4x = 12$
$x = 3$

13. $\dfrac{5b}{2} - 10 = \dfrac{b}{3} + 3$
$6\left(\dfrac{5b}{2} - 10\right) = 6\left(\dfrac{b}{3} + 3\right)$
$3(5b) - 60 = 2b + 18$
$15b - 60 = 2b + 18$
$13b = 78$
$b = 6$

14. $2a - 5 = -2a + 4(a-2) + 1$
$2a - 5 = -2a + 4a - 8 + 1$
$2a - 5 = 2a - 7$
$-5 = -7$
contradiction

15. $3x + 2y = 12$ $\qquad$ $2x - 3y = 5$
$\quad 2y = -3x + 12$ $\qquad -3y = -2x + 5$
$\quad\ y = -\frac{3}{2}x + 6$ $\qquad\ y = \frac{2}{3}x - \frac{5}{3}$
$m = -\frac{3}{2}$ $\qquad\qquad m = \frac{2}{3}$
$\qquad\qquad$ perpendicular

16. $3x = y + 4$ $\qquad y = 3(x-4) - 1$
$-y = -3x + 4$ $\qquad y = 3x - 12 - 1$
$\ \ y = 3x - 4$ $\qquad\ y = 3x - 13$
$\ \ m = 3$ $\qquad\qquad\ m = 3$
$\qquad\qquad$ parallel

17. $3x + y = 8$
$\quad y = -3x + 8$
$m = -3$
Use $m = \frac{1}{3}$ (the perpendicular slope):
$y - y_1 = m(x - x_1)$
$y - 3 = \frac{1}{3}(x + 2)$
$\quad y = \frac{1}{3}x + \frac{11}{3}$

18. $A = \dfrac{1}{2}h(b_1 + b_2)$
$2A = h(b_1 + b_2)$
$\dfrac{2A}{b_1 + b_2} = h$, or $h = \dfrac{2A}{b_1 + b_2}$

19. $f(2) = 3(2)^2 - 2 = 3(4) - 2$
$= 12 - 2 = 10$

20. $f(-2) = 3(-2)^2 - (-2) = 3(4) + 2$
$= 12 + 2 = 14$

21. $\begin{cases} 2x + y = 5 \\ x - 2y = 0 \end{cases}$

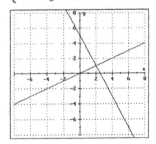

solution: $(2, 1)$

22. $\begin{cases} 3x + y = 4 \quad (1) \\ 2x - 3y = -1 \quad (2) \end{cases}$

Substitute $y = -3x + 4$ from (1) into (2):

$$2x - 3\mathbf{y} = -1$$
$$2x - 3(\mathbf{-3x + 4}) = -1$$
$$2x + 9x - 12 = -1$$
$$11x = 11$$
$$x = 1$$

Substitute this and solve for y:

$$y = -3x + 4$$
$$y = -3(1) + 4$$
$$y = 1$$

Solution: $(1, 1)$

23.
$$\begin{array}{l} x + 2y = -2 \\ 2x - y = 6 \end{array} \Rightarrow \times 2 \quad \begin{array}{r} x + 2y = -2 \\ 4x - 2y = 12 \\ \hline 5x = 10 \\ x = 2 \end{array}$$

Substitute and solve for y:

$$x + 2y = -2$$
$$2 + 2y = -2$$
$$2y = -4$$
$$y = -2$$

The solution is $(2, -2)$.

24.
$$\begin{array}{l} \dfrac{x}{10} + \dfrac{y}{5} = \dfrac{1}{2} \Rightarrow \times 10 \quad x + 2y = 5 \\ \dfrac{x}{2} - \dfrac{y}{5} = \dfrac{13}{10} \Rightarrow \times 10 \quad 5x - 2y = 13 \\ \hline 6x = 18 \\ x = 3 \end{array}$$

Substitute and solve for y:

$$x + 2y = 5$$
$$3 + 2y = 5$$
$$2y = 2$$
$$y = 1$$

The solution is $(3, 1)$.

25.
$$\begin{array}{ll} (1) & x + y + z = 1 \\ (2) & 2x - y - z = -4 \\ (3) & x - 2y + z = 4 \end{array} \qquad \begin{array}{ll} (1) & x + y + z = 1 \\ (2) & 2x - y - z = -4 \\ (4) & 3x = -3 \\ \hline & x = -1 \end{array} \qquad \begin{array}{ll} (2) & 2x - y - z = -4 \\ (3) & x - 2y + z = 4 \\ (5) & \overline{3x - 3y = 0} \end{array}$$

$$\begin{array}{r} 3x - 3y = 0 \\ 3(-1) - 3y = 0 \\ -3 - 3y = 0 \\ -3y = 3 \\ y = -1 \end{array} \qquad \begin{array}{r} x + y + z = 1 \\ -1 + (-1) + z = 1 \\ -2 + z = 1 \\ z = 3 \end{array}$$

Solution:

$\boxed{(-1, -1, 3)}$

26.

$$
\begin{array}{lll}
(1)\ x+2y+3z=1 & (1)\quad x+2y+3z=1 & (1)\quad x+2y+3z=1 \\
(2)\ 3x+2y+z=-1 & -3\cdot(2)\ -9x-6y-3z=3 & -3\cdot(3)\ -6x-9y-3z=6 \\
(3)\ 2x+3y+z=-2 & (4)\ \overline{-8x-4y\ \ \ \ \ =4} & (5)\ \overline{-5x-7y\ \ \ \ \ =7}
\end{array}
$$

$$
\begin{array}{ll}
-8x-4y=4 \Rightarrow \times 7 & -56x-28y=\ \ 28 \\
-5x-7y=7 \Rightarrow \times(-4) & \underline{20x+28y=-28} \\
& -36x\ \ \ \ \ \ =\ \ \ 0 \\
& \quad\ x\ \ \ \ \ \ \ =\ \ \ 0
\end{array}
\qquad
\begin{array}{l}
-5x-7y=7 \\
-5(0)-7y=7 \\
\quad y=-1
\end{array}
$$

$$
2x+3y+z=-2
$$
$$
2(0)+3(-1)+z=-2
$$
$$
-3+z=-2
$$
$$
z=1 \quad \boxed{\text{The solution is } (0,-1,1).}
$$

27. $\begin{vmatrix} 3 & -2 \\ 1 & -1 \end{vmatrix} = 3(-1)-(-2)(1) = -3-(-2) = -3+2 = -1$

28. $\begin{vmatrix} 2 & 3 & -1 \\ -1 & -1 & 2 \\ 4 & 1 & -1 \end{vmatrix} = 2\begin{vmatrix} -1 & 2 \\ 1 & -1 \end{vmatrix} - 3\begin{vmatrix} -1 & 2 \\ 4 & -1 \end{vmatrix} + (-1)\begin{vmatrix} -1 & -1 \\ 4 & 1 \end{vmatrix} = 2(-1)-3(-7)-1(3) = 16$

29. $x = \dfrac{\begin{vmatrix} -1 & -3 \\ -7 & 4 \end{vmatrix}}{\begin{vmatrix} 4 & -3 \\ 3 & 4 \end{vmatrix}} = \dfrac{-4-(21)}{16-(-9)} = \dfrac{-25}{25} = -1;\ y = \dfrac{\begin{vmatrix} 4 & -1 \\ 3 & -7 \end{vmatrix}}{\begin{vmatrix} 4 & -3 \\ 3 & 4 \end{vmatrix}} = \dfrac{-28-(-3)}{25} = \dfrac{-25}{25} = -1$

solution: $(-1,-1)$

30. $D = \begin{vmatrix} 1 & -2 & -1 \\ 3 & 1 & -1 \\ 2 & -1 & 1 \end{vmatrix} = 1\begin{vmatrix} 1 & -1 \\ -1 & 1 \end{vmatrix} - (-2)\begin{vmatrix} 3 & -1 \\ 2 & 1 \end{vmatrix} + (-1)\begin{vmatrix} 3 & 1 \\ 2 & -1 \end{vmatrix} = 1(0)+2(5)-1(-5)$

$$= 15$$

$N_x = \begin{vmatrix} -2 & -2 & -1 \\ 6 & 1 & -1 \\ -1 & -1 & 1 \end{vmatrix} = -2\begin{vmatrix} 1 & -1 \\ -1 & 1 \end{vmatrix} - (-2)\begin{vmatrix} 6 & -1 \\ -1 & 1 \end{vmatrix} + (-1)\begin{vmatrix} 6 & 1 \\ -1 & -1 \end{vmatrix} = 15$

$N_y = \begin{vmatrix} 1 & -2 & -1 \\ 3 & 6 & -1 \\ 2 & -1 & 1 \end{vmatrix} = 1\begin{vmatrix} 6 & -1 \\ -1 & 1 \end{vmatrix} - (-2)\begin{vmatrix} 3 & -1 \\ 2 & 1 \end{vmatrix} + (-1)\begin{vmatrix} 3 & 6 \\ 2 & -1 \end{vmatrix} = 30$

$N_z = \begin{vmatrix} 1 & -2 & -2 \\ 3 & 1 & 6 \\ 2 & -1 & -1 \end{vmatrix} = 1\begin{vmatrix} 1 & 6 \\ -1 & -1 \end{vmatrix} - (-2)\begin{vmatrix} 3 & 6 \\ 2 & -1 \end{vmatrix} + (-2)\begin{vmatrix} 3 & 1 \\ 2 & -1 \end{vmatrix} = -15$

$x = \dfrac{N_x}{D} = \dfrac{15}{15} = 1;\ y = \dfrac{N_y}{D} = \dfrac{30}{15} = 2;\ z = \dfrac{N_z}{D} = \dfrac{-15}{15} = -1 \Rightarrow$ solution: $(1,2,-1)$

31. $-3(x-4) \geq x - 32$
$-3x + 12 \geq x - 32$
$-4x \geq -44$
$x \leq 11$

32. $-8 < -3x + 1 < 10$
$-9 < -3x < 9$
$3 > x > -3$, or $-3 < x < 3$

33. $|4x - 3| = 9$
$4x - 3 = 9$ **or** $4x - 3 = -9$
$4x = 12$ $4x = -6$
$x = 3$ $x = -\frac{3}{2}$

34. $|2x - 1| = |3x + 4|$
$2x - 1 = 3x + 4$ **or** $2x - 1 = -(3x + 4)$
$-x = 5$ $2x - 1 = -3x - 4$
$x = -5$ $5x = -3$
                    $x = -\frac{3}{5}$

35. $|3x - 2| \leq 4$
$-4 \leq 3x - 2 \leq 4$
$-2 \leq 3x \leq 6$
$-\dfrac{2}{3} \leq x \leq 2$

36. $|2x + 3| - 1 > 4$
$|2x + 3| > 5$
$2x + 3 < -5$ **or** $2x + 3 > 5$
$2x < -8$ $2x > 2$
$x < -4$ $x > 1$

37. $2x - 3y \leq 12$

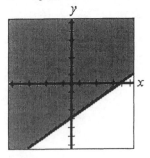

38. $3 > x \geq -2$, or $-2 \leq x < 3$

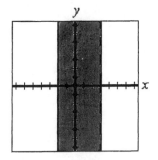

39. $\begin{cases} 3x - 2y < 6 \\ y < -x + 2 \end{cases}$

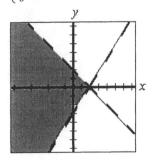

40. $\begin{cases} y < x + 2 \\ 3x + y \leq 6 \end{cases}$

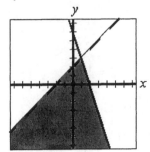

41. Let x = the number of mice used, and let y = the number of rats used.
Form the following system of constraints:
$$\begin{cases} x \geq 0 \\ y \geq 0 \\ 12x + 8y \leq 240 \quad \text{(Maze 1)} \\ 10x + 15y \leq 300 \quad \text{(Maze 2)} \end{cases}$$

Let the objective function be the total number of animals: $P = x + y$.

Vertex	$P = x + y$
$(0,0)$	$= 0$
$(20,0)$	$= 20$
$(12,12)$	$= 24$
$(0,20)$	$= 20$

The maximum number of animals that can be used in both mazes is 24, resulting from using 12 mice and 12 rats.

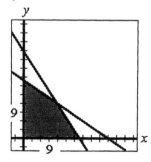

Exercise 5.1 (page 272)

1. $a^3 a^2 = a^5$

3. $\dfrac{3(y^3)^{10}}{y^3 y^4} = \dfrac{3y^{30}}{y^7} = 3y^{23}$

5. $114{,}000{,}000 = 1.14 \times 10^8$

7. sum; whole

9. binomial

11. one

13. monomial

15. trinomial

17. binomial

19. monomial

21. deg $= 2$

23. deg $= 8$

25. deg $= 3 + 3 + 4 = 10$

27. deg $= 0$

29. $3x - 2x^4 + 7 - 5x^2 = -2x^4 - 5x^2 + 3x + 7$

31. $a^2 x - ax^3 + 7a^3 x^5 - 5a^3 x^2 = 7a^3 x^5 - ax^3 - 5a^3 x^2 + a^2 x$

33. $4y^2 - 2y^5 + 7y - 5y^3 = 7y + 4y^2 - 5y^3 - 2y^5$

35. $5x^3 y^6 + 2x^4 y - 5x^3 y^3 + x^5 y^7 - 2y^4 = 2x^4 y - 5x^3 y^3 - 2y^4 + 5x^3 y^6 + x^5 y^7$

37. $P(0) = 2(0)^2 + 0 + 2 = 2(0) + 2 = 2$

39. $P(-2) = 2(-2)^2 + (-2) + 2 = 2(4) + 0 = 8$

41. $h = f(t) = f(0) = -16(0)^2 + 64(0) = 0 + 0 = 0$ ft

43. $h = f(t) = f(2) = -16(2)^2 + 64(2) = -16(4) + 128 = -64 + 128 = 64$ ft

45. $x^2 + y^2 = 2^2 + (-3)^2 = 4 + 9 = 13$

47. $x^3 - y^3 = 2^3 - (-3)^3 = 8 - (-27) = 35$

49. $3x^2 y + xy^3 = 3(2)^2(-3) + 2(-3)^3 = 3(4)(-3) + 2(-27) = -36 - 54 = -90$

SECTION 5.1

51. $-2xy^2 + x^2y = -2(2)(-3)^2 + (2)^2(-3) = -2(2)(9) + 4(-3) = -36 - 12 = -48$

Problems 53-57 are to be solved using a calculator. The keystrokes needed to solve each problem using a TI-83 graphing calculator appear in each solution. There may be other solutions. Keystrokes for other calculators may be slightly different.

53. $x^2y = (3.7)^2(-2.5) \Rightarrow$ [3] [.] [7] [x^2] [×] [(-)] [2] [.] [5] [ENTER] $\{-34.225\}$

55. $\dfrac{x^2}{z^2} = \dfrac{(3.7)^2}{(8.9)^2} \Rightarrow$ [3] [.] [7] [x^2] [÷] [8] [.] [9] [x^2] [ENTER] $\{0.17283171\}$

57. $\dfrac{x+y+z}{xyz} = \dfrac{3.7+(-2.5)+8.9}{(3.7)(-2.5)(8.9)} \Rightarrow$ [(] [3] [.] [7] [+] [(-)] [2] [.] [5] [+] [8] [.] [9] [)]

[÷] [(] [3] [.] [7] [×] [(-)] [2] [.] [5] [×] [8] [.] [9] [)] [ENTER] $\{-0.12268448\}$

59. $f(x) = x^2 + 2$
Shift $y = x^2$ up 2.

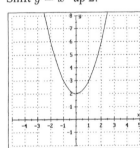

61. $f(x) = -x^3$
Reflect $y = x^3$ about x-axis.

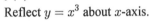
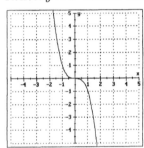

63. $f(x) = -x^3 + x$

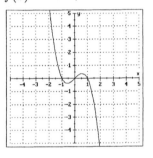

65. $f(x) = x^2 - 2x + 1$

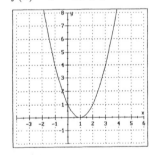

67. $f(x) = 2.75x^2 - 4.7x + 1.5$

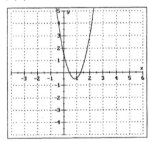

69. $f(x) = 0.25x^2 - 0.5x - 2.5$

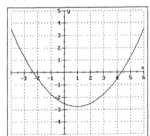

71. $d = f(v) = f(30) = 0.04(30)^2 + 0.9(30) = 0.04(900) + 27 = 36 + 27 = 63$ ft

73. $d = f(v) = f(60) = 0.04(60)^2 + 0.9(60) = 0.04(3600) + 54 = 144 + 54 = 198$ ft

75. $c = f(x) = f(1) = -2(1)^2 + 12(1) = -2(1) + 12 = -2 + 12 = 10$ in.^{2}

77. $c = f(x) = f(3) = -2(3)^2 + 12(3) = -2(9) + 36 = -18 + 36 = 18$ in.^{2}

122

79. a. $h(0) = 0.001(0)^3 - 0.12(0)^2 + 3.6(0) + 10 = 0 - 0 + 0 + 10 = 10$ ft

b. $h(10) = 0.001(10)^3 - 0.12(10)^2 + 3.6(10) + 10$
$= 0.001(1000) - 0.12(100) + 36 + 10 = 1 - 12 + 36 + 10 = 35$ ft

c. $h(40) = 0.001(40)^3 - 0.12(40)^2 + 3.6(40) + 10$
$= 0.001(64000) - 0.12(1600) + 144 + 10 = 64 - 192 + 144 + 10 = 26$ ft

81. Answers may vary.

83. $P(2) + P(3) = [2^2 - 5(2)] + [3^2 - 5(3)] = [4 - 10] + [9 - 15] = -6 + (-6) = -12$
$P(2 + 3) = P(5) = 5^2 - 5(5) = 0 \Rightarrow$ They are not equal.

85. $P(P(0)) = P(0^2 - 2(0) - 3) = P(-3) = (-3)^2 - 2(-3) - 3 = 9 + 6 - 3 = 12$

87. Answers may vary.

Exercise 5.2 (page 278)

1. $2x + 3 \le 11$
$2x \le 8$
$x \le 4$
solution set: $(-\infty, 4]$

3. $|x - 4| < 5$
$-5 < x - 4 < 5$
$-1 < x < 9$
solution set: $(-1, 9)$

5. exponents

7. coefficients

9. like terms, $3x + 7x = 10x$

11. unlike terms

13. like terms, $3r^2t^3 - 8r^2t^3 = -5r^2t^3$

15. unlike terms

17. $8x + 4x = 12x$

19. $5x^3y^2z - 3x^3y^2z = 2x^3y^2z$

21. $-2x^2y^3 + 3xy^4 - 5x^2y^3 = -7x^2y^3 + 3xy^4$

23. $(3x^2y)^2 + 2x^4y^2 - x^4y^2 = 9x^4y^2 + 2x^4y^2 - x^4y^2 = 10x^4y^2$

25. $(3x^2 + 2x + 1) + (-2x^2 - 7x + 5) = 3x^2 + 2x + 1 - 2x^2 - 7x + 5 = x^2 - 5x + 6$

27. $(-a^2 + 2a + 3) - (4a^2 - 2a - 1) = -a^2 + 2a + 3 - 4a^2 + 2a + 1 = -5a^2 + 4a + 4$

29. $(7y^3 + 4y^2 + y + 3) + (-8y^3 - y + 3) = 7y^3 + 4y^2 + y + 3 - 8y^3 - y + 3 = -y^3 + 4y^2 + 6$

31. $(3x^2 + 4x - 3) + (2x^2 - 3x - 1) - (x^2 + x + 7) = 3x^2 + 4x - 3 + 2x^2 - 3x - 1 - x^2 - x - 7$
$= 4x^2 - 11$

33. $\left(3x^3 - 2x + 3\right) + \left(4x^3 + 3x^2 - 2\right) + \left(-4x^3 - 3x^2 + x + 12\right)$
$$= 3x^3 - 2x + 3 + 4x^3 + 3x^2 - 2 - 4x^3 - 3x^2 + x + 12$$
$$= 3x^3 - x + 13$$

35. $\left(3y^2 - 2y + 4\right) + \left[\left(2y^2 - 3y + 2\right) - \left(y^2 + 4y + 3\right)\right]$
$$= 3y^2 - 2y + 4 + \left[2y^2 - 3y + 2 - y^2 - 4y - 3\right]$$
$$= 3y^2 - 2y + 4 + \left[y^2 - 7y - 1\right]$$
$$= 3y^2 - 2y + 4 + y^2 - 7y - 1 = 4y^2 - 9y + 3$$

37.
$$\begin{array}{r} 3x^3 - 2x^2 + 4x - 3 \\ -2x^3 + 3x^2 + 3x - 2 \\ +\quad 5x^3 - 7x^2 + 7x - 12 \\ \hline 6x^3 - 6x^2 + 14x - 17 \end{array}$$

39.
$$\begin{array}{r} -2y^4 - 2y^3 + 4y^2 - 3y + 10 \\ -3y^4 + 7y^3 - y^2 + 14y - 3 \\ -3y^3 - 5y^2 - 5y + 7 \\ +\quad -4y^4 + y^3 - 13y^2 + 14y - 2 \\ \hline -9y^4 + 3y^3 - 15y^2 + 20y + 12 \end{array}$$

41.
$$\begin{array}{r} 3x^2 - 4x + 17 \\ -\quad 2x^2 + 4x - 5 \\ \hline 3x^2 - 4x + 17 \\ +\quad -2x^2 - 4x + 5 \\ \hline x^2 - 8x + 22 \end{array}$$

43.
$$\begin{array}{r} -5y^3 + 4y^2 - 11y + 3 \\ -\quad -2y^3 - 14y^2 + 17y - 32 \\ \hline -5y^3 + 4y^2 - 11y + 3 \\ +\quad 2y^3 + 14y^2 - 17y + 32 \\ \hline -3y^3 + 18y^2 - 28y + 35 \end{array}$$

45. $3(x + 2) + 2(x - 5) = 3x + 6 + 2x - 10$
$$= 5x - 4$$

47. $-6(t - 4) - 5(t - 1) = -6t + 24 - 5t + 5$
$$= -11t + 29$$

49. $2\left(x^3 + x^2\right) + 3\left(2x^3 - x^2\right) = 2x^3 + 2x^2 + 6x^3 - 3x^2 = 8x^3 - x^2$

51. $-3(2m - n) + 2(m - 3n) = -6m + 3n + 2m - 6n = -4m - 3n$

53. $-5\left(2x^3 + 7x^2 + 4x\right) - 2\left(3x^3 - 4x^2 - 4x\right) = -10x^3 - 35x^2 - 20x - 6x^3 + 8x^2 + 8x$
$$= -16x^3 - 27x^2 - 12x$$

55. $4\left(3z^2 - 4z + 5\right) + 6\left(-2z^2 - 3z + 4\right) - 2\left(4z^2 + 3z - 5\right)$
$$= 12z^2 - 16z + 20 - 12z^2 - 18z + 24 - 8z^2 - 6z + 10$$
$$= -8z^2 - 40z + 54$$

57. $5\left(2a^2 + 4a - 2\right) - 2\left(-3a^2 - a + 12\right) - 2\left(a^2 + 3a - 5\right)$
$$= 10a^2 + 20a - 10 + 6a^2 + 2a - 24 - 2a^2 - 6a + 10$$
$$= 14a^2 + 16a - 24$$

59. $x + (x + 1) + (x + 2) = x + x + 1 + x + 2 = 3x + 3$

61. $\left(2x^2 + 5x + 1\right) - \left(x^2 - 5\right) = 2x^2 + 5x + 1 - x^2 + 5 = x^2 + 5x + 6$ meters

63. $f(x) = 1100x + 125000;\ f(10) = 1100(10) + 125000 = 11000 + 125000 = \$136{,}000$

65. $y = (1100x + 125000) + (1400x + 150000) = 2500x + 275000$

67. $y = -2100x + 16600$

69. $y = (-2100x + 16600) + (-2700x + 19200)$
$= -4800x + 35800$

71. Answers may vary.

73. $\left[(-2x^2 - x + 7) + (5x^2 + 3x - 1)\right] - (3x^2 + 4x - 3)$
$= \left[-2x^2 - x + 7 + 5x^2 + 3x - 1\right] - 3x^2 - 4x + 3$
$= -2x^2 - x + 7 + 5x^2 + 3x - 1 - 3x^2 - 4x + 3$
$= -2x + 9$

75. $\left[(2x^2 - 4x + 3) - (8x^2 + 5x - 3)\right] + (-2x^2 + 7x - 4)$
$= \left[2x^2 - 4x + 3 - 8x^2 - 5x + 3\right] - 2x^2 + 7x - 4$
$= 2x^2 - 4x + 3 - 8x^2 - 5x + 3 - 2x^2 + 7x - 4$
$= -8x^2 - 2x + 2$

Exercise 5.3 (page 288)

1. $|3a - b| = |3(-2) - 4| = |-6 - 4| = |-10| = 10$

3. $-|a^2b - b^0| = -|(-2)^2(4) - 1| = -|4(4) - 1| = -|16 - 1| = -|15| = -15$

5. The current value of ABC $= \$126.5$ per share, and the current value of WD $= \$73.5$ per share. Total value $= 200(126.5) + 350(73.5) = 25300 + 25725 = \$51,025$.

7. variable

9. term

11. $x^2 + 2xy + y^2$

13. $x^2 - y^2$

15. $(2a^2)(-3ab) = -6a^3b$

17. $(-3ab^2c)(5ac^2) = -15a^2b^2c^3$

19. $(4a^2b)(-5a^3b^2)(6a^4) = -120a^9b^3$

21. $(5x^3y^2)^4\left(\frac{1}{5}x^{-2}\right)^2 = (625x^{12}y^8)\left(\frac{1}{25}x^{-4}\right)$
$= 25x^8y^8$

23. $(-5xx^2)(-3xy)^4 = (-5x^3)(81x^4y^4)$
$= -405x^7y^4$

25. $3(x + 2) = 3x + 3(2) = 3x + 6$

27. $-a(a - b) = -a(a) + (-a)(-b)$
$= -a^2 + ab$

29. $3x(x^2 + 3x) = 3x(x^2) + 3x(3x) = 3x^3 + 9x^2$

31. $-2x(3x^2 - 3x + 2) = -2x(3x^2) + (-2x)(-3x) + (-2x)(2) = -6x^3 + 6x^2 - 4x$

33. $5a^2b^3(2a^4b - 5a^0b^3) = 5a^2b^3(2a^4b) + 5a^2b^3(-5a^0b^3) = 10a^6b^4 - 25a^2b^6$

35. $7rst(r^2 + s^2 - t^2) = 7rst(r^2) + 7rst(s^2) + 7rst(-t^2) = 7r^3st + 7rs^3t - 7rst^3$

37. $4m^2n(-3mn)(m + n) = -12m^3n^2(m + n) = -12m^4n^2 - 12m^3n^3$

39. $(x + 2)(x + 3) = x^2 + 3x + 2x + 6$
$= x^2 + 5x + 6$

41. $(z - 7)(z - 2) = z^2 - 2z - 7z + 14$
$= z^2 - 9z + 14$

43. $(2a + 1)(a - 2) = 2a^2 - 4a + a - 2$
$= 2a^2 - 3a - 2$

45. $(3t - 2)(2t + 3) = 6t^2 + 9t - 4t - 6$
$= 6t^2 + 5t - 6$

47. $(3y - z)(2y - z) = 6y^2 - 3yz - 2yz + z^2 = 6y^2 - 5yz + z^2$

49. $(2x - 3y)(x + 2y) = 2x^2 + 4xy - 3xy - 6y^2 = 2x^2 + xy - 6y^2$

51. $(3x + y)(3x - 3y) = 9x^2 - 9xy + 3xy - 3y^2 = 9x^2 - 6xy - 3y^2$

53. $(4a - 3b)(2a + 5b) = 8a^2 + 20ab - 6ab - 15b^2 = 8a^2 + 14ab - 15b^2$

55. $(x + 2)^2 = (x + 2)(x + 2) = x^2 + 2x + 2x + 4 = x^2 + 4x + 4$

57. $(a - 4)^2 = (a - 4)(a - 4) = a^2 - 4a - 4a + 16 = a^2 - 8a + 16$

59. $(2a + b)^2 = (2a + b)(2a + b) = 4a^2 + 2ab + 2ab + b^2 = 4a^2 + 4ab + b^2$

61. $(2x - y)^2 = (2x - y)(2x - y) = 4x^2 - 2xy - 2xy + y^2 = 4x^2 - 4xy + y^2$

63. $(x + 2)(x - 2) = x^2 - 2x + 2x - 4$
$= x^2 - 4$

65. $(a + b)(a - b) = a^2 - ab + ab - b^2$
$= a^2 - b^2$

67. $(2x + 3y)(2x - 3y) = 4x^2 - 6xy + 6xy - 9y^2 = 4x^2 - 9y^2$

69. $(x - y)(x^2 + xy + y^2) = x(x^2 + xy + y^2) - y(x^2 + xy + y^2)$
$= x^3 + x^2y + xy^2 - x^2y - xy^2 - y^3 = x^3 - y^3$

71. $(3y + 1)(2y^2 + 3y + 2) = 3y(2y^2 + 3y + 2) + 1(2y^2 + 3y + 2)$
$= 6y^3 + 9y^2 + 6y + 2y^2 + 3y + 2 = 6y^3 + 11y^2 + 9y + 2$

73. $(2a - b)(4a^2 + 2ab + b^2) = 2a(4a^2 + 2ab + b^2) - b(4a^2 + 2ab + b^2)$
$= 8a^3 + 4a^2b + 2ab^2 - 4a^2b - 2ab^2 - b^3 = 8a^3 - b^3$

75. $(2x - 1)[2x^2 - 3(x + 2)] = (2x - 1)(2x^2 - 3x - 6) = 2x(2x^2 - 3x - 6) - 1(2x^2 - 3x - 6)$
$= 4x^3 - 6x^2 - 12x - 2x^2 + 3x + 6$
$= 4x^3 - 8x^2 - 9x + 6$

77. $(a + b)(a - b)(a - 3b) = (a^2 - b^2)(a - 3b) = a^3 - 3a^2b - ab^2 + 3b^3$

SECTION 5.3

79. $x^3(2x^2 + x^{-2}) = 2x^5 + x$

81. $x^3 y^{-6} z^{-2}(3x^{-2}y^2 z - x^3 y^{-4}) = 3xy^{-4}z^{-1} - x^6 y^{-10}z^{-2} = \dfrac{3x}{y^4 z} - \dfrac{x^6}{y^{10}z^2}$

83. $(x^{-1} + y)(x^{-1} - y) = x^{-2} - x^{-1}y + x^{-1}y - y^2 = \dfrac{1}{x^2} - y^2$

85. $(2x^{-3} + y^3)(2x^3 - y^{-3}) = 4x^0 - 2x^{-3}y^{-3} + 2x^3 y^3 - y^0 = 4 - \dfrac{2}{x^3 y^3} + 2x^3 y^3 - 1$

$$= 2x^3 y^3 - \dfrac{2}{x^3 y^3} + 3$$

87. $x^n(x^{2n} - x^n) = x^{3n} - x^{2n}$

89. $(x^n + 1)(x^n - 1) = x^{2n} - x^n + x^n - 1$
$$= x^{2n} - 1$$

91. $(x^n - y^n)(x^n - y^{-n}) = x^{2n} - x^n y^{-n} - x^n y^n + y^0 = x^{2n} - \dfrac{x^n}{y^n} - x^n y^n + 1$

93. $(x^{2n} + y^{2n})(x^{2n} - y^{2n}) = x^{4n} - x^{2n}y^{2n} + x^{2n}y^{2n} - y^{4n} = x^{4n} - y^{4n}$

95. $(x^n + y^n)(x^n + 1) = x^{2n} + x^n + x^n y^n + y^n$

97. $3x(2x + 4) - 3x^2 = 6x^2 + 12x - 3x^2$
$$= 3x^2 + 12x$$

99. $3pq - p(p - q) = 3pq - p^2 + pq$
$$= -p^2 + 4pq$$

101. $2m(m - n) - (m + n)(m - 2n) = 2m^2 - 2mn - (m^2 - 2mn + mn - 2n^2)$
$$= 2m^2 - 2mn - m^2 + 2mn - mn + 2n^2$$
$$= m^2 - mn + 2n^2$$

103. $(x + 3)(x - 3) + (2x - 1)(x + 2) = x^2 - 3x + 3x - 9 + 2x^2 + 4x - x - 2$
$$= 3x^2 + 3x - 11$$

105. $(3x - 4)^2 - (2x + 3)^2 = (3x - 4)(3x - 4) - (2x + 3)(2x + 3)$
$$= 9x^2 - 12x - 12x + 16 - (4x^2 + 6x + 6x + 9)$$
$$= 9x^2 - 24x + 16 - 4x^2 - 6x - 6x - 9$$
$$= 5x^2 - 36x + 7$$

107. $3(x - 3y)^2 + 2(3x + y)^2 = 3(x - 3y)(x - 3y) + 2(3x + y)(3x + y)$
$$= 3(x^2 - 3xy - 3xy + 9y^2) + 2(9x^2 + 3xy + 3xy + y^2)$$
$$= 3x^2 - 9xy - 9xy + 27y^2 + 18x^2 + 6xy + 6xy + 2y^2$$
$$= 21x^2 - 6xy + 29y^2$$

109. $5(2y - z)^2 + 4(y + 2z)^2 = 5(2y - z)(2y - z) + 4(y + 2z)(y + 2z)$
$$= 5(4y^2 - 2yz - 2yz + z^2) + 4(y^2 + 2yz + 2yz + 4z^2)$$
$$= 20y^2 - 10yz - 10yz + 5z^2 + 4y^2 + 8yz + 8yz + 16z^2$$
$$= 24y^2 - 4yz + 21z^2$$

111. $(3.21x - 7.85)(2.87x + 4.59) = 9.2127x^2 + 14.7339x - 22.5295x - 36.0315$
$$= 9.2127x^2 - 7.7956x - 36.0315$$

113. $(-17.3y + 4.35)^2 = (-17.3y + 4.35)(-17.3y + 4.35)$
$$= 299.29y^2 - 75.255y - 75.255y + 18.9225$$
$$= 299.29y^2 - 150.51y + 18.9225$$

115. a. Area of large square $= x^2$ **b.** Area I $= (x - y)^2$
   c. Area II $= y(x - y) = xy - y^2$ **d.** Area III $= y(x - y) = xy - y^2$
   e. Area IV $= y^2$
   f.   Area I = Area of large square − Area II − Area III − Area IV
$$(x - y)^2 = x^2 - (xy - y^2) - (xy - y^2) - y^2$$
$$(x - y)^2 = x^2 - xy + y^2 - xy + y^2 - y^2$$
$$(x - y)^2 = x^2 - 2xy + y^2$$

117. $x = -\frac{1}{5}(375) + 90 = -75 + 90 = 15$ **119.** $r = p\left(-\frac{1}{5}p + 90\right) = -\frac{1}{5}p^2 + 90p$

121. $A = lw = (2x - 3)(2x - 3) = 4x^2 - 6x - 6x + 9 = 4x^2 - 12x + 9 \text{ ft}^2$

123. $A = \frac{1}{2}bh = \frac{1}{2}(b + 5)(b - 2) = \frac{1}{2}(b^2 - 2b + 5b - 10) = \frac{1}{2}(b^2 + 3b - 10) \text{ in.}^2$

125. Answers may vary.

127. $0.35 \times 10^7 + 1.96 \times 10^7 = (0.35 + 1.96) \times 10^7 = 2.31 \times 10^7$

Exercise 5.4   (page 297)

1. $(a + 4)(a - 4) = a^2 - 16$ **3.** $(4r^2 + 3s)(4r^2 - 3s) = 16r^4 - 9s^2$

5. $(m + 4)(m^2 - 4m + 16) = m^3 - 4m^2 + 16m + 4m^2 - 16m + 64 = m^3 + 64$

7.   factoring   **9.**   greatest common factor   **11.** $6 = 2 \cdot 3$

13. $135 = 3^3 \cdot 5$ **15.** $128 = 2^7$ **17.** $325 = 5^2 \cdot 13$

19. $36 = 2^2 \cdot 3^2; 48 = 2^4 \cdot 3; \text{gcf} = 2^2 \cdot 3 = 12$ **21.** $42 = 2 \cdot 3 \cdot 7; 36 = 2^2 \cdot 3^2; 98 = 2 \cdot 7^2$
                                                                           $\text{gcf} = 2$

23. $4a^2b = 2^2 \cdot a^2b; 8a^3c = 2^3 \cdot a^3c; \text{gcf} = 2^2 \cdot a^2 = 4a^2$

25. $18x^4y^3z^2 = 2 \cdot 3^2 \cdot x^4y^3z^2$; $-12xy^2z^3 = -1 \cdot 2^2 \cdot 3xy^2z^2$; $\gcf = 2 \cdot 3 \cdot x \cdot y^2 \cdot z^2 = 6xy^2z^2$

27. $3a - 12 = 3(a - \underline{4})$ **29.** $8z^2 + 2z = 2z(4z + \underline{1})$

31. $2x + 8 = \mathbf{2} \cdot x + \mathbf{2} \cdot 4 = \mathbf{2}(x + 4)$ **33.** $2x^2 - 6x = \mathbf{2x} \cdot x - \mathbf{2x} \cdot 3 = \mathbf{2x}(x - 3)$

35. $5xy + 12ab^2 \Rightarrow$ prime **37.** $15x^2y - 10x^2y^2 = \mathbf{5x^2y} \cdot 3 - \mathbf{5x^2y} \cdot 2y$
$= \mathbf{5x^2y}(3 - 2y)$

39. $63x^3y^2 + 81x^2y^4 = \mathbf{9x^2y^2} \cdot 7x + \mathbf{9x^2y^2} \cdot 9y^2 = \mathbf{9x^2y^2}(7x + 9y^2)$

41. $14r^2s^3 + 15t^6 \Rightarrow$ prime

43. $27z^3 + 12z^2 + 3z = \mathbf{3z} \cdot 9z^2 + \mathbf{3z} \cdot 4z + \mathbf{3z} \cdot 1 = \mathbf{3z}(9z^2 + 4z + 1)$

45. $24s^3 - 12s^2t + 6st^2 = \mathbf{6s}(4s^2) - \mathbf{6s}(2st) + \mathbf{6s}(t^2) = \mathbf{6s}(4s^2 - 2st + t^2)$

47. $45x^{10}y^3 - 63x^7y^7 + 81x^{10}y^{10} = \mathbf{9x^7y^3}(5x^3) - \mathbf{9x^7y^3}(7y^4) + \mathbf{9x^7y^3}(9x^3y^7)$
$= \mathbf{9x^7y^3}(5x^3 - 7y^4 + 9x^3y^7)$

49. $25x^3 - 14y^3 + 36x^3y^3 \Rightarrow$ prime

51. $-3a - 6 = (\mathbf{-3})(a) + (\mathbf{-3})(2)$ **53.** $-3x^2 - x = (\mathbf{-x})(3x) + (\mathbf{-x})(1)$
$= \mathbf{-3}(a + 2)$ $= \mathbf{-x}(3x + 1)$

55. $-6x^2 - 3xy = (\mathbf{-3x})(2x) + (\mathbf{-3x})(y) = \mathbf{-3x}(2x + y)$

57. $-18a^2b - 12ab^2 = (\mathbf{-6ab})(3a) + (\mathbf{-6ab})(2b) = \mathbf{-6ab}(3a + 2b)$

59. $-63u^3v^6z^9 + 28u^2v^7z^2 - 21u^3v^3z^4 = -7u^2v^3z^2(9uv^3z^7 - 4v^4 + 3uz^2)$

61. $x^{n+2} + x^{n+3} = x^2(x^{n+2-2} + x^{n+3-2})$ **63.** $2y^{n+2} - 3y^{n+3} = y^n(2y^{n+2-n} - 3y^{n+3-n})$
$= x^2(x^n + x^{n+1})$ $= y^n(2y^2 - 3y^3)$

65. $x^4 - 5x^6 = x^{-2}\left(x^{4-(-2)} - 5x^{6-(-2)}\right)$ **67.** $t^5 + 4t^{-6} = t^{-3}\left(t^{5-(-3)} + 4t^{-6-(-3)}\right)$
$= x^{-2}\left(x^6 - 5x^8\right)$ $= t^{-3}(t^8 + 4t^{-3})$

69. $8y^{2n} + 12 + 16y^{-2n} = 4y^{-2n}\left(2y^{2n-(-2n)} + 3y^{0-(-2n)} + 4y^{-2n-(-2n)}\right) = 4y^{-2n}(2y^{4n} + 3y^{2n} + 4)$

71. $4(\mathbf{x+y}) + t(\mathbf{x+y}) = (\mathbf{x+y})(4+t)$ **73.** $(\mathbf{a-b})r - (\mathbf{a-b})s = (\mathbf{a-b})(r-s)$

75. $3(\mathbf{m+n+p}) + x(\mathbf{m+n+p}) = (\mathbf{m+n+p})(3+x)$

77. $(x+y)(\mathbf{x+y}) + z(\mathbf{x+y}) = (\mathbf{x+y})[(x+y) + z] = (x+y)(x+y+z)$

79. $(u+v)^2 - (u+v) = (\mathbf{u+v})(\mathbf{u+v}) - 1(\mathbf{u+v}) = (\mathbf{u+v})[(u+v) - 1] = (u+v)(u+v-1)$

81. $-a(x+y)+b(x+y)=(x+y)(-a+b)=-(x+y)(a-b)$

83. $ax+bx+ay+by=x(a+b)+y(a+b)$
$=(a+b)(x+y)$

85. $x^2+yx+2x+2y=x(x+y)+2(x+y)$
$=(x+y)(x+2)$

87. $3c-cd+3d-c^2=3c+3d-c^2-cd=3(c+d)-c(c+d)=(c+d)(3-c)$

89. $a^2-4b+ab-4a=a^2+ab-4a-4b=a(a+b)-4(a+b)=(a+b)(a-4)$

91. $ax+bx-a-b=x(a+b)-1(a+b)=(a+b)(x-1)$

93. $x^2+xy+xz+xy+y^2+zy=x(x+y+z)+y(z+y+z)=(x+y+z)(x+y)$

95. $mpx+mqx+npx+nqx=x(mp+mq+np+nq)=x[m(p+q)+n(p+q)]$
$=x(p+q)(m+n)$

97. $x^2y+xy^2+2xyz+xy^2+y^3+2y^2z=y(x^2+xy+2xz+xy+y^2+2yz)$
$=y[x(x+y+2z)+y(x+y+2z)]$
$=y(x+y+2z)(x+y)$

99. $2n^4p-2n^2-n^3p^2+np+2mn^3p-2mn=n(2n^3p-2n-n^2p^2+p+2mn^2p-2m)$
$=n(2n^3p-n^2p^2+2mn^2p-2n+p-2m)$
$=n[n^2p(2n-p+2m)-1(2n-p+2m)]$
$=n(2n-p+2m)(n^2p-1)$

101.
$$r_1r_2=rr_2+rr_1$$
$$r_1r_2-rr_1=rr_2$$
$$r_1(r_2-r)=rr_2$$
$$r_1=\frac{rr_2}{r_2-r}$$

103.
$$d_1d_2=fd_2+fd_1$$
$$d_1d_2=f(d_2+d_1)$$
$$\frac{d_1d_2}{d_2+d_1}=f,\text{ or }f=\frac{d_1d_2}{d_2+d_1}$$

105. $b^2x^2+a^2y^2=a^2b^2$
$$b^2x^2=a^2b^2-a^2y^2$$
$$b^2x^2=a^2(b^2-y^2)$$
$$\frac{b^2x^2}{b^2-y^2}=a^2,\text{ or }a^2=\frac{b^2x^2}{b^2-y^2}$$

107. $S(1-r)=a-lr$
$$S-Sr=a-lr$$
$$S-a=Sr-lr$$
$$S-a=r(S-l)$$
$$\frac{S-a}{S-l}=r,\text{ or }r=\frac{S-a}{S-l}$$

109. $H(a+b)=2ab$
$$Ha+Hb=2ab$$
$$Hb=2ab-Ha$$
$$Hb=a(2b-H)$$
$$\frac{Hb}{2b-H}=a,\text{ or }a=\frac{Hb}{2b-H}$$

111. $3xy-x=2y+3$
$$3xy-2y=x+3$$
$$y(3x-2)=x+3$$
$$y=\frac{x+3}{3x-2}$$

113. $2x^3 + 5x^2 - 2x + 8 = x(2x^2 + 5x - 2) + 8 = x[x(2x + 5) - 2] + 8$

115. a. $(2x^2)(6x) = 12x^3$ in.^{2}
b. $(5x)(4x) = 20x^2$ in.^{2}
c. $12x^3 - 20x^2 = 4x^2(3x - 5)$ in.^{2}

117. Answers may vary.

119. Answers may vary.

121. $14 = 2 \cdot 7; 45 = 3^2 \cdot 5; \text{gcf} = 1 \Rightarrow$ relatively prime

123. $60 = 2^2 \cdot 3 \cdot 5; 28 = 2^2 \cdot 7; 36 = 2^2 \cdot 3^2; \text{gcf} = 2^2 \Rightarrow$ not relatively prime

125. $12x^2y = 2^2 \cdot 3 \cdot x^2y; 5ab^3 = 5 \cdot ab^3; 35x^2b^3 = 5 \cdot 7 \cdot x^2b^3; \text{gcf} = 1 \Rightarrow$ relatively prime

Exercise 5.5 (page 305)

1. $(x + 1)(x + 1) = x^2 + x + x + 1 = x^2 + 2x + 1$

3. $(2m + n)(2m + n) = 4m^2 + 2mn + 2mn + n^2 = 4m^2 + 4mn + n^2$

5. $(a + 4)(a + 3) = a^2 + 3a + 4a + 12 = a^2 + 7a + 12$

7. $(4r - 3s)(2r - s) = 8r^2 - 4rs - 6rs + 3s^2 = 8r^2 - 10rs + 3s^2$

9. $1, 4, 9, 16, 25, 36, 49, 64, 81, 100$

11. cannot

13. $(p^2 - pq + q^2)$

15. $x^2 - 4 = x^2 - 2^2 = (x + 2)(x - 2)$

17. $9y^2 - 64 = (3y)^2 - 8^2 = (3y + 8)(3y - 8)$

19. $x^2 + 25 \Rightarrow$ prime (sum of two squares)

21. $625a^2 - 169b^4 = (25a)^2 - (13b^2)^2$
$= (25a + 13b^2)(25a - 13b^2)$

23. $81a^4 - 49b^2 = (9a^2)^2 - (7b)^2$
$= (9a^2 + 7b)(9a^2 - 7b)$

25. $36x^4y^2 - 49z^4 = (6x^2y)^2 - (7z^2)^2 = (6x^2y + 7z^2)(6x^2y - 7z^2)$

27. $(x + y)^2 - z^2 = [(x + y) + z][(x + y) - z]$
$= (x + y + z)(x + y - z)$

29. $(a - b)^2 - c^2 = [(a - b) + c][(a - b) - c]$
$= (a - b + c)(a - b - c)$

31. $x^4 - y^4 = (x^2 + y^2)(x^2 - y^2) = (x^2 + y^2)(x + y)(x - y)$

33. $256x^4y^4 - z^8 = (16x^2y^2 + z^4)(16x^2y^2 - z^4) = (16x^2y^2 + z^4)(4xy + z^2)(4xy - z^2)$

35. $2x^2 - 288 = 2(x^2 - 144)$
$= 2(x + 12)(x - 12)$

37. $2x^3 - 32x = 2x(x^2 - 16)$
$= 2x(x + 4)(x - 4)$

39. $5x^3 - 125x = 5x(x^2 - 25)$
$= 5x(x + 5)(x - 5)$

41. $r^2s^2t^2 - t^2x^4y^2 = t^2(r^2s^2 - x^4y^2)$
$= t^2(rs + x^2y)(rs - x^2y)$

43. $r^3 + s^3 = (r + s)(r^2 - rs + s^2)$

45. $x^3 - 8y^3 = x^3 - (2y)^3 = (x - 2y)[x^2 + x(2y) + (2y)^2] = (x - 2y)(x^2 + 2xy + 4y^2)$

47. $64a^3 - 125b^6 = (4a)^3 - (5b^2)^3 = (4a - 5b^2)[(4a)^2 + (4a)(5b^2) + (5b^2)^2]$
$$= (4a - 5b^2)(16a^2 + 20ab^2 + 25b^4)$$

49. $125x^3y^6 + 216z^9 = (5xy^2)^3 + (6z^3)^3 = (5xy^2 + 6z^3)[(5xy^2)^2 - (5xy^2)(6z^3) + (6z^3)^2]$
$$= (5xy^2 + 6z^3)(25x^2y^4 - 30xy^2z^3 + 36z^6)$$

51. $x^6 + y^6 = (x^2)^3 + (y^2)^3 = (x^2 + y^2)[(x^2)^2 - x^2y^2 + (y^2)^2] = (x^2 + y^2)(x^4 - x^2y^2 + y^4)$

53. $5x^3 + 625 = 5(x^3 + 125) = 5(x^3 + 5^3) = 5(x + 5)(x^2 - 5x + 25)$

55. $4x^5 - 256x^2 = 4x^2(x^3 - 64) = 4x^2(x^3 - 4^3) = 4x^2(x - 4)(x^2 + 4x + 16)$

57. $128u^2v^3 - 2t^3u^2 = 2u^2(64v^3 - t^3) = 2u^2[(4v)^3 - t^3] = 2u^2(4v - t)[(4v)^2 + 4vt + t^2]$
$$= 2u^2(4v - t)(16v^2 + 4vt + t^2)$$

59. $(a + b)x^3 + 27(a + b) = (a + b)(x^3 + 27) = (a + b)(x^3 + 3^3) = (a + b)(x + 3)(x^2 - 3x + 9)$

61. $x^{2m} - y^{4n} = (x^m)^2 - (y^{2n})^2 = (x^m + y^{2n})(x^m - y^{2n})$

63. $100a^{4m} - 81b^{2n} = (10a^{2m})^2 - (9b^n)^2 = (10a^{2m} + 9b^n)(10a^{2m} - 9b^n)$

65. $x^{3n} - 8 = (x^n)^3 - 2^3 = (x^n - 2)[(x^n)^2 + 2x^n + 2^2] = (x^n - 2)(x^{2n} + 2x^n + 4)$

67. $a^{3m} + b^{3n} = (a^m)^3 + (b^n)^3 = (a^m + b^n)[(a^m)^2 - a^mb^n + (b^n)^2] = (a^m + b^n)(a^{2m} - a^mn^n + b^{2n})$

69. $2x^{6m} + 16y^{3m} = 2(x^{6m} + 8y^{3m}) = 2[(x^{2m})^3 + (2y^m)^3]$
$$= 2(x^{2m} + 2y^m)[(x^{2m})^2 - 2x^{2m}y^m + (2y^m)^2]$$
$$= 2(x^{2m} + 2y^m)(x^{4m} - 2x^{2m}y^m + 4y^{2m})$$

71. $a^2 - b^2 + a + b = (a^2 - b^2) + (a + b) = (a + b)(a - b) + 1(a + b)$
$$= (a + b)(a - b + 1)$$

73. $a^2 - b^2 + 2a - 2b = (a^2 - b^2) + 2(a - b) = (a + b)(a - b) + 2(a - b)$
$$= (a - b)(a + b + 2)$$

75. $2x + y + 4x^2 - y^2 = (2x + y) + (4x^2 - y^2) = (2x + y)(1) + (2x + y)(2x - y)$
$$= (2x + y)(1 + 2x - y)$$

77. $0.5gt_1^2 - 0.5gt_2^2 = 0.5g(t_1^2 - t_2^2) = 0.5g(t_1 + t_2)(t_1 - t_2)$

79. $V = \dfrac{4}{3}\pi r_1^3 - \dfrac{4}{3}\pi r_2^3 = \dfrac{4}{3}\pi\left(r_1^3 - r_2^3\right) = \dfrac{4}{3}\pi(r_1 - r_2)\left(r_1^2 + r_1r_2 + r_2^2\right)$

81. Answers may vary.

83. $x^{32} - y^{32} = (x^{16} + y^{16})(x^{16} - y^{16}) = (x^{16} + y^{16})(x^8 + y^8)(x^8 - y^8)$
$$= (x^{16} + y^{16})(x^8 + y^8)(x^4 + y^4)(x^4 - y^4)$$
$$= (x^{16} + y^{16})(x^8 + y^8)(x^4 + y^4)(x^2 + y^2)(x^2 - y^2)$$
$$= (x^{16} + y^{16})(x^8 + y^8)(x^4 + y^4)(x^2 + y^2)(x + y)(x - y)$$

Exercise 5.6 (page 316)

1. $\dfrac{2 + x}{11} = 3$
$2 + x = 33$
$x = 31$

3. $\dfrac{2}{3}(5t - 3) = 38$
$2(5t - 3) = 114$
$10t - 6 = 114$
$10t = 120$
$t = 12$

5. $11r + 6(3 - r) = 3$
$11r + 18 - 6r = 3$
$5r = -15$
$r = -3$

7. $2xy + y^2$

9. $x^2 - y^2$

11. $x + 2$

13. $x - 3$

15. $2a + 1$

17. $2m + 3n$

19. $x^2 + 2x + 1 = (x + 1)(x + 1) = (x + 1)^2$

21. $a^2 - 18a + 81 = (a - 9)(a - 9) = (a - 9)^2$

23. $4y^2 + 4y + 1 = (2y + 1)(2y + 1)$
$= (2y + 1)^2$

25. $9b^2 - 12b + 4 = (3b - 2)(3b - 2)$
$= (3b - 2)^2$

27. $9z^2 + 24z + 16 = (3z + 4)(3z + 4)$
$= (3z + 4)^2$

29. $x^2 + 9x + 8 = (x + 1)(x + 8)$

31. $x^2 - 7x + 10 = (x - 5)(x - 2)$

33. $b^2 + 8b + 18 \Rightarrow$ prime

35. $x^2 - x - 30 = (x - 6)(x + 5)$

37. $a^2 + 5a - 50 = (a + 10)(a - 5)$

39. $y^2 - 4y - 21 = (y - 7)(y + 3)$

41. $3x^2 + 12x - 63 = 3(x^2 + 4x - 21)$
$= 3(x + 7)(x - 3)$

43. $a^2 b^2 - 13ab^2 + 22b^2 = b^2(a^2 - 13a + 22) = b^2(a - 11)(a - 2)$

45. $b^2 x^2 - 12bx^2 + 35x^2 = x^2(b^2 - 12b + 35)$
$= x^2(b - 5)(b - 7)$

47. $-a^2 + 4a + 32 = -(a^2 - 4a - 32)$
$= -(a - 8)(a + 4)$

49. $-3x^2 + 15x - 18 = -3(x^2 - 5x + 6)$
$= -3(x - 2)(x - 3)$

51. $-4x^2 + 4x + 80 = -4(x^2 - x - 20)$
$= -4(x - 5)(x + 4)$

53. $6y^2 + 7y + 2 = (2y + 1)(3y + 2)$

55. $8a^2 + 6a - 9 = (4a - 3)(2a + 3)$

57. $6x^2 - 5x - 4 = (3x - 4)(2x + 1)$

59. $5x^2 + 4x + 1 \Rightarrow$ prime

61. $8x^2 - 10x + 3 = (4x - 3)(2x - 1)$

63. $a^2 - 3ab - 4b^2 = (a - 4b)(a + b)$

65. $2y^2 + yt - 6t^2 = (2y - 3t)(y + 2t)$

67. $3x^3 - 10x^2 + 3x = x(3x^2 - 10x + 3)$
$$= x(3x - 1)(x - 3)$$

69. $-3a^2 + ab + 2b^2 = -(3a^2 - ab - 2b^2) = -(3a + 2b)(a - b)$

71. $-4x^2 - 9 + 12x = -4x^2 + 12x - 9 = -(4x^2 - 12x + 9) = -(2x - 3)(2x - 3) = -(2x - 3)^2$

73. $5a^2 + 45b^2 - 30ab = 5a^2 - 30ab + 45b^2 = 5(a^2 - 6ab + 9b^2) = 5(a - 3b)(a - 3b) = 5(a - 3b)^2$

75. $8x^2z + 6xyz + 9y^2z = z(8x^2 + 6xy + 9y^2)$ NOTE: $8x^2 + 6xy + 9y^2$ is a prime trinomial.

77. $21x^4 - 10x^3 - 16x^2 = x^2(21x^2 - 10x - 16) = x^2(7x - 8)(3x + 2)$

79. $x^4 + 8x^2 + 15 = (x^2 + 5)(x^2 + 3)$

81. $y^4 - 13y^2 + 30 = (y^2 - 10)(y^2 - 3)$

83. $a^4 - 13a^2 + 36 = (a^2 - 4)(a^2 - 9) = (a + 2)(a - 2)(a + 3)(a - 3)$

85. $z^4 - z^2 - 12 = (z^2 - 4)(z^2 + 3) = (z + 2)(z - 2)(z^2 + 3)$

87. $4x^3 + x^6 + 3 = x^6 + 4x^3 + 3 = (x^3 + 1)(x^3 + 3) = (x + 1)(x^2 - x + 1)(x^3 + 3)$

89. $x^{2n} + 2x^n + 1 = (x^n + 1)(x^n + 1)$
$$= (x^n + 1)^2$$

91. $2a^{6n} - 3a^{3n} - 2 = (2a^{3n} + 1)(a^{3n} - 2)$

93. $x^{4n} + 2x^{2n}y^{2n} + y^{4n} = (x^{2n} + y^{2n})(x^{2n} + y^{2n}) = (x^{2n} + y^{2n})^2$

95. $6x^{2n} + 7x^n - 3 = (3x^n - 1)(2x^n + 3)$

97. $(x + 1)^2 + 2(x + 1) + 1 = [(x + 1) + 1][(x + 1) + 1] = (x + 2)(x + 2) = (x + 2)^2$

99. $(a + b)^2 - 2(a + b) - 24 = [(a + b) - 6][(a + b) + 4] = (a + b - 6)(a + b + 4)$

101. $6(x + y)^2 - 7(x + y) - 20 = [(3(x + y) + 4][2(x + y) - 5] = (3x + 3y + 4)(2x + 2y - 5)$

103. $x^2 + 4x + 4 - y^2 = (x^2 + 4x + 4) - y^2 = (x + 2)(x + 2) - y^2 = (x + 2)^2 - y^2$
$$= [(x + 2) + y][(x + 2) - y]$$
$$= (x + 2 + y)(x + 2 - y)$$

105. $x^2 + 2x + 1 - 9z^2 = (x^2 + 2x + 1) - 9z^2 = (x + 1)(x + 1) - 9z^2$
$$= (x + 1)^2 - (3z)^2$$
$$= [(x + 1) + 3z][(x + 1) - 3z]$$
$$= (x + 1 + 3z)(x + 1 - 3z)$$

107. $c^2 - 4a^2 + 4ab - b^2 = c^2 - (4a^2 - 4ab + b^2) = c^2 - (2a - b)(2a - b)$
$$= c^2 - (2a - b)^2$$
$$= [c + (2a - b)][c - (2a - b)]$$
$$= (c + 2a - b)(c - 2a + b)$$

109. $a^2 - b^2 + 8a + 16 = (a^2 + 8a + 16) - b^2 = (a + 4)(a + 4) - b^2 = (a + 4)^2 - b^2$
$$= [(a + 4) + b][(a + 4) - b]$$
$$= (a + 4 + b)(a + 4 - b)$$

111. $4x^2 - z^2 + 4xy + y^2 = (4x^2 + 4xy + y^2) - z^2 = (2x + y)(2x + y) - z^2$
$$= (2x + y)^2 - z^2$$
$$= [(2x + y) + z][(2x + y) - z]$$
$$= (2x + y + z)(2x + y - z)$$

113. $a^2 - 17a + 16$: $a = 1, b = -17, c = 16 \Rightarrow$ key $\# = ac = 1(16) = 16$.
Find two factors of 16 which add to equal -17: -1 and -16.
Rewrite and factor: $a^2 - 17a + 16 = a^2 - a - 16a + 16$
$$= a(a - 1) - 16(a - 1) = (a - 1)(a - 16)$$

115. $2u^2 + 5u + 3$: $a = 2, b = 5, c = 3 \Rightarrow$ key $\# = ac = 2(3) = 6$.
Find two factors of 6 which add to equal 5: $+2$ and $+3$.
Rewrite and factor: $2u^2 + 5u + 3 = 2u^2 + 2u + 3u + 3$
$$= 2u(u + 1) + 3(u + 1) = (u + 1)(2u + 3)$$

117. $20r^2 - 7rs - 6s^2$: $a = 20, b = -7, c = -6 \Rightarrow$ key $\# = ac = 20(-6) = -120$.
Find two factors of -120 which add to equal -7: -15 and $+8$.
Rewrite and factor: $20r^2 - 7rs - 6s^2 = 20r^2 - 15rs + 8rs - 6s^2$
$$= 5r(4r - 3s) + 2s(4r - 3s) = (4r - 3s)(5r + 2s)$$

119. $20u^2 + 19uv + 3v^2$: $a = 20, b = 19, c = 3 \Rightarrow$ key $\# = ac = 20(3) = 60$.
Find two factors of 60 which add to equal 19: $+15$ and $+4$.
Rewrite and factor: $20u^2 + 19uv + 3v^2 = 20u^2 + 15uv + 4uv + 3v^2$
$$= 5u(4u + 3v) + v(4u + 3v) = (4u + 3v)(5u + v)$$

121. $4x^2 + 20x - 11 = (2x + 11)(2x - 1)$ difference $= (2x + 11) - (2x - 1)$
length $= 2x + 11$ in., width $= 2x - 1$ in. $= 2x + 11 - 2x + 1 = 12$
The difference is 12 inches.

123. Answers may vary.

125. $x^2 - q^2 = x^2 + 0x - q^2$: $a = 1, b = 0, c = -q^2 \Rightarrow b^2 - 4ac = 0^2 - 4(1)(-q^2) = 4q^2 = (2q)^2$.
Since $4q^2$ is a perfect square, the binomial is factorable, and the test works.

Exercise 5.7 (page 321)

1. $(3a^2 + 4a - 2) + (4a^2 - 3a - 5) = 3a^2 + 4a - 2 + 4a^2 - 3a - 5 = 7a^2 + a - 7$

3. $5(2y^2 - 3y + 3) - 2(3y^2 - 2y + 6) = 10y^2 - 15y + 15 - 6y^2 + 4y - 12 = 4y^2 - 11y + 3$

5. $(m + 4)(m - 2) = m^2 - 2m + 4m - 8$
$= m^2 + 2m - 8$

7. common factors

9. trinomial

11. $x^2 + 8x + 16 = (x + 4)(x + 4) = (x + 4)^2$

13. $8x^3y^3 - 27 = (2xy)^3 - 3^3 = (2xy - 3)[(2xy)^2 + (2xy)(3) + 3^2] = (2xy - 3)(4x^2y^2 + 6xy + 9)$

15. $xy - ty + xs - ts = y(x - t) + s(x - t)$
$= (x - t)(y + s)$

17. $25x^2 - 16y^2 = (5x)^2 - (4y)^2$
$= (5x + 4y)(5x - 4y)$

19. $12x^2 + 52x + 35 = (6x + 5)(2x + 7)$

21. $6x^2 - 14x + 8 = 2(3x^2 - 7x + 4)$
$= 2(3x - 4)(x - 1)$

23. $56x^2 - 15x + 1 = (8x - 1)(7x - 1)$

25. $4x^2y^2 + 4xy^2 + y^2 = y^2(4x^2 + 4x + 1)$
$= y^2(2x + 1)(2x + 1)$

27. $x^3 + (a^2y)^3 = (x + a^2y)[x^2 - xa^2y + (a^2y)^2] = (x + a^2y)(x^2 - a^2xy + a^4y^2)$

29. $2x^3 - 54 = 2(x^3 - 27) = 2(x^3 - 3^3) = 2(x - 3)(x^2 + 3x + 3^2) = 2(x - 3)(x^2 + 3x + 9)$

31. $ae + bf + af + be = ae + af + be + bf = a(e + f) + b(e + f) = (e + f)(a + b)$

33. $2(x + y)^2 + (x + y) - 3 = [2(x + y) + 3][(x + y) - 1] = (2x + 2y + 3)(x + y - 1)$

35. $625x^4 - 256y^4 = (25x^2)^2 - (16y^2)^2 = (25x^2 + 16y^2)(25x^2 - 16y^2)$
$= (25x^2 + 16y^2)(5x + 4y)(5x - 4y)$

37. $36x^4 - 36 = 36(x^4 - 1) = 36(x^2 + 1)(x^2 - 1) = 36(x^2 + 1)(x + 1)(x - 1)$

39. $2x^6 + 2y^6 = 2(x^6 + y^6) = 2[(x^2)^3 + (y^2)^3] = 2(x^2 + y^2)(x^4 - x^2y^2 + y^4)$

41. $a^4 - 13a^2 + 36 = (a^2 - 4)(a^2 - 9) = (a + 2)(a - 2)(a + 3)(a - 3)$

43. $x^2 + 6x + 9 - y^2 = (x + 3)^2 - y^2 = (x + 3 + y)(x + 3 - y)$

45. $4x^2 + 4x + 1 - 4y^2 = (2x + 1)^2 - (2y)^2 = (2x + 1 + 2y)(2x + 1 - 2y)$

47. $x^2 - y^2 - 2y - 1 = x^2 - (y^2 + 2y + 1) = x^2 - (y + 1)^2 = [x + (y + 1)][x - (y + 1)]$
$= (x + y + 1)(x - y - 1)$

SECTION 5.7

49. $x^5 + x^2 - x^3 - 1 = x^2(x^3 + 1) - 1(x^3 + 1) = (x^3 + 1)(x^2 - 1)$
$$= (x + 1)(x^2 - x + 1)(x + 1)(x - 1)$$

51. $x^5 - 9x^3 + 8x^2 - 72 = x^3(x^2 - 9) + 8(x^2 - 9) = (x^2 - 9)(x^3 + 8)$
$$= (x + 3)(x - 3)(x + 2)(x^2 - 2x + 4)$$

53. $2x^5 z - 2x^2 y^3 z - 2x^3 y^2 z + 2y^5 z = 2z(x^5 - x^2 y^3 - x^3 y^2 + y^5)$
$$= 2z[x^2(x^3 - y^3) - y^2(x^3 - y^3)]$$
$$= 2z(x^3 - y^3)(x^2 - y^2)$$
$$= 2z(x - y)(x^2 + xy + y^2)(x + y)(x - y)$$
$$= 2z(x + y)(x - y)^2(x^2 + xy + y^2)$$

55. $x^{2m} - x^m - 6 = (x^m - 3)(x^m + 2)$

57. $a^{3n} - b^{3n} = (a^n)^3 - (b^n)^3$
$$= (a^n - b^n)(a^{2n} + a^n b^n + b^{2n})$$

59. $x^{-2} + 2x^{-1} + 1 = (x^{-1} + 1)(x^{-1} + 1) = (x^{-1} + 1)^2 = \left(\dfrac{1}{x} + 1\right)^2$

61. $6x^{-2} - 5x^{-1} - 6 = (3x^{-1} + 2)(2x^{-1} - 3) = \left(\dfrac{3}{x} + 2\right)\left(\dfrac{2}{x} - 3\right)$

63. Answers may vary.

65. Answers may vary.

67. $x^4 + x^2 + 1 = x^4 + x^2 + x^2 + 1 - x^2 = x^4 + 2x^2 + 1 - x^2 = (x^2 + 1)^2 - x^2$
$$= (x^2 + 1 + x)(x^2 + 1 - x)$$

Exercise 5.8 (page 329)

1. 2, 3, 5, 7

3. $V = \frac{4}{3}\pi r^3 = \frac{4}{3}\pi(21.23)^3 = 40{,}081.00$ cm^3

5. $ax^2 + bx + c = 0$

7.
$$4x^2 + 8x = 0$$
$$4x(x + 2) = 0$$
$$4x = 0 \quad \text{or} \quad x + 2 = 0$$
$$x = 0 \qquad\qquad x = -2$$

9.
$$y^2 - 16 = 0$$
$$(y + 4)(y - 4) = 0$$
$$y + 4 = 0 \quad \text{or} \quad y - 4 = 0$$
$$y = -4 \qquad\qquad y = 4$$

11.
$$x^2 + x = 0$$
$$x(x + 1) = 0$$
$$x = 0 \quad \text{or} \quad x + 1 = 0$$
$$x = -1$$

13.
$$5y^2 - 25 = 0$$
$$5y(y - 5) = 0$$
$$5y = 0 \quad \text{or} \quad y - 5 = 0$$
$$y = 0 \qquad\qquad y = 5$$

15.
$$z^2 + 8z + 15 = 0$$
$$(z + 5)(z + 3) = 0$$
$$z + 5 = 0 \quad \textbf{or} \quad z + 3 = 0$$
$$z = -5 \qquad\qquad z = -3$$

17.
$$y^2 - 7y + 6 = 0$$
$$(y - 6)(y - 1) = 0$$
$$y - 6 = 0 \quad \textbf{or} \quad y - 1 = 0$$
$$y = 6 \qquad\qquad y = 1$$

19.
$$y^2 - 7y + 12 = 0$$
$$(y - 4)(y - 3) = 0$$
$$y - 4 = 0 \quad \textbf{or} \quad y - 3 = 0$$
$$y = 4 \qquad\qquad y = 3$$

21.
$$x^2 + 6x + 8 = 0$$
$$(x + 4)(x + 2) = 0$$
$$x + 4 = 0 \quad \textbf{or} \quad x + 2 = 0$$
$$x = -4 \qquad\qquad x = -2$$

23.
$$3m^2 + 10m + 3 = 0$$
$$(3m + 1)(m + 3) = 0$$
$$3m + 1 = 0 \quad \textbf{or} \quad m + 3 = 0$$
$$3m = -1 \qquad\qquad m = -3$$
$$m = -\tfrac{1}{3}$$

25.
$$2y^2 - 5y + 2 = 0$$
$$(2y - 1)(y - 2) = 0$$
$$2y - 1 = 0 \quad \textbf{or} \quad y - 2 = 0$$
$$2y = 1 \qquad\qquad y = 2$$
$$y = \tfrac{1}{2}$$

27.
$$2x^2 - x - 1 = 0$$
$$(2x + 1)(x - 1) = 0$$
$$2x + 1 = 0 \quad \textbf{or} \quad x - 1 = 0$$
$$2x = -1 \qquad\qquad x = 1$$
$$x = -\tfrac{1}{2}$$

29.
$$3s^2 - 5s - 2 = 0$$
$$(3s + 1)(s - 2) = 0$$
$$3s + 1 = 0 \quad \textbf{or} \quad s - 2 = 0$$
$$3s = -1 \qquad\qquad s = 2$$
$$s = -\tfrac{1}{3}$$

31.
$$x(x - 6) + 9 = 0$$
$$x^2 - 6x + 9 = 0$$
$$(x - 3)(x - 3) = 0$$
$$x - 3 = 0 \quad \textbf{or} \quad x - 3 = 0$$
$$x = 3 \qquad\qquad x = 3$$

33.
$$8a^2 = 3 - 10a$$
$$8a^2 + 10a - 3 = 0$$
$$(4a - 1)(2a + 3) = 0$$
$$4a - 1 = 0 \quad \textbf{or} \quad 2a + 3 = 0$$
$$4a = 1 \qquad\qquad 2a = -3$$
$$a = \tfrac{1}{4} \qquad\qquad a = -\tfrac{3}{2}$$

35.
$$b(6b - 7) = 10$$
$$6b^2 - 7b - 10 = 0$$
$$(6b + 5)(b - 2) = 0$$
$$6b + 5 = 0 \quad \textbf{or} \quad b - 2 = 0$$
$$6b = -5 \qquad\qquad b = 2$$
$$b = -\tfrac{5}{6}$$

37.
$$\frac{3a^2}{2} = \frac{1}{2} - a$$
$$3a^2 = 1 - 2a$$
$$3a^2 + 2a - 1 = 0$$
$$(3a - 1)(a + 1) = 0$$
$$3a - 1 = 0 \quad \textbf{or} \quad a + 1 = 0$$
$$3a = 1 \qquad\qquad a = -1$$
$$a = \tfrac{1}{3}$$

39.
$$\frac{1}{2}x^2 - \frac{5}{4}x = -\frac{1}{2}$$
$$2x^2 - 5x = -2$$
$$2x^2 - 5x + 2 = 0$$
$$(2x - 1)(x - 2) = 0$$
$$2x - 1 = 0 \quad \text{or} \quad x - 2 = 0$$
$$2x = 1 \qquad\qquad x = 2$$
$$x = \tfrac{1}{2}$$

41.
$$x\left(3x + \frac{22}{5}\right) = 1$$
$$3x^2 + \frac{22}{5}x = 1$$
$$15x^2 + 22x = 5$$
$$15x^2 + 22x - 5 = 0$$
$$(5x - 1)(3x + 5) = 0$$
$$5x - 1 = 0 \quad \text{or} \quad 3x + 5 = 0$$
$$5x = 1 \qquad\qquad 3x = -5$$
$$x = \tfrac{1}{5} \qquad\qquad x = -\tfrac{5}{3}$$

43.
$$x^3 + x^2 = 0$$
$$x^2(x + 1) = 0$$
$$x^2 = 0 \quad \text{or} \quad x + 1 = 0$$
$$x = 0 \qquad\qquad x = -1$$

45.
$$y^3 - 49y = 0$$
$$y(y^2 - 49) = 0$$
$$y(y + 7)(y - 7) = 0$$
$$y = 0 \quad \text{or} \quad y + 7 = 0 \quad \text{or} \quad y - 7 = 0$$
$$y = -7 \qquad\qquad y = 7$$

47.
$$x^3 - 4x^2 - 21x = 0$$
$$x(x^2 - 4x - 21) = 0$$
$$x(x + 3)(x - 7) = 0$$
$$x = 0 \quad \text{or} \quad x + 3 = 0 \quad \text{or} \quad x - 7 = 0$$
$$x = -3 \qquad\qquad x = 7$$

49.
$$z^4 - 13z^2 + 36 = 0$$
$$(z^2 - 4)(z^2 - 9) = 0$$
$$(z + 2)(z - 2)(z + 3)(z - 3) = 0$$
$$z + 2 = 0 \quad \text{or} \quad z - 2 = 0 \quad \text{or} \quad z + 3 = 0 \quad \text{or} \quad z - 3 = 0$$
$$z = -2 \qquad\qquad z = 2 \qquad\qquad z = -3 \qquad\qquad z = 3$$

51.
$$3a(a^2 + 5a) = -18a$$
$$3a^3 + 15a^2 = -18a$$
$$3a^3 + 15a^2 + 18a = 0$$
$$3a(a^2 + 5a + 6) = 0$$
$$3a(a + 2)(a + 3) = 0$$
$$3a = 0 \quad \text{or} \quad a + 2 = 0 \quad \text{or} \quad a + 3 = 0$$
$$a = 0 \qquad\qquad a = -2 \qquad\qquad a = -3$$

53.
$$\frac{x^2(6x + 37)}{35} = x$$
$$x^2(6x + 37) = 35x$$
$$6x^3 + 37x^2 = 35x$$
$$6x^3 + 37x^2 - 35x = 0$$
$$x(6x^2 + 37x - 35) = 0$$
$$x(6x - 5)(x + 7) = 0$$
$$x = 0 \quad \text{or} \quad 6x - 5 = 0 \quad \text{or} \quad x + 7 = 0$$
$$6x = 5 \qquad\qquad x = -7$$
$$x = \tfrac{5}{6}$$

55.
$$f(x) = 0$$
$$x^2 - 49 = 0$$
$$(x + 7)(x - 7) = 0$$
$$x + 7 = 0 \quad \text{or} \quad x - 7 = 0$$
$$x = -7 \qquad\qquad x = 7$$

57.
$$f(x) = 0$$
$$2x^2 + 5x - 3 = 0$$
$$(2x - 1)(x + 3) = 0$$
$$2x - 1 = 0 \quad \text{or} \quad x + 3 = 0$$
$$2x = 1 \qquad\qquad x = -3$$
$$x = \tfrac{1}{2}$$

59.
$$f(x) = 0$$
$$5x^3 + 3x^2 - 2x = 0$$
$$x(5x^2 + 3x - 2) = 0$$
$$x(5x - 2)(x + 1) = 0$$
$$x = 0 \quad \text{or} \quad 5x - 2 = 0 \quad \text{or} \quad x + 1 = 0$$
$$5x = 2 \qquad\qquad x = -1$$
$$x = \tfrac{2}{5}$$

61.
$$x^3 + 3x^2 - x - 3 = 0$$
$$x^2(x + 3) - 1(x + 3) = 0$$
$$(x + 3)(x^2 - 1) = 0$$
$$(x + 3)(x + 1)(x - 1) = 0$$
$$x + 3 = 0 \quad \text{or} \quad x + 1 = 0 \quad \text{or} \quad x - 1 = 0$$
$$x = -3 \qquad\qquad x = -1 \qquad\qquad x = 1$$

63.
$$2r^3 + 3r^2 - 18r - 27 = 0$$
$$r^2(2r + 3) - 9(2r + 3) = 0$$
$$(2r + 3)(r^2 - 9) = 0$$
$$(2r + 3)(r + 3)(r - 3) = 0$$
$$2r + 3 = 0 \quad \text{or} \quad r + 3 = 0 \quad \text{or} \quad r - 3 = 0$$
$$2r = -3 \qquad\qquad r = -3 \qquad\qquad r = 3$$
$$r = -\tfrac{3}{2}$$

65.

$$3y^3 + y^2 = 4(3y + 1)$$
$$3y^3 + y^2 = 12y + 4$$
$$3y^3 + y^2 - 12y - 4 = 0$$
$$y^2(3y + 1) - 4(3y + 1) = 0$$
$$(3y + 1)(y^2 - 4) = 0$$
$$(3y + 1)(y + 2)(y - 2) = 0$$

$3y + 1 = 0 \quad$ **or** $\quad y + 2 = 0 \quad$ **or** $\quad y - 2 = 0$
$3y = -1 \qquad\qquad y = -2 \qquad\qquad y = 2$
$y = -\frac{1}{3}$

67. Let $x =$ the first even integer and $x + 2 =$ the second.

$$\boxed{\text{First}} \cdot \boxed{\text{Second}} = 288$$
$$x(x + 2) = 288$$
$$x^2 + 2x - 288 = 0$$
$$(x + 18)(x - 16) = 0$$

$x + 18 = 0 \quad$ **or** $\quad x - 16 = 0$
$x = -18 \qquad\qquad x = 16$

The integers are -18 and -16, or 16 and 18.

69. Let $x =$ the first positive integer and $x + 1 =$ the second.

$$\boxed{\text{First}^2} + \boxed{\text{Second}^2} = 85$$
$$x^2 + (x + 1)^2 = 85$$
$$x^2 + (x + 1)(x + 1) = 85$$
$$x^2 + x^2 + 2x + 1 = 85$$
$$2x^2 + 2x - 84 = 0$$
$$2(x^2 + x - 42) = 0$$
$$2(x + 7)(x - 6) = 0$$

$x + 7 = 0 \quad$ **or** $\quad x - 6 = 0$
$x = -7 \qquad\qquad x = 6$
(not positive)

The integers are 6 and 7.

71.

$$\boxed{\text{Length}} \cdot \boxed{\text{Width}} = \boxed{\text{Area}}$$
$$(w + 4)w = 96$$
$$w^2 + 4w = 96$$
$$w^2 + 4w - 96 = 0$$
$$(w + 12)(w - 8) = 0$$

$w + 12 = 0 \quad$ **or** $\quad w - 8 = 0$
$x = -12 \qquad\qquad w = 8$
(not positive)

The dimensions are 8 m by 12 m, so the perimeter is 40 meters.

73.

$$\boxed{\text{Length}} \cdot \boxed{\text{Width}} = \boxed{\text{Area}}$$
$$(2x - 5)x = 375$$
$$2x^2 - 5x = 375$$
$$2x^2 - 5x - 375 = 0$$
$$(2x + 25)(x - 15) = 0$$

$2x + 25 = 0 \quad$ **or** $\quad x - 15 = 0$
$2x = -25 \qquad\qquad x = 15$
(x is not positive)

The dimensions are 15 ft by 25 ft.

75. The length of the mural is $18 - 2w$, while the width is $11 - 2w$.

$$\boxed{\text{Length}} \cdot \boxed{\text{Width}} = \boxed{\text{Area}}$$
$$(18 - 2w)(11 - 2w) = 60$$
$$198 - 36w - 22w + 4w^2 = 60$$
$$4w^2 - 58w + 138 = 0$$
$$2(2w^2 - 29w + 69) = 0$$
$$2(2w - 23)(w - 3) = 0$$
$$2w - 23 = 0 \quad \text{or} \quad w - 3 = 0$$
$$2w = 23 \qquad\qquad w = 3$$
$$w = \tfrac{23}{2}$$
(makes width < 0)

The dimensions are 12 ft by 5 ft.

77. Let $w = $ the width of the room and let $2w = $ the length of the room.

$$\boxed{\text{Length}} \cdot \boxed{\text{Width}} = \boxed{\text{Area}}$$
$$(2w - 12)w = 560$$
$$2w^2 - 12w - 560 = 0$$
$$2(w^2 - 6w - 280) = 0$$
$$2(w + 14)(w - 20) = 0$$
$$w + 14 = 0 \quad \text{or} \quad w - 20 = 0$$
$$w = -14 \qquad\qquad w = 20$$
(impossible)

The dimensions are 20 feet by 40 feet.

79.
$$h = vt - 16t^2$$
$$0 = 160t - 16t^2$$
$$0 = 16t(10 - t)$$
$$16t = 0 \quad \text{or} \quad 10 - t = 0$$
$$t = 0 \qquad\qquad 10 = t$$

The object will hit the ground after **10 sec.**

81.
$$h = vt - 16t^2$$
$$3344 = 480t - 16t^2$$
$$16t^2 - 480t + 3344 = 0$$
$$16(t^2 - 30t + 209) = 0$$
$$16(t - 11)(t - 19) = 0$$
$$t - 11 = 0 \quad \text{or} \quad t - 19 = 0$$
$$t = 11 \qquad\qquad t = 19$$

The cannonball will be at that height after **11 and 19 seconds.**

83. Let $w =$ the width and $2w + 20 =$ the length.

$$\boxed{\text{Length}} \cdot \boxed{\text{Width}} = \boxed{\text{Area}}$$
$$(2w + 20)w = 6000$$
$$2w^2 + 20w - 6000 = 0$$
$$2(w^2 + 10w - 3000) = 0$$
$$2(w - 50)(w + 60) = 0$$
$$w - 50 = 0 \quad \textbf{or} \quad w + 60 = 0$$
$$w = 50 \qquad\qquad w = -60$$
$$\qquad\qquad\qquad \text{(impossible)}$$

The width is 50 meters.

85. Let $x =$ the width of the pool and $2x - 10 =$ the length of the pool.

$$\boxed{\text{Length}} \cdot \boxed{\text{Width}} = \boxed{\text{Area of pool}}$$
$$(2x - 10)x = 1500$$
$$2x^2 - 10x - 1500 = 0$$
$$2(x^2 - 5x - 750) = 0$$
$$2(x + 25)(x - 30) = 0$$
$$x + 25 = 0 \quad \textbf{or} \quad x - 30 = 0$$
$$x = -25 \qquad\qquad x = 30$$
$$\text{(impossible)}$$

The dimensions of the pool are 30 feet by 50 feet.
Then let $w =$ the width of the walkway.

$$30w + 30w + 50w + 50w + 4w^2 = \boxed{\text{Area of walkway}}$$
$$4w^2 + 160w = 516$$
$$4w^2 + 160w - 516 = 0$$
$$4(w^2 + 40w - 129) = 0$$
$$4(w + 43)(w - 3) = 0$$
$$w + 43 = 0 \quad \textbf{or} \quad w - 3 = 0$$
$$w = -43 \qquad\qquad w = 3$$
$$\text{(impossible)}$$

The walkway should be 3 feet wide.

87. $x^2 - 4x + 7$

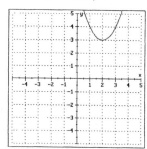

x-intercepts: none $\Rightarrow$ no solution

89. $-3x^3 - 2x^2 + 5 = 0$

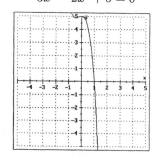

x-intercept: 1.00

91. **Answers may vary.**

93. $(x - 3)(x - 5) = 0$
$$x^2 - 8x + 15 = 0$$

95. $(x - 0)(x + 5) = 0$
$$x^2 + 5x = 0$$

Chapter 5 Summary (page 333)

1. degree $= 5$

2. degree $= 4 + 4 = 8$

3. $P(0) = -0^2 + 4(0) + 6 = 6$

4. $P(1) = -1^2 + 4(1) + 6$
$= -1 + 4 + 6 = 9$

5. $P(-t) = -(-t)^2 + 4(-t) + 6$
$= -t^2 - 4t + 6$

6. $P(z) = -z^2 + 4z + 6$

7. $f(x) = x^3 - 1$

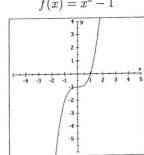

8. $f(x) = x^2 - 2x$

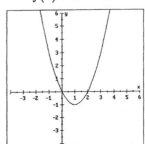

9. $(3x^2 + 4x + 9) + (2x^2 - 2x + 7) = 3x^2 + 4x + 9 + 2x^2 - 2x + 7 = 5x^2 + 2x + 16$

10. $(4x^3 + 4x^2 + 7) - (-2x^3 - x - 2) = 4x^3 + 4x^2 + 7 + 2x^3 + x + 2 = 6x^3 + 4x^2 + x + 9$

11. $(2x^2 - 5x + 9) - (x^2 - 3) - (-3x^2 + 4x - 7)$
$= 2x^2 - 5x + 9 - x^2 + 3 + 3x^2 - 4x + 7 = 4x^2 - 9x + 19$

12. $2(7x^3 - 6x^2 + 4x - 3) - 3(7x^3 + 6x^2 + 4x - 3)$
$= 14x^3 - 12x^2 + 8x - 6 - 21x^3 - 18x^2 - 12x + 9 = -7x^3 - 30x^2 - 4x + 3$

13. $(8a^2b^2)(-2abc) = -16a^3b^3c$

14. $(-3xy^2z)(2xz^3) = -6x^2y^2z^4$

15. $2xy^2(x^3y - 4xy^5) = 2x^4y^3 - 8x^2y^7$

16. $a^2b(a^2 + 2ab + b^2) = a^4b + 2a^3b^2 + a^2b^3$

17. $(8x - 5)(2x + 3) = 16x^2 + 24x - 10x - 15 = 16x^2 + 14x - 15$

18. $(3x + 2)(2x - 4) = 6x^2 - 12x + 4x - 8$
$= 6x^2 - 8x - 8$

19. $(5x - 4)(3x - 2) = 15x^2 - 10x - 12x + 8$
$= 15x^2 - 22x + 8$

20. $(3x^2 - 2)(x^2 - x + 2) = 3x^4 - 3x^3 + 6x^2 - 2x^2 + 2x - 4 = 3x^4 - 3x^3 + 4x^2 + 2x - 4$

21. $4x + 8 = 4(x + 2)$

22. $3x^2 - 6x = 3x(x - 2)$

23. $5x^2y^3 - 10xy^2 = 5xy^2(xy - 2)$

24. $7a^4b^2 + 49a^3b = 7a^3b(ab + 7)$

25. $-8x^2y^3z^4 - 12x^4y^3z^2 = -4x^2y^3z^2(2z^2 + 3x^2)$

26. $12a^6b^4c^2 + 15a^2b^4c^6 = 3a^2b^4c^2(4a^4 + 5c^4)$

27. $27x^3y^3z^3 + 81x^4y^5z^2 - 90x^2y^3z^7 = 9x^2y^3z^2(3xz + 9x^2y^2 - 10z^5)$

28. $-36a^5b^4c^2 + 60a^7b^5c^3 - 24a^2b^3c^7 = -12a^2b^3c^2(3a^3b - 5a^5b^2c + 2c^5)$

29. $x^{2n} + x^n = x^n(x^{2n-n} + 1) = x^n(x^n + 1)$ 30. $y^{2n} - y^{4n} = y^{2n}(1 - y^{4n-2n}) = y^{2n}(1 - y^{2n})$

31. $\begin{aligned} x^{-4} - x^{-2} &= x^{-2}(x^{-4-(-2)} - 1) \\ &= x^{-2}(x^{-2} - 1) \end{aligned}$ 32. $\begin{aligned} a^6 + 1 &= a^{-3}(a^{6-(-3)} + a^{0-(-3)}) \\ &= a^{-3}(a^9 + a^3) \end{aligned}$

33. $5x^2(x + y)^3 - 15x^3(x + y)^4 = 5x^2(x + y)^3[1 - 3x(x + y)] = 5x^2(x + y)^3(1 - 3x^2 - 3xy)$

34. $\begin{aligned} -49a^3b^2(a - b)^4 + 63a^2b^4(a - b)^3 &= -7a^2b^2(a - b)^3[7a(a - b) - 9b^2] \\ &= -7a^2b^2(a - b)^3(7a^2 - 7ab - 9b^2) \end{aligned}$

35. $\begin{aligned} xy + 2y + 4x + 8 &= y(x + 2) + 4(x + 2) \\ &= (x + 2)(y + 4) \end{aligned}$ 36. $\begin{aligned} ac + bc + 3a + 3b &= c(a + b) + 3(a + b) \\ &= (a + b)(c + 3) \end{aligned}$

37. $x^4 + 4y + 4x^2 + x^2y = x^4 + x^2y + 4x^2 + 4y = x^2(x^2 + y) + 4(x^2 + y) = (x^2 + y)(x^2 + 4)$

38. $a^5 + b^2c + a^2c + a^3b^2 = a^5 + a^3b^2 + a^2c + b^2c = a^3(a^2 + b^2) + c(a^2 + b^2) = (a^2 + b^2)(a^3 + c)$

39. $\begin{aligned} S &= 2wh + 2wl + 2lh \\ S - 2wl &= 2wh + 2lh \\ S - 2wl &= h(2w + 2l) \\ \frac{S - 2wl}{2w + 2l} &= h, \text{ or } h = \frac{S - 2wl}{2w + 2l} \end{aligned}$ 40. $\begin{aligned} S &= 2wh + 2wl + 2lh \\ S - 2wh &= 2wl + 2lh \\ S - 2wh &= l(2w + 2h) \\ \frac{S - 2wh}{2w + 2h} &= l, \text{ or } l = \frac{S - 2wh}{2w + 2h} \end{aligned}$

41. $z^2 - 16 = z^2 - 4^2 = (z + 4)(z - 4)$ 42. $y^2 - 121 = y^2 - 11^2 = (y + 11)(y - 11)$

43. $\begin{aligned} x^2y^4 - 64z^6 &= (xy^2)^2 - (8z^3)^2 \\ &= (xy^2 + 8z^3)(xy^2 - 8z^3) \end{aligned}$ 44. $a^2b^2 + c^2 \Rightarrow \text{prime}$

45. $\begin{aligned} (x + z)^2 - t^2 &= [(x + z) + t][(x + z) - t] \\ &= (x + z + t)(x + z - t) \end{aligned}$ 46. $\begin{aligned} c^2 - (a + b)^2 &= [c + (a + b)][c - (a + b)] \\ &= (c + a + b)(c - a - b) \end{aligned}$

47. $\begin{aligned} 2x^4 - 98 &= 2(x^4 - 49) \\ &= 2(x^2 + 7)(x^2 - 7) \end{aligned}$ 48. $\begin{aligned} 3x^6 - 300x^2 &= 3x^2(x^4 - 100) \\ &= 3x^2(x^2 + 10)(x^2 - 10) \end{aligned}$

49. $\begin{aligned} x^3 + 343 &= x^3 + 7^3 \\ &= (x + 7)(x^2 - 7x + 49) \end{aligned}$ 50. $\begin{aligned} a^3 - 125 &= a^3 - 5^3 \\ &= (a - 5)(a^2 + 5a + 25) \end{aligned}$

51. $8y^3 - 512 = 8(y^3 - 64) = 8(y^3 - 4^3) = 8(y - 4)(y^2 + 4y + 16)$

52. $4x^3y + 108yz^3 = 4y(x^3 + 27z^3) = 4y[x^3 + (3z)^3] = 4y(x + 3z)(x^2 - 3xz + 9z^2)$

53. $x^2 + 10x + 25 = (x+5)(x+5)$

54. $a^2 - 14a + 49 = (a-7)(a-7)$

55. $y^2 + 21y + 20 = (y+20)(y+1)$

56. $z^2 - 11z + 30 = (z-5)(z-6)$

57. $-x^2 - 3x + 28 = -(x^2 + 3x - 28)$
$\qquad\qquad\qquad = -(x+7)(x-4)$

58. $y^2 - 5y - 24 = (y-8)(y+3)$

59. $4a^2 - 5a + 1 = (4a-1)(a-1)$

60. $3b^2 + 2b + 1 \Rightarrow$ prime

61. $7x^2 + x + 2 \Rightarrow$ prime

62. $-15x^2 + 14x + 8 = -(15x^2 - 14x - 8)$
$\qquad\qquad\qquad = -(5x+2)(3x-4)$

63. $y^3 + y^2 - 2y = y(y^2 + y - 2)$
$\qquad\qquad\quad = y(y+2)(y-1)$

64. $2a^4 + 4a^3 - 6a^2 = 2a^2(a^2 + 2a - 3)$
$\qquad\qquad\qquad = 2a^2(a+3)(a-1)$

65. $-3x^2 - 9x - 6 = -3(x^2 + 3x + 2)$
$\qquad\qquad\quad = -3(x+2)(x+1)$

66. $8x^2 - 4x - 24 = 4(2x^2 - x - 6)$
$\qquad\qquad\quad = 4(2x+3)(x-2)$

67. $15x^2 - 57xy - 12y^2 = 3(5x^2 - 19xy - 4y^2)$
$\qquad\qquad\qquad\quad = 3(5x+y)(x-4y)$

68. $30x^2 + 65xy + 10y^2 = 5(6x^2 + 13xy + 2y^2)$
$\qquad\qquad\qquad\quad = 5(6x+y)(x+2y)$

69. $24x^2 - 23xy - 12y^2 = (8x+3y)(3x-4y)$

70. $14x^2 + 13xy - 12y^2 = (2x+3y)(7x-4y)$

71. $x^3 + 5x^2 - 6x = x(x^2 + 5x - 6)$
$\qquad\qquad\quad = x(x-1)(x+6)$

72. $3x^2y - 12xy - 63y = 3y(x^2 - 4x - 21)$
$\qquad\qquad\qquad = 3y(x-7)(x+3)$

73. $z^2 - 4 + zx - 2x = (z^2 - 4) + zx - 2x = (z+2)(z-2) + x(z-2) = (z-2)(z+2+x)$

74. $x^2 + 2x + 1 - p^2 = (x^2 + 2x + 1) - p^2 = (x+1)^2 - p^2 = (x+1+p)(x+1-p)$

75. $x^2 + 4x + 4 - 4p^4 = (x^2 + 4x + 4) - 4p^4 = (x+2)^2 - (2p^2)^2$
$\qquad\qquad\qquad\qquad = (x+2+2p^2)(x+2-2p^2)$

76. $y^2 + 3y + 2 + 2x + xy = (y^2 + 3y + 2) + 2x + xy = (y+1)(y+2) + x(2+y)$
$\qquad\qquad\qquad\qquad = (y+2)(y+1+x)$

77. $x^{2m} + 2x^m - 3 = (x^m + 3)(x^m - 1)$

78. $x^{-2} - x^{-1} - 2 = (x^{-1} - 2)(x^{-1} + 1)$
$\qquad\qquad\qquad = \left(\frac{1}{x} - 2\right)\left(\frac{1}{x} + 1\right)$

79.
$$4x^2 - 3x = 0$$
$$x(4x-3) = 0$$
$$x = 0 \quad \text{or} \quad 4x - 3 = 0$$
$$4x = 3$$
$$x = \tfrac{3}{4}$$

80.
$$x^2 - 36 = 0$$
$$(x+6)(x-6) = 0$$
$$x + 6 = 0 \quad \text{or} \quad x - 6 = 0$$
$$x = -6 \qquad\qquad x = 6$$

81.
$$12x^2 + 4x - 5 = 0$$
$$(2x - 1)(6x + 5) = 0$$
$$2x - 1 = 0 \quad \text{or} \quad 6x + 5 = 0$$
$$2x = 1 \qquad\qquad 6x = -5$$
$$x = \tfrac{1}{2} \qquad\qquad x = -\tfrac{5}{6}$$

82.
$$7y^2 - 37y + 10 = 0$$
$$(7y - 2)(y - 5) = 0$$
$$7y - 2 = 0 \quad \text{or} \quad y - 5 = 0$$
$$7y = 2 \qquad\qquad y = 5$$
$$y = \tfrac{2}{7}$$

83.
$$t^2(15t - 2) = 8t$$
$$15t^3 - 2t^2 = 8t$$
$$15t^3 - 2t^2 - 8t = 0$$
$$t(15t^2 - 2t - 8) = 0$$
$$t(3t + 2)(5t - 4) = 0$$
$$t = 0 \quad \text{or} \quad 3t + 2 = 0 \quad \text{or} \quad 5t - 4 = 0$$
$$\qquad\qquad 3t = -2 \qquad\qquad 5t = 4$$
$$\qquad\qquad t = -\tfrac{2}{3} \qquad\qquad t = \tfrac{4}{5}$$

84.
$$3u^3 = u(19u + 14)$$
$$3u^3 = 19u^2 + 14u$$
$$3u^3 - 19u^2 - 14u = 0$$
$$u(3u^2 - 19u - 14) = 0$$
$$u(3u + 2)(u - 7) = 0$$
$$u = 0 \quad \text{or} \quad 3u + 2 = 0 \quad \text{or} \quad u - 7 = 0$$
$$\qquad\qquad 3u = -2 \qquad\qquad u = 7$$
$$\qquad\qquad u = -\tfrac{2}{3}$$

85. Let $h =$ the height and $h + 3 =$ the width
$$lwh = V$$
$$12(h + 3)h = 840$$
$$12h^2 + 36h - 840 = 0$$
$$3(4h^2 + 12h - 280) = 0$$
$$3(4h + 40)(h - 7) = 0$$
$$4h + 40 = 0 \quad \text{or} \quad h - 7 = 0$$
$$4h = -40 \qquad\qquad h = 7$$
$$h = -10$$
$$(\text{impossible}) \qquad\qquad \text{The height is 7 cm.}$$

86. Let $x =$ one side of the base and $x + 3 =$ the other side of the base.
$$\tfrac{Bh}{3} = V$$
$$\tfrac{x(x+3)9}{3} = 1020$$
$$\tfrac{9x^2 + 27x}{3} = 1020$$
$$9x^2 + 27x = 3060$$
$$9x^2 + 27x - 3060 = 0$$
$$9(x^2 + 3x - 340) = 0$$
$$9(x + 20)(x - 17) = 0$$
$$x + 20 = 0 \quad \text{or} \quad x - 17 = 0$$
$$x = -20 \qquad\qquad x = 17$$
$$(\text{impossible}) \qquad\qquad \text{The dimensions are}$$
$$17 \text{ m by } 20 \text{ m.}$$

Chapter 5 Test (page 338)

1. degree $= 5$

2. degree $= 9 + 4 = 13$

3. $P(2) = -3(2)^2 + 2(2) - 1 = -3(4) + 4 - 1 = -12 + 4 - 1 = -9$

4. $P(-1) = -3(-1)^2 + 2(-1) - 1 = -3(1) - 2 - 1 = -3 - 2 - 1 = -6$

CHAPTER 5 TEST

5. $f(x) = x^2 + 2x$

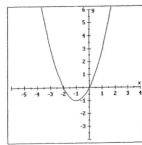

6. $(2y^2 + 4y + 3) + (3y^2 - 3y - 4) = 2y^2 + 4y + 3 + 3y^2 - 3y - 4 = 5y^2 + y - 1$

7. $(-3u^2 + 2u - 7) - (u^2 + 7) = -3u^2 + 2u - 7 - u^2 - 7 = -4u^2 + 2u - 14$

8. $3(2a^2 - 4a + 2) - 4(-a^2 - 3a - 4) = 6a^2 - 12a + 6 + 4a^2 + 12a + 16 = 10a^2 + 22$

9. $-2(2x^2 - 2) + 3(x^2 + 5x - 2) = -4x^2 + 4 + 3x^2 + 15x - 6 = -x^2 + 15x - 2$

10. $(3x^3y^2z)(-2xy^{-1}z^3) = -6x^4yz^4$

11. $-5a^2b(3ab^3 - 2ab^4) = -15a^3b^4 + 10a^3b^5$

12. $(z + 4)(z - 4) = z^2 - 4z + 4z - 16$
$ = z^2 - 16$

13. $(3x - 2)(4x + 3) = 12x^2 + 9x - 8x - 6$
$ = 12x^2 + x - 6$

14. $3xy^2 + 6x^2y = 3xy(y + 2x)$

15. $12a^3b^2c - 3a^2b^2c^2 + 6abc^3 = 3abc(4a^2b - abc + 2c^2)$

16. $x^2y^{n+2} + y^n = y^n(x^2y^{n+2-n} + y^{n-n}) = y^n(x^2y^2 + 1)$

17. $a^nb^n - ab^{-n} = b^n(a^nb^{n-n} - ab^{-n-n}) = b^n(a^n - ab^{-2n})$

18. $(u - v)r + (u - v)s = (u - v)(r + s)$

19. $ax - xy + ay - y^2 = x(a - y) + y(a - y)$
$ = (a - y)(x + y)$

20. $x^2 - 49 = x^2 - 7^2 = (x + 7)(x - 7)$

21. $2x^2 - 32 = 2(x^2 - 16) = 2(x + 4)(x - 4)$

22. $4y^4 - 64 = 4(y^4 - 16) = 4(y^2 + 4)(y^2 - 4) = 4(y^2 + 4)(y + 2)(y - 2)$

23. $b^3 + 125 = b^3 + 5^3$
$ = (b + 5)(b^2 - 5b + 25)$

24. $b^3 - 27 = b^3 - 3^3$
$ = (b - 3)(b^2 + 3b + 9)$

25. $3u^3 - 24 = 3(u^3 - 8) = 3(u^3 - 2^3) = 3(u - 2)(u^2 + 2u + 4)$

26. $a^2 - 5a - 6 = (a - 6)(a + 1)$

27. $6b^2 + b - 2 = (3b + 2)(2b - 1)$

28. $6u^2 + 9u - 6 = 3(2u^2 + 3u - 2)$
$$= 3(2u - 1)(u + 2)$$

29. $20r^2 - 15r - 5 = 5(4r^2 - 3r - 1)$
$$= 5(4r + 1)(r - 1)$$

30. $x^{2n} + 2x^n + 1 = (x^n + 1)(x^n + 1) = (x^n + 1)^2$

31. $x^2 + 6x + 9 - y^2 = (x^2 + 6x + 9) - y^2 = (x + 3)^2 - y^2 = (x + 3 + y)(x + 3 - y)$

32. $r_1 r_2 - r_2 r = r_1 r$
$$r_1 r_2 = r_1 r + r_2 r$$
$$r_1 r_2 = r(r_1 + r_2)$$
$$\frac{r_1 r_2}{r_1 + r_2} = r, \text{ or } r = \frac{r_1 r_2}{r_1 + r_2}$$

33. $x^2 - 5x - 6 = 0$
$$(x + 1)(x - 6) = 0$$
$$x + 1 = 0 \quad \text{or} \quad x - 6 = 0$$
$$x = -1 \qquad\qquad x = 6$$

34. Let x and $x + 1$ represent the integers.
$$x(x + 1) = 156$$
$$x^2 + x - 156 = 0$$
$$(x + 13)(x - 12) = 0$$
$$x + 13 = 0 \quad \text{or} \quad x - 12 = 0$$
$$x = -13 \qquad\qquad x = 12$$
(impossible)
The integers are 12 and 13 (sum of 25).

35. $\boxed{\text{Area of border}} = 70$
$$2w(1) + 2w(1) + w(1) + w(1) + 4(1) = 70$$
$$6w + 4 = 70$$
$$6w = 66$$
$$w = 11$$
The dimensions are 11 ft by 22 ft.

Exercise 6.1 (page 349)

1. $3x^2 - 9x = 3x(x - 3)$

3. $27x^6 + 64y^3 = (3x^2)^3 + (4y)^3 = (3x^2 + 4y)(9x^4 - 12x^2y + 16y^2)$

5. rational

7. asymptote

9. a

11. $\frac{a}{b}; 0$

13. $t = f(r) = \dfrac{600}{r}$
$$f(30) = \frac{600}{30}$$
$$= 20$$
$$t = 20 \text{ hours}$$

15. $t = f(r) = \dfrac{600}{r}$
$$f(50) = \frac{600}{50}$$
$$= 12$$
$$t = 12 \text{ hours}$$

17. $C = f(p) = \dfrac{50{,}000p}{100 - p}$
$$f(10) = \frac{50{,}000(10)}{100 - 10}$$
$$= \frac{500{,}000}{90}$$
$$\approx 5555.56$$
$$C = \$5555.56$$

19. $C = f(p) = \dfrac{50{,}000p}{100 - p}$
$$f(50) = \frac{50{,}000(50)}{100 - 50}$$
$$= \frac{2{,}500{,}000}{50}$$
$$= 50{,}000.00$$
$$C = \$50{,}000$$

21. $c = f(x) = 1.25x + 700$

23. Refer to #21.
$$c = f(500) = 1.25(500) + 700 = \$1325$$

25. Refer to #22. $\bar{c} = f(1000) = \dfrac{1.25(1000) + 700}{1000} = \dfrac{1950}{1000} = \$1.95$

27. $c = f(n) = 0.09n + 7.50$

29. Refer to #27. $c = f(775) = 0.09(775) + 7.50 = 69.75 + 7.50 = \$77.25$

31. Refer to #28. $\bar{c} = f(1000) = \dfrac{0.09(1000) + 7.50}{1000} = \dfrac{97.50}{1000} = \$0.0975 = 9.75¢$

33. $f(t) = \dfrac{t^2 + 2t}{2t + 2}$

$f(15) = \dfrac{15^2 + 2(15)}{2(15) + 2}$

$= \dfrac{225 + 30}{30 + 2} = \dfrac{255}{32} \approx 7.96875$

It will take them almost 8 days.

35. $f(t) = \dfrac{t^2 + 3t}{2t + 3}$

Small pipe: 7 hrs $\Rightarrow$ Large pipe: 4 hrs

$f(4) = \dfrac{4^2 + 3(4)}{2(4) + 3}$

$= \dfrac{16 + 12}{8 + 3} = \dfrac{28}{11} = 2.5455$

It will take the pipes about 2.55 hours.

37. $f(x) = \dfrac{x}{x - 2}$

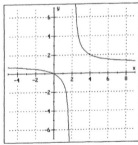

domain: $(-\infty, 2) \cup (2, \infty)$

39. $f(x) = \dfrac{x + 1}{x^2 - 4}$

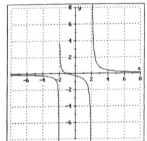

domain: $(-\infty, -2) \cup (-2, 2) \cup (2, \infty)$

41. $\dfrac{12}{18} = \dfrac{\cancel{6} \cdot 2}{\cancel{6} \cdot 3} = \dfrac{2}{3}$

43. $-\dfrac{112}{36} = -\dfrac{\cancel{4} \cdot 28}{\cancel{4} \cdot 9} = -\dfrac{28}{9}$

45. $\dfrac{288}{312} = \dfrac{\cancel{24} \cdot 12}{\cancel{24} \cdot 13} = \dfrac{12}{13}$

47. $-\dfrac{244}{74} = -\dfrac{\cancel{2} \cdot 122}{\cancel{2} \cdot 37} = -\dfrac{122}{37}$

49. $\dfrac{12x^3}{3x} = \dfrac{12}{3} \cdot \dfrac{x^3}{x} = 4x^2$

51. $\dfrac{-24x^3y^4}{18x^4y^3} = -\dfrac{24}{18} \cdot \dfrac{x^3y^4}{x^4y^3} = -\dfrac{4y}{3x}$

53. $\dfrac{(3x^3)^2}{9x^4} = \dfrac{9x^6}{9x^4} = x^2$

55. $-\dfrac{11x(x - y)}{22(x - y)} = -\dfrac{x}{2}$

57. $\dfrac{9y^2(y - z)}{21y(y - z)^2} = \dfrac{3y}{7(y - z)}$

59. $\dfrac{(a-b)(c-d)}{(c-d)(a-b)} = 1$

61. $\dfrac{x+y}{x^2-y^2} = \dfrac{x+y}{(x+y)(x-y)} = \dfrac{1}{x-y}$

63. $\dfrac{5x-10}{x^2-4x+4} = \dfrac{5(x-2)}{(x-2)(x-2)} = \dfrac{5}{x-2}$

65. $\dfrac{12-3x^2}{x^2-x-2} = \dfrac{3(4-x^2)}{(x-2)(x+1)} = \dfrac{3(2+x)(2-x)}{(x-2)(x+1)} = \dfrac{-3(x+2)(x-2)}{(x-2)(x+1)} = \dfrac{-3(x+2)}{x+1}$

67. $\dfrac{3x+6y}{x+2y} = \dfrac{3(x+2y)}{x+2y} = 3$

69. $\dfrac{x^3+8}{x^2-2x+4} = \dfrac{(x+2)(x^2-2x+4)}{x^2-2x+4}$
$= x+2$

71. $\dfrac{x^2+2x+1}{x^2+4x+3} = \dfrac{(x+1)(x+1)}{(x+1)(x+3)} = \dfrac{x+1}{x+3}$

73. $\dfrac{3m-6n}{3n-6m} = \dfrac{3(m-2n)}{3(n-2m)} = \dfrac{m-2n}{n-2m}$

75. $\dfrac{4x^2+24x+32}{16x^2+8x-48} = \dfrac{4(x^2+6x+8)}{8(2x^2+x-6)} = \dfrac{4(x+4)(x+2)}{8(2x-3)(x+2)} = \dfrac{x+4}{2(2x-3)}$

77. $\dfrac{3x^2-3y^2}{x^2+2y+2x+yx} = \dfrac{3(x^2-y^2)}{x^2+2x+yx+2y} = \dfrac{3(x+y)(x-y)}{(x+2)(x+y)} = \dfrac{3(x-y)}{x+2}$

79. $\dfrac{4x^2+8x+3}{6+x-2x^2} = \dfrac{4x^2+8x+3}{-(2x^2-x-6)} = \dfrac{(2x+3)(2x+1)}{-(2x+3)(x-2)} = \dfrac{2x+1}{-(x-2)} = \dfrac{2x+1}{2-x}$

81. $\dfrac{a^3+27}{4a^2-36} = \dfrac{a^3+27}{4(a^2-9)} = \dfrac{(a+3)(a^2-3a+9)}{4(a+3)(a-3)} = \dfrac{a^2-3a+9}{4(a-3)}$

83. $\dfrac{2x^2-3x-9}{2x^2+3x-9} = \dfrac{(2x+3)(x-3)}{(2x-3)(x+3)}$
$\Rightarrow$ lowest terms

85. $\dfrac{(m+n)^3}{m^2+2mn+n^2} = \dfrac{(m+n)^3}{(m+n)(m+n)}$
$= \dfrac{(m+n)^3}{(m+n)^2} = m+n$

87. $\dfrac{m^3-mn^2}{mn^2+m^2n-2m^3} = \dfrac{m(m^2-n^2)}{m(n^2+mn-2m^2)} = \dfrac{m(m+n)(m-n)}{m(n+2m)(n-m)} = \dfrac{m(m+n)(m-n)}{-m(n+2m)(m-n)}$
$= -\dfrac{m+n}{2m+n}$

89. $\dfrac{x^4-y^4}{(x^2+2xy+y^2)(x^2+y^2)} = \dfrac{(x^2+y^2)(x^2-y^2)}{(x+y)^2(x^2+y^2)} = \dfrac{(x^2+y^2)(x+y)(x-y)}{(x+y)^2(x^2+y^2)} = \dfrac{x-y}{x+y}$

91. $\dfrac{4a^2-9b^2}{2a^2-ab-6b^2} = \dfrac{(2a+3b)(2a-3b)}{(2a+3b)(a-2b)} = \dfrac{2a-3b}{a-2b}$

93. $\dfrac{x-y}{x^3-y^3-x+y} = \dfrac{x-y}{(x^3-y^3)-(x-y)} = \dfrac{x-y}{(x-y)(x^2+xy+y^2)-1(x-y)}$

$$= \dfrac{x-y}{(x-y)(x^2+xy+y^2-1)}$$

$$= \dfrac{1}{x^2+xy+y^2-1}$$

95. $\dfrac{px-py+qx-qy}{px+qx+py+qy} = \dfrac{p(x-y)+q(x-y)}{x(p+q)+y(p+q)} = \dfrac{(x-y)(p+q)}{(p+q)(x+y)} = \dfrac{x-y}{x+y}$

97. $\dfrac{(x^2-1)(x+1)}{(x^2-2x+1)^2} = \dfrac{(x+1)(x-1)(x+1)}{[(x-1)^2]^2} = \dfrac{(x+1)^2(x-1)}{(x-1)^4} = \dfrac{(x+1)^2}{(x-1)^3}$

99. $\dfrac{(2x^2+3xy+y^2)(3a+b)}{(x+y)(2xy+2bx+y^2+by)} = \dfrac{(2x+y)(x+y)(3a+b)}{(x+y)[2x(y+b)+y(y+b)]} = \dfrac{(2x+y)(x+y)(3a+b)}{(x+y)(y+b)(2x+y)}$

$$= \dfrac{3a+b}{y+b}$$

101. $P(E) = \dfrac{s}{n} = \dfrac{1}{6}$

103. $P(E) = \dfrac{s}{n} = \dfrac{0}{6} = 0$

105. $P(E) = \dfrac{s}{n} = \dfrac{26}{52} = \dfrac{1}{2}$

107. $P(E) = \dfrac{s}{n} = \dfrac{4}{52} = \dfrac{1}{13}$

109. a. $C = f(p) = \dfrac{50{,}000p}{100-p}$ **b.** $C = f(p) = \dfrac{50{,}000p}{100-p}$

$f(40) = \dfrac{50{,}000(40)}{100-40}$ $f(70) = \dfrac{50{,}000(70)}{100-70}$

$\qquad\quad = \dfrac{2{,}000{,}000}{60}$ $= \dfrac{3{,}500{,}000}{30}$

$\qquad\quad \approx 33333.33$ ≈ 116666.67

$\quad C = \$33333.33$ $C = \$116666.67$

111. $P(E) = \dfrac{s}{n} = \dfrac{1}{6{,}000{,}000}$

113. Sample space: $\{BBB, BBG, BGB, GBB, BGG, GBG, GGB, GGG\}$; $P(E) = \dfrac{s}{n} = \dfrac{1}{8}$

115. Answers may vary.

117. $\dfrac{a-3b}{2b-a} = \dfrac{-(3b-a)}{-(a-2b)} = \dfrac{3b-a}{a-2b}$

The two answers are the same.

119. You can divide out the 4's in parts a and d.   **121.** It is impossible for a man to be 20 feet tall.

123. Since an impossible event has a probability of 0, and an event certain to occur has a probability of 1, all events must have a probability between 0 and 1 (including the endpoints).

Exercise 6.2 (page 361)

1. $(x^2x^3)^2 = (x^5)^2 = x^{10}$

3. $\dfrac{b^0 - 2b^0}{b^0} = \dfrac{1-2}{1} = \dfrac{-1}{1} = -1$

5. $35{,}000 = 3.5 \times 10^4$

7. $2.5 \times 10^{-3} = 0.0025$

9. unit costs; rates

11. extremes; means

13. direct

15. rational

17. joint

19. direct

21. neither

23.
$$\frac{x}{5} = \frac{15}{25}$$
$$25x = 75$$
$$x = 3$$

25.
$$\frac{r-2}{3} = \frac{r}{5}$$
$$5(r-2) = 3r$$
$$5r - 10 = 3r$$
$$2r = 10$$
$$r = 5$$

27.
$$\frac{3}{n} = \frac{2}{n+1}$$
$$3(n+1) = 2n$$
$$3n + 3 = 2n$$
$$n = -3$$

29.
$$\frac{5}{5z+3} = \frac{2z}{2z^2+6}$$
$$5(2z^2+6) = 2z(5z+3)$$
$$10z^2 + 30 = 10z^2 + 6z$$
$$30 = 6z$$
$$5 = z$$

31.
$$\frac{2}{c} = \frac{c-3}{2}$$
$$4 = c^2 - 3c$$
$$0 = c^2 - 3c - 4$$
$$0 = (c+1)(c-4)$$
$$c + 1 = 0 \quad \text{or} \quad c - 4 = 0$$
$$c = -1 \qquad\qquad c = 4$$

33.
$$\frac{2}{3x} = \frac{6x}{36}$$
$$72 = 18x^2$$
$$0 = 18x^2 - 72$$
$$0 = 18(x+2)(x-2)$$
$$x + 2 = 0 \quad \text{or} \quad x - 2 = 0$$
$$x = -2 \qquad\qquad x = 2$$

35.
$$\frac{2(x+3)}{3} = \frac{4(x-4)}{5}$$
$$10(x+3) = 12(x-4)$$
$$10x + 30 = 12x - 48$$
$$-2x = -78$$
$$x = 39$$

37.
$$\frac{1}{x+3} = \frac{-2x}{x+5}$$
$$x + 5 = -2x^2 - 6x$$
$$2x^2 + 7x + 5 = 0$$
$$(2x+5)(x+1)$$
$$2x + 5 = 0 \quad \text{or} \quad x + 1 = 0$$
$$x = -\tfrac{5}{2} \qquad\qquad x = -1$$

39.
$$\frac{a-4}{a+2} = \frac{a-5}{a+1}$$
$$(a-4)(a+1) = (a+2)(a-5)$$
$$a^2 - 3a - 4 = a^2 - 3a - 10$$
$$-3a - 4 = -3a - 10$$
$$-4 \neq -10$$
$$\text{no solution}$$

41. $A = kp^2$

43. $v = \dfrac{k}{r^3}$

45. $B = kmn$

47. $P = \dfrac{ka^2}{j^3}$

49. L varies jointly with m and n.

51. E varies jointly with a and the square of b.

53. X varies directly with x^2 and inversely with y^2.

55. R varies directly with L and inversely with d^2.

57. Let c = the cost of 5 shirts.
$$\frac{2}{25} = \frac{5}{c}$$
$$2c = 125$$
$$c = 62.5$$
5 shirts will cost $62.50.

59. Let g = gallons of gas for 315 miles.
$$\frac{42}{1} = \frac{315}{g}$$
$$42g = 315$$
$$g = \frac{315}{42} = 7.5$$
7.5 gallons of gas are needed.

61. Let w = the width if it were a real house.
$$\frac{1 \text{ in.}}{1 \text{ ft}} = \frac{32 \text{ in.}}{w \text{ ft}}$$
$$w = 32$$
The width would be 32 feet.

63. Let h = the actual height of the building.
$$\frac{7 \text{ in.}}{280 \text{ ft}} = \frac{2 \text{ in.}}{h \text{ ft}}$$
$$7h = 560$$
$$h = 80$$
The building is 80 feet tall.

65. Let d = the dosage required.
$$\frac{0.006}{1} = \frac{d}{30}$$
$$0.18 = d$$
The dosage is 0.18 grams.

67. Let h = the height of the tree.
$$\frac{6}{4} = \frac{h}{28}$$
$$168 = 4h$$
$$42 = h$$
The tree is 42 feet tall.

69. $$\frac{20}{32} = \frac{w}{75}$$
$$1500 = 32w$$
$$w = \frac{1500}{32} = 46\frac{7}{8}$$
The width of the river is $46\frac{7}{8}$ feet.

71. $$\frac{1350}{1} = \frac{x}{5}$$
$$6750 = x$$
The plane will descend 6750 feet.

73. $A = kr^2$
$A = \pi r^2$
$A = \pi (6 \text{ in.})^2$
$\boxed{A = 36\pi \text{ in.}^2}$

75. $d = kg$ $d = 24g$
$288 = k(12)$ $d = 24(18)$
$24 = k$ $\boxed{d = 432 \text{ mi}}$

77. $t = \dfrac{k}{n}$ $t = \dfrac{250}{n}$
$10 = \dfrac{k}{25}$ $t = \dfrac{250}{10}$
$250 = k$ $\boxed{t = 25 \text{ days}}$

79. $V = \dfrac{k}{P}$ $V = \dfrac{120}{P}$
$20 = \dfrac{k}{6}$ $V = \dfrac{120}{10}$
$120 = k$ $\boxed{V = 12 \text{ in.}^3}$

81. $f = \dfrac{k}{l}$ $f = \dfrac{512}{l}$

$256 = \dfrac{k}{2}$ $f = \dfrac{512}{6}$

$512 = k$ $\boxed{f = 85\frac{1}{3}}$

83. $V_1 = klwh$

$V_2 = k(2l)(3w)(2h) = 12klwh = 12V_1$

The volume is multiplied by 12.

85. $g = khr^2$

$g = 23.5hr^2$

$g = 23.5(20)(7.5)^2$

$\boxed{g = 26{,}437.5 \text{ gallons}}$

87. $V = kC$

$6 = k(2)$

$3 = k \Rightarrow$ The resistance is 3 ohms.

89. $D = \dfrac{k}{wd^3}$ $D = \dfrac{281.6}{wd^3}$

$1.1 = \dfrac{k}{4(4)^3}$ $D = \dfrac{281.6}{2(8)^3}$

$1.1 = \dfrac{k}{256}$ $D = \dfrac{281.6}{1024}$

$281.6 = k$ $\boxed{D = 0.275 \text{ in.}}$

91. $P = \dfrac{kT}{V}$ $P = \dfrac{\frac{1}{273}T}{V}$

$1 = \dfrac{k(273)}{1}$ $1 = \dfrac{\frac{1}{273}T}{2}$

$\dfrac{1}{273} = k$ $2 = \dfrac{1}{273}T$

$\boxed{546 \text{ K} = T}$

93. Answers may vary.

95. Answers may vary.

97. This is not direct variation. For this to be direct variation, one temperature would have to be a constant multiple of the other.

Exercise 6.3 (page 371)

1. $-2a^2(3a^3 - a^2) = -6a^5 + 2a^4$

3. $(m^n + 2)(m^n - 2) = m^{2n} - 4$

5. $\dfrac{ac}{bd}$

7. 0

9. $\dfrac{3}{4} \cdot \dfrac{5}{3} \cdot \dfrac{8}{7} = \dfrac{\cancel{3}}{\cancel{4}} \cdot \dfrac{5}{\cancel{3}} \cdot \dfrac{\cancel{4} \cdot 2}{7} = \dfrac{10}{7}$

11. $-\dfrac{6}{11} \div \dfrac{36}{55} = -\dfrac{6}{11} \cdot \dfrac{55}{36} = -\dfrac{\cancel{6}}{\cancel{11}} \cdot \dfrac{\cancel{11} \cdot 5}{\cancel{6} \cdot 6}$

$= -\dfrac{5}{6}$

13. $\dfrac{x^2y^2}{cd} \cdot \dfrac{c^{-2}d^2}{x} = x^{2-1}y^2 c^{-2-1}d^{2-1} = xy^2 c^{-3}d = \dfrac{xy^2 d}{c^3}$

15. $\dfrac{-x^2y^{-2}}{x^{-1}y^{-3}} \div \dfrac{x^{-3}y^2}{x^4 y^{-1}} = \dfrac{-x^2 y^{-2}}{x^{-1}y^{-3}} \cdot \dfrac{x^4 y^{-1}}{x^{-3}y^2} = \dfrac{-x^6 y^{-3}}{x^{-4}y^{-1}} = -x^{6-(-4)}y^{-3-(-1)} = -x^{10}y^{-2} = -\dfrac{x^{10}}{y^2}$

17. $\dfrac{x^2 + 2x + 1}{x} \cdot \dfrac{x^2 - x}{x^2 - 1} = \dfrac{(x+1)(x+1)}{x} \cdot \dfrac{x(x-1)}{(x+1)(x-1)} = x+1$

19. $\dfrac{2x^2 - x - 3}{x^2 - 1} \cdot \dfrac{x^2 + x - 2}{2x^2 + x - 6} = \dfrac{(2x-3)(x+1)}{(x+1)(x-1)} \cdot \dfrac{(x+2)(x-1)}{(2x-3)(x+2)} = 1$

21. $\dfrac{x^2 - 16}{x^2 - 25} \div \dfrac{x+4}{x-5} = \dfrac{x^2 - 16}{x^2 - 25} \cdot \dfrac{x-5}{x+4} = \dfrac{(x+4)(x-4)}{(x+5)(x-5)} \cdot \dfrac{x-5}{x+4} = \dfrac{x-4}{x+5}$

23. $\dfrac{a^2 + 2a - 35}{12x} \div \dfrac{ax - 3x}{a^2 + 4a - 21} = \dfrac{a^2 + 2a - 35}{12x} \cdot \dfrac{a^2 + 4a - 21}{ax - 3x}$

$$= \dfrac{(a+7)(a-5)}{12x} \cdot \dfrac{(a+7)(a-3)}{x(a-3)} = \dfrac{(a+7)^2(a-5)}{12x^2}$$

25. $\dfrac{3t^2 - t - 2}{6t^2 - 5t - 6} \cdot \dfrac{4t^2 - 9}{2t^2 + 5t + 3} = \dfrac{(3t+2)(t-1)}{(3t+2)(2t-3)} \cdot \dfrac{(2t+3)(2t-3)}{(2t+3)(t+1)} = \dfrac{t-1}{t+1}$

27. $\dfrac{3n^2 + 5n - 2}{12n^2 - 13n + 3} \div \dfrac{n^2 + 3n + 2}{4n^2 + 5n - 6} = \dfrac{3n^2 + 5n - 2}{12n^2 - 13n + 3} \cdot \dfrac{4n^2 + 5n - 6}{n^2 + 3n + 2}$

$$= \dfrac{(3n-1)(n+2)}{(3n-1)(4n-3)} \cdot \dfrac{(4n-3)(n+2)}{(n+2)(n+1)} = \dfrac{n+2}{n+1}$$

29. $(x+1) \cdot \dfrac{1}{x^2 + 2x + 1} = \dfrac{x+1}{1} \cdot \dfrac{1}{(x+1)(x+1)} = \dfrac{1}{x+1}$

31. $(x^2 - x - 2) \cdot \dfrac{x^2 + 3x + 2}{x^2 - 4} = \dfrac{x^2 - x - 2}{1} \cdot \dfrac{x^2 + 3x + 2}{x^2 - 4} = \dfrac{(x-2)(x+1)}{1} \cdot \dfrac{(x+2)(x+1)}{(x+2)(x-2)}$

$$= (x+1)^2$$

33. $(2x^2 - 15x + 25) \div \dfrac{2x^2 - 3x - 5}{x+1} = \dfrac{2x^2 - 15x + 25}{1} \cdot \dfrac{x+1}{2x^2 - 3x - 5}$

$$= \dfrac{(2x-5)(x-5)}{1} \cdot \dfrac{x+1}{(2x-5)(x+1)} = x - 5$$

35. $\dfrac{x^3 + y^3}{x^3 - y^3} \div \dfrac{x^2 - xy + y^2}{x^2 + xy + y^2} = \dfrac{x^3 + y^3}{x^3 - y^3} \cdot \dfrac{x^2 + xy + y^2}{x^2 - xy + y^2} = \dfrac{(x+y)(x^2 - xy + y^2)}{(x-y)(x^2 + xy + y^2)} \cdot \dfrac{x^2 + xy + y^2}{x^2 - xy + y^2}$

$$= \dfrac{x+y}{x-y}$$

37. $\dfrac{m^2 - n^2}{2x^2 + 3x - 2} \cdot \dfrac{2x^2 + 5x - 3}{n^2 - m^2} = \dfrac{(m+n)(m-n)}{(2x-1)(x+2)} \cdot \dfrac{(2x-1)(x+3)}{(n+m)(n-m)} = -\dfrac{x+3}{x+2}$

39. $\dfrac{ax + ay + bx + by}{x^3 - 27} \cdot \dfrac{x^2 + 3x + 9}{xc + xd + yc + yd} = \dfrac{a(x+y) + b(x+y)}{(x-3)(x^2 + 3x + 9)} \cdot \dfrac{x^2 + 3x + 9}{x(c+d) + y(c+d)}$

$$= \dfrac{(x+y)(a+b)}{(x-3)(x^2 + 3x + 9)} \cdot \dfrac{x^2 + 3x + 9}{(c+d)(x+y)}$$

$$= \dfrac{a+b}{(x-3)(c+d)}$$

41. $\dfrac{x^2-x-6}{x^2-4}\cdot\dfrac{x^2-x-2}{9-x^2}=\dfrac{(x-3)(x+2)}{(x+2)(x-2)}\cdot\dfrac{(x-2)(x+1)}{(3+x)(3-x)}=-\dfrac{x+1}{x+3}$

43. $\dfrac{2x^2+3xy+y^2}{y^2-x^2}\div\dfrac{6x^2+5xy+y^2}{2x^2-xy-y^2}=\dfrac{2x^2+3xy+y^2}{y^2-x^2}\cdot\dfrac{2x^2-xy-y^2}{6x^2+5xy+y^2}$

$$=\dfrac{(2x+y)(x+y)}{(y+x)(y-x)}\cdot\dfrac{(2x+y)(x-y)}{(2x+y)(3x+y)}=-\dfrac{2x+y}{3x+y}$$

45. $\dfrac{3x^2y^2}{6x^3y}\cdot\dfrac{-4x^7y^{-2}}{18x^{-2}y}\div\dfrac{36x}{18y^{-2}}=\dfrac{3x^2y^2}{6x^3y}\cdot\dfrac{-4x^7y^{-2}}{18x^{-2}y}\cdot\dfrac{18y^{-2}}{36x}=\dfrac{-2^3\cdot3^3x^9y^{-2}}{2^4\cdot3^5x^2y^2}=-\dfrac{x^7}{18y^4}$

47. $(4x+12)\cdot\dfrac{x^2}{2x-6}\div\dfrac{2}{x-3}=\dfrac{4x+12}{1}\cdot\dfrac{x^2}{2x-6}\cdot\dfrac{x-3}{2}=\dfrac{4(x+3)}{1}\cdot\dfrac{x^2}{2(x-3)}\cdot\dfrac{x-3}{2}$

$$=x^2(x+3)$$

49. $\dfrac{2x^2-2x-4}{x^2+2x-8}\cdot\dfrac{3x^2+15x}{x+1}\div\dfrac{4x^2-100}{x^2-x-20}=\dfrac{2x^2-2x-4}{x^2+2x-8}\cdot\dfrac{3x^2+15x}{x+1}\cdot\dfrac{x^2-x-20}{4x^2-100}$

$$=\dfrac{2(x-2)(x+1)}{(x+4)(x-2)}\cdot\dfrac{3x(x+5)}{x+1}\cdot\dfrac{(x-5)(x+4)}{4(x+5)(x-5)}$$

$$=\dfrac{3x}{2}$$

51. $\dfrac{2t^2+5t+2}{t^2-4t+16}\div\dfrac{t+2}{t^3+64}\div\dfrac{2t^3+9t^2+4t}{t+1}=\dfrac{2t^2+5t+2}{t^2-4t+16}\cdot\dfrac{t^3+64}{t+2}\cdot\dfrac{t+1}{2t^3+9t^2+4t}$

$$=\dfrac{(2t+1)(t+2)}{t^2-4t+16}\cdot\dfrac{(t+4)(t^2-4t+16)}{t+2}\cdot\dfrac{t+1}{t(2t+1)(t+4)}=\dfrac{t+1}{t}$$

53. $\dfrac{x^4-3x^2-4}{x^4-1}\cdot\dfrac{x^2+3x+2}{x^2+4x+4}=\dfrac{(x^2-4)(x^2+1)}{(x^2+1)(x^2-1)}\cdot\dfrac{(x+1)(x+2)}{(x+2)(x+2)}$

$$=\dfrac{(x+2)(x-2)}{(x+1)(x-1)}\cdot\dfrac{(x+1)(x+2)}{(x+2)(x+2)}=\dfrac{x-2}{x-1}$$

55. $(x^2-x-6)\div(x-3)\div(x-2)=\dfrac{x^2-x-6}{1}\div\dfrac{x-3}{1}\div\dfrac{x-2}{1}$

$$=\dfrac{(x-3)(x+2)}{1}\cdot\dfrac{1}{x-3}\cdot\dfrac{1}{x-2}=\dfrac{x+2}{x-2}$$

57. $\dfrac{3x^2-2x}{3x+2}\div(3x-2)\div\dfrac{3x}{3x-3}=\dfrac{3x^2-2x}{3x+2}\cdot\dfrac{1}{3x-2}\cdot\dfrac{3x-3}{3x}$

$$=\dfrac{x(3x-2)}{3x+2}\cdot\dfrac{1}{3x-2}\cdot\dfrac{3(x-1)}{3x}=\dfrac{x-1}{3x+2}$$

59. $\dfrac{2x^2 + 5x - 3}{x^2 + 2x - 3} \div \left(\dfrac{x^2 + 2x - 35}{x^2 - 6x + 5} \div \dfrac{x^2 - 9x + 14}{2x^2 - 5x + 2} \right) =$

$\quad = \dfrac{2x^2 + 5x - 3}{x^2 + 2x - 3} \div \left(\dfrac{x^2 + 2x - 35}{x^2 - 6x + 5} \cdot \dfrac{2x^2 - 5x + 2}{x^2 - 9x + 14} \right)$

$\quad = \dfrac{2x^2 + 5x - 3}{x^2 + 2x - 3} \div \left(\dfrac{(x+7)(x-5)}{(x-5)(x-1)} \cdot \dfrac{(2x-1)(x-2)}{(x-7)(x-2)} \right)$

$\quad = \dfrac{2x^2 + 5x - 3}{x^2 + 2x - 3} \div \dfrac{(x+7)(2x-1)}{(x-1)(x-7)} = \dfrac{(2x-1)(x+3)}{(x+3)(x-1)} \cdot \dfrac{(x-1)(x-7)}{(x+7)(2x-1)} = \dfrac{x-7}{x+7}$

61. $\dfrac{x^2 - x - 12}{x^2 + x - 2} \div \dfrac{x^2 - 6x + 8}{x^2 - 3x - 10} \cdot \dfrac{x^2 - 3x + 2}{x^2 - 2x - 15} = \dfrac{x^2 - x - 12}{x^2 + x - 2} \cdot \dfrac{x^2 - 3x - 10}{x^2 - 6x + 8} \cdot \dfrac{x^2 - 3x + 2}{x^2 - 2x - 15}$

$\quad = \dfrac{(x-4)(x+3)}{(x+2)(x-1)} \cdot \dfrac{(x-5)(x+2)}{(x-4)(x-2)} \cdot \dfrac{(x-2)(x-1)}{(x-5)(x+3)} = 1$

63. $A = \dfrac{1}{2}bh = \dfrac{1}{2} \cdot \dfrac{b^2 - 4}{b+3} \cdot \dfrac{b^2 - 9}{b+2} = \dfrac{1}{2} \cdot \dfrac{(b+2)(b-2)}{b+3} \cdot \dfrac{(b+3)(b-3)}{b+2} = \dfrac{(b-2)(b-3)}{2}$ cm^2

65. $d = rt = \dfrac{k^2 - k - 6}{k - 4} \cdot \dfrac{k^2 - 16}{k^2 - 2k - 3} = \dfrac{(k-3)(k+2)}{k-4} \cdot \dfrac{(k+4)(k-4)}{(k-3)(k+1)} = \dfrac{(k+2)(k+4)}{k+1}$ miles

67. **Answers may vary.**

69. $\dfrac{x^2}{y} \boxed{\div} \dfrac{x}{y^2} \boxed{\times} \dfrac{x^2}{y^2} = \dfrac{x^3}{y}$

Exercise 6.4 (page 380)

1. $(-1, 4] \Rightarrow$

3. $P = 2l + 2w$

$\quad P - 2l = 2w$

$\quad \dfrac{P - 2l}{2} = w$, or $w = \dfrac{P - 2l}{2}$

5. $\dfrac{a+c}{b}$

7. subtract; keep

9. LCD

11. $\dfrac{3}{4} + \dfrac{7}{4} = \dfrac{10}{4} = \dfrac{5}{2}$

13. $\dfrac{10}{33} - \dfrac{21}{33} = \dfrac{-11}{33} = -\dfrac{1}{3}$

15. $\dfrac{3}{4y} + \dfrac{8}{4y} = \dfrac{11}{4y}$

17. $\dfrac{3}{a+b} - \dfrac{a}{a+b} = \dfrac{3-a}{a+b}$

19. $\dfrac{3x}{2x+2} + \dfrac{x+4}{2x+2} = \dfrac{4x+4}{2x+2} = \dfrac{4(x+1)}{2(x+1)} = 2$

21. $\dfrac{3x}{x-3} - \dfrac{9}{x-3} = \dfrac{3x-9}{x-3} = \dfrac{3(x-3)}{x-3} = 3$

23. $\dfrac{5x}{x+1} + \dfrac{3}{x+1} - \dfrac{2x}{x+1} = \dfrac{3x+3}{x+1}$

$\quad = \dfrac{3(x+1)}{x+1} = 3$

25. $\dfrac{3(x^2 + x)}{x^2 - 5x + 6} + \dfrac{-3(x^2 - x)}{x^2 - 5x + 6} = \dfrac{3x^2 + 3x}{x^2 - 5x + 6} + \dfrac{-3x^2 + 3x}{x^2 - 5x + 6} = \dfrac{6x}{(x - 3)(x - 2)}$

27. $8 = 2^3;\ 12 = 2^2 \cdot 3;\ 18 = 2 \cdot 3^2 \Rightarrow \text{LCD} = 2^3 \cdot 3^2 = 72$

29. $x^2 + 3x = x(x + 3);\ x^2 - 9 = (x + 3)(x - 3) \Rightarrow \text{LCD} = x(x + 3)(x - 3)$

31. $x^3 + 27 = (x + 3)(x^2 - 3x + 9);\ x^2 + 6x + 9 = (x + 3)^2 \Rightarrow \text{LCD} = (x + 3)^2(x^2 - 3x + 9)$

33. $2x^2 + 5x + 3 = (2x + 3)(x + 1)$
$4x^2 + 12x + 9 = (2x + 3)^2$
$x^2 + 2x + 1 = (x + 1)^2$
$\text{LCD} = (2x + 3)^2(x + 1)^2$

35. $\dfrac{1}{2} + \dfrac{1}{3} = \dfrac{1 \cdot 3}{2 \cdot 3} + \dfrac{1 \cdot 2}{3 \cdot 2} = \dfrac{3}{6} + \dfrac{2}{6} = \dfrac{5}{6}$

37. $\dfrac{7}{15} - \dfrac{17}{25} = \dfrac{7 \cdot 5}{15 \cdot 5} - \dfrac{17 \cdot 3}{25 \cdot 3} = \dfrac{35}{75} - \dfrac{51}{75}$
$= -\dfrac{16}{75}$

39. $\dfrac{a}{2} + \dfrac{2a}{5} = \dfrac{a \cdot 5}{2 \cdot 5} + \dfrac{2a \cdot 2}{5 \cdot 2} = \dfrac{5a}{10} + \dfrac{4a}{10}$
$= \dfrac{9a}{10}$

41. $\dfrac{3a}{2} - \dfrac{4b}{7} = \dfrac{3a \cdot 7}{2 \cdot 7} - \dfrac{4b \cdot 2}{7 \cdot 2} = \dfrac{21a}{14} - \dfrac{8b}{14}$
$= \dfrac{21a - 8b}{14}$

43. $\dfrac{3}{4x} + \dfrac{2}{3x} = \dfrac{3 \cdot 3}{4x \cdot 3} + \dfrac{2 \cdot 4}{3x \cdot 4} = \dfrac{9}{12x} + \dfrac{8}{12x}$
$= \dfrac{17}{12x}$

45. $\dfrac{3a}{2b} - \dfrac{2b}{3a} = \dfrac{3a \cdot 3a}{2b \cdot 3a} - \dfrac{2b \cdot 2b}{3a \cdot 2b} = \dfrac{9a^2}{6ab} - \dfrac{4b^2}{6ab} = \dfrac{9a^2 - 4b^2}{6ab}$

47. $\dfrac{a + b}{3} + \dfrac{a - b}{7} = \dfrac{(a + b)7}{3(7)} + \dfrac{(a - b)3}{7(3)} = \dfrac{7a + 7b}{21} + \dfrac{3a - 3b}{21} = \dfrac{10a + 4b}{21}$

49. $\dfrac{3}{x + 2} + \dfrac{5}{x - 4} = \dfrac{3(x - 4)}{(x + 2)(x - 4)} + \dfrac{5(x + 2)}{(x - 4)(x + 2)} = \dfrac{3x - 12}{(x + 2)(x - 4)} + \dfrac{5x + 10}{(x + 2)(x - 4)}$
$= \dfrac{8x - 2}{(x + 2)(x - 4)} = \dfrac{2(4x - 1)}{(x + 2)(x - 4)}$

51. $\dfrac{x + 2}{x + 5} - \dfrac{x - 3}{x + 7} = \dfrac{(x + 2)(x + 7)}{(x + 5)(x + 7)} - \dfrac{(x - 3)(x + 5)}{(x + 7)(x + 5)} = \dfrac{x^2 + 9x + 14}{(x + 5)(x + 7)} - \dfrac{x^2 + 2x - 15}{(x + 5)(x + 7)}$
$= \dfrac{7x + 29}{(x + 5)(x + 7)}$

53. $x + \dfrac{1}{x} = \dfrac{x}{1} + \dfrac{1}{x} = \dfrac{x(x)}{1(x)} + \dfrac{1}{x} = \dfrac{x^2}{x} + \dfrac{1}{x} = \dfrac{x^2 + 1}{x}$

55. $\dfrac{x + 8}{x - 3} - \dfrac{x - 14}{3 - x} = \dfrac{x + 8}{x - 3} - \dfrac{-x + 14}{x - 3} = \dfrac{2x - 6}{x - 3} = \dfrac{2(x - 3)}{x - 3} = 2$

57. $\dfrac{2a+1}{3a+2} - \dfrac{a-4}{2-3a} = \dfrac{2a+1}{3a+2} - \dfrac{-a+4}{3a-2} = \dfrac{(2a+1)(3a-2)}{(3a+2)(3a-2)} + \dfrac{(a-4)(3a+2)}{(3a-2)(3a+2)}$

$$= \dfrac{6a^2 - a - 2}{(3a+2)(3a-2)} + \dfrac{3a^2 - 10a - 8}{(3a+2)(3a-2)}$$

$$= \dfrac{9a^2 - 11a - 10}{(3a+2)(3a-2)}$$

59. $\dfrac{x}{x^2 + 5x + 6} + \dfrac{x}{x^2 - 4} = \dfrac{x}{(x+2)(x+3)} + \dfrac{x}{(x+2)(x-2)}$

$$= \dfrac{x(x-2)}{(x+2)(x+3)(x-2)} + \dfrac{x(x+3)}{(x+2)(x-2)(x+3)}$$

$$= \dfrac{x^2 - 2x}{(x+2)(x+3)(x-2)} + \dfrac{x^2 + 3x}{(x+2)(x+3)(x-2)}$$

$$= \dfrac{2x^2 + x}{(x+2)(x+3)(x-2)}$$

61. $\dfrac{4}{x^2 - 2x - 3} - \dfrac{x}{3x^2 - 7x - 6} = \dfrac{4}{(x-3)(x+1)} - \dfrac{x}{(3x+2)(x-3)}$

$$= \dfrac{4(3x+2)}{(x-3)(x+1)(3x+2)} - \dfrac{x(x+1)}{(3x+2)(x-3)(x+1)}$$

$$= \dfrac{12x + 8}{(x-3)(x+1)(3x+2)} - \dfrac{x^2 + x}{(x-3)(x+1)(3x+2)}$$

$$= \dfrac{-x^2 + 11x + 8}{(x-3)(x+1)(3x+2)}$$

63. $\dfrac{8}{x^2 - 9} + \dfrac{2}{x-3} - \dfrac{6}{x} = \dfrac{8}{(x+3)(x-3)} + \dfrac{2}{x-3} - \dfrac{6}{x}$

$$= \dfrac{8x}{x(x+3)(x-3)} + \dfrac{2x(x+3)}{x(x+3)(x-3)} - \dfrac{6(x+3)(x-3)}{x(x+3)(x-3)}$$

$$= \dfrac{8x}{x(x+3)(x-3)} + \dfrac{2x^2 + 6x}{x(x+3)(x-3)} - \dfrac{6(x^2 - 9)}{x(x+3)(x-3)}$$

$$= \dfrac{8x}{x(x+3)(x-3)} + \dfrac{2x^2 + 6x}{x(x+3)(x-3)} - \dfrac{6x^2 - 54}{x(x+3)(x-3)}$$

$$= \dfrac{-4x^2 + 14x + 54}{x(x+3)(x-3)}$$

65.
$$\frac{x}{x+1} - \frac{x}{1-x^2} + \frac{1}{x} = \frac{x}{x+1} + \frac{x}{x^2-1} + \frac{1}{x}$$
$$= \frac{x}{x+1} + \frac{x}{(x+1)(x-1)} + \frac{1}{x}$$
$$= \frac{x(x)(x-1)}{x(x+1)(x-1)} + \frac{x(x)}{x(x+1)(x-1)} + \frac{1(x+1)(x-1)}{x(x+1)(x-1)}$$
$$= \frac{x^2(x-1)}{x(x+1)(x-1)} + \frac{x^2}{x(x+1)(x-1)} + \frac{x^2-1}{x(x+1)(x-1)}$$
$$= \frac{x^3-x^2}{x(x+1)(x-1)} + \frac{x^2}{x(x+1)(x-1)} + \frac{x^2-1}{x(x+1)(x-1)}$$
$$= \frac{x^3+x^2-1}{x(x+1)(x-1)}$$

67.
$$2x+3+\frac{1}{x+1} = \frac{2x+3}{1} + \frac{1}{x+1} = \frac{(2x+3)(x+1)}{1(x+1)} + \frac{1}{x+1}$$
$$= \frac{2x^2+5x+3}{x+1} + \frac{1}{x+1} = \frac{2x^2+5x+4}{x+1}$$

69.
$$1+x-\frac{x}{x-5} = \frac{x+1}{1} - \frac{x}{x-5} = \frac{(x+1)(x-5)}{1(x-5)} - \frac{x}{x-5}$$
$$= \frac{x^2-4x-5}{x-5} - \frac{x}{x-5} = \frac{x^2-5x-5}{x-5}$$

71.
$$\frac{3x}{x-1} - 2x - x^2 = \frac{3x}{x-1} - \frac{x^2+2x}{1} = \frac{3x}{x-1} - \frac{(x^2+2x)(x-1)}{1(x-1)}$$
$$= \frac{3x}{x-1} - \frac{x^3+x^2-2x}{x-1} = \frac{-x^3-x^2+5x}{x-1}$$

73.
$$\frac{y+4}{y^2+7y+12} - \frac{y-4}{y+3} + \frac{47}{y+4} = \frac{y+4}{(y+4)(y+3)} - \frac{y-4}{y+3} + \frac{47}{y+4}$$
$$= \frac{y+4}{(y+4)(y+3)} - \frac{(y-4)(y+4)}{(y+4)(y+3)} + \frac{47(y+3)}{(y+4)(y+3)}$$
$$= \frac{y+4}{(y+4)(y+3)} - \frac{y^2-16}{(y+4)(y+3)} + \frac{47y+141}{(y+4)(y+3)}$$
$$= \frac{-y^2+48y+161}{(y+4)(y+3)}$$

75. $\dfrac{3}{x+1} - \dfrac{2}{x-1} + \dfrac{x+3}{x^2-1} = \dfrac{3}{x+1} - \dfrac{2}{x-1} + \dfrac{x+3}{(x+1)(x-1)}$

$$= \dfrac{3(x-1)}{(x+1)(x-1)} - \dfrac{2(x+1)}{(x+1)(x-1)} + \dfrac{x+3}{(x+1)(x-1)}$$

$$= \dfrac{3x-3}{(x+1)(x-1)} - \dfrac{2x+2}{(x+1)(x-1)} + \dfrac{x+3}{(x+1)(x-1)}$$

$$= \dfrac{2x-2}{(x+1)(x-1)} = \dfrac{2(x-1)}{(x+1)(x-1)} = \dfrac{2}{x+1}$$

77. $\dfrac{x-2}{x^2-3x} + \dfrac{2x-1}{x^2+3x} - \dfrac{2}{x^2-9} = \dfrac{x-2}{x(x-3)} + \dfrac{2x-1}{x(x+3)} - \dfrac{2}{(x+3)(x-3)}$

$$= \dfrac{(x-2)(x+3)}{x(x-3)(x+3)} + \dfrac{(2x-1)(x-3)}{x(x-3)(x+3)} - \dfrac{2x}{x(x-3)(x+3)}$$

$$= \dfrac{x^2+x-6}{x(x-3)(x+3)} + \dfrac{2x^2-7x+3}{x(x-3)(x+3)} - \dfrac{2x}{x(x-3)(x+3)}$$

$$= \dfrac{3x^2-8x-3}{x(x-3)(x+3)} = \dfrac{(3x+1)(x-3)}{x(x-3)(x+3)} = \dfrac{3x+1}{x(x+3)}$$

79. $\dfrac{5}{x^2-25} - \dfrac{3}{2x^2-9x-5} + 1 = \dfrac{5}{(x+5)(x-5)} - \dfrac{3}{(2x+1)(x-5)} + \dfrac{1}{1}$

$$= \dfrac{5(2x+1)}{(x+5)(x-5)(2x+1)} - \dfrac{3(x+5)}{(x+5)(x-5)(2x+1)} + \dfrac{(x+5)(x-5)(2x+1)}{(x+5)(x-5)(2x+1)}$$

$$= \dfrac{10x+5}{(x+5)(x-5)(2x+1)} - \dfrac{3x+15}{(x+5)(x-5)(2x+1)} + \dfrac{2x^3+x^2-50x-25}{(x+5)(x-5)(2x+1)}$$

$$= \dfrac{2x^3+x^2-43x-35}{(x+5)(x-5)(2x+1)}$$

81. $\dfrac{3x}{x-3} + \dfrac{4}{x-2} - \dfrac{5x}{x^3-5x^2+6x} = \dfrac{3x}{x-3} + \dfrac{4}{x-2} - \dfrac{5x}{x(x-3)(x-2)}$

$$= \dfrac{3x(x)(x-2)}{x(x-3)(x-2)} + \dfrac{4x(x-3)}{x(x-3)(x-2)} - \dfrac{5x}{x(x-3)(x-2)}$$

$$= \dfrac{3x^3-6x^2}{x(x-3)(x-2)} + \dfrac{4x^2-12x}{x(x-3)(x-2)} - \dfrac{5x}{x(x-3)(x-2)}$$

$$= \dfrac{3x^3-2x^2-17x}{x(x-3)(x-2)} = \dfrac{x(3x^2-2x-17)}{x(x-3)(x-2)} = \dfrac{3x^2-2x-17}{(x-3)(x-2)}$$

83. $2 + \dfrac{4a}{a^2 - 1} - \dfrac{2}{a+1} = \dfrac{2}{1} + \dfrac{4a}{(a+1)(a-1)} - \dfrac{2}{(a+1)}$

$\qquad\qquad = \dfrac{2(a+1)(a-1)}{(a+1)(a-1)} + \dfrac{4a}{(a+1)(a-1)} - \dfrac{2(a-1)}{(a+1)(a-1)}$

$\qquad\qquad = \dfrac{2a^2 - 2}{(a+1)(a-1)} + \dfrac{4a}{(a+1)(a-1)} - \dfrac{2a - 2}{(a+1)(a-1)}$

$\qquad\qquad = \dfrac{2a^2 + 2a}{(a+1)(a-1)} = \dfrac{2a(a+1)}{(a+1)(a-1)} = \dfrac{2a}{a-1}$

85. $\dfrac{x+5}{2x^2 - 2} + \dfrac{x}{2x+2} - \dfrac{3}{x-1} = \dfrac{x+5}{2(x+1)(x-1)} + \dfrac{x}{2(x+1)} - \dfrac{3}{x-1}$

$\qquad\qquad = \dfrac{x+5}{2(x+1)(x-1)} + \dfrac{x(x-1)}{2(x+1)(x-1)} - \dfrac{3(2)(x+1)}{2(x+1)(x-1)}$

$\qquad\qquad = \dfrac{x+5}{2(x+1)(x-1)} + \dfrac{x^2 - x}{2(x+1)(x-1)} - \dfrac{6x+6}{2(x+1)(x-1)}$

$\qquad\qquad = \dfrac{x^2 - 6x - 1}{2(x+1)(x-1)}$

87. $\dfrac{a}{a-b} + \dfrac{b}{a+b} + \dfrac{a^2 + b^2}{b^2 - a^2} = \dfrac{-a}{b-a} + \dfrac{b}{b+a} + \dfrac{a^2 + b^2}{(b+a)(b-a)}$

$\qquad\qquad = \dfrac{-a(b+a)}{(b+a)(b-a)} + \dfrac{b(b-a)}{(b+a)(b-a)} + \dfrac{a^2 + b^2}{(b+a)(b-a)}$

$\qquad\qquad = \dfrac{-ab - a^2}{(b+a)(b-a)} + \dfrac{b^2 - ab}{(b+a)(b-a)} + \dfrac{a^2 + b^2}{(b+a)(b-a)}$

$\qquad\qquad = \dfrac{2b^2 - 2ab}{(b+a)(b-a)} = \dfrac{2b(b-a)}{(b+a)(b-a)} = \dfrac{2b}{b+a}$

89. $\dfrac{7n^2}{m-n} + \dfrac{3m}{n-m} - \dfrac{3m^2 - n}{m^2 - 2mn + n^2} = \dfrac{7n^2}{m-n} - \dfrac{3m}{m-n} - \dfrac{3m^2 - n}{(m-n)^2}$

$\qquad\qquad = \dfrac{7n^2(m-n)}{(m-n)^2} - \dfrac{3m(m-n)}{(m-n)^2} - \dfrac{3m^2 - n}{(m-n)^2}$

$\qquad\qquad = \dfrac{7mn^2 - 7n^3}{(m-n)^2} - \dfrac{3m^2 - 3mn}{(m-n)^2} - \dfrac{3m^2 - n}{(m-n)^2}$

$\qquad\qquad = \dfrac{7mn^2 - 7n^3 - 6m^2 + 3mn + n}{(m-n)^2}$

91.
$$\frac{m+1}{m^2+2m+1}+\frac{m-1}{m^2-2m+1}+\frac{2}{m^2-1}=\frac{m+1}{(m+1)^2}+\frac{m-1}{(m-1)^2}+\frac{2}{(m+1)(m-1)}$$
$$=\frac{\cdot\ 1}{m+1}+\frac{1}{m-1}+\frac{2}{(m+1)(m-1)}$$
$$=\frac{1(m-1)}{(m+1)(m-1)}+\frac{1(m+1)}{(m+1)(m-1)}+\frac{2}{(m+1)(m-1)}$$
$$=\frac{2m+2}{(m+1)(m-1)}=\frac{2(m+1)}{(m+1)(m-1)}=\frac{2}{m-1}$$

93.
$$\left(\frac{1}{x-1}+\frac{1}{1-x}\right)^2=\left(\frac{1}{x-1}+\frac{-1}{x-1}\right)^2=\left(\frac{0}{x-1}\right)^2=0^2=0$$

95.
$$\left(\frac{x}{x-3}+\frac{3}{3-x}\right)^3=\left(\frac{x}{x-3}+\frac{-3}{x-3}\right)^3=\left(\frac{x-3}{x-3}\right)^3=1^3=1$$

97.
$$\frac{a}{b}+\frac{c}{d}=\frac{ad}{bd}+\frac{cb}{db}=\frac{ad+bc}{bd}$$

99.
$$h=\frac{9x-2}{2}+x+\frac{3x+6}{x}=\frac{(9x-2)x}{2x}+\frac{x(2x)}{1(2x)}+\frac{(3x+6)2}{x(2)}$$
$$=\frac{9x^2-2x}{2x}+\frac{2x^2}{2x}+\frac{6x+12}{12}=\frac{11x^2+4x+12}{2x}\ \text{ft}$$

101.
$$w=\frac{7x+5}{2}-\frac{8x-6}{x}=\frac{(7x+5)x}{2x}-\frac{(8x-6)2}{2x}=\frac{7x^2+5x}{2x}-\frac{16x-12}{2x}=\frac{7x^2-11x+12}{2x}$$

103. Answers may vary.

105. In the second line, the **whole** numerator must be subtracted.
The second line should have $\dfrac{8x+2-3x-8}{5}$.

Exercise 6.5 (page 390)

1.
$$\frac{8(a-5)}{3}=2(a-4)$$
$$8(a-5)=6(a-4)$$
$$8a-40=6a-24$$
$$2a=16$$
$$a=8$$

3.
$$a^4 - 13a^2 + 36 = 0$$
$$(a^2 - 4)(a^2 - 9) = 0$$
$$(a + 2)(a - 2)(a + 3)(a - 3) = 0$$

$a + 2 = 0$ **or** $a - 2 = 0$ **or** $a + 3 = 0$ **or** $a - 3 = 0$

       $a = -2$ $a = 2$ $a = -3$ $a = 3$

5.   complex

7. $\dfrac{\frac{1}{2}}{\frac{3}{4}} = \dfrac{1}{2} \div \dfrac{3}{4} = \dfrac{1}{2} \cdot \dfrac{4}{3} = \dfrac{2}{3}$

9. $\dfrac{-\frac{2}{3}}{\frac{6}{9}} = -\dfrac{2}{3} \div \dfrac{6}{9} = -\dfrac{2}{3} \cdot \dfrac{9}{6} = -1$

11. $\dfrac{\frac{1}{2} + \frac{1}{3}}{\frac{1}{4}} = \dfrac{\frac{3}{6} + \frac{2}{6}}{\frac{1}{4}} = \dfrac{\frac{5}{6}}{\frac{1}{4}} = \dfrac{5}{6} \div \dfrac{1}{4} = \dfrac{5}{6} \cdot \dfrac{4}{1} = \dfrac{10}{3}$

13. $\dfrac{\frac{1}{2} - \frac{2}{3}}{\frac{2}{3} + \frac{1}{2}} = \dfrac{\frac{3}{6} - \frac{4}{6}}{\frac{4}{6} + \frac{3}{6}} = \dfrac{-\frac{1}{6}}{\frac{7}{6}} = -\dfrac{1}{6} \div \dfrac{7}{6} = -\dfrac{1}{6} \cdot \dfrac{6}{7} = -\dfrac{1}{7}$

15. $\dfrac{\frac{4x}{y}}{\frac{6xz}{y^2}} = \dfrac{4x}{y} \div \dfrac{6xz}{y^2} = \dfrac{4x}{y} \cdot \dfrac{y^2}{6xz} = \dfrac{2y}{3z}$

17. $\dfrac{5ab^2}{\frac{ab}{25}} = 5ab^2 \div \dfrac{ab}{25} = \dfrac{5ab^2}{1} \cdot \dfrac{25}{ab} = \dfrac{125b}{1} = 125b$

19. $\dfrac{\frac{x-y}{xy}}{\frac{y-x}{x}} = \dfrac{x-y}{xy} \div \dfrac{y-x}{x} = \dfrac{x-y}{xy} \cdot \dfrac{x}{y-x} = \dfrac{x-y}{xy} \cdot \dfrac{x}{-(x-y)} = -\dfrac{1}{y}$

21. $\dfrac{\frac{1}{x} - \frac{1}{y}}{xy} = \dfrac{\left(\frac{1}{x} - \frac{1}{y}\right) \cdot xy}{(xy) \cdot xy} = \dfrac{\frac{1}{x} \cdot xy - \frac{1}{y} \cdot xy}{x^2 y^2} = \dfrac{y - x}{x^2 y^2}$

23. $\dfrac{\frac{1}{a} + \frac{1}{b}}{\frac{1}{a}} = \dfrac{\left(\frac{1}{a} + \frac{1}{b}\right) \cdot ab}{\frac{1}{a} \cdot ab} = \dfrac{\frac{1}{a} \cdot ab + \frac{1}{b} \cdot ab}{b} = \dfrac{b + a}{b}$

25. $\dfrac{1 + \frac{x}{y}}{1 - \frac{x}{y}} = \dfrac{\left(1 + \frac{x}{y}\right) \cdot y}{\left(1 - \frac{x}{y}\right) \cdot y} = \dfrac{1 \cdot y + \frac{x}{y} \cdot y}{1 \cdot y - \frac{x}{y} \cdot y} = \dfrac{y + x}{y - x}$

27. $\dfrac{\frac{y}{x} - \frac{x}{y}}{\frac{1}{x} + \frac{1}{y}} = \dfrac{\left(\frac{y}{x} - \frac{x}{y}\right) \cdot xy}{\left(\frac{1}{x} + \frac{1}{y}\right) \cdot xy} = \dfrac{\frac{y}{x} \cdot xy - \frac{x}{y} \cdot xy}{\frac{1}{x} \cdot xy + \frac{1}{y} \cdot xy} = \dfrac{y^2 - x^2}{y + x} = \dfrac{(y + x)(y - x)}{y + x} = y - x$

29. $\dfrac{\frac{1}{a} - \frac{1}{b}}{\frac{a}{b} - \frac{b}{a}} = \dfrac{\left(\frac{1}{a} - \frac{1}{b}\right) \cdot ab}{\left(\frac{a}{b} - \frac{b}{a}\right) \cdot ab} = \dfrac{\frac{1}{a} \cdot ab - \frac{1}{b} \cdot ab}{\frac{a}{b} \cdot ab - \frac{b}{a} \cdot ab} = \dfrac{b - a}{a^2 - b^2} = \dfrac{b - a}{(a + b)(a - b)} = -\dfrac{1}{a + b}$

31. $\dfrac{x + 1 - \frac{6}{x}}{\frac{1}{x}} = \dfrac{\left(x + 1 - \frac{6}{x}\right) \cdot x}{\frac{1}{x} \cdot x} = \dfrac{x^2 + x - 6}{1} = x^2 + x - 6$

33. $\dfrac{5xy}{1 + \frac{1}{xy}} = \dfrac{(5xy) \cdot xy}{\left(1 + \frac{1}{xy}\right) \cdot xy} = \dfrac{5x^2y^2}{xy + 1}$

35. $\dfrac{1 + \frac{6}{x} + \frac{8}{x^2}}{1 + \frac{1}{x} - \frac{12}{x^2}} = \dfrac{\left(1 + \frac{6}{x} + \frac{8}{x^2}\right) \cdot x^2}{\left(1 + \frac{1}{x} - \frac{12}{x^2}\right) \cdot x^2} = \dfrac{x^2 + 6x + 8}{x^2 + x - 12} = \dfrac{(x + 4)(x + 2)}{(x + 4)(x - 3)} = \dfrac{x + 2}{x - 3}$

37. $\dfrac{\frac{1}{a+1} + 1}{\frac{3}{a-1} + 1} = \dfrac{\left(\frac{1}{a+1} + 1\right)(a + 1)(a - 1)}{\left(\frac{3}{a-1} + 1\right)(a + 1)(a - 1)} = \dfrac{1(a - 1) + 1(a + 1)(a - 1)}{3(a + 1) + 1(a + 1)(a - 1)} = \dfrac{a - 1 + a^2 - 1}{3a + 3 + a^2 - 1}$

$$= \dfrac{a^2 + a - 2}{a^2 + 3a + 2}$$
$$= \dfrac{(a + 2)(a - 1)}{(a + 2)(a + 1)}$$
$$= \dfrac{a - 1}{a + 1}$$

39. $\dfrac{x^{-1} + y^{-1}}{x} = \dfrac{\frac{1}{x} + \frac{1}{y}}{x} = \dfrac{\left(\frac{1}{x} + \frac{1}{y}\right)(xy)}{x(xy)} = \dfrac{y + x}{x^2 y}$

41. $\dfrac{y}{x^{-1} - y^{-1}} = \dfrac{y}{\frac{1}{x} - \frac{1}{y}} = \dfrac{y(xy)}{\left(\frac{1}{x} - \frac{1}{y}\right)(xy)} = \dfrac{xy^2}{y - x}$

43. $\dfrac{x^{-1} + y^{-1}}{x^{-1} - y^{-1}} = \dfrac{\frac{1}{x} + \frac{1}{y}}{\frac{1}{x} - \frac{1}{y}} = \dfrac{\left(\frac{1}{x} + \frac{1}{y}\right)(xy)}{\left(\frac{1}{x} - \frac{1}{y}\right)(xy)} = \dfrac{y + x}{y - x}$

45. $\dfrac{x + y}{x^{-1} + y^{-1}} = \dfrac{x + y}{\frac{1}{x} + \frac{1}{y}} = \dfrac{(x + y)(xy)}{\left(\frac{1}{x} + \frac{1}{y}\right)(xy)} = \dfrac{(x + y)xy}{y + x} = xy$

47. $\dfrac{x-y^{-2}}{y-x^{-2}} = \dfrac{x-\frac{1}{y^2}}{y-\frac{1}{x^2}} = \dfrac{\left(x-\frac{1}{y^2}\right)(x^2y^2)}{\left(y-\frac{1}{x^2}\right)(x^2y^2)} = \dfrac{x^3y^2-x^2}{x^2y^3-y^2} = \dfrac{x^2(xy^2-1)}{y^2(x^2y-1)}$

49. $\dfrac{1+\frac{a}{b}}{1-\frac{a}{1-\frac{a}{b}}} = \dfrac{1+\frac{a}{b}}{1-\frac{a(b)}{\left(1-\frac{a}{b}\right)(b)}} = \dfrac{1+\frac{a}{b}}{1-\frac{ab}{b-a}} = \dfrac{\left(1+\frac{a}{b}\right)(b)(b-a)}{\left(1-\frac{ab}{b-a}\right)(b)(b-a)} = \dfrac{b(b-a)+a(b-a)}{b(b-a)-ab(b)}$

$= \dfrac{b^2-ab+ab-a^2}{b^2-ab-ab^2}$

$= \dfrac{b^2-a^2}{b(b-a-ab)}$

$= \dfrac{(b+a)(b-a)}{b(b-a-ab)}$

51. $\dfrac{x-\frac{1}{x}}{1+\frac{1}{\frac{1}{x}}} = \dfrac{x-\frac{1}{x}}{1+\frac{1(x)}{\frac{1}{x}(x)}} = \dfrac{x-\frac{1}{x}}{1+\frac{x}{1}} = \dfrac{x-\frac{1}{x}}{1+x} = \dfrac{\left(x-\frac{1}{x}\right)x}{(1+x)x} = \dfrac{x^2-1}{x(x+1)} = \dfrac{(x+1)(x-1)}{x(x+1)}$

$= \dfrac{x-1}{x}$

53. $\dfrac{b}{b+\frac{2}{2+\frac{1}{2}}} = \dfrac{b}{b+\frac{2(2)}{\left(2+\frac{1}{2}\right)(2)}} = \dfrac{b}{b+\frac{4}{4+1}} = \dfrac{b}{b+\frac{4}{5}} = \dfrac{b(5)}{\left(b+\frac{4}{5}\right)(5)} = \dfrac{5b}{5b+4}$

55. $a+\dfrac{a}{1+\frac{a}{a+1}} = a+\dfrac{a(a+1)}{\left(1+\frac{a}{a+1}\right)(a+1)} = a+\dfrac{a(a+1)}{a+1+a} = a+\dfrac{a^2+a}{2a+1} = \dfrac{a(2a+1)}{2a+1}+\dfrac{a^2+a}{2a+1}$

$= \dfrac{2a^2+a+a^2+a}{2a+1}$

$= \dfrac{3a^2+2a}{2a+1}$

57. $\dfrac{x-\frac{1}{1-\frac{x}{2}}}{\frac{3}{x+\frac{2}{3}}-x} = \dfrac{x-\frac{1(2)}{\left(1-\frac{x}{2}\right)2}}{\frac{3(3)}{\left(x+\frac{2}{3}\right)3}-x} = \dfrac{x-\frac{2}{2-x}}{\frac{9}{3x+2}-x} = \dfrac{\left(x-\frac{2}{2-x}\right)(2-x)(3x+2)}{\left(\frac{9}{3x+2}-x\right)(2-x)(3x+2)}$

$= \dfrac{x(2-x)(3x+2)-2(3x+2)}{9(2-x)-x(2-x)(3x+2)}$

$= \dfrac{(2x-x^2)(3x+2)-2(3x+2)}{9(2-x)-(3x^2+2x)(2-x)}$

$= \dfrac{(3x+2)(-x^2+2x-2)}{(2-x)(-3x^2-2x+9)}$

59.
$$\dfrac{2x + \frac{1}{2-\frac{x}{2}}}{\frac{4}{\frac{x}{2}-2} - x} = \dfrac{2x + \frac{1(2)}{(2-\frac{x}{2})2}}{\frac{4(2)}{(\frac{x}{2}-2)2} - x} = \dfrac{2x + \frac{2}{4-x}}{\frac{8}{x-4} - x} = \dfrac{\left(2x + \frac{2}{4-x}\right)(x-4)}{\left(\frac{8}{x-4} - x\right)(x-4)}$$

$$= \dfrac{2x^2 - 8x - 2}{8 - x^2 + 4x} = \dfrac{2(x^2 - 4x - 1)}{-x^2 + 4x + 8}$$

61.
$$\dfrac{\frac{1}{x^2+3x+2} + \frac{1}{x^2+x-2}}{\frac{3x}{x^2-1} - \frac{x}{x+2}} = \dfrac{\frac{1}{(x+2)(x+1)} + \frac{1}{(x+2)(x-1)}}{\frac{3x}{(x+1)(x-1)} - \frac{x}{x+2}}$$

$$= \dfrac{\left(\frac{1}{(x+2)(x+1)} + \frac{1}{(x+2)(x-1)}\right)(x+2)(x+1)(x-1)}{\left(\frac{3x}{(x+1)(x-1)} - \frac{x}{x+2}\right)(x+2)(x+1)(x-1)}$$

$$= \dfrac{x-1+x+1}{3x(x+2) - x(x+1)(x-1)}$$

$$= \dfrac{2x}{3x^2 + 6x - x^3 + x}$$

$$= \dfrac{2x}{-x^3 + 3x^2 + 7x} = \dfrac{2x}{-x(x^2 - 3x - 7)} = \dfrac{-2}{x^2 - 3x - 7}$$

63.
$$\dfrac{1}{\frac{1}{k_1} + \frac{1}{k_2}} = \dfrac{1 k_1 k_2}{\left(\frac{1}{k_1} + \frac{1}{k_2}\right)k_1 k_2} = \dfrac{k_1 k_2}{k_2 + k_1}$$

65. average $= \dfrac{\frac{k}{2} + \frac{k}{3} + \frac{k}{2}}{3} = \dfrac{\frac{3k}{6} + \frac{2k}{6} + \frac{3k}{6}}{3} = \dfrac{\frac{8k}{6}}{\frac{3}{1}} = \dfrac{8k}{6} \div \dfrac{3}{1} = \dfrac{8k}{6} \cdot \dfrac{1}{3} = \dfrac{8k}{18} = \dfrac{4k}{9}$

67. Answers may vary.

69. $(x^{-1}y^{-1})(x^{-1} + y^{-1})^{-1} = \dfrac{1}{x} \cdot \dfrac{1}{y} \cdot \dfrac{1}{\frac{1}{x} + \frac{1}{y}} = \dfrac{1}{xy} \cdot \dfrac{1(xy)}{\left(\frac{1}{x} + \frac{1}{y}\right)xy} = \dfrac{1}{xy} \cdot \dfrac{xy}{y+x} = \dfrac{1}{y+x}$

Exercise 6.6 (page 400)

1. $(m^2 n^{-3})^{-2} = m^{-4} n^6 = \dfrac{n^6}{m^4}$

3. $\dfrac{a^0 + 2a^0 - 3a^0}{(a-b)^0} = \dfrac{1+2-3}{1} = 0$

5. rational

7.
$$\frac{1}{4} + \frac{9}{x} = 1$$
$$\left(\frac{1}{4} + \frac{9}{x}\right)4x = 1(4x)$$
$$x + 36 = 4x$$
$$-3x = -36$$
$$x = 12$$
The answer checks.

9.
$$\frac{34}{x} - \frac{3}{2} = -\frac{13}{20}$$
$$\left(\frac{34}{x} - \frac{3}{2}\right)20x = -\left(\frac{13}{20}\right)20x$$
$$680 - 30x = -13x$$
$$-17x = -680$$
$$x = 40$$
The answer checks.

11.
$$\frac{3}{y} + \frac{7}{2y} = 13$$
$$\left(\frac{3}{y} + \frac{7}{2y}\right)2y = (13)2y$$
$$6 + 7 = 26y$$
$$-26y = -13$$
$$y = \frac{-13}{-26} = \frac{1}{2}$$
The answer checks.

13.
$$\frac{x+1}{x} - \frac{x-1}{x} = 0$$
$$\left(\frac{x+1}{x} - \frac{x-1}{x}\right)x = (0)x$$
$$x + 1 - (x - 1) = 0$$
$$x + 1 - x + 1 = 0$$
$$2 \neq 0$$
There is no solution.

15.
$$\frac{7}{5x} - \frac{1}{2} = \frac{5}{6x} + \frac{1}{3}$$
$$\left(\frac{7}{5x} - \frac{1}{2}\right)30x = \left(\frac{5}{6x} + \frac{1}{3}\right)30x$$
$$42 - 15x = 25 + 10x$$
$$-25x = -17$$
$$x = \frac{-17}{-25} = \frac{17}{25}$$
The answer checks.

17.
$$\frac{y-3}{y+2} = 3 - \frac{1-2y}{y+2}$$
$$\frac{y-3}{y+2} \cdot (y+2) = \left(3 - \frac{1-2y}{y+2}\right)(y+2)$$
$$y - 3 = 3(y + 2) - (1 - 2y)$$
$$y - 3 = 3y + 6 - 1 + 2y$$
$$y - 3 = 5y + 5$$
$$-4y = 8$$
$$y = -2$$
The answer does not check.
There is no solution.

19.
$$\frac{3-5y}{2+y} = \frac{3+5y}{2-y}$$
$$\left(\frac{3-5y}{2+y}\right)(2+y)(2-y) = \left(\frac{3+5y}{2-y}\right)(2+y)(2-y)$$
$$(3 - 5y)(2 - y) = (3 + 5y)(2 + y)$$
$$5y^2 - 13y + 6 = 5y^2 + 13y + 6$$
$$-26y = 0$$
$$y = 0 \qquad \text{The answer checks.}$$

21.
$$\frac{a+2}{a+1} = \frac{a-4}{a-3}$$
$$\left(\frac{a+2}{a+1}\right)(a+1)(a-3) = \left(\frac{a-4}{a-3}\right)(a+1)(a-3)$$
$$(a+2)(a-3) = (a-4)(a+1)$$
$$a^2 - a - 6 = a^2 - 3a - 4$$
$$2a = 2$$
$$a = 1 \qquad \text{The answer checks.}$$

23.
$$\frac{x+2}{x+3} - 1 = \frac{1}{3-2x-x^2}$$
$$\frac{x+2}{x+3} - 1 = \frac{-1}{x^2+2x-3}$$
$$\frac{x+2}{x+3} - 1 = \frac{-1}{(x+3)(x-1)}$$
$$\left(\frac{x+2}{x+3} - 1\right)(x+3)(x-1) = \left(\frac{-1}{(x+3)(x-1)}\right)(x+3)(x-1)$$
$$(x+2)(x-1) - (x+3)(x-1) = -1$$
$$x^2 + x - 2 - (x^2 + 2x - 3) = -1$$
$$x^2 + x - 2 - x^2 - 2x + 3 = -1$$
$$-x = -2$$
$$x = 2 \qquad \text{The answer checks.}$$

25.
$$\frac{x}{x+2} = 1 - \frac{3x+2}{x^2+4x+4}$$
$$\frac{x}{x+2} = 1 - \frac{3x+2}{(x+2)(x+2)}$$
$$\left(\frac{x}{x+2}\right)(x+2)(x+2) = \left(1 - \frac{3x+2}{(x+2)(x+2)}\right)(x+2)(x+2)$$
$$x(x+2) = (x+2)(x+2) - (3x+2)$$
$$x^2 + 2x = x^2 + 4x + 4 - 3x - 2$$
$$x^2 + 2x = x^2 + x + 2$$
$$x = 2 \qquad \text{The answer checks.}$$

27.
$$\frac{2}{x-2} + \frac{1}{x+1} = \frac{1}{x^2-x-2}$$
$$\left(\frac{2}{x-2} + \frac{1}{x+1}\right)(x-2)(x+1) = \left(\frac{1}{(x-2)(x+1)}\right)(x-2)(x+1)$$
$$2(x+1) + 1(x-2) = 1$$
$$2x + 2 + x - 2 = 1$$
$$3x = 1$$
$$x = \tfrac{1}{3} \qquad \text{The answer checks.}$$

29.
$$\frac{a-1}{a+3} - \frac{1-2a}{3-a} = \frac{2-a}{a-3}$$
$$\frac{a-1}{a+3} - \frac{2a-1}{a-3} = \frac{2-a}{a-3}$$
$$\left(\frac{a-1}{a+3} - \frac{2a-1}{a-3}\right)(a+3)(a-3) = \left(\frac{2-a}{a-3}\right)(a+3)(a-3)$$
$$(a-1)(a-3) - (2a-1)(a+3) = (2-a)(a+3)$$
$$a^2 - 4a + 3 - (2a^2 + 5a - 3) = -a^2 - a + 6$$
$$a^2 - 4a + 3 - 2a^2 - 5a + 3 = -a^2 - a + 6$$
$$-a^2 - 9a + 6 = -a^2 - a + 6$$
$$-8a = 0$$
$$a = 0 \qquad \text{The answer checks.}$$

31.
$$\frac{5}{x+4} + \frac{1}{x+4} = x - 1$$
$$\left(\frac{5}{x+4} + \frac{1}{x+4}\right)(x+4) = (x-1)(x+4)$$
$$5 + 1 = x^2 + 3x - 4$$
$$0 = x^2 + 3x - 10$$
$$0 = (x+5)(x-2)$$
$$x + 5 = 0 \quad \textbf{or} \quad x - 2 = 0$$
$$x = -5 \qquad\qquad x = 2 \qquad \text{Both answers check.}$$

33.
$$\frac{3}{x+1} - \frac{x-2}{2} = \frac{x-2}{x+1}$$
$$\left(\frac{3}{x+1} - \frac{x-2}{2}\right)(2)(x+1) = \left(\frac{x-2}{x+1}\right)(2)(x+1)$$
$$6 - (x-2)(x+1) = 2x - 4$$
$$6 - (x^2 - x - 2) = 2x - 4$$
$$6 - x^2 + x + 2 = 2x - 4$$
$$0 = x^2 + x - 12$$
$$0 = (x+4)(x-3)$$
$$x + 4 = 0 \quad \textbf{or} \quad x - 3 = 0$$
$$x = -4 \qquad\qquad x = 3 \qquad \text{Both answers check.}$$

35.
$$\frac{2}{x-3} + \frac{3}{4} = \frac{17}{2x}$$
$$\left(\frac{2}{x-3} + \frac{3}{4}\right)(4x)(x-3) = \left(\frac{17}{2x}\right)(4x)(x-3)$$
$$8x + 3x(x-3) = 34(x-3)$$
$$8x + 3x^2 - 9x = 34x - 102$$
$$3x^2 - 35x + 102 = 0$$
$$(3x - 17)(x - 6) = 0$$
$$3x - 17 = 0 \quad \textbf{or} \quad x - 6 = 0$$
$$3x = 17 \qquad\qquad x = 6$$
$$x = \tfrac{17}{3} \qquad\qquad \text{Both answers check.}$$

37.
$$\frac{x+4}{x+7} - \frac{x}{x+3} = \frac{3}{8}$$
$$\left(\frac{x+4}{x+7} - \frac{x}{x+3}\right)(8)(x+7)(x+3) = \left(\frac{3}{8}\right)(8)(x+7)(x+3)$$
$$8(x+4)(x+3) - 8x(x+7) = 3(x+7)(x+3)$$
$$8(x^2 + 7x + 12) - 8x^2 - 56x = 3(x^2 + 10x + 21)$$
$$8x^2 + 56x + 96 - 8x^2 - 56x = 3x^2 + 30x + 63$$
$$96 = 3x^2 + 30x + 63$$
$$0 = 3x^2 + 30x - 33$$
$$0 = 3(x^2 + 10x - 11)$$
$$0 = 3(x+11)(x-1)$$
$$x + 11 = 0 \quad \textbf{or} \quad x - 1 = 0$$
$$x = -11 \qquad\qquad x = 1 \qquad \text{Both answers check.}$$

39.
$$\frac{1}{p} + \frac{1}{q} = \frac{1}{f}$$
$$\left(\frac{1}{p} + \frac{1}{q}\right)pqf = \left(\frac{1}{f}\right)pqf$$
$$qf + pf = pq$$
$$f(q + p) = pq$$
$$f = \frac{pq}{q+p}$$

41.
$$S = \frac{a - lr}{1 - r}$$
$$S(1 - r) = \left(\frac{a - lr}{1 - r}\right)(1 - r)$$
$$S - Sr = a - lr$$
$$S - a = Sr - lr$$
$$S - a = r(S - l)$$
$$\frac{S - a}{S - l} = r, \text{ or } r = \frac{S - a}{S - l}$$

43.
$$\frac{1}{R} = \frac{1}{r_1} + \frac{1}{r_2} + \frac{1}{r_3}$$
$$\left(\frac{1}{R}\right)Rr_1r_2r_3 = \left(\frac{1}{r_1} + \frac{1}{r_2} + \frac{1}{r_3}\right)Rr_1r_2r_3$$
$$r_1r_2r_3 = Rr_2r_3 + Rr_1r_3 + Rr_1r_2$$
$$r_1r_2r_3 = R(r_2r_3 + r_1r_3 + r_1r_2)$$
$$\frac{r_1r_2r_3}{r_2r_3 + r_1r_3 + r_1r_2} = R, \text{ or } R = \frac{r_1r_2r_3}{r_2r_3 + r_1r_3 + r_1r_2}$$

45.
$$\frac{1}{f} = \frac{1}{s_1} + \frac{1}{s_2}$$
$$\left(\frac{1}{f}\right) f s_1 s_2 = \left(\frac{1}{s_1} + \frac{1}{s_2}\right) f s_1 s_2$$
$$s_1 s_2 = f s_2 + f s_1$$
$$s_1 s_2 = f(s_2 + s_1)$$
$$\frac{s_1 s_2}{s_2 + s_1} = f, \text{ or } f = \frac{s_1 s_2}{s_2 + s_1}$$
$$f = \frac{s_1 s_2}{s_2 + s_1} = \frac{(60 \text{ in.})(5 \text{ in.})}{60 \text{ in.} + 5 \text{ in.}} = \frac{300 \text{ in.}^2}{65 \text{ in.}} = \frac{60}{13} \text{ in.} = 4\frac{8}{13} \text{ inches}$$

47. The first painter paints $\frac{1}{5}$ of a house in 1 day, while the second paints $\frac{1}{3}$ of a house in 1 day. Let $x =$ the number of days it takes them to paint the house together. Then they can paint $\frac{1}{x}$ of a house together in 1 day.

Amount 1st paints in 1 day		Amount 2nd paints in 1 day		Amount painted by both in 1 day
	+		=	

$$\frac{1}{5} + \frac{1}{3} = \frac{1}{x}$$
$$\left(\frac{1}{5} + \frac{1}{3}\right) 15x = \frac{1}{x} \cdot 15x$$
$$3x + 5x = 15$$
$$8x = 15$$
$$x = \frac{15}{8} = 1\frac{7}{8}$$

They can paint the house together in $1\frac{7}{8}$ days.

49. The first belt can move $\frac{1}{10}$ of the 1000 bushels in 1 minute, while the second can move $\frac{1}{14}$ of the 1000 bushels in 1 minute. Let $x =$ the number of minutes for them to move the 1000 bushels together. Then $\frac{1}{x} =$ the amount of the 1000 bushels they can move together in 1 minute.

Amount 1st moves in 1 minute		Amount 2nd moves in 1 minute		Amount moved together in 1 minute
	+		=	

$$\frac{1}{10} + \frac{1}{14} = \frac{1}{x}$$
$$\left(\frac{1}{10} + \frac{1}{14}\right)(70x) = \frac{1}{x} \cdot 70x$$
$$7x + 5x = 70$$
$$12x = 70$$
$$x = \frac{70}{12} = \frac{35}{6} = 5\frac{5}{6}$$

It will take them $5\frac{5}{6}$ minutes to move 1000 bushels working together.

SECTION 6.6

51. The first drain can drain $\frac{1}{3}$ of the pool in 1 day, while the second can drain $\frac{1}{2}$ of the pool in 1 day. Let x = the number of days needed for them to drain the pool together. Then $\frac{1}{x}$ = the amount of the pool they can drain together in 1 day.

$$\boxed{\begin{array}{c}\text{Amount 1st drains}\\\text{in 1 day}\end{array}} + \boxed{\begin{array}{c}\text{Amount 2nd drains}\\\text{in 1 day}\end{array}} = \boxed{\begin{array}{c}\text{Amount drained together}\\\text{in 1 day}\end{array}}$$

$$\frac{1}{3} + \frac{1}{2} = \frac{1}{x}$$
$$\left(\frac{1}{3} + \frac{1}{2}\right)6x = \frac{1}{x} \cdot 6x$$
$$2x + 3x = 6$$
$$5x = 6$$
$$x = \frac{6}{5} = 1\frac{1}{5}$$

It will take the two drains $1\frac{1}{5}$ days to drain the pool.

53. The first pipe fills $\frac{1}{3}$ of the pond in 1 week, while the second fills $\frac{1}{5}$ of the pond in a week and evaporation empties $\frac{1}{10}$ of the pond in a week. Let x = the number of weeks needed to fill the pond with both pipes, considering evaporation. Then $\frac{1}{x}$ of the pond is filled in 1 week.

$$\boxed{\begin{array}{c}\text{Amount 1st pipe}\\\text{fills in 1 week}\end{array}} + \boxed{\begin{array}{c}\text{Amount 2nd pipe}\\\text{fills in 1 week}\end{array}} - \boxed{\begin{array}{c}\text{Amount emptied}\\\text{in 1 week}\end{array}} = \boxed{\begin{array}{c}\text{Total amount filled}\\\text{in 1 week}\end{array}}$$

$$\frac{1}{3} + \frac{1}{5} - \frac{1}{10} = \frac{1}{x}$$
$$\left(\frac{1}{3} + \frac{1}{5} - \frac{1}{10}\right)30x = \frac{1}{x} \cdot 30x$$
$$10x + 6x - 3x = 30$$
$$13x = 30$$
$$x = \frac{30}{13} = 2\frac{4}{13}$$

It will take $2\frac{4}{13}$ weeks to fill the pond.

55. Let $w =$ the rate at which he walks. Then $w + 5 =$ the rate at which he bicycles.

	Rate	Time	Dist.
Walk	w	$\frac{24}{w}$	24
Bicycle	$w + 5$	$\frac{24}{w+5}$	24

$$\boxed{\begin{array}{c}\text{Time}\\\text{walking}\end{array}} + \boxed{\begin{array}{c}\text{Time}\\\text{bicycling}\end{array}} = \boxed{\begin{array}{c}\text{Total}\\\text{time}\end{array}}$$

$$\frac{24}{w} + \frac{24}{w + 5} = 11$$

$$\left(\frac{24}{w} + \frac{24}{w + 5}\right)(w)(w + 5) = 11w(w + 5)$$

$$24(w + 5) + 24w = 11w^2 + 55w$$

$$0 = 11w^2 + 7w - 120$$

$$0 = (11w + 40)(w - 3)$$

$11w + 40 = 0$ **or** $w - 3 = 0$

$w = -\frac{40}{11}$ $w = 3$ $w = 3$ is the only answer that makes sense.

He walks at a rate of 3 miles per hour.

57. Let $r =$ the first rate. Then $r - 20 =$ the second rate.

	Rate	Time	Dist.
First	r	$\frac{120}{r}$	120
Second	$r - 20$	$\frac{120}{r-20}$	120

$$\boxed{\begin{array}{c}\text{First}\\\text{time}\end{array}} + \boxed{\begin{array}{c}\text{Second}\\\text{time}\end{array}} = 5$$

$$\frac{120}{r} + \frac{120}{r - 20} = 5$$

$$\left(\frac{120}{r} + \frac{120}{r - 20}\right)(r)(r - 20) = 5r(r - 20)$$

$$120(r - 20) + 120r = 5r^2 - 100r$$

$$0 = 5r^2 - 340r + 2400$$

$$0 = 5(r - 60)(r - 8)$$

$r - 60 = 0$ **or** $r - 8 = 0$

$r = 60$ $r = 8$ $r = 60$ is the only answer that makes sense.

The train traveled at rates of 60 and 40 miles per hour.

59. Let c = the speed of the current.

	Rate	Time	Dist.
Downstream	$12 + c$	$\frac{45}{12+c}$	45
Upstream	$12 - c$	$\frac{27}{12-c}$	27

$$\boxed{\text{Time downstream}} = \boxed{\text{Time upstream}}$$

$$\frac{45}{12+c} = \frac{27}{12-c}$$

$$\frac{45}{12+c}(12+c)(12-c) = \frac{27}{12-c}(12+c)(12-c)$$

$$45(12-c) = 27(12+c)$$

$$540 - 45c = 324 + 27c$$

$$-72c = -216$$

$$c = \frac{-216}{-72} = 3$$

The speed of the current is 3 miles per hour.

61. Let w = the speed of the wind.

	Rate	Time	Dist.
Downwind	$340 + w$	$\frac{200}{340+w}$	200
Upwind	$340 - w$	$\frac{140}{340-w}$	140

$$\boxed{\text{Time downwind}} = \boxed{\text{Time upwind}}$$

$$\frac{200}{340+w} = \frac{140}{340-w}$$

$$\left(\frac{200}{340+w}\right)(340+w)(340-w) = \left(\frac{140}{340-w}\right)(340+w)(340-w)$$

$$200(340-w) = 140(340+w)$$

$$68000 - 200w = 47600 + 140w$$

$$-340w = -20400$$

$$w = 60 \qquad \text{The speed of the wind is 60 miles per hour.}$$

63. Let x = the number purchased.

Then the unit cost $= \frac{224}{x}$.

$$\boxed{\begin{array}{c}\text{New}\\\text{unit cost}\end{array}}\cdot\boxed{\begin{array}{c}\text{New \#}\\\text{motors}\end{array}} = 224$$

$$\left(\frac{224}{x} - 4\right)(x+1) = 224$$

$$\left(\frac{224}{x} - 4\right)(x+1)x = 224x$$

$$224(x+1) - 4x(x+1) = 224x$$

$$224x + 224 - 4x^2 - 4x = 224x$$

$$-4x^2 - 4x + 224 = 0$$

$$-4(x+8)(x-7) = 0$$

$$x + 8 = 0 \quad \text{or} \quad x - 7 = 0$$

$$x = -8 \qquad\qquad x = 7$$

He originally bought 7 motors.

65. Let x = the number of days.

Then the daily cost $= \frac{1200}{x}$.

$$\boxed{\begin{array}{c}\text{New}\\\text{daily cost}\end{array}}\cdot\boxed{\begin{array}{c}\text{New \#}\\\text{days}\end{array}} = 1200$$

$$\left(\frac{1200}{x} - 20\right)(x+3) = 1200$$

$$\left(\frac{1200}{x} - 20\right)(x+3)x = 1200x$$

$$1200(x+3) - 20x(x+3) = 1200x$$

$$1200x + 3600 - 20x^2 - 60x = 1200x$$

$$-20x^2 - 60x + 3600 = 0$$

$$-20(x+15)(x-12) = 0$$

$$x + 15 = 0 \quad \text{or} \quad x - 12 = 0$$

$$x = -15 \qquad\qquad x = 12$$

Her original vacation was 12 days.

67. **Answers may vary.**

69. $\dfrac{x}{x-3} + \dfrac{3}{x-3} = \dfrac{6}{x-3}$

Exercise 6.7 (page 408)

1. $2(x^2 + 4x - 1) + 3(2x^2 - 2x + 2) = 2x^2 + 8x - 2 + 6x^2 - 6x + 6 = 8x^2 + 2x + 4$

3. $-2(3y^3 - 2y + 7) - 3(y^2 + 2y - 4) + 4(y^3 + 2y - 1)$
$\qquad = -6y^3 + 4y - 14 - 3y^2 - 6y + 12 + 4y^3 + 8y - 4 = -2y^3 - 3y^2 + 6y - 6$

5. $\dfrac{1}{b}$

7. quotient

9. $\dfrac{4x^2y^3}{8x^5y^2} = \dfrac{y}{2x^3}$

11. $\dfrac{33a^{-2}b^2}{44a^2b^{-2}} = \dfrac{3b^4}{4a^4}$

13. $\dfrac{45x^{-2}y^{-3}t^0}{-63x^{-1}y^4t^2} = -\dfrac{5}{7xy^7t^2}$

15. $\dfrac{-65a^{2n}b^nc^{3n}}{-15a^nb^{-n}c} = \dfrac{13a^nb^{2n}c^{3n-1}}{3}$

17. $\dfrac{4x^2 - x^3}{6x} = \dfrac{4x^2}{6x} - \dfrac{x^3}{6x} = \dfrac{2x}{3} - \dfrac{x^2}{6}$

19. $\dfrac{4x^2y^3 + x^3y^2}{6xy} = \dfrac{4x^2y^3}{6xy} + \dfrac{x^3y^2}{6xy}$
$\qquad\qquad = \dfrac{2xy^2}{3} + \dfrac{x^2y}{6}$

21. $\dfrac{24x^6y^7 - 12x^5y^{12} + 36xy}{48x^2y^3} = \dfrac{24x^6y^7}{48x^2y^3} - \dfrac{12x^5y^{12}}{48x^2y^3} + \dfrac{36xy}{48x^2y^3} = \dfrac{x^4y^4}{2} - \dfrac{x^3y^9}{4} + \dfrac{3}{4xy^2}$

23. $\dfrac{3a^{-2}b^3 - 6a^2b^{-3} + 9a^{-2}}{12a^{-1}b} = \dfrac{3a^{-2}b^3}{12a^{-1}b} - \dfrac{6a^2b^{-3}}{12a^{-1}b} + \dfrac{9a^{-2}}{12a^{-1}b} = \dfrac{b^2}{4a} - \dfrac{a^3}{2b^4} + \dfrac{3}{4ab}$

25. $\dfrac{x^ny^n - 3x^{2n}y^{2n} + 6x^{3n}y^{3n}}{x^ny^n} = \dfrac{x^ny^n}{x^ny^n} - \dfrac{3x^{2n}y^{2n}}{x^ny^n} + \dfrac{6x^{3n}y^{3n}}{x^ny^n} = 1 - 3x^ny^n + 6x^{2n}y^{2n}$

27.
$$\begin{array}{r} x+\ 2 \\ x+3\ \overline{\smash{\big)}\ x^2+5x+6} \\ \underline{x^2+3x} \\ 2x+6 \\ \underline{2x+6} \\ 0 \end{array}$$

29.
$$\begin{array}{r} x+\ 7 \\ x+3\ \overline{\smash{\big)}\ x^2+10x+21} \\ \underline{x^2+\ 3x} \\ 7x+21 \\ \underline{7x+21} \\ 0 \end{array}$$

31.
$$\begin{array}{r} 3x-\ 5+\frac{3}{2x+3} \\ 2x+3\ \overline{\smash{\big)}\ 6x^2-\ \ x-12} \\ \underline{6x^2+\ 9x} \\ -10x-12 \\ \underline{-10x-15} \\ 3 \end{array}$$

33.
$$\begin{array}{r} 3x^2+\ \ x+\ 2+\frac{8}{x-1} \\ x-1\ \overline{\smash{\big)}\ 3x^3-2x^2+\ \ x+\ \ 6} \\ \underline{3x^3-3x^2} \\ x^2+\ x \\ \underline{x^2-\ x} \\ 2x+\ 6 \\ \underline{2x-\ 2} \\ 8 \end{array}$$

35.
$$\begin{array}{r} 2x^2+\ \ 5x+\ \ 3+\frac{4}{3x-2} \\ 3x-2\ \overline{\smash{\big)}\ 6x^3+11x^2-\ \ x-\ \ 2} \\ \underline{6x^3-\ 4x^2} \\ 15x^2-\ \ x \\ \underline{15x^2-10x} \\ 9x-\ 2 \\ \underline{9x-\ 6} \\ 4 \end{array}$$

37.
$$\begin{array}{r} 3x^2+\ 4x+\ \ 3 \\ 2x-3\ \overline{\smash{\big)}\ 6x^3-\ \ x^2-\ 6x-9} \\ \underline{6x^3-9x^2} \\ 8x^2-\ 6x \\ \underline{8x^2-12x} \\ 6x-9 \\ \underline{6x-9} \\ 0 \end{array}$$

39.
$$\begin{array}{r} a+\ 1 \\ a+1\ \overline{\smash{\big)}\ a^2+2a+1} \\ \underline{a^2+\ a} \\ a+1 \\ \underline{a+1} \\ 0 \end{array}$$

41.
$$\begin{array}{r} 2y+\ \ 2 \\ 5y-2\ \overline{\smash{\big)}\ 10y^2+\ 6y-4} \\ \underline{10y^2-\ 4y} \\ 10y-4 \\ \underline{10y-4} \\ 0 \end{array}$$

43.
$$\begin{array}{r} 6x-\ 12 \\ x-1\ \overline{\smash{\big)}\ 6x^2-18x+12} \\ \underline{6x^2-\ 6x} \\ -12x+12 \\ \underline{-12x+12} \\ 0 \end{array}$$

45.
$$\begin{array}{r} 3x^2-\ \ x+\ 2 \\ 3x-2\ \overline{\smash{\big)}\ 9x^3-9x^2+8x-4} \\ \underline{9x^3-6x^2} \\ -3x^2+8x \\ \underline{-3x^2+2x} \\ 6x-4 \\ \underline{6x-4} \\ 0 \end{array}$$

47.
$$\begin{array}{r} 4x^3-\ 3x^2+\ \ 3x+\ \ 1 \\ 4x+3\ \overline{\smash{\big)}\ 16x^4+\ 0x^3+\ 3x^2+13x+3} \\ \underline{16x^4+12x^3} \\ -12x^3+\ 3x^2 \\ \underline{-12x^3-\ 9x^2} \\ 12x^2+13x \\ \underline{12x^2+\ 9x} \\ 4x+3 \\ \underline{4x+3} \\ 0 \end{array}$$

49.
$$\begin{array}{r} a^2+\ \ a+\ 1+\frac{2}{a-1} \\ a-1\ \overline{\smash{\big)}\ a^3+0a^2+0a+\ \ 1} \\ \underline{a^3-\ a^2} \\ a^2+0a \\ \underline{a^2-\ a} \\ a+\ 1 \\ \underline{a-\ 1} \\ 2 \end{array}$$

51.
$$\begin{array}{r} 5a^2-\ \ 3a-\ \ 4 \\ 3a-4\ \overline{\smash{\big)}\ 15a^3-29a^2+\ 0a+16} \\ \underline{15a^3-20a^2} \\ -\ 9a^2+\ 0a \\ \underline{-\ 9a^2+12a} \\ -12a+16 \\ \underline{-12a+16} \\ 0 \end{array}$$

53.

$$\begin{array}{r} 6y - 12 \\ y - 2 \enclose{longdiv}{6y^2 - 24y + 24} \\ \underline{6y^2 - 12y} \\ -12y + 24 \\ \underline{-12y + 24} \\ 0 \end{array}$$

55.

$$\begin{array}{r} 16x^4 - 8x^3y + 4x^2y^2 - 2xy^3 + y^4 \\ 2x + y \enclose{longdiv}{32x^5 + 0x^4y + 0x^3y^2 + 0x^2y^3 + 0xy^4 + y^5} \\ \underline{32x^5 + 16x^4y} \\ -16x^4y + 0x^3y^2 \\ \underline{-16x^4y - 8x^3y^2} \\ 8x^3y^2 + 0x^2y^3 \\ \underline{8x^3y^2 + 4x^2y^3} \\ -4x^2y^3 + 0xy^4 \\ \underline{-4x^2y^3 - 2xy^4} \\ 2xy^4 + y^5 \\ \underline{2xy^4 + y^5} \\ 0 \end{array}$$

57.

$$\begin{array}{r} x^4 + x^2 + 4 \\ x^2 - 2 \enclose{longdiv}{x^6 - x^4 + 2x^2 - 8} \\ \underline{x^6 - 2x^4} \\ x^4 + 2x^2 \\ \underline{x^4 - 2x^2} \\ 4x^2 - 8 \\ \underline{4x^2 - 8} \\ 0 \end{array}$$

59.

$$\begin{array}{r} x^2 + x + 1 \\ x^2 + x + 2 \enclose{longdiv}{x^4 + 2x^3 + 4x^2 + 3x + 2} \\ \underline{x^4 + x^3 + 2x^2} \\ x^3 + 2x^2 + 3x \\ \underline{x^3 + x^2 + 2x} \\ x^2 + x + 2 \\ \underline{x^2 + x + 2} \\ 0 \end{array}$$

61.

$$\begin{array}{r} x^2 + x + 2 \\ x^2 + 0x + 3 \enclose{longdiv}{x^4 + x^3 + 5x^2 + 3x + 6} \\ \underline{x^4 + 0x^3 + 3x^2} \\ x^3 + 2x^2 + 3x \\ \underline{x^3 + 0x^2 + 3x} \\ 2x^2 + 0x + 6 \\ \underline{2x^2 + 0x + 6} \\ 0 \end{array}$$

63.

$$\begin{array}{r} 9.8x + 16.4 - \frac{36.5}{x-2} \\ x - 2 \enclose{longdiv}{9.8x^2 - 3.2x - 69.3} \\ \underline{9.8x^2 - 19.6x} \\ 16.4x - 69.3 \\ \underline{16.4x - 32.8} \\ -36.5 \end{array}$$

65.

$$A = lw$$
$$\frac{A}{w} = l$$
$$\frac{9x^2 + 21x + 10}{3x + 2} = l$$

$$\begin{array}{r} 3x + 5 \\ 3x + 2 \enclose{longdiv}{9x^2 + 21x + 10} \\ \underline{9x^2 + 6x} \\ 15x + 10 \\ \underline{15x + 10} \\ 0 \end{array}$$

The length is $3x + 5$.

67. Answers may vary.

69.
$$2x - 3 \overline{\smash{\big)}\, \begin{array}{r} 5x + 7 \\ 10x^2 - x - 21 \end{array}}$$
$$\underline{10x^2 - 15x}$$
$$14x - 21$$
$$\underline{14x - 21}$$
$$0 \Rightarrow \text{It is a factor.}$$

Exercise 6.8 (page 415)

1. $f(1) = 3(1)^2 + 2(1) - 1 = 4$

3. $f(2a) = 3(2a)^2 + 2(2a) - 1$
$= 12a^2 + 4a - 1$

5. $2(x^2 + 4x - 1) + 3(2x^2 - 2x + 2) = 2x^2 + 8x - 2 + 6x^2 - 6x + 6 = 8x^2 + 2x + 4$

7. $P(r)$

9.
$$\begin{array}{r|rrr} 1 & 1 & 1 & -2 \\ & & 1 & 2 \\ \hline & 1 & 2 & 0 \end{array} \Rightarrow \boxed{x + 2}$$

11.
$$\begin{array}{r|rrr} 4 & 1 & -7 & 12 \\ & & 4 & -12 \\ \hline & 1 & -3 & 0 \end{array} \Rightarrow \boxed{x - 3}$$

13.
$$\begin{array}{r|rrr} -4 & 1 & 6 & 8 \\ & & -4 & -8 \\ \hline & 1 & 2 & 0 \end{array} \Rightarrow \boxed{x + 2}$$

15.
$$\begin{array}{r|rrr} -2 & 1 & -5 & 14 \\ & & -2 & 14 \\ \hline & 1 & -7 & 28 \end{array} \Rightarrow \boxed{x - 7 + \frac{28}{x+2}}$$

17.
$$\begin{array}{r|rrrr} 3 & 3 & -10 & 5 & -6 \\ & & 9 & -3 & 6 \\ \hline & 3 & -1 & 2 & 0 \end{array}$$
$$\Rightarrow \boxed{3x^2 - x + 2}$$

19.
$$\begin{array}{r|rrrr} 2 & 2 & 0 & -5 & -6 \\ & & 4 & 8 & 6 \\ \hline & 2 & 4 & 3 & 0 \end{array}$$
$$\Rightarrow \boxed{2x^2 + 4x + 3}$$

21.
$$\begin{array}{r|rrrr} -1 & 6 & 5 & 0 & 4 \\ & & -6 & 1 & -1 \\ \hline & 6 & -1 & 1 & 3 \end{array}$$
$$\Rightarrow \boxed{6x^2 - x + 1 + \frac{3}{x+1}}$$

23.
$$\begin{array}{r|rrr} 0.2 & 7.2 & -2.1 & 0.5 \\ & & 1.44 & -0.132 \\ \hline & 7.2 & -0.66 & 0.368 \end{array}$$
$$\Rightarrow \boxed{7.2x - 0.66 + \frac{0.368}{x-0.2}}$$

25.
$$\begin{array}{r|rrr} -1.7 & 2.7 & 1.0 & -5.2 \\ & & -4.59 & 6.103 \\ \hline & 2.7 & -3.59 & 0.903 \end{array}$$
$$\Rightarrow \boxed{2.7x - 3.59 + \frac{0.903}{x+1.7}}$$

27.
$$\begin{array}{r|rrrr} -57 & 9 & 0 & 0 & -25 \\ & & -513 & 29241 & -1666737 \\ \hline & 9 & -513 & 29241 & -1666762 \end{array}$$
$$\Rightarrow \boxed{9x^2 - 513x + 29{,}241 - \frac{1{,}666{,}762}{x+57}}$$

SECTION 6.8

29. $P(1) = 2(1)^3 - 4(1)^2 + 2(1) - 1 = \boxed{-1}$

$$\begin{array}{r|rrrr} 1 & 2 & -4 & 2 & -1 \\ & & 2 & -2 & 0 \\ \hline & 2 & -2 & 0 & \boxed{-1} \end{array}$$

31. $P(-2) = 2(-2)^3 - 4(-2)^2 + 2(-2) - 1$
$= \boxed{-37}$

$$\begin{array}{r|rrrr} -2 & 2 & -4 & 2 & -1 \\ & & -4 & 16 & -36 \\ \hline & 2 & -8 & 18 & \boxed{-37} \end{array}$$

33. $P(3) = 2(3)^3 - 4(3)^2 + 2(3) - 1 = \boxed{23}$

$$\begin{array}{r|rrrr} 3 & 2 & -4 & 2 & -1 \\ & & 6 & 6 & 24 \\ \hline & 2 & 2 & 8 & \boxed{23} \end{array}$$

35. $P(0) = 2(0)^3 - 4(0)^2 + 2(0) - 1 = \boxed{-1}$

$$\begin{array}{r|rrrr} 0 & 2 & -4 & 2 & -1 \\ & & 0 & 0 & 0 \\ \hline & 2 & -4 & 2 & \boxed{-1} \end{array}$$

37. $Q(-1) = (-1)^4 - 3(-1)^3 + 2(-1)^2 + (-1) - 3 = \boxed{2}$

$$\begin{array}{r|rrrrr} -1 & 1 & -3 & 2 & 1 & -3 \\ & & -1 & 4 & -6 & 5 \\ \hline & 1 & -4 & 6 & -5 & \boxed{2} \end{array}$$

39. $Q(2) = (2)^4 - 3(2)^3 + 2(2)^2 + (2) - 3 = \boxed{-1}$

$$\begin{array}{r|rrrrr} 2 & 1 & -3 & 2 & 1 & -3 \\ & & 2 & -2 & 0 & 2 \\ \hline & 1 & -1 & 0 & 1 & \boxed{-1} \end{array}$$

41. $Q(3) = (3)^4 - 3(3)^3 + 2(3)^2 + (3) - 3 = \boxed{18}$

$$\begin{array}{r|rrrrr} 3 & 1 & -3 & 2 & 1 & -3 \\ & & 3 & 0 & 6 & 21 \\ \hline & 1 & 0 & 2 & 7 & \boxed{18} \end{array}$$

43. $Q(-3) = (-3)^4 - 3(-3)^3 + 2(-3)^2 + (-3) - 3 = \boxed{174}$

$$\begin{array}{r|rrrrr} -3 & 1 & -3 & 2 & 1 & -3 \\ & & -3 & 18 & -60 & 177 \\ \hline & 1 & -6 & 20 & -59 & \boxed{174} \end{array}$$

45.
$$\begin{array}{r|rrrr} 2 & 1 & -4 & 1 & -2 \\ & & 2 & -4 & -6 \\ \hline & 1 & -2 & -3 & \boxed{-8} \end{array}$$

47.
$$\begin{array}{r|rrrr} 3 & 2 & 0 & 1 & 2 \\ & & 6 & 18 & 57 \\ \hline & 2 & 6 & 19 & \boxed{59} \end{array}$$

49.
$$\begin{array}{r|rrrrr} -2 & 1 & -2 & 1 & -3 & 2 \\ & & -2 & 8 & -18 & 42 \\ \hline & 1 & -4 & 9 & -21 & \boxed{44} \end{array}$$

51.
$$\begin{array}{r|rrrrrr} -\frac{1}{2} & 3 & 0 & 0 & 0 & 0 & 1 \\ & & -\frac{3}{2} & \frac{3}{4} & -\frac{3}{8} & \frac{3}{16} & -\frac{3}{32} \\ \hline & 3 & -\frac{3}{2} & \frac{3}{4} & -\frac{3}{8} & \frac{3}{16} & \boxed{\frac{29}{32}} \end{array}$$

53.
$$\begin{array}{r|rrrr} 3 & 1 & -3 & 5 & -15 \\ & & 3 & 0 & 15 \\ \hline & 1 & 0 & 5 & \boxed{0} \end{array} \Rightarrow \text{factor}$$

55.
$$\begin{array}{r|rrr} -2 & 3 & -7 & 4 \\ & & -6 & 26 \\ \hline & 3 & -13 & \boxed{30} \end{array} \Rightarrow \text{not a factor}$$

57.
$$\begin{array}{r|rrrrrrr} 2 & 1 & 0 & 0 & 0 & 0 & 0 & 0 \\ & & 2 & 4 & 8 & 16 & 32 & 64 \\ \hline & 1 & 2 & 4 & 8 & 16 & 32 & \boxed{64} \end{array}$$

59. Answers may vary.

61. remainder $= P(1) = 1^{100} - 1^{99} + 1^{98} - 1^{97} + \cdots + 1^2 - 1 + 1 = 1$

Chapter 6 Summary (page 418)

1.
$$f(x) = \frac{3x + 2}{x}$$

Horizontal: $y = 3$

Vertical: $x = 0$

2. $\dfrac{248x^2y}{576xy^2} = \dfrac{8 \cdot 31x^2y}{8 \cdot 72xy^2} = \dfrac{31x}{72y}$

3. $\dfrac{212m^3n}{588m^2n^3} = \dfrac{4 \cdot 53m^3n}{4 \cdot 147m^2n^3} = \dfrac{53m}{147n^2}$

4. $\dfrac{x^2 - 49}{x^2 + 14x + 49} = \dfrac{(x+7)(x-7)}{(x+7)(x+7)} = \dfrac{x-7}{x+7}$

5. $\dfrac{x^2 + 6x + 36}{x^3 - 216} = \dfrac{x^2 + 6x + 36}{(x-6)(x^2 + 6x + 36)}$
$$= \dfrac{1}{x - 6}$$

6. $\dfrac{x^2 - 2x + 4}{2x^3 + 16} = \dfrac{x^2 - 2x + 4}{2(x^3 + 8)} = \dfrac{x^2 - 2x + 4}{2(x+2)(x^2 - 2x + 4)} = \dfrac{1}{2(x+2)} = \dfrac{1}{2x + 4}$

7. $\dfrac{x - y}{y - x} = \dfrac{x - y}{-1(x - y)} = -1$

8. $\dfrac{2m - 2n}{n - m} = \dfrac{2(m - n)}{-1(m - n)} = -2$

9. $\dfrac{ac - ad + bc - bd}{d^2 - c^2} = \dfrac{a(c - d) + b(c - d)}{(d + c)(d - c)} = \dfrac{(c - d)(a + b)}{(d + c)(d - c)} = \dfrac{a + b}{c + d} = \dfrac{-a - b}{c + d}$

10. $P(E) = \dfrac{s}{n} = \dfrac{2}{36} = \dfrac{1}{18}$

11. $P(E) = \dfrac{s}{n} = \dfrac{1}{8}$

CHAPTER 6 SUMMARY

12.
$$\frac{x+1}{8} = \frac{4x-2}{24}$$
$$24(x+1) = 8(4x-2)$$
$$24x + 24 = 32x - 16$$
$$-8x = -40$$
$$x = 5$$

13.
$$\frac{1}{x+6} = \frac{x+10}{12}$$
$$12 = (x+6)(x+10)$$
$$12 = x^2 + 16x + 60$$
$$0 = x^2 + 16x + 48$$
$$0 = (x+12)(x+4)$$
$$x + 12 = 0 \quad \textbf{or} \quad x + 4 = 0$$
$$x = -12 \qquad\qquad x = -4$$

14. Let h = the height of the tree.
$$\frac{44}{2.5} = \frac{h}{4}$$
$$176 = 2.5h$$
$$70.4 = h \Rightarrow \text{The tree is 70.4 feet tall.}$$

15.
$$x = ky \qquad x = 6y$$
$$12 = k(2) \qquad x = 6(12)$$
$$6 = k \qquad \boxed{x = 72}$$

16.
$$x = \frac{k}{y} \qquad x = \frac{72}{y}$$
$$24 = \frac{k}{3} \qquad 12 = \frac{72}{y}$$
$$72 = k \qquad 12y = 72$$
$$\boxed{y = 6}$$

17.
$$x = kyz$$
$$24 = k(3)(4)$$
$$24 = 12k$$
$$\boxed{2 = k}$$

18.
$$x = \frac{kt}{y}$$
$$2 = \frac{k(8)}{64}$$
$$128 = 8k$$
$$\boxed{16 = k}$$

19.
$$\frac{x^2 + 4x + 4}{x^2 - x - 6} \cdot \frac{x^2 - 9}{x^2 + 5x + 6} = \frac{(x+2)(x+2)}{(x+2)(x-3)} \cdot \frac{(x+3)(x-3)}{(x+2)(x+3)} = 1$$

20.
$$\frac{x^3 - 64}{x^2 + 4x + 16} \div \frac{x^2 - 16}{x+4} = \frac{x^3 - 64}{x^2 + 4x + 16} \cdot \frac{x+4}{x^2 - 16} = \frac{(x-4)(x^2 + 4x + 16)}{x^2 + 4x + 16} \cdot \frac{x+4}{(x+4)(x-4)}$$
$$= 1$$

21.
$$\frac{x^2 + 3x + 2}{x^2 - x - 6} \cdot \frac{3x^2 - 3x}{x^2 - 3x - 4} \div \frac{x^2 + 3x + 2}{x^2 - 2x - 8}$$
$$= \frac{x^2 + 3x + 2}{x^2 - x - 6} \cdot \frac{3x^2 - 3x}{x^2 - 3x - 4} \cdot \frac{x^2 - 2x - 8}{x^2 + 3x + 2}$$
$$= \frac{(x+2)(x+1)}{(x-3)(x+2)} \cdot \frac{3x(x-1)}{(x-4)(x+1)} \cdot \frac{(x-4)(x+2)}{(x+2)(x+1)}$$
$$= \frac{3x(x-1)}{(x-3)(x+1)}$$

22.
$$\frac{x^2 - x - 6}{x^2 - 3x - 10} \div \frac{x^2 - x}{x^2 - 5x} \cdot \frac{x^2 - 4x + 3}{x^2 - 6x + 9} = \frac{x^2 - x - 6}{x^2 - 3x - 10} \cdot \frac{x^2 - 5x}{x^2 - x} \cdot \frac{x^2 - 4x + 3}{x^2 - 6x + 9}$$
$$= \frac{(x-3)(x+2)}{(x-5)(x+2)} \cdot \frac{x(x-5)}{x(x-1)} \cdot \frac{(x-3)(x-1)}{(x-3)(x-3)} = 1$$

23. $\dfrac{5y}{x-y} - \dfrac{3}{x-y} = \dfrac{5y-3}{x-y}$

24. $\dfrac{3x-1}{x^2+2} + \dfrac{3(x-2)}{x^2+2} = \dfrac{3x-1+3x-6}{x^2+2}$

$$= \dfrac{6x-7}{x^2+2}$$

25. $\dfrac{3}{x+2} + \dfrac{2}{x+3} = \dfrac{3(x+3)}{(x+2)(x+3)} + \dfrac{2(x+2)}{(x+2)(x+3)} = \dfrac{3x+9+2x+4}{(x+2)(x+3)} = \dfrac{5x+13}{(x+2)(x+3)}$

26. $\dfrac{4x}{x-4} - \dfrac{3}{x+3} = \dfrac{4x(x+3)}{(x-4)(x+3)} - \dfrac{3(x-4)}{(x-4)(x+3)} = \dfrac{4x^2+12x-3x+12}{(x-4)(x+3)} = \dfrac{4x^2+9x+12}{(x-4)(x+3)}$

27. $\dfrac{2x}{x+1} + \dfrac{3x}{x+2} + \dfrac{4x}{x^2+3x+2} = \dfrac{2x}{x+1} + \dfrac{3x}{x+2} + \dfrac{4x}{(x+2)(x+1)}$

$$= \dfrac{2x(x+2)}{(x+1)(x+2)} + \dfrac{3x(x+1)}{(x+1)(x+2)} + \dfrac{4x}{(x+2)(x+1)}$$

$$= \dfrac{2x^2+4x+3x^2+3x+4x}{(x+1)(x+2)} = \dfrac{5x^2+11x}{(x+1)(x+2)}$$

28. $\dfrac{5x}{x-3} + \dfrac{5}{x^2-5x+6} + \dfrac{x+3}{x-2} = \dfrac{5x}{x-3} + \dfrac{5}{(x-3)(x-2)} + \dfrac{x+3}{x-2}$

$$= \dfrac{5x(x-2)}{(x-3)(x-2)} + \dfrac{5}{(x-3)(x-2)} + \dfrac{(x+3)(x-3)}{(x-3)(x-2)}$$

$$= \dfrac{5x^2-10x+5+x^2-9}{(x-3)(x-2)}$$

$$= \dfrac{6x^2-10x-4}{(x-3)(x-2)} = \dfrac{2(3x+1)(x-2)}{(x-3)(x-2)} = \dfrac{2(3x+1)}{x-3}$$

29. $\dfrac{3(x+2)}{x^2-1} - \dfrac{2}{x+1} + \dfrac{4(x+3)}{x^2-2x+1} = \dfrac{3(x+2)}{(x+1)(x-1)} - \dfrac{2}{(x+1)} + \dfrac{4(x+3)}{(x-1)(x-1)}$

$$= \dfrac{3(x+2)(x-1)}{(x+1)(x-1)(x-1)} - \dfrac{2(x-1)(x-1)}{(x+1)(x-1)(x-1)} + \dfrac{4(x+3)(x+1)}{(x+1)(x-1)(x-1)}$$

$$= \dfrac{3(x^2+x-2) - 2(x^2-2x+1) + 4(x^2+4x+3)}{(x+1)(x-1)(x-1)}$$

$$= \dfrac{3x^2+3x-6-2x^2+4x-2+4x^2+16x+12}{(x+1)(x-1)(x-1)} = \dfrac{5x^2+23x+4}{(x+1)(x-1)(x-1)}$$

30. $\dfrac{-2(3+x)}{x^2+6x+9} + \dfrac{3(x+2)}{x^2-6x+9} - \dfrac{1}{x^2-9}$

$$= \dfrac{-2(3+x)}{(x+3)(x+3)} + \dfrac{3(x+2)}{(x-3)(x-3)} - \dfrac{1}{(x+3)(x-3)}$$

$$= \dfrac{-2}{x+3} + \dfrac{3(x+2)}{(x-3)(x-3)} - \dfrac{1}{(x+3)(x-3)}$$

$$= \dfrac{-2(x-3)(x-3)}{(x+3)(x-3)(x-3)} + \dfrac{3(x+2)(x+3)}{(x+3)(x-3)(x-3)} - \dfrac{1(x-3)}{(x+3)(x-3)(x-3)}$$

$$= \dfrac{-2(x^2-6x+9)+3(x^2+5x+6)-x+3}{(x+3)(x-3)(x-3)} = \dfrac{x^2+26x+3}{(x+3)(x-3)(x-3)}$$

31. $\dfrac{\frac{3}{x}-\frac{2}{y}}{xy} = \dfrac{\left(\frac{3}{x}-\frac{2}{y}\right)xy}{xy(xy)} = \dfrac{3y-2x}{x^2y^2}$

32. $\dfrac{\frac{1}{x}+\frac{2}{y}}{\frac{2}{x}-\frac{1}{y}} = \dfrac{\left(\frac{1}{x}+\frac{2}{y}\right)xy}{\left(\frac{2}{x}-\frac{1}{y}\right)xy} = \dfrac{y+2x}{2y-x}$

33. $\dfrac{2x+3+\frac{1}{x}}{x+2+\frac{1}{x}} = \dfrac{\left(2x+3+\frac{1}{x}\right)x}{\left(x+2+\frac{1}{x}\right)x} = \dfrac{2x^2+3x+1}{x^2+2x+1} = \dfrac{(2x+1)(x+1)}{(x+1)(x+1)} = \dfrac{2x+1}{x+1}$

34. $\dfrac{6x+13+\frac{6}{x}}{6x+5-\frac{6}{x}} = \dfrac{\left(6x+13+\frac{6}{x}\right)x}{\left(6x+5-\frac{6}{x}\right)x} = \dfrac{6x^2+13x+6}{6x^2+5x-6} = \dfrac{(2x+3)(3x+2)}{(2x+3)(3x-2)} = \dfrac{3x+2}{3x-2}$

35. $\dfrac{1-\frac{1}{x}-\frac{2}{x^2}}{1+\frac{4}{x}+\frac{3}{x^2}} = \dfrac{\left(1-\frac{1}{x}-\frac{2}{x^2}\right)x^2}{\left(1+\frac{4}{x}+\frac{3}{x^2}\right)x^2} = \dfrac{x^2-x-2}{x^2+4x+3} = \dfrac{(x-2)(x+1)}{(x+3)(x+1)} = \dfrac{x-2}{x+3}$

36. $\dfrac{x^{-1}+1}{x+1} = \dfrac{\frac{1}{x}+1}{x+1} = \dfrac{\left(\frac{1}{x}+1\right)x}{(x+1)x} = \dfrac{1+x}{(x+1)x} = \dfrac{1}{x}$

37. $\dfrac{x^{-1}-y^{-1}}{x^{-1}+y^{-1}} = \dfrac{\frac{1}{x}-\frac{1}{y}}{\frac{1}{x}+\frac{1}{y}} = \dfrac{\left(\frac{1}{x}-\frac{1}{y}\right)xy}{\left(\frac{1}{x}+\frac{1}{y}\right)xy} = \dfrac{y-x}{y+x}$

38. $\dfrac{(x-y)^{-2}}{x^{-2}-y^{-2}} = \dfrac{\frac{1}{(x-y)^2}}{\frac{1}{x^2}-\frac{1}{y^2}} = \dfrac{\left(\frac{1}{(x-y)^2}\right)x^2y^2(x-y)^2}{\left(\frac{1}{x^2}-\frac{1}{y^2}\right)x^2y^2(x-y)^2} = \dfrac{x^2y^2}{y^2(x-y)^2-x^2(x-y)^2}$

$$= \dfrac{x^2y^2}{(x-y)^2(y^2-x^2)}$$

39.

$$\frac{4}{x} - \frac{1}{10} = \frac{7}{2x}$$

$$\left(\frac{4}{x} - \frac{1}{10}\right)10x = \frac{7}{2x} \cdot 10x$$

$$40 - x = 35$$

$$5 = x \qquad \text{The answer checks.}$$

40.

$$\frac{2}{x+5} - \frac{1}{6} = \frac{1}{x+4}$$

$$\left(\frac{2}{x+5} - \frac{1}{6}\right)6(x+5)(x+4) = \left(\frac{1}{x+4}\right)6(x+5)(x+4)$$

$$12(x+4) - (x+5)(x+4) = 6(x+5)$$

$$12x + 48 - x^2 - 9x - 20 = 6x + 30$$

$$0 = x^2 + 3x + 2$$

$$0 = (x+2)(x+1)$$

$$x + 2 = 0 \quad \textbf{or} \quad x + 1 = 0$$

$$x = -2 \qquad\qquad x = -1 \qquad \text{Both answers check.}$$

41.

$$\frac{2(x-5)}{x-2} = \frac{6x+12}{4-x^2}$$

$$\frac{2(x-5)}{x-2} = \frac{6(x+2)}{(2+x)(2-x)}$$

$$\frac{2(x-5)}{x-2} = \frac{-6(x+2)}{(x+2)(x-2)}$$

$$2(x-5)(x+2) = -6(x+2)$$

$$2x^2 - 6x - 20 = -6x - 12$$

$$2x^2 - 8 = 0$$

$$2(x+2)(x-2) = 0$$

$$x = 2, \text{ or } x = -2 \qquad \text{Neither answer checks, so there is no solution.}$$

42.

$$\frac{7}{x+9} - \frac{x+2}{2} = \frac{x+4}{x+9}$$

$$\left(\frac{7}{x+9} - \frac{x+2}{2}\right)2(x+9) = \left(\frac{x+4}{x+9}\right)2(x+9)$$

$$14 - (x+2)(x+9) = 2(x+4)$$

$$14 - x^2 - 11x - 18 = 2x + 8$$

$$0 = x^2 + 13x + 12$$

$$0 = (x+12)(x+1)$$

$$x + 12 = 0 \quad \textbf{or} \quad x + 1 = 0$$

$$x = -12 \qquad\qquad x = -1 \qquad \text{Both answers check.}$$

CHAPTER 6 SUMMARY

43.

$$\frac{x^2}{a^2} - \frac{y^2}{b^2} = 1$$

$$\left(\frac{x^2}{a^2} - \frac{y^2}{b^2}\right)a^2b^2 = a^2b^2$$

$$x^2b^2 - y^2a^2 = a^2b^2$$

$$x^2b^2 - a^2b^2 = y^2a^2$$

$$\frac{x^2b^2 - a^2b^2}{a^2} = y^2$$

44.

$$H = \frac{2ab}{a+b}$$

$$H(a+b) = 2ab$$

$$Ha + Hb = 2ab$$

$$Ha = 2ab - Hb$$

$$Ha = b(2a - H)$$

$$\frac{Ha}{2a - H} = b$$

45. Let r = the usual rate. Then $r - 10$ = the slower rate.

	Rate	Time	Dist.
Usual	r	$\frac{200}{r}$	200
Slower	$r - 10$	$\frac{200}{r-10}$	200

$$\boxed{\text{Usual time}} + 1 = \boxed{\text{Slower time}}$$

$$\frac{200}{r} + 1 = \frac{200}{r - 10}$$

$$\left(\frac{200}{r} + 1\right)(r)(r-10) = \frac{200}{r-10} \cdot r(r-10)$$

$$200(r - 10) + r(r - 10) = 200r$$

$$r^2 - 10r - 2000 = 0$$

$$(r + 40)(r - 50) = 0$$

$$r + 40 = 0 \quad \textbf{or} \quad r - 50 = 0$$

$$r = -40 \qquad\qquad r = 50$$

$r = 50$ is the only answer that makes sense. The usual rate is 50 miles per hour.

46. Let r = the usual rate. Then $r + 40$ = the faster rate.

	Rate	Time	Dist.
Usual	r	$\frac{600}{r}$	600
Faster	$r + 40$	$\frac{600}{r+40}$	600

$$\boxed{\text{Usual time}} - \frac{1}{2} = \boxed{\text{Faster time}}$$

$$\frac{600}{r} - \frac{1}{2} = \frac{600}{r + 40}$$

$$\left(\frac{600}{r} - \frac{1}{2}\right)(2r)(r+40) = \frac{600}{r+40} \cdot 2r(r+40)$$

$$1200(r + 40) + r(r + 40) = 1200r$$

$$r^2 + 40r - 48000 = 0$$

$$(r + 240)(r - 200) = 0$$

$$r + 240 = 0 \quad \textbf{or} \quad r - 200 = 0$$

$$r = -240 \qquad\qquad r = 200$$

$r = 200$ is the only answer that makes sense. The usual rate is 200 miles per hour.

47. The first pipe can drain $\frac{1}{24}$ of the tank in 1 hour, while the second can drain $\frac{1}{36}$ of the pool in 1 hour. Let $x =$ the number of hours needed for them to drain the tank together. Then $\frac{1}{x} =$ the amount of the tank they can drain together in 1 hour.

Amount 1st drains in 1 hour		Amount 2nd drains in 1 hour		Amount drained together in 1 hour
	$+$		$=$	

$$\frac{1}{24} + \frac{1}{36} = \frac{1}{x}$$

$$\left(\frac{1}{24} + \frac{1}{36}\right)72x = \frac{1}{x} \cdot 72x$$

$$3x + 2x = 72$$

$$5x = 72$$

$$x = \frac{72}{5} = 14\frac{2}{5}$$

It will take the two pipes $14\frac{2}{5}$ hours to drain the tank.

48. The first man can side $\frac{1}{14}$ of the house in 1 day. Let $x =$ the number of days needed for the second man to side the house alone. Then he can side $\frac{1}{x}$ of the house alone. Since both men can side the house in 8 days, they side $\frac{1}{8}$ of the house in 1 day.

Amount 1st sides in 1 day		Amount 2nd sides in 1 day		Amount sided together in 1 day
	$+$		$=$	

$$\frac{1}{14} + \frac{1}{x} = \frac{1}{8}$$

$$\left(\frac{1}{14} + \frac{1}{x}\right)(56x) = \frac{1}{8} \cdot 56x$$

$$4x + 56 = 7x$$

$$56 = 3x, \text{ or } x = \frac{56}{3} = 18\frac{2}{3}$$

It takes the second man $18\frac{2}{3}$ days to side the house alone.

49. $\dfrac{-5x^6y^3}{10x^3y^6} = -\dfrac{x^3}{2y^3}$

50. $\dfrac{30x^3y^2 - 15x^2y - 10xy^2}{-10xy} = \dfrac{30x^3y^2}{-10xy} + \dfrac{-15x^2y}{-10xy} + \dfrac{-10xy^2}{-10xy} = -3x^2y + \dfrac{3x}{2} + y$

51.
```
            x +   5y
3x - 2y | 3x² + 13xy - 10y²
          3x² -  2xy
              15xy - 10y²
              15xy - 10y²
                        0
```

52.
```
             x² + 2x - 1 + 6/(2x+3)
2x + 3 | 2x³ + 7x² + 4x +     3
         2x³ + 3x²
              4x² + 4x
              4x² + 6x
                  - 2x +     3
                  - 2x -     3
                              6
```

188

53.

$$5 \;\big|\; \begin{array}{rrrr} 1 & -3 & -8 & -10 \\ & 5 & 10 & 10 \end{array}$$
$$\overline{\quad\;\; 1 \quad\; 2 \quad\; 2 \quad \boxed{0}} \;\Rightarrow \text{factor}$$

54.

$$-5 \;\big|\; \begin{array}{rrrr} 1 & 4 & -5 & 5 \\ & -5 & 5 & 0 \end{array}$$
$$\overline{\quad\;\; 1 \;\; -1 \quad\; 0 \quad \boxed{5}} \;\Rightarrow \text{not a factor}$$

Chapter 6 Test (page 421)

1. $\dfrac{-12x^2y^3z^2}{18x^3y^4z^2} = -\dfrac{2}{3xy}$

2. $\dfrac{2x+4}{x^2-4} = \dfrac{2(x+2)}{(x+2)(x-2)} = \dfrac{2}{x-2}$

3. $\dfrac{3y-6z}{2z-y} = \dfrac{3(y-2z)}{2z-y} = -3$

4. $\dfrac{2x^2+7x+3}{4x+12} = \dfrac{(2x+1)(x+3)}{4(x+3)} = \dfrac{2x+1}{4}$

5. $P(E) = \dfrac{s}{n} = \dfrac{1}{8}$

6. Let h = the height of the tree.
$$\dfrac{12}{2} = \dfrac{h}{3}$$
$$36 = 2h$$
$$18 = h$$
The tree is 18 feet tall.

7.
$$\dfrac{3}{x-2} = \dfrac{x+3}{2x}$$
$$6x = (x+3)(x-2)$$
$$6x = x^2+x-6$$
$$0 = x^2-5x-6$$
$$0 = (x-6)(x+1)$$
$$x-6=0 \quad\text{or}\quad x+1=0$$
$$x=6 \qquad\qquad x=-1$$

8.
$$V = \dfrac{k}{t} \qquad V = \dfrac{1100}{t}$$
$$55 = \dfrac{k}{20} \qquad 75 = \dfrac{1100}{t}$$
$$1100 = k \qquad 75t = 1100$$
$$t = \dfrac{1100}{75}$$
$$\boxed{t = \dfrac{44}{3}}$$

9. $\dfrac{x^2y^{-2}}{x^3z^2} \cdot \dfrac{x^2z^4}{y^2z} = \dfrac{x^4y^{-2}z^4}{x^3y^2z^3} = \dfrac{xz}{y^4}$

10. $\dfrac{(x+1)(x+2)}{10} \cdot \dfrac{5}{x+2} = \dfrac{x+1}{2}$

11. $\dfrac{u^2+5u+6}{u^2-4} \cdot \dfrac{u^2-5u+6}{u^2-9} = \dfrac{(u+2)(u+3)}{(u+2)(u-2)} \cdot \dfrac{(u-2)(u-3)}{(u+3)(u-3)} = 1$

12. $\dfrac{x^3+y^3}{4} \div \dfrac{x^2-xy+y^2}{2x+2y} = \dfrac{x^3+y^3}{4} \cdot \dfrac{2x+2y}{x^2-xy+y^2} = \dfrac{(x+y)(x^2-xy+y^2)}{4} \cdot \dfrac{2(x+y)}{x^2-xy+y^2}$
$$= \dfrac{(x+y)^2}{2}$$

13. $\dfrac{xu+2u+3x+6}{u^2-9} \cdot \dfrac{2u-6}{x^2+3x+2} = \dfrac{u(x+2)+3(x+2)}{(u+3)(u-3)} \cdot \dfrac{2(u-3)}{(x+2)(x+1)}$
$$= \dfrac{(x+2)(u+3)}{(u+3)(u-3)} \cdot \dfrac{2(u-3)}{(x+2)(x+1)} = \dfrac{2}{x+1}$$

14. $\dfrac{a^2 + 7a + 12}{a + 3} \div \dfrac{16 - a^2}{a - 4} = \dfrac{a^2 + 7a + 12}{a + 3} \cdot \dfrac{a - 4}{16 - a^2} = \dfrac{(a + 4)(a + 3)}{a + 3} \cdot \dfrac{a - 4}{(4 + a)(4 - a)} = -1$

15. $\dfrac{3t}{t + 3} + \dfrac{9}{t + 3} = \dfrac{3t + 9}{t + 3} = \dfrac{3(t + 3)}{t + 3)} = 3$

16. $\dfrac{3w}{w - 5} + \dfrac{w + 10}{5 - w} = \dfrac{3w}{w - 5} - \dfrac{w + 10}{w - 5} = \dfrac{2w - 10}{w - 5} = \dfrac{2(w - 5)}{w - 5} = 2$

17. $\dfrac{2}{r} + \dfrac{r}{s} = \dfrac{2s}{rs} + \dfrac{rr}{rs} = \dfrac{2s + r^2}{rs}$

18. $\dfrac{x + 2}{x + 1} - \dfrac{x + 1}{x + 2} = \dfrac{(x + 2)(x + 2)}{(x + 1)(x + 2)} - \dfrac{(x + 1)(x + 1)}{(x + 1)(x + 2)} = \dfrac{x^2 + 4x + 4 - (x^2 + 2x + 1)}{(x + 1)(x + 2)}$

$$= \dfrac{x^2 + 4x + 4 - x^2 - 2x - 1}{(x + 1)(x + 2)}$$

$$= \dfrac{2x + 3}{(x + 1)(x + 2)}$$

19. $\dfrac{\frac{2u^2w^3}{v^2}}{\frac{4uw^4}{uv}} = \dfrac{2u^2w^3}{v^2} \div \dfrac{4uw^4}{uv} = \dfrac{2u^2w^3}{v^2} \cdot \dfrac{uv}{4uw^4} = \dfrac{2u^3w^3v}{4uv^2w^4} = \dfrac{u^2}{2vw}$

20. $\dfrac{\frac{x}{y} + \frac{1}{2}}{\frac{x}{2} - \frac{1}{y}} = \dfrac{\left(\frac{x}{y} + \frac{1}{2}\right)2y}{\left(\frac{x}{2} - \frac{1}{y}\right)2y} = \dfrac{2x + y}{xy - 2}$

21.
$$\dfrac{2}{x - 1} + \dfrac{5}{x + 2} = \dfrac{11}{x + 2}$$
$$\left(\dfrac{2}{x - 1} + \dfrac{5}{x + 2}\right)(x - 1)(x + 2) = \left(\dfrac{11}{x + 2}\right)(x - 1)(x + 2)$$
$$2(x + 2) + 5(x - 1) = 11(x - 1)$$
$$2x + 4 + 5x - 5 = 11x - 11$$
$$7x - 1 = 11x - 11$$
$$-4x = -10$$
$$x = \dfrac{-10}{-4} = \dfrac{5}{2} \quad \text{The answer checks.}$$

22.
$$\frac{u-2}{u-3} + 3 = u + \frac{u-4}{3-u}$$
$$\frac{u-2}{u-3} + 3 = u + \frac{4-u}{u-3}$$
$$\left(\frac{u-2}{u-3} + 3\right)(u-3) = \left(u + \frac{4-u}{u-3}\right)(u-3)$$
$$u - 2 + 3(u-3) = u(u-3) + 4 - u$$
$$4u - 11 = u^2 - 4u + 4$$
$$0 = u^2 - 8u + 15$$
$$0 = (u-5)(u-3)$$

$u - 5 = 0$ **or** $u - 3 = 0$
$u = 5$ $\qquad\qquad u = 3$ $\qquad$ $u = 3$ does not check and is not a solution. $u = 5$ is a solution.

23.
$$\frac{x^2}{a^2} + \frac{y^2}{b^2} = 1$$
$$\left(\frac{x^2}{a^2} + \frac{y^2}{b^2}\right)a^2b^2 = a^2b^2$$
$$x^2b^2 + y^2a^2 = a^2b^2$$
$$x^2b^2 = a^2b^2 - y^2a^2$$
$$x^2b^2 = a^2(b^2 - y^2)$$
$$\frac{x^2b^2}{b^2 - y^2} = a^2$$

24.
$$\frac{1}{r} = \frac{1}{r_1} + \frac{1}{r_2}$$
$$\frac{1}{r} \cdot rr_1r_2 = \left(\frac{1}{r_1} + \frac{1}{r_2}\right)rr_1r_2$$
$$r_1r_2 = rr_2 + rr_1$$
$$r_1r_2 - rr_2 = rr_1$$
$$r_2(r_1 - r) = rr_1$$
$$r_2 = \frac{rr_1}{r_1 - r}$$

25. Let r = the usual rate (in nautical miles per day). Then $r + 11$ = the faster rate.

	Rate	Time	Dist.
Usual	r	$\frac{440}{r}$	440
Faster	$r+11$	$\frac{440}{r+11}$	440

$$\boxed{\begin{array}{c}\text{Usual}\\\text{time}\end{array}} - 2 = \boxed{\begin{array}{c}\text{Faster}\\\text{time}\end{array}}$$
$$\frac{440}{r} - 2 = \frac{440}{r+11}$$
$$\left(\frac{440}{r} - 2\right)(r)(r+11) = \frac{440}{r+11} \cdot r(r+11)$$
$$440(r+11) - 2r(r+11) = 440r$$
$$-2r^2 - 22r + 4840 = 0$$
$$-2(r+55)(r-44) = 0$$

$r + 55 = 0$ $\qquad$ **or** $\qquad r - 44 = 0$
$r = -55$ $\qquad\qquad\quad r = 44$ $\quad$ $r = 44$ is the only answer that makes sense.
The usual rate is 44 nautical miles per day.
The usual time is $\frac{440}{44} = 10$ days.

26. Let $r =$ the usual rate. Then $r + 0.04 =$ the higher rate.

$$\text{Interest} = \text{Principal} \cdot \text{Rate} (\cdot \text{ Time}) \Rightarrow \text{Principal} = \frac{\text{Interest}}{\text{Rate}}$$

$$\boxed{\begin{array}{c}\text{Original}\\\text{principal}\end{array}} - 2000 = \boxed{\begin{array}{c}\text{New}\\\text{principal}\end{array}}$$

$$\frac{300}{r} - 2000 = \frac{300}{r + 0.04}$$

$$\left(\frac{300}{r} - 2000\right)r(r + 0.04) = \left(\frac{300}{r + 0.04}\right)r(r + 0.04)$$

$$300(r + 0.04) - 2000r(r + 0.04) = 300r$$

$$-2000r^2 - 80r + 12 = 0$$

$$-4(500r^2 + 20r - 3) = 0$$

$$-4(50r - 3)(10r + 1) = 0$$

$$50r - 3 = 0 \qquad \text{or} \qquad 10r + 1 = 0$$

$$r = 0.06 \qquad\qquad\qquad r = -0.10 \quad r = 0.06 \text{ is the only answer that makes sense.}$$

She would invest \$5000 at 6% or \$3000 at 10%.

27. $\dfrac{18x^2y^3 - 12x^3y^2 + 9xy}{-3xy^4} = \dfrac{18x^2y^3}{-3xy^4} + \dfrac{-12x^3y^2}{-3xy^4} + \dfrac{9xy}{-3xy^4} = -\dfrac{6x}{y} + \dfrac{4x^2}{y^2} - \dfrac{3}{y^3}$

28.
$$\require{enclose}\begin{array}{r}3x^2 + 4x + 2\\2x - 1 \enclose{longdiv}{6x^3 + 5x^2 + 0x - 2}\\\underline{6x^3 - 3x^2}\\8x^2 + 0x\\\underline{8x^2 - 4x}\\4x - 2\\\underline{4x - 2}\\0\end{array}$$

29.
$$\begin{array}{r}x^2 - 5x + 10\\x + 1 \enclose{longdiv}{x^3 - 4x^2 + 5x + 3}\\\underline{x^3 + x^2}\\-5x^2 + 5x\\\underline{-5x^2 - 5x}\\10x + 3\\\underline{10x + 10}\\-7\end{array}$$

30.
$$\begin{array}{r|rrrr}2 & 4 & 3 & 2 & -1\\ & & 8 & 22 & 48\\\hline & 4 & 11 & 24 & \boxed{47}\end{array}$$

Cumulative Review Exercises (page 422)

1. $a^3b^2a^5b^2 = a^8b^4$

2. $\dfrac{a^3b^6}{a^7b^2} = \dfrac{b^4}{a^4}$

3. $\left(\dfrac{2a^2}{3b^4}\right)^{-4} = \left(\dfrac{3b^4}{2a^2}\right)^4 = \dfrac{81b^{16}}{16a^8}$

4. $\left(\dfrac{x^{-2}y^3}{x^2x^3y^4}\right)^{-3} = \left(\dfrac{x^2x^3y^4}{x^{-2}y^3}\right)^3$

$$= (x^7y)^3 = x^{21}y^3$$

5. $4.25 \times 10^4 = 42{,}500$

6. $7.12 \times 10^{-4} = 0.000712$

7.
$$\frac{a+2}{5} - \frac{8}{5} = 4a - \frac{a+9}{2}$$
$$\left(\frac{a+2}{5} - \frac{8}{5}\right)10 = \left(4a - \frac{a+9}{2}\right)10$$
$$2(a+2) - 16 = 40a - 5(a+9)$$
$$2a + 4 - 16 = 40a - 5a - 45$$
$$33 = 33a$$
$$1 = a$$

8.
$$\frac{3x-4}{6} - \frac{x-2}{2} = \frac{-2x-3}{3}$$
$$\left(\frac{3x-4}{6} - \frac{x-2}{2}\right)6 = \left(\frac{-2x-3}{3}\right)6$$
$$3x - 4 - 3(x-2) = 2(-2x-3)$$
$$3x - 4 - 3x + 6 = -4x - 6$$
$$4x = -8$$
$$x = -2$$

9.
$$m = \frac{\Delta y}{\Delta x} = \frac{5-10}{-2-4} = \frac{-5}{-6} = \frac{5}{6}$$

10.
$$3x + 4y = 13$$
$$4y = -3x + 13$$
$$y = -\tfrac{3}{4}x + \tfrac{13}{4}$$
$$m = -\tfrac{3}{4}$$

11. $y = 3x + 2 \Rightarrow m = 3$
A parallel line will also have $m = 3$.

12. $y = 3x + 2 \Rightarrow m = 3$
A perpendicular line will have $m = -\frac{1}{3}$.

13. $f(0) = 0^2 - 2(0) = 0$

14. $f(-2) = (-2)^2 - 2(-2) = 8$

15. $f\left(\frac{2}{5}\right) = \left(\frac{2}{5}\right)^2 - 2\left(\frac{2}{5}\right) = \frac{4}{25} - \frac{4}{5} = -\frac{16}{25}$

16. $f(t-1) = (t-1)^2 - 2(t-1) = t^2 - 2t + 1 - 2t + 2 = t^2 - 4t + 3$

17. $y = \dfrac{kxz}{r}$

18. Since the graph does not pass the vertical line test, it is not the graph of a function.

19.
$$x - 2 \leq 3x + 1 \leq 5x - 4$$

$$x - 2 \leq 3x + 1 \qquad \text{and} \qquad 3x + 1 \leq 5x - 4$$
$$x - 2 - 1 \leq 3x + 1 - 1 \qquad\qquad 3x + 1 + 4 \leq 5x - 4 + 4$$
$$x - x - 3 \leq 3x - x \qquad\qquad 3x - 3x + 5 \leq 5x - 3x$$
$$-3 \leq 2x \qquad\qquad 5 \leq 2x$$
$$\frac{-3}{2} \leq x \qquad\qquad \frac{5}{2} \leq x$$
$$x \geq -\frac{3}{2} \qquad\qquad x \geq \frac{5}{2}$$

If $x \geq -\dfrac{3}{2}$ **and** $x \geq \dfrac{5}{2}$, then $x \geq \dfrac{5}{2}$. The solution set is $\left[\dfrac{5}{2}, \infty\right)$

$\xleftarrow{\qquad} [\xrightarrow{\qquad\qquad}$
$\frac{5}{2}$

CUMULATIVE REVIEW EXERCISES

20.
$$\left|\frac{3a}{5} - 2\right| + 1 \geq \frac{6}{5}$$

$$\left|\frac{3a}{5} - 2\right| \geq \frac{1}{5}$$

$$\frac{3a}{5} - 2 \leq -\frac{1}{5} \quad \text{or} \quad \frac{3a}{5} - 2 \geq \frac{1}{5}$$

$$\frac{3a}{5} \leq \frac{9}{5} \qquad\qquad \frac{3a}{5} \geq \frac{11}{5}$$

$$3a \leq 9 \qquad\qquad 3a \geq 11$$

$$a \leq 3 \qquad\qquad a \geq \frac{11}{3}$$

solution set: $(-\infty, 3] \cup \left[\frac{11}{3}, \infty\right)$

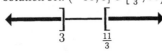

21. trinomial

22. degree $= 3 + 4 = 7$

23. $f(-2) = -3(-2)^3 + (-2) - 4 = 18$

24. $y = f(x) = 2x^2 - 3$

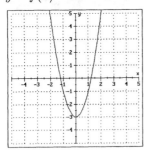

25. $(3x^2 - 2x + 7) + (-2x^2 + 2x + 5) + (3x^2 - 4x + 2)$
$= 3x^2 - 2x + 7 - 2x^2 + 2x + 5 + 3x^2 - 4x + 2 = 4x^2 - 4x + 14$

26. $(-5x^2 + 3x + 4) - (-2x^2 + 3x + 7) = -5x^2 + 3x + 4 + 2x^2 - 3x - 7 = -3x^2 - 3$

27. $(3x + 4)(2x - 5) = 6x^2 - 15x + 8x - 20$
$= 6x^2 - 7x - 20$

28. $(2x^n - 1)(x^n + 2) = 2x^{2n} + 4x^n - x^n - 2$
$= 2x^{2n} + 3x^n - 2$

29. $3r^2s^3 - 6rs^4 = 3rs^3(r - 2s)$

30. $5(x - y) - a(x - y) = (x - y)(5 - a)$

31. $xu + yv + xv + yu = xu + xv + yu + yv = x(u + v) + y(u + v) = (u + v)(x + y)$

32. $81x^4 - 16y^4 = (9x^2 + 4y^2)(9x^2 - 4y^2) = (9x^2 + 4y^2)(3x + 2y)(3x - 2y)$

33. $8x^3 - 27y^6 = (2x)^3 - (3y^2)^3 = (2x - 3y^2)(4x^2 + 6xy^2 + 9y^4)$

34. $6x^2 + 5x - 6 = (2x + 3)(3x - 2)$

35. $9x^2 - 30x + 25 = (3x - 5)(3x - 5)$

36. $15x^2 - x - 6 = (5x + 3)(3x - 2)$

37. $27a^3 + 8b^3 = (3a + 2b)(9a^2 - 6ab + 4b^2)$

CUMULATIVE REVIEW EXERCISES

38. $6x^2 + x - 35 = (3x - 7)(2x + 5)$

39. $x^2 + 10x + 25 - y^4 = (x + 5)^2 - y^4 = (x + 5 + y^2)(x + 5 - y^2)$

40. $y^2 - x^2 + 4x - 4 = y^2 - (x^2 - 4x + 4) = y^2 - (x - 2)^2 = (y + x - 2)(y - x + 2)$

41.
$$x^3 - 4x = 0$$
$$x(x^2 - 4) = 0$$
$$x(x + 2)(x - 2) = 0$$
$$x = 0 \quad \text{or} \quad x + 2 = 0 \quad \text{or} \quad x - 2 = 0$$
$$x = -2 \qquad x = 2$$

42.
$$6x^2 + 7 = -23x$$
$$6x^2 + 23x + 7 = 0$$
$$(2x + 7)(3x + 1) = 0$$
$$2x + 7 = 0 \quad \text{or} \quad 3x + 1 = 0$$
$$x = -\tfrac{7}{2} \qquad x = -\tfrac{1}{3}$$

43. $\dfrac{2x^2y + xy - 6y}{3x^2y + 5xy - 2y} = \dfrac{y(2x^2 + x - 6)}{y(3x^2 + 5x - 2)} = \dfrac{y(2x - 3)(x + 2)}{y(3x - 1)(x + 2)} = \dfrac{2x - 3}{3x - 1}$

44. $\dfrac{x^2 - 4}{x^2 + 9x + 20} \div \dfrac{x^2 + 5x + 6}{x^2 + 4x - 5} \cdot \dfrac{x^2 + 3x - 4}{(x - 1)^2} = \dfrac{x^2 - 4}{x^2 + 9x + 20} \cdot \dfrac{x^2 + 4x - 5}{x^2 + 5x + 6} \cdot \dfrac{x^2 + 3x - 4}{(x - 1)^2}$

$$= \dfrac{(x + 2)(x - 2)}{(x + 4)(x + 5)} \cdot \dfrac{(x + 5)(x - 1)}{(x + 2)(x + 2)} \cdot \dfrac{(x + 4)(x - 1)}{(x - 1)(x - 1)}$$

$$= \dfrac{x - 2}{x + 3}$$

45. $\dfrac{2}{x + y} + \dfrac{3}{x - y} - \dfrac{x - 3y}{x^2 - y^2} = \dfrac{2}{x + y} + \dfrac{3}{x - y} - \dfrac{x - 3y}{(x + y)(x - y)}$

$$= \dfrac{2(x - y)}{(x + y)(x - y)} + \dfrac{3(x + y)}{(x + y)(x - y)} - \dfrac{x - 3y}{(x + y)(x - y)}$$

$$= \dfrac{2x - 2y + 3x + 3y - x + 3y}{(x + y)(x - y)}$$

$$= \dfrac{4x + 4y}{(x + y)(x - y)} = \dfrac{4(x + y)}{(x + y)(x - y)} = \dfrac{4}{x - y}$$

46. $\dfrac{\frac{a}{b} + b}{a - \frac{b}{a}} = \dfrac{\left(\frac{a}{b} + b\right)ab}{\left(a - \frac{b}{a}\right)ab} = \dfrac{a^2 + ab^2}{a^2b - b^2}$

47.
$$\dfrac{5x - 3}{x + 2} = \dfrac{5x + 3}{x - 2}$$
$$(5x - 3)(x - 2) = (5x + 3)(x + 2)$$
$$5x^2 - 13x + 6 = 5x^2 = 13x + 6$$
$$-26x = 0$$
$$x = 0 \quad \text{The answer checks.}$$

195

48.
$$\frac{3}{x-2} + \frac{x^2}{(x+3)(x-2)} = \frac{x+4}{x+3}$$

$$\left(\frac{3}{x-2} + \frac{x^2}{(x+3)(x-2)}\right)(x+3)(x-2) = \left(\frac{x+4}{x+3}\right)(x+3)(x-2)$$

$$3(x+3) + x^2 = (x+4)(x-2)$$

$$x^2 + 3x + 9 = x^2 + 2x - 8$$

$$x = -17$$

49.
$$\begin{array}{r} x + 4 \\ x+5 \overline{\smash{\big)}\ x^2 + 9x + 20} \\ \underline{x^2 + 5x} \\ 4x + 20 \\ \underline{4x + 20} \\ 0 \end{array}$$

50.
$$\begin{array}{r} -x^2 + x + 5 + \frac{8}{x-1} \\ x-1 \overline{\smash{\big)}\ -x^3 + 2x^2 + 4x + 3} \\ \underline{-x^3 + x^2} \\ x^2 + 4x \\ \underline{x^2 - x} \\ 5x + 3 \\ \underline{5x - 5} \\ 8 \end{array}$$

Exercise 7.1 (page 435)

1. $\dfrac{x^2 + 7x + 12}{x^2 - 16} = \dfrac{(x+4)(x+3)}{(x+4)(x-4)} = \dfrac{x+3}{x-4}$

3. $\dfrac{x^2 - x - 6}{x^2 - 2x - 3} \cdot \dfrac{x^2 - 1}{x^2 + x - 2} = \dfrac{(x-3)(x+2)}{(x-3)(x+1)} \cdot \dfrac{(x+1)(x-1)}{(x+2)(x-1)} = 1$

5. $\dfrac{3}{m+1} + \dfrac{3m}{m-1} = \dfrac{3(m-1)}{(m+1)(m-1)} + \dfrac{3m(m+1)}{(m+1)(m-1)} = \dfrac{3m^2 + 6m - 3}{(m+1)(m-1)} = \dfrac{3(m^2 + 2m - 1)}{(m+1)(m-1)}$

7. $(5x^2)^2$ **9.** positive **11.** 3; up **13.** x

15. odd **17.** 0 **19.** $3x^2$ **21.** $a^2 + b^3$

23. $\sqrt{121} = \sqrt{11^2} = 11$ **25.** $-\sqrt{64} = -\sqrt{8^2} = -8$ **27.** $\sqrt{\frac{1}{9}} = \sqrt{\left(\frac{1}{3}\right)^2} = \frac{1}{3}$

29. $-\sqrt{\frac{25}{49}} = -\sqrt{\left(\frac{5}{7}\right)^2} = -\frac{5}{7}$ **31.** $\sqrt{-25}$: not a real number **33.** $\sqrt{0.16} = 0.4$

35. $\sqrt{(-4)^2} = \sqrt{16} = 4$ **37.** $\sqrt{-36}$: not a real number **39.** $\sqrt{12} \approx 3.4641$

41. $\sqrt{679.25} \approx 26.0624$ **43.** $\sqrt{4x^2} = \sqrt{(2x)^2} = |2x| = 2|x|$

45. $\sqrt{9a^4} = \sqrt{(3a^2)^2} = |3a^2| = 3a^2$ **47.** $\sqrt{(t+5)^2} = |t+5|$

49. $\sqrt{(-5b)^2} = |-5b| = 5|b|$ **51.** $\sqrt{a^2 + 6a + 9} = \sqrt{(a+3)^2} = |a+3|$

53. $\sqrt{t^2 + 24t + 144} = \sqrt{(t + 12)^2} = |t + 12|$ **55.** $\sqrt[3]{1} = \sqrt[3]{1^3} = 1$

57. $\sqrt[3]{-125} = \sqrt[3]{(-5)^3} = -5$ **59.** $\sqrt[3]{-\frac{8}{27}} = \sqrt[3]{\left(-\frac{2}{3}\right)^3} = -\frac{2}{3}$

61. $\sqrt[3]{0.064} = 0.4$ **63.** $\sqrt[3]{8a^3} = \sqrt[3]{(2a)^3} = 2a$

65. $\sqrt[3]{-1000p^3q^3} = \sqrt[3]{(-10pq)^3} = -10pq$ **67.** $\sqrt[3]{-\frac{1}{8}m^6n^3} = \sqrt[3]{\left(-\frac{1}{2}m^2n\right)^3} = -\frac{1}{2}m^2n$

69. $\sqrt[3]{0.008z^9} = \sqrt[3]{(0.2z^3)^3} = 0.2z^3$ **71.** $\sqrt[4]{81} = \sqrt[4]{3^4} = 3$

73. $-\sqrt[5]{243} = -\sqrt[5]{3^5} = -3$ **75.** $\sqrt[5]{-32} = \sqrt[5]{(-2)^5} = -2$

77. $\sqrt[4]{\frac{16}{625}} = \sqrt[4]{\left(\frac{2}{5}\right)^4} = \frac{2}{5}$ **79.** $-\sqrt[5]{-\frac{1}{32}} = -\sqrt[5]{\left(-\frac{1}{2}\right)^5} = -\left(-\frac{1}{2}\right) = \frac{1}{2}$

81. $\sqrt[4]{-256} \Rightarrow$ not a real number   **83.** $\sqrt[4]{16x^4} = \sqrt[4]{(2x)^4} = |2x| = 2|x|$

85. $\sqrt[3]{8a^3} = \sqrt[3]{(2a)^3} = 2a$ **87.** $\sqrt[4]{\frac{1}{16}x^4} = \sqrt[4]{\left(\frac{1}{2}x\right)^4} = \left|\frac{1}{2}x\right| = \frac{1}{2}|x|$

89. $\sqrt[4]{x^{12}} = \sqrt[4]{(x^3)^4} = |x^3|$ **91.** $\sqrt[5]{-x^5} = \sqrt[5]{(-x)^5} = -x$

93. $\sqrt[3]{-27a^6} = \sqrt[3]{(-3a^2)^3} = -3a^2$ **95.** $\sqrt[25]{(x + 2)^{25}} = x + 2$

97. $\sqrt[8]{0.00000001x^{16}y^8} = \sqrt[8]{(0.1x^2y)^8} = |0.1x^2y| = 0.1x^2|y|$

99. $f(4) = \sqrt{4 - 4} = \sqrt{0} = 0$ **101.** $f(20) = \sqrt{20 - 4} = \sqrt{16} = 4$

103. $g(9) = \sqrt{9 - 8} = \sqrt{1} = 1$ **105.** $g(8.25) = \sqrt{8.25 - 8} = \sqrt{0.25} = 0.5$

107. $f(4) = \sqrt{4^2 + 1} = \sqrt{17} \approx 4.1231$ **109.** $f(2.35) = \sqrt{(2.35)^2 + 1} = \sqrt{6.5225}$
$$\approx 2.5539$$

111. $f(x) = \sqrt{x+4}$; Shift $y = \sqrt{x}$ left 4.

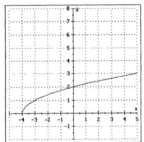

$D = [-4, \infty); R = [0, \infty)$

113. $f(x) = -\sqrt{x} - 3$; Reflect $y = \sqrt{x}$ about the x-axis and shift down 3.

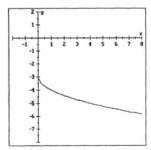

$D = [0, \infty); R = (-\infty, -3]$

115. mean $= \dfrac{2+5+5+6+7}{5} = \dfrac{25}{5} = 5$

Original term	Mean	Difference (term−mean)	Square of difference
2	5	−3	9
5	5	0	0
5	5	0	0
6	5	1	1
7	5	2	4

st. dev. $= \sqrt{\dfrac{9+0+0+1+4}{5}} \approx 1.67$

117. $s_{\overline{x}} = \dfrac{s}{\sqrt{N}} = \dfrac{65}{\sqrt{30}} \approx 11.8673$

119. $r = \sqrt{\dfrac{A}{\pi}} = \sqrt{\dfrac{9\pi}{\pi}} = \sqrt{9} = 3$ units

121. $t = \dfrac{\sqrt{s}}{4} = \dfrac{\sqrt{256}}{4} = \dfrac{16}{4} = 4$ seconds

123. $I = \sqrt{\dfrac{P}{18}} = \sqrt{\dfrac{980}{18}} \approx \sqrt{54.44} \approx 7.4$ amps

125. Answers may vary.

127. $\sqrt{x^2 - 4x + 4} = \sqrt{(x-2)^2} = |x - 2|$. $|x - 2| = x - 2$ when $x - 2 \geq 0$, or $x \geq 2$.

Exercise 7.2 (page 442)

1. $(4x + 2)(3x - 5) = 12x^2 - 14x - 10$

3. $(5t + 4s)(3t - 2s) = 15t^2 + 2ts - 8s^2$

5. hypotenuse

7. $a^2 + b^2 = c^2$

9. distance

SECTION 7.2

11.
$$c^2 = a^2 + b^2$$
$$c^2 = 6^2 + 8^2$$
$$c^2 = 36 + 64$$
$$c^2 = 100$$
$$c = \sqrt{100}$$
$$c = 10 \text{ ft}$$

13.
$$c^2 = a^2 + b^2$$
$$82^2 = a^2 + 18^2$$
$$6724 = a^2 + 324$$
$$6400 = a^2$$
$$\sqrt{6400} = a$$
$$80 \text{ m} = a$$

15.
$$c^2 = a^2 + b^2$$
$$50^2 = 14^2 + b^2$$
$$2500 = 196 + b^2$$
$$2304 = b^2$$
$$\sqrt{2304} = b$$
$$48 \text{ in.} = b$$

17. Let $x =$ the length of the diagonal.
$$7^2 + 7^2 = x^2$$
$$49 + 49 = x^2$$
$$98 = x^2$$
$$\sqrt{98} = x$$
$$9.9 \text{ cm} \approx x$$

19. $d = \sqrt{(x_2 - x_1)^2 + (y_2 - y_1)^2} = \sqrt{(0 - 3)^2 + [0 - (-4)]^2} = \sqrt{(-3)^2 + (4)^2} = \sqrt{9 + 16}$
$$= \sqrt{25} = 5$$

21. $d = \sqrt{(x_2 - x_1)^2 + (y_2 - y_1)^2} = \sqrt{(2 - 5)^2 + (4 - 8)^2} = \sqrt{(-3)^2 + (-4)^2} = \sqrt{9 + 16}$
$$= \sqrt{25} = 5$$

23. $d = \sqrt{(x_2 - x_1)^2 + (y_2 - y_1)^2} = \sqrt{(-2 - 3)^2 + (-8 - 4)^2} = \sqrt{(-5)^2 + (-12)^2} = \sqrt{25 + 144}$
$$= \sqrt{169} = 13$$

25. $d = \sqrt{(x_2 - x_1)^2 + (y_2 - y_1)^2} = \sqrt{(6 - 12)^2 + (8 - 16)^2} = \sqrt{(-6)^2 + (-8)^2} = \sqrt{36 + 64}$
$$= \sqrt{100} = 10$$

27. $d = \sqrt{(x_2 - x_1)^2 + (y_2 - y_1)^2} = \sqrt{[-3 - (-5)]^2 + [5 - (-5)]^2} = \sqrt{2^2 + 10^2} = \sqrt{4 + 100}$
$$= \sqrt{104} \approx 10.2$$

29. Let the points be represented by $A(5, 1)$, $B(7, 0)$ and $C(3, 0)$. Find the length of $\overline{AB}$ and $\overline{AC}$:
$\overline{AB}: \sqrt{(5 - 7)^2 + (1 - 0)^2} = \sqrt{(-2)^2 + 1^2} = \sqrt{5}$
$\overline{AC}: \sqrt{(5 - 3)^2 + (1 - 0)^2} = \sqrt{(2)^2 + 1^2} = \sqrt{5}$
Since $\overline{AB}$ and $\overline{AC}$ have the same length, $(5, 1)$ is equidistant from $(7, 0)$ and $(3, 0)$.

31. Let the points be represented by $A(-2, 4)$, $B(2, 8)$ and $C(6, 4)$. Find the length of each side:
$\overline{AB}: \sqrt{(-2 - 2)^2 + (4 - 8)^2} = \sqrt{(-4)^2 + (-4)^2} = \sqrt{32}$
$\overline{AC}: \sqrt{(-2 - 6)^2 + (4 - 4)^2} = \sqrt{(-8)^2 + 0^2} = \sqrt{64}$
$\overline{BC}: \sqrt{(2 - 6)^2 + (8 - 4)^2} = \sqrt{(-4)^2 + 4^2} = \sqrt{32}$
Since $\overline{AB}$ and $\overline{BC}$ have the same length, the triangle is isosceles.

33. $d^2 = 5^2 + 12^2$
$d^2 = 25 + 144$
$d^2 = 169$
$d = 13$ ft

35. Let $x =$ distance to 2nd
$x^2 = 90^2 + 90^2$
$x^2 = 8100 + 8100$
$x^2 = 16200$
$x \approx 127$ ft

37. Refer to the diagram provided. The 3rd baseman is at B, so $\overline{BC}$ has a length of 10 ft. Let $\overline{AC}$ and $\overline{AB}$ both have a length x.

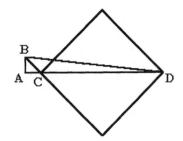

$x^2 + x^2 = 10^2$
$2x^2 = 100$
$x^2 = 50$
$x = \sqrt{50} \approx 7.1$ ft

From **#35**, the length of $\overline{CD} \approx 127.3$ ft, so $\overline{AD}$ has a length of about $127.3 + 7.1 = 134.4$ ft. Let $y =$ the length of $\overline{BD}$.
$y^2 = (7.1)^2 + (134.4)^2$
$y^2 = 18113.77$
$y = \sqrt{18113.77} \approx 135$ ft

39. $d = \sqrt{a^2 + b^2 + c^2} = \sqrt{12^2 + 24^2 + 17^2} = \sqrt{1009} = 31.76 \Rightarrow$ The racket will not fit.

41. $d = \sqrt{a^2 + b^2 + c^2} = \sqrt{21^2 + 21^2 + 21^2} = \sqrt{1323} = 36.4 \Rightarrow$ The femur will fit.

43. Let $x =$ direct distance from A to D.
$x^2 = 52^2 + (105 + 60)^2$
$x^2 = 52^2 + 165^2$
$x^2 = 29929$
$x = 173$ yd

45. Let $x =$ half the length of the stretched wire.
$x^2 = 20^2 + 1^2$
$x^2 = 401$
$x = 20.025$ ft
The stretched wire has a length of 40.05 ft. It has been stretched by 0.05 ft.

47. $A = 6\sqrt[3]{V^2} = 6\sqrt[3]{8^2} = 6\sqrt[3]{64} = 6(4) = 24$ cm^2

49. **Answers may vary.**

51. $I = \dfrac{703w}{h^2} = \dfrac{703(104)}{(54.1)^2} = \dfrac{73112}{2926.81} \approx 25$

Exercise 7.3 (page 451)

1. $5x - 4 < 11$
$5x < 15$
$x < 3$

3. $\frac{4}{5}(r - 3) > \frac{2}{3}(r + 2)$
$15 \cdot \frac{4}{5}(r - 3) > 15 \cdot \frac{2}{3}(r + 2)$
$12(r - 3) > 10(r + 2)$
$2r > 56$
$r > 28$

SECTION 7.3

5. Let x = pints of water added (0% alcohol).

$$\boxed{\substack{\text{Alcohol} \\ \text{at start}}} + \boxed{\substack{\text{Alcohol} \\ \text{added}}} = \boxed{\substack{\text{Alcohol} \\ \text{at end}}}$$

$$0.20(5) + 0(x) = 0.15(5 + x)$$
$$1 + 0 = 0.75 + 0.15x$$
$$0.25 = 0.15x$$
$$x = \frac{0.25}{0.15} = \frac{5}{3} \Rightarrow 1\tfrac{2}{3} \text{ pints of water should be added.}$$

7. $a \cdot a \cdot a \cdot a$ 　　　　**9.** a^{mn} 　　　　**11.** $\dfrac{a^n}{b^n}$ 　　　　**13.** $\dfrac{1}{a^n}; 0$

15. $\left(\dfrac{b}{a}\right)^n$ 　　　　**17.** $|x|$ 　　　　**19.** $7^{1/3} = \sqrt[3]{7}$ 　　　　**21.** $8^{1/5} = \sqrt[5]{8}$

23. $(3x)^{1/4} = \sqrt[4]{3x}$ 　　　　**25.** $\left(\tfrac{1}{2}x^3y\right)^{1/4} = \sqrt[4]{\tfrac{1}{2}x^3y}$ 　　　　**27.** $(4a^2b^3)^{1/5} = \sqrt[5]{4a^2b^3}$

29. $(x^2 + y^2)^{1/2} = \sqrt{x^2 + y^2}$ 　　**31.** $\sqrt{11} = 11^{1/2}$ 　　　　**33.** $\sqrt[4]{3a} = (3a)^{1/4}$

35. $3\sqrt[5]{a} = 3a^{1/5}$ 　　　　**37.** $\sqrt[6]{\tfrac{1}{7}abc} = \left(\tfrac{1}{7}abc\right)^{1/6}$ 　　　**39.** $\sqrt[5]{\tfrac{1}{2}mn} = \left(\tfrac{1}{2}mn\right)^{1/5}$

41. $\sqrt[3]{a^2 - b^2} = (a^2 - b^2)^{1/3}$ 　　**43.** $4^{1/2} = 2$ 　　　　　**45.** $27^{1/3} = 3$

47. $16^{1/4} = 2$ 　　**49.** $32^{1/5} = 2$ 　　**51.** $\left(\tfrac{1}{4}\right)^{1/2} = \tfrac{1}{2}$ 　　**53.** $\left(\tfrac{1}{8}\right)^{1/3} = \tfrac{1}{2}$

55. $-16^{1/4} = -2$ 　　**57.** $(-27)^{1/3} = -3$ 　　**59.** $(-64)^{1/2}$ not a real number 　　**61.** $0^{1/3} = 0$

63. $(25y^2)^{1/2} = \left[(5y)^2\right]^{1/2} = |5y| = 5|y|$ 　　　**65.** $(16x^4)^{1/4} = \left[(2x)^4\right]^{1/4} = |2x| = 2|x|$

67. $(243x^5)^{1/5} = \left[(3x)^5\right]^{1/5} = 3x$ 　　　**69.** $(-64x^8)^{1/4} \Rightarrow$ not a real number

71. $36^{3/2} = \left(36^{1/2}\right)^3 = 6^3 = 216$ 　　　**73.** $81^{3/4} = \left(81^{1/4}\right)^3 = 3^3 = 27$

75. $144^{3/2} = \left(144^{1/2}\right)^3 = 12^3 = 1728$ 　　**77.** $\left(\tfrac{1}{8}\right)^{2/3} = \left[\left(\tfrac{1}{8}\right)^{1/3}\right]^2 = \left(\tfrac{1}{2}\right)^2 = \tfrac{1}{4}$

79. $\left(25x^4\right)^{3/2} = \left[\left(25x^4\right)^{1/2}\right]^3 = \left(5x^2\right)^3$
$$= 125x^6$$

81. $\left(\dfrac{8x^3}{27}\right)^{2/3} = \left[\left(\dfrac{8x^3}{27}\right)^{1/3}\right]^2 = \left(\dfrac{2x}{3}\right)^2$
$$= \dfrac{4x^2}{9}$$

83. $4^{-1/2} = \dfrac{1}{4^{1/2}} = \dfrac{1}{2}$

85. $4^{-3/2} = \dfrac{1}{4^{3/2}} = \dfrac{1}{\left(4^{1/2}\right)^3} = \dfrac{1}{2^3} = \dfrac{1}{8}$

87. $\left(16x^2\right)^{-3/2} = \dfrac{1}{\left(16x^2\right)^{3/2}} = \dfrac{1}{\left[\left(16x^2\right)^{1/2}\right]^3} = \dfrac{1}{(4x)^3} = \dfrac{1}{64x^3}$

89. $\left(-27y^3\right)^{-2/3} = \dfrac{1}{\left(-27y^3\right)^{2/3}} = \dfrac{1}{\left[\left(-27y^3\right)^{1/3}\right]^2} = \dfrac{1}{(-3y)^2} = \dfrac{1}{9y^2}$

91. $\left(-32p^5\right)^{-2/5} = \dfrac{1}{\left(-32p^5\right)^{2/5}} = \dfrac{1}{\left[\left(-32p^5\right)^{1/5}\right]^2} = \dfrac{1}{(-2p)^2} = \dfrac{1}{4p^2}$

93. $\left(\dfrac{1}{4}\right)^{-3/2} = \left(\dfrac{4}{1}\right)^{3/2} = 4^{3/2} = \left(4^{1/2}\right)^3 = 2^3 = 8$

95. $\left(\dfrac{27}{8}\right)^{-4/3} = \left(\dfrac{8}{27}\right)^{4/3} = \left[\left(\dfrac{8}{27}\right)^{1/3}\right]^4 = \left(\dfrac{2}{3}\right)^4 = \dfrac{16}{81}$

97. $\left(-\dfrac{8x^3}{27}\right)^{-1/3} = \left(-\dfrac{27}{8x^3}\right)^{1/3} = -\dfrac{3}{2x}$

99. $5^{4/9}5^{4/9} = 5^{4/9+4/9} = 5^{8/9}$ **101.** $\left(4^{1/5}\right)^3 = 4^{(1/5)\cdot 3} = 4^{3/5}$ **103.** $\dfrac{9^{4/5}}{9^{3/5}} = 9^{4/5-3/5} = 9^{1/5}$

105. $\dfrac{7^{1/2}}{7^0} = 7^{1/2-0} = 7^{1/2}$

107. $6^{-2/3}6^{-4/3} = 6^{-6/3} = 6^{-2} = \dfrac{1}{36}$

109. $\dfrac{2^{5/6}2^{1/3}}{2^{1/2}} = \dfrac{2^{7/6}}{2^{1/2}} = 2^{4/6} = 2^{2/3}$

111. $a^{2/3}a^{1/3} = a^{3/3} = a^1 = a$

113. $\left(a^{2/3}\right)^{1/3} = a^{(2/3)(1/3)} = a^{2/9}$

115. $\left(a^{1/2}b^{1/3}\right)^{3/2} = a^{3/4}b^{1/2}$

117. $\left(mn^{-2/3}\right)^{-3/5} = m^{-3/5}n^{2/5} = \dfrac{n^{2/5}}{m^{3/5}}$

119. $\dfrac{\left(4x^3y\right)^{1/2}}{(9xy)^{1/2}} = \dfrac{2x^{3/2}y^{1/2}}{3x^{1/2}y^{1/2}} = \dfrac{2x}{3}$

121. $\left(27x^{-3}\right)^{-1/3} = (27)^{-1/3}x = \dfrac{1}{3}x$

123. $y^{1/3}\left(y^{2/3} + y^{5/3}\right) = y^{3/3} + y^{6/3} = y + y^2$

125. $x^{3/5}\left(x^{7/5} - x^{2/5} + 1\right) = x^{10/5} - x^{5/5} + x^{3/5} = x^2 - x + x^{3/5}$

127. $\left(x^{1/2} + 2\right)\left(x^{1/2} - 2\right) = x^{2/2} - 2x^{1/2} + 2x^{1/2} - 4 = x - 4$

SECTION 7.3

129. $\left(x^{2/3} - x\right)\left(x^{2/3} + x\right) = x^{4/3} + x^{5/3} - x^{5/3} - x^2 = x^{4/3} - x^2$

131. $\left(x^{2/3} + y^{2/3}\right)^2 = \left(x^{2/3} + y^{2/3}\right)\left(x^{2/3} + y^{2/3}\right) = x^{4/3} + x^{2/3}y^{2/3} + x^{2/3}y^{2/3} + y^{4/3}$
$$= x^{4/3} + 2x^{2/3}y^{2/3} + y^{4/3}$$

133. $\left(a^{3/2} - b^{3/2}\right)^2 = \left(a^{3/2} - b^{3/2}\right)\left(a^{3/2} - b^{3/2}\right) = a^{6/2} - a^{3/2}b^{3/2} - a^{3/2}b^{3/2} + b^{6/2}$
$$= a^3 - 2a^{3/2}b^{3/2} + b^3$$

135. $\sqrt[6]{p^3} = \left(p^3\right)^{1/6} = p^{3/6} = p^{1/2} = \sqrt{p}$ **137.** $\sqrt[4]{25b^2} = \left(5^2 b^2\right)^{1/4} = 5^{1/2}b^{1/2} = \sqrt{5b}$

139. Answers may vary. **141.** $16^{2/4} = 2^2 = 4;\ 16^{1/2} = 4$

Exercise 7.4 (page 461)

1. $3x^2 y^3(-5x^3 y^{-4}) = -15x^5 y^{-1} = \dfrac{-15x^5}{y}$ **3.** $(3t + 2)^2 = (3t + 2)(3t + 2) = 9t^2 + 12t + 4$

5.
$$\begin{array}{r} 3p + \ \ 4 + \frac{-5}{2p-5} \\ 2p - 5 \overline{\smash{\big)}\ 6p^2 - \ 7p - \ 25} \\ \underline{6p^2 - 15p} \\ 8p - \ 25 \\ \underline{8p - \ 20} \\ -5 \end{array}$$

7. $\sqrt[n]{a}\sqrt[n]{b}$

9. $\sqrt{6}\sqrt{6} = \sqrt{36} = 6$ **11.** $\sqrt{t}\sqrt{t} = \sqrt{t^2} = t$

13. $\sqrt[3]{5x^2}\sqrt[3]{25x} = \sqrt[3]{125x^3} = 5x$ **15.** $\dfrac{\sqrt{500}}{\sqrt{5}} = \sqrt{\dfrac{500}{5}} = \sqrt{100} = 10$

17. $\dfrac{\sqrt{98x^3}}{\sqrt{2x}} = \sqrt{\dfrac{98x^3}{2x}} = \sqrt{49x^2} = 7x$ **19.** $\dfrac{\sqrt{180ab^4}}{\sqrt{5ab^2}} = \sqrt{\dfrac{180ab^4}{5ab^2}} = \sqrt{36b^2} = 6b$

21. $\dfrac{\sqrt[3]{48}}{\sqrt[3]{6}} = \sqrt[3]{\dfrac{48}{6}} = \sqrt[3]{8} = 2$ **23.** $\dfrac{\sqrt[3]{189a^4}}{\sqrt[3]{7a}} = \sqrt[3]{\dfrac{189a^4}{7a}} = \sqrt[3]{27a^3} = 3a$

25. $\sqrt{20} = \sqrt{4 \cdot 5} = \sqrt{4}\sqrt{5} = 2\sqrt{5}$ **27.** $-\sqrt{200} = -\sqrt{100 \cdot 2} = -\sqrt{100}\sqrt{2}$
$$= -10\sqrt{2}$$

29. $\sqrt[3]{80} = \sqrt[3]{8 \cdot 10} = \sqrt[3]{8}\sqrt[3]{10} = 2\sqrt[3]{10}$ **31.** $\sqrt[3]{-81} = \sqrt[3]{-27 \cdot 3} = \sqrt[3]{-27}\sqrt[3]{3} = -3\sqrt[3]{3}$

33. $\sqrt[4]{32} = \sqrt[4]{16 \cdot 2} = \sqrt[4]{16}\sqrt[4]{2} = 2\sqrt[4]{2}$ **35.** $\sqrt[5]{96} = \sqrt[5]{32 \cdot 3} = \sqrt[5]{32}\sqrt[5]{3} = 2\sqrt[5]{3}$

SECTION 7.4

37. $\sqrt{\dfrac{7}{9}} = \dfrac{\sqrt{7}}{\sqrt{9}} = \dfrac{\sqrt{7}}{3}$

39. $\sqrt[3]{\dfrac{7}{64}} = \dfrac{\sqrt[3]{7}}{\sqrt[3]{64}} = \dfrac{\sqrt[3]{7}}{4}$

41. $\sqrt[4]{\dfrac{3}{10,000}} = \dfrac{\sqrt[4]{3}}{\sqrt[4]{10,000}} = \dfrac{\sqrt[4]{3}}{10}$

43. $\sqrt[5]{\dfrac{3}{32}} = \dfrac{\sqrt[5]{3}}{\sqrt[5]{32}} = \dfrac{\sqrt[5]{3}}{2}$

45. $\sqrt{50x^2} = \sqrt{25x^2 \cdot 2} = \sqrt{25x^2}\sqrt{2} = 5x\sqrt{2}$ **47.** $\sqrt{32b} = \sqrt{16 \cdot 2b} = \sqrt{16}\sqrt{2b} = 4\sqrt{2b}$

49. $-\sqrt{112a^3} = -\sqrt{16a^2 \cdot 7a} = -\sqrt{16a^2}\sqrt{7a} = -4a\sqrt{7a}$

51. $\sqrt{175a^2b^3} = \sqrt{25a^2b^2 \cdot 7b} = \sqrt{25a^2b^2}\sqrt{7b} = 5ab\sqrt{7b}$

53. $-\sqrt{300xy} = -\sqrt{100 \cdot 3xy} = -\sqrt{100}\sqrt{3xy} = -10\sqrt{3xy}$

55. $\sqrt[3]{-54x^6} = \sqrt[3]{-27x^6 \cdot 2} = \sqrt[3]{-27x^6}\sqrt[3]{2}$ **57.** $\sqrt[3]{16x^{12}y^3} = \sqrt[3]{8x^{12}y^3 \cdot 2} = \sqrt[3]{8x^{12}y^3}\sqrt[3]{2}$
$\qquad\qquad = -3x^2\sqrt[3]{2}$ $\qquad\qquad\qquad\qquad = 2x^4y\sqrt[3]{2}$

59. $\sqrt[4]{32x^{12}y^4} = \sqrt[4]{16x^{12}y^4 \cdot 2} = \sqrt[4]{16x^{12}y^4}\sqrt[4]{2}$ **61.** $\sqrt{\dfrac{z^2}{16x^2}} = \dfrac{\sqrt{z^2}}{\sqrt{16x^2}} = \dfrac{z}{4x}$
$\qquad\qquad\qquad = 2x^3y\sqrt[4]{2}$

63. $\sqrt[4]{\dfrac{5x}{16z^4}} = \dfrac{\sqrt[4]{5x}}{\sqrt[4]{16z^4}} = \dfrac{\sqrt[4]{5x}}{2z}$

65. $4\sqrt{2x} + 6\sqrt{2x} = 10\sqrt{2x}$

67. $8\sqrt[5]{7a^2} - 7\sqrt[5]{7a^2} = \sqrt[5]{7a^2}$

69. $\sqrt{3} + \sqrt{27} = \sqrt{3} + \sqrt{9}\sqrt{3}$
$\qquad\qquad = \sqrt{3} + 3\sqrt{3} = 4\sqrt{3}$

71. $\sqrt{2} - \sqrt{8} = \sqrt{2} - \sqrt{4}\sqrt{2}$
$\qquad\quad = \sqrt{2} - 2\sqrt{2} = -\sqrt{2}$

73. $\sqrt{98} - \sqrt{50} = \sqrt{49}\sqrt{2} - \sqrt{25}\sqrt{2}$
$\qquad\qquad = 7\sqrt{2} - 5\sqrt{2} = 2\sqrt{2}$

75. $3\sqrt{24} + \sqrt{54} = 3\sqrt{4}\sqrt{6} + \sqrt{9}\sqrt{6} = 3(2)\sqrt{6} + 3\sqrt{6} = 6\sqrt{6} + 3\sqrt{6} = 9\sqrt{6}$

77. $\sqrt[3]{24} + \sqrt[3]{3} = \sqrt[3]{8}\sqrt[3]{3} + \sqrt[3]{3} = 2\sqrt[3]{3} + \sqrt[3]{3} = 3\sqrt[3]{3}$

79. $\sqrt[3]{32} - \sqrt[3]{108} = \sqrt[3]{8}\sqrt[3]{4} - \sqrt[3]{27}\sqrt[3]{4} = 2\sqrt[3]{4} - 3\sqrt[3]{4} = -\sqrt[3]{4}$

81. $2\sqrt[3]{125} - 5\sqrt[3]{64} = 2(5) - 5(4) = 10 - 20 = -10$

83. $14\sqrt[4]{32} - 15\sqrt[4]{162} = 14\sqrt[4]{16}\sqrt[4]{2} - 15\sqrt[4]{81}\sqrt[4]{2} = 14(2)\sqrt[4]{2} - 15(3)\sqrt[4]{2} = 28\sqrt[4]{2} - 45\sqrt[4]{2}$
$\qquad\qquad\qquad\qquad\qquad\qquad\qquad\qquad\qquad\qquad\qquad = -17\sqrt[4]{2}$

85. $3\sqrt[4]{512} + 2\sqrt[4]{32} = 3\sqrt[4]{256}\sqrt[4]{2} + 2\sqrt[4]{16}\sqrt[4]{2} = 3(4)\sqrt[4]{2} + 2(2)\sqrt[4]{2} = 12\sqrt[4]{2} + 4\sqrt[4]{2} = 16\sqrt[4]{2}$

87. $\sqrt{98} - \sqrt{50} - \sqrt{72} = \sqrt{49}\sqrt{2} - \sqrt{25}\sqrt{2} - \sqrt{36}\sqrt{2} = 7\sqrt{2} - 5\sqrt{2} - 6\sqrt{2} = -4\sqrt{2}$

89. $\sqrt{18} + \sqrt{300} - \sqrt{243} = \sqrt{9}\sqrt{2} + \sqrt{100}\sqrt{3} - \sqrt{81}\sqrt{3} = 3\sqrt{2} + 10\sqrt{3} - 9\sqrt{3} = 3\sqrt{2} + \sqrt{3}$

91. $2\sqrt[3]{16} - \sqrt[3]{54} - 3\sqrt[3]{128} = 2\sqrt[3]{8}\sqrt[3]{2} - \sqrt[3]{27}\sqrt[3]{2} - 3\sqrt[3]{64}\sqrt[3]{2} = 2(2)\sqrt[3]{2} - 3\sqrt[3]{2} - 3(4)\sqrt[3]{2}$
$$= 4\sqrt[3]{2} - 3\sqrt[3]{2} - 12\sqrt[3]{2} = -11\sqrt[3]{2}$$

93. $\sqrt{25y^2z} - \sqrt{16y^2z} = \sqrt{25y^2}\sqrt{z} - \sqrt{16y^2}\sqrt{z} = 5y\sqrt{z} - 4y\sqrt{z} = y\sqrt{z}$

95. $\sqrt{36xy^2} + \sqrt{49xy^2} = \sqrt{36y^2}\sqrt{x} + \sqrt{49y^2}\sqrt{x} = 6y\sqrt{x} + 7y\sqrt{x} = 13y\sqrt{x}$

97. $2\sqrt[3]{64a} + 2\sqrt[3]{8a} = 2\sqrt[3]{64}\sqrt[3]{a} + 2\sqrt[3]{8}\sqrt[3]{a} = 2(4)\sqrt[3]{a} + 2(2)\sqrt[3]{a} = 8\sqrt[3]{a} + 4\sqrt[3]{a} = 12\sqrt[3]{a}$

99. $\sqrt{y^5} - \sqrt{9y^5} - \sqrt{25y^5} = \sqrt{y^4}\sqrt{y} - \sqrt{9y^4}\sqrt{y} - \sqrt{25y^4}\sqrt{y} = y^2\sqrt{y} - 3y^2\sqrt{y} - 5y^2\sqrt{y}$
$$= -7y^2\sqrt{y}$$

101. $\sqrt[5]{x^6y^2} + \sqrt[5]{32x^6y^2} + \sqrt[5]{x^6y^2} = \sqrt[5]{x^5}\sqrt[5]{xy^2} + \sqrt[5]{32x^5}\sqrt[5]{xy^2} + \sqrt[5]{x^5}\sqrt[5]{xy^2}$
$$= x\sqrt[5]{xy^2} + 2x\sqrt[5]{xy^2} + x\sqrt[5]{xy^2} = 4x\sqrt[5]{xy^2}$$

103. $\sqrt{x^2 + 2x + 1} + \sqrt{x^2 + 2x + 1} = \sqrt{(x+1)^2} + \sqrt{(x+1)^2} = x + 1 + x + 1 = 2x + 2$

105. $x = 2.00$;
$h = 2\sqrt{2} \approx 2.83$

107. $h = 2(5) = 10.00$;
$x = 5\sqrt{3} \approx 8.66$

109. $x = \frac{9.37}{2} \approx 4.69$;
$y \approx 4.69\sqrt{3} \approx 8.11$

111. $x = y = \frac{17.12}{\sqrt{2}} \approx 12.11$ **113. Answers may vary.**

115. $\sqrt{a+b} = \sqrt{a} + \sqrt{b}$ if either a or b equals 0 and the other is nonnegative.

Exercise 7.5 (page 469)

1. $\dfrac{2}{3-a} = 1$
$2 = 3 - a$
$a = 1$

3. $\dfrac{8}{b-2} + \dfrac{3}{2-b} = -\dfrac{1}{b}$
$\dfrac{8}{b-2} + \dfrac{-3}{b-2} = -\dfrac{1}{b}$
$\dfrac{5}{b-2} = \dfrac{-1}{b}$
$5b = -b + 2$
$6b = 2$
$b = \dfrac{2}{6} = \dfrac{1}{3}$

5. $2; \sqrt{7}; \sqrt{5}$ **7.** FOIL   **9.** conjugate

11. $\sqrt{2}\sqrt{8} = \sqrt{16} = 4$

13. $\sqrt{5}\sqrt{10} = \sqrt{50} = \sqrt{25}\sqrt{2} = 5\sqrt{2}$

15. $2\sqrt{3}\sqrt{6} = 2\sqrt{18} = 2\sqrt{9}\sqrt{2} = 2(3)\sqrt{2}$
$= 6\sqrt{2}$

17. $\sqrt[3]{5}\sqrt[3]{25} = \sqrt[3]{125} = 5$

19. $\left(3\sqrt[3]{9}\right)\left(2\sqrt[3]{3}\right) = 6\sqrt[3]{27} = 6(3) = 18$

21. $\sqrt[3]{2}\sqrt[3]{12} = \sqrt[3]{24} = \sqrt[3]{8}\sqrt[3]{3} = 2\sqrt[3]{3}$

23. $\sqrt{ab^3}\sqrt{ab} = \sqrt{a^2b^4} = ab^2$

25. $\sqrt{5ab}\sqrt{5a} = \sqrt{25a^2b} = \sqrt{25a^2}\sqrt{b}$
$= 5a\sqrt{b}$

27. $\sqrt[3]{5r^2s}\sqrt[3]{2r} = \sqrt[3]{10r^3s} = \sqrt[3]{r^3}\sqrt[3]{10s}$
$= r\sqrt[3]{10s}$

29. $\sqrt[3]{a^5b}\sqrt[3]{16ab^5} = \sqrt[3]{16a^6b^6} = \sqrt[3]{8a^6b^6}\sqrt[3]{2}$
$= 2a^2b^2\sqrt[3]{2}$

31. $\sqrt{x(x+3)}\sqrt{x^3(x+3)} = \sqrt{x^4(x+3)^2} = x^2(x+3)$

33. $\sqrt[3]{6x^2(y+z)^2}\sqrt[3]{18x(y+z)} = \sqrt[3]{108x^3(y+z)^3} = \sqrt[3]{27x^3(y+z)^3}\sqrt[3]{4} = 3x(y+z)\sqrt[3]{4}$

35. $3\sqrt{5}\left(4-\sqrt{5}\right) = 12\sqrt{5} - 3\sqrt{25} = 12\sqrt{5} - 3(5) = 12\sqrt{5} - 15$

37. $3\sqrt{2}\left(4\sqrt{3} + 2\sqrt{7}\right) = 12\sqrt{6} + 6\sqrt{14}$

39. $-2\sqrt{5x}\left(4\sqrt{2x} - 3\sqrt{3}\right) = -8\sqrt{10x^2} + 6\sqrt{15x} = -8x\sqrt{10} + 6\sqrt{15x}$

41. $\left(\sqrt{2}+1\right)\left(\sqrt{2}-3\right) = \sqrt{4} - 3\sqrt{2} + \sqrt{2} - 3 = 2 - 2\sqrt{2} - 3 = -1 - 2\sqrt{2}$

43. $\left(4\sqrt{x}+3\right)\left(2\sqrt{x}-5\right) = 8\sqrt{x^2} - 20\sqrt{x} + 6\sqrt{x} - 15 = 8x - 14\sqrt{x} - 15$

45. $\left(\sqrt{5z}+\sqrt{3}\right)\left(\sqrt{5z}+\sqrt{3}\right) = \sqrt{25z^2} + \sqrt{15z} + \sqrt{15z} + \sqrt{9} = 5z + 2\sqrt{15z} + 3$

47. $\left(\sqrt{3x}-\sqrt{2y}\right)\left(\sqrt{3x}+\sqrt{2y}\right) = \sqrt{9x^2} + \sqrt{6xy} - \sqrt{6xy} - \sqrt{4y^2} = 3x - 2y$

49. $\left(2\sqrt{3a}-\sqrt{b}\right)\left(\sqrt{3a}+3\sqrt{b}\right) = 2\sqrt{9a^2} + 6\sqrt{3ab} - \sqrt{3ab} - 3\sqrt{b^2} = 6a + 5\sqrt{3ab} - 3b$

51. $\left(3\sqrt{2r}-2\right)^2 = \left(3\sqrt{2r}-2\right)\left(3\sqrt{2r}-2\right) = 9\sqrt{4r^2} - 6\sqrt{2r} - 6\sqrt{2r} + 4 = 18r - 12\sqrt{2r} + 4$

53. $-2\left(\sqrt{3x}+\sqrt{3}\right)^2 = -2\left(\sqrt{3x}+\sqrt{3}\right)\left(\sqrt{3x}+\sqrt{3}\right)$
$= -2\left(\sqrt{9x^2} + \sqrt{9x} + \sqrt{9x} + \sqrt{9}\right)$
$= -2\left(3x + 3\sqrt{x} + 3\sqrt{x} + 3\right) = -2(3x + 6\sqrt{x} + 3) = -6x - 12\sqrt{x} - 6$

55. $\sqrt{\dfrac{1}{7}} = \dfrac{\sqrt{1}}{\sqrt{7}} = \dfrac{1\sqrt{7}}{\sqrt{7}\sqrt{7}} = \dfrac{\sqrt{7}}{7}$

57. $\sqrt{\dfrac{2}{3}} = \dfrac{\sqrt{2}}{\sqrt{3}} = \dfrac{\sqrt{2}\sqrt{3}}{\sqrt{3}\sqrt{3}} = \dfrac{\sqrt{6}}{3}$

59. $\dfrac{\sqrt{5}}{\sqrt{8}} = \dfrac{\sqrt{5}\sqrt{2}}{\sqrt{8}\sqrt{2}} = \dfrac{\sqrt{10}}{\sqrt{16}} = \dfrac{\sqrt{10}}{4}$

61. $\dfrac{\sqrt{8}}{\sqrt{2}} = \sqrt{\dfrac{8}{2}} = \sqrt{4} = 2$

63. $\dfrac{1}{\sqrt[3]{2}} = \dfrac{1\sqrt[3]{4}}{\sqrt[3]{2}\sqrt[3]{4}} = \dfrac{\sqrt[3]{4}}{\sqrt[3]{8}} = \dfrac{\sqrt[3]{4}}{2}$

65. $\dfrac{3}{\sqrt[3]{9}} = \dfrac{3\sqrt[3]{3}}{\sqrt[3]{9}\sqrt[3]{3}} = \dfrac{3\sqrt[3]{3}}{\sqrt[3]{27}} = \dfrac{3\sqrt[3]{3}}{3} = \sqrt[3]{3}$

67. $\dfrac{\sqrt[3]{2}}{\sqrt[3]{9}} = \dfrac{\sqrt[3]{2}\sqrt[3]{3}}{\sqrt[3]{9}\sqrt[3]{3}} = \dfrac{\sqrt[3]{6}}{\sqrt[3]{27}} = \dfrac{\sqrt[3]{6}}{3}$

69. $\dfrac{\sqrt{8x^2y}}{\sqrt{xy}} = \sqrt{\dfrac{8x^2y}{xy}} = \sqrt{8x} = 2\sqrt{2x}$

71. $\dfrac{\sqrt{10xy^2}}{\sqrt{2xy^3}} = \sqrt{\dfrac{10xy^2}{2xy^3}} = \sqrt{\dfrac{5}{y}} = \dfrac{\sqrt{5}\sqrt{y}}{\sqrt{y}\sqrt{y}}$
$= \dfrac{\sqrt{5y}}{y}$

73. $\dfrac{\sqrt[3]{4a^2}}{\sqrt[3]{2ab}} = \sqrt[3]{\dfrac{4a^2}{2ab}} = \sqrt[3]{\dfrac{2a}{b}} = \dfrac{\sqrt[3]{2a}\sqrt[3]{b^2}}{\sqrt[3]{b}\sqrt[3]{b^2}}$
$= \dfrac{\sqrt[3]{2ab^2}}{b}$

75. $\dfrac{1}{\sqrt[4]{4}} = \dfrac{1\sqrt[4]{4}}{\sqrt[4]{4}\sqrt[4]{4}} = \dfrac{\sqrt[4]{4}}{\sqrt[4]{16}} = \dfrac{\sqrt[4]{4}}{2}$

77. $\dfrac{1}{\sqrt[5]{16}} = \dfrac{1\sqrt[5]{2}}{\sqrt[5]{16}\sqrt[5]{2}} = \dfrac{\sqrt[5]{2}}{\sqrt[5]{32}} = \dfrac{\sqrt[5]{2}}{2}$

79. $\dfrac{1}{\sqrt{2}-1} = \dfrac{1(\sqrt{2}+1)}{(\sqrt{2}-1)(\sqrt{2}+1)} = \dfrac{\sqrt{2}+1}{\sqrt{4}-1} = \dfrac{\sqrt{2}+1}{2-1} = \dfrac{\sqrt{2}+1}{1} = \sqrt{2}+1$

81. $\dfrac{\sqrt{2}}{\sqrt{5}+3} = \dfrac{\sqrt{2}(\sqrt{5}-3)}{(\sqrt{5}+3)(\sqrt{5}-3)} = \dfrac{\sqrt{2}(\sqrt{5}-3)}{\sqrt{25}-9} = \dfrac{\sqrt{2}(\sqrt{5}-3)}{5-9} = \dfrac{\sqrt{2}(\sqrt{5}-3)}{-4}$
$= \dfrac{\sqrt{10}-3\sqrt{2}}{-4}$
$= \dfrac{3\sqrt{2}-\sqrt{10}}{4}$

83. $\dfrac{\sqrt{3}+1}{\sqrt{3}-1} = \dfrac{(\sqrt{3}+1)(\sqrt{3}+1)}{(\sqrt{3}-1)(\sqrt{3}+1)} = \dfrac{\sqrt{9}+\sqrt{3}+\sqrt{3}+1}{\sqrt{9}-1} = \dfrac{4+2\sqrt{3}}{2} = \dfrac{2(2+\sqrt{3})}{2}$
$= 2+\sqrt{3}$

85. $\dfrac{\sqrt{7}-\sqrt{2}}{\sqrt{2}+\sqrt{7}} = \dfrac{(\sqrt{7}-\sqrt{2})(\sqrt{2}-\sqrt{7})}{(\sqrt{2}+\sqrt{7})(\sqrt{2}-\sqrt{7})} = \dfrac{\sqrt{14}-7-2+\sqrt{14}}{\sqrt{4}-\sqrt{49}} = \dfrac{2\sqrt{14}-9}{-5} = \dfrac{9-2\sqrt{14}}{5}$

87. $\dfrac{2}{\sqrt{x}+1} = \dfrac{2(\sqrt{x}-1)}{(\sqrt{x}+1)(\sqrt{x}-1)} = \dfrac{2(\sqrt{x}-1)}{\sqrt{x^2}-1} = \dfrac{2(\sqrt{x}-1)}{x-1} = \dfrac{2\sqrt{x}-2}{x-1}$

89. $\dfrac{x}{\sqrt{x}-4} = \dfrac{x(\sqrt{x}+4)}{(\sqrt{x}-4)(\sqrt{x}+4)} = \dfrac{x(\sqrt{x}+4)}{\sqrt{x^2}-16} = \dfrac{x(\sqrt{x}+4)}{x-16} = \dfrac{x\sqrt{x}+4x}{x-16}$

91. $\dfrac{2z-1}{\sqrt{2z}-1} = \dfrac{(2z-1)\left(\sqrt{2z}+1\right)}{\left(\sqrt{2z}-1\right)\left(\sqrt{2z}+1\right)} = \dfrac{(2z-1)\left(\sqrt{2z}+1\right)}{\sqrt{4z^2}-1} = \dfrac{(2z-1)\left(\sqrt{2z}+1\right)}{2z-1}$

$$= \sqrt{2z}+1$$

93. $\dfrac{\sqrt{x}-\sqrt{y}}{\sqrt{x}+\sqrt{y}} = \dfrac{(\sqrt{x}-\sqrt{y})(\sqrt{x}-\sqrt{y})}{(\sqrt{x}+\sqrt{y})(\sqrt{x}-\sqrt{y})} = \dfrac{\sqrt{x^2}-\sqrt{xy}-\sqrt{xy}+\sqrt{y^2}}{\sqrt{x^2}-\sqrt{y^2}} = \dfrac{x-2\sqrt{xy}+y}{x-y}$

95. $\dfrac{\sqrt{3}+1}{2} = \dfrac{\left(\sqrt{3}+1\right)\left(\sqrt{3}-1\right)}{2\left(\sqrt{3}-1\right)} = \dfrac{\sqrt{9}-1}{2\left(\sqrt{3}-1\right)} = \dfrac{2}{2\left(\sqrt{3}-1\right)} = \dfrac{1}{\sqrt{3}-1}$

97. $\dfrac{\sqrt{x}+3}{x} = \dfrac{(\sqrt{x}+3)(\sqrt{x}-3)}{x(\sqrt{x}-3)} = \dfrac{\sqrt{x^2}-9}{x(\sqrt{x}-3)} = \dfrac{x-9}{x\sqrt{x}-3x}$

99. $\dfrac{\sqrt{x}+\sqrt{y}}{\sqrt{x}} = \dfrac{(\sqrt{x}+\sqrt{y})(\sqrt{x}-\sqrt{y})}{\sqrt{x}(\sqrt{x}-\sqrt{y})} = \dfrac{\sqrt{x^2}-\sqrt{y^2}}{\sqrt{x^2}-\sqrt{xy}} = \dfrac{x-y}{x-\sqrt{xy}}$

101. If the area of the aperture is again cut in half, the area will equal $9\pi/4$ cm^2.

$$A = \pi r^2 \qquad\qquad f\text{-number} = \dfrac{f}{d}$$
$$\dfrac{9\pi}{4} = \pi\left(\dfrac{d}{2}\right)^2 \qquad\qquad = \dfrac{12}{3}$$
$$\dfrac{9\pi}{4} = \dfrac{\pi d^2}{4} \qquad\qquad = 4$$
$$36\pi = 4\pi d^2$$
$$\dfrac{36\pi}{4\pi} = d^2$$
$$9 = d^2$$
$$3 = d$$

103. Answers may vary.

105. $\dfrac{\sqrt{x}-3}{4} = \dfrac{(\sqrt{x}-3)(\sqrt{x}+3)}{4(\sqrt{x}+3)} = \dfrac{x-9}{4(\sqrt{x}+3)}$

Exercise 7.6 (page 477)

1. $f(0) = 3(0)^2 - 4(0) + 2 = 2$

3. $f(2) = 3(2)^2 - 4(2) + 2 = 6$

5. $x^n = y^n$

7. square

9. extraneous

11.
$$\sqrt{5x - 6} = 2$$
$$\left(\sqrt{5x - 6}\right)^2 = 2^2$$
$$5x - 6 = 4$$
$$5x = 10$$
$$x = 2$$
The answer checks.

13.
$$\sqrt{6x + 1} + 2 = 7$$
$$\sqrt{6x + 1} = 5$$
$$\left(\sqrt{6x + 1}\right)^2 = 5^2$$
$$6x + 1 = 25$$
$$6x = 24$$
$$x = 4$$
The answer checks.

15.
$$2\sqrt{4x + 1} = \sqrt{x + 4}$$
$$\left(2\sqrt{4x + 1}\right)^2 = \left(\sqrt{x + 4}\right)^2$$
$$4(4x + 1) = x + 4$$
$$16x + 4 = x + 4$$
$$15x = 0$$
$$x = 0$$
The answer checks.

17.
$$\sqrt[3]{7n - 1} = 3$$
$$\left(\sqrt[3]{7n - 1}\right)^3 = 3^3$$
$$7n - 1 = 27$$
$$7n = 28$$
$$n = 4$$
The answer checks.

19.
$$\sqrt[4]{10p + 1} = \sqrt[4]{11p - 7}$$
$$\left(\sqrt[4]{10p + 1}\right)^4 = \left(\sqrt[4]{11p - 7}\right)^4$$
$$10p + 1 = 11p - 7$$
$$-p = -8$$
$$p = 8$$
The answer checks.

21.
$$x = \frac{\sqrt{12x - 5}}{2}$$
$$2x = \sqrt{12x - 5}$$
$$(2x)^2 = \left(\sqrt{12x - 5}\right)^2$$
$$4x^2 = 12x - 5$$
$$4x^2 - 12x + 5 = 0$$
$$(2x - 1)(2x - 5)$$
$$2x - 1 = 0 \quad \text{or} \quad 2x - 5 = 0$$
$$2x = 1 \qquad\qquad 2x = 5$$
$$x = \tfrac{1}{2} \qquad\qquad x = \tfrac{5}{2}$$
Both answers check.

23.
$$\sqrt{x+2} = \sqrt{4-x}$$
$$\left(\sqrt{x+2}\right)^2 = \left(\sqrt{4-x}\right)^2$$
$$x + 2 = 4 - x$$
$$2x = 2$$
$$x = 1$$
The answer checks.

25.
$$2\sqrt{x} = \sqrt{5x-16}$$
$$\left(2\sqrt{x}\right)^2 = \left(\sqrt{5x-16}\right)^2$$
$$4x = 5x - 16$$
$$-x = -16$$
$$x = 16$$
The answer checks.

27.
$$r - 9 = \sqrt{2r-3}$$
$$(r-9)^2 = \left(\sqrt{2r-3}\right)^2$$
$$r^2 - 18r + 81 = 2r - 3$$
$$r^2 - 20r + 84 = 0$$
$$(r-14)(r-6) = 0$$
$$r - 14 = 0 \quad \textbf{or} \quad r - 6 = 0$$
$$r = 14 \qquad\qquad r = 6$$
solution $\qquad$ not a solution

29.
$$\sqrt{-5x+24} = 6 - x$$
$$\left(\sqrt{-5x+24}\right)^2 = (6-x)^2$$
$$-5x + 24 = 36 - 12x + x^2$$
$$0 = x^2 - 7x + 12$$
$$0 = (x-3)(x-4)$$
$$x - 3 = 0 \quad \textbf{or} \quad x - 4 = 0$$
$$x = 3 \qquad\qquad x = 4$$
solution $\qquad$ solution

31.
$$\sqrt{y+2} = 4 - y$$
$$\left(\sqrt{y+2}\right) = (4-y)^2$$
$$y + 2 = 16 - 8y + y^2$$
$$0 = y^2 - 9y + 14$$
$$0 = (y-2)(y-7)$$
$$y - 2 = 0 \quad \textbf{or} \quad y - 7 = 0$$
$$y = 2 \qquad\qquad y = 7$$
solution $\qquad$ not a solution

33.
$$\sqrt{x}\sqrt{x+16} = 15$$
$$\left(\sqrt{x}\sqrt{x+16}\right)^2 = 15^2$$
$$x(x+16) = 225$$
$$x^2 + 16x - 225 = 0$$
$$(x-9)(x+25) = 0$$
$$x - 9 = 0 \quad \textbf{or} \quad x + 25 = 0$$
$$x = 9 \qquad\qquad x = -25$$
solution $\qquad$ not a solution

35.
$$\sqrt[3]{x^3-7} = x - 1$$
$$\left(\sqrt[3]{x^3-7}\right)^3 = (x-1)^3$$
$$x^3 - 7 = x^3 - 3x^2 + 3x - 1$$
$$3x^2 - 3x - 6 = 0$$
$$3(x-2)(x+1) = 0$$
$$x - 2 = 0 \quad \textbf{or} \quad x + 1 = 0$$
$$x = 2 \qquad\qquad x = -1$$
solution $\qquad$ solution

37.
$$\sqrt[4]{x^4+4x^2-4} = -x$$
$$\left(\sqrt[4]{x^4+4x^2-4}\right)^4 = (-x)^4$$
$$x^4 + 4x^2 - 4 = x^4$$
$$4x^2 - 4 = 0$$
$$4(x+1)(x-1) = 0$$
$$x + 1 = 0 \quad \textbf{or} \quad x - 1 = 0$$
$$x = -1 \qquad\qquad x = 1$$
solution $\qquad$ not a solution

SECTION 7.6

39.
$$\sqrt[4]{12t+4}+2=0$$
$$\sqrt[4]{12t+4}=-2$$
$$\left(\sqrt[4]{12t+4}\right)^4=(-2)^4$$
$$12t+4=16$$
$$12t=12$$
$$t=1$$
The answer does not check. $\Rightarrow$ no solution

41.
$$\sqrt{2y+1}=1-2\sqrt{y}$$
$$\left(\sqrt{2y+1}\right)^2=\left(1-2\sqrt{y}\right)^2$$
$$2y+1=1-4\sqrt{y}+4y$$
$$4\sqrt{y}=2y$$
$$\left(4\sqrt{y}\right)^2=(2y)^2$$
$$16y=4y^2$$
$$0=4y^2-16y$$
$$0=4y(y-4)$$
$$4y=0\quad\text{or}\quad y-4=0$$
$$y=0\qquad\qquad y=4$$
$$\text{solution}\qquad\text{not a solution}$$

43.
$$\sqrt{y+7}+3=\sqrt{y+4}$$
$$\left(\sqrt{y+7}+3\right)^2=\left(\sqrt{y+4}\right)^2$$
$$y+7+6\sqrt{y+7}+9=y+4$$
$$6\sqrt{y+7}=-12$$
$$\left(6\sqrt{y+7}\right)^2=(-12)^2$$
$$36(y+7)=144$$
$$y+7=4$$
$$y=-3$$
The answer does not check. $\Rightarrow$ no solution

45.
$$\sqrt{v}+\sqrt{3}=\sqrt{v+3}$$
$$\left(\sqrt{v}+\sqrt{3}\right)^2=\left(\sqrt{v+3}\right)^2$$
$$v+2\sqrt{3v}+3=v+3$$
$$2\sqrt{3v}=0$$
$$\left(2\sqrt{3v}\right)^2=0^2$$
$$12v=0$$
$$v=0$$
The answer checks.

47.
$$2+\sqrt{u}=\sqrt{2u+7}$$
$$\left(2+\sqrt{u}\right)^2=\left(\sqrt{2u+7}\right)^2$$
$$4+4\sqrt{u}+u=2u+7$$
$$4\sqrt{u}=u+3$$
$$\left(4\sqrt{u}\right)^2=(u+3)^2$$
$$16u=u^2+6u+9$$
$$0=u^2-10u+9$$
$$0=(u-9)(u-1)$$
$$u-9=0\quad\text{or}\quad u-1=0$$
$$u=9\qquad\qquad u=1$$
$$\text{solution}\qquad\text{solution}$$

49.
$$\sqrt{6t+1}-3\sqrt{t}=-1$$
$$\sqrt{6t+1}=3\sqrt{t}-1$$
$$\left(\sqrt{6t+1}\right)^2=\left(3\sqrt{t}-1\right)^2$$
$$6t+1=9t-6\sqrt{t}+1$$
$$6\sqrt{t}=3t$$
$$\left(6\sqrt{t}\right)^2=(3t)^2$$
$$36t=9t^2$$
$$0=9t^2-36t$$
$$0=9t(t-4)$$
$$9t=0\quad\text{or}\quad t-4=0$$
$$t=0\qquad\qquad t=4$$
$$\text{not a solution}\qquad\text{solution}$$

51. $\sqrt{2x+5} + \sqrt{x+2} = 5$

$$\sqrt{2x+5} = 5 - \sqrt{x+2}$$

$$\left(\sqrt{2x+5}\right)^2 = \left(5 - \sqrt{x+2}\right)^2$$

$$2x+5 = 25 - 10\sqrt{x+2} + x + 2$$

$$10\sqrt{x+2} = -x + 22$$

$$\left(10\sqrt{x+2}\right)^2 = (-x+22)^2$$

$$100(x+2) = x^2 - 44x + 484$$

$$0 = x^2 - 144x + 284$$

$$0 = (x-142)(x-2)$$

$x - 142 = 0$ **or** $x - 2 = 0$

$\quad x = 142 \qquad\qquad x = 2$

not a solution   solution

53. $\sqrt{z-1} + \sqrt{z+2} = 3$

$$\sqrt{z-1} = 3 - \sqrt{z+2}$$

$$\left(\sqrt{z-1}\right)^2 = \left(3 - \sqrt{z+2}\right)^2$$

$$z - 1 = 9 - 6\sqrt{z+2} + z + 2$$

$$6\sqrt{z+2} = 12$$

$$\sqrt{z+2} = 2$$

$$\left(\sqrt{z+2}\right)^2 = 2^2$$

$$z + 2 = 4$$

$z = 2$: The answer checks.

55. $\sqrt{x-5} - \sqrt{x+3} = 4$

$$\sqrt{x-5} = \sqrt{x+3} + 4$$

$$\left(\sqrt{x-5}\right)^2 = \left(\sqrt{x+3} + 4\right)^2$$

$$x - 5 = x + 3 + 8\sqrt{x+3} + 16$$

$$-24 = 8\sqrt{x+3}$$

$$-3 = \sqrt{x+3}$$

$$(-3)^2 = \left(\sqrt{x+3}\right)^2$$

$$9 = x + 3$$

$$6 = x$$

The answer does not check. $\Rightarrow$ no solution

57.
$$\sqrt{x+1} + \sqrt{3x} = \sqrt{5x+1}$$
$$\left(\sqrt{x+1} + \sqrt{3x}\right)^2 = \left(\sqrt{5x+1}\right)^2$$
$$x + 1 + 2\sqrt{3x(x+1)} + 3x = 5x + 1$$
$$2\sqrt{3x^2 + 3x} = x$$
$$\left(2\sqrt{3x^2 + 3x}\right)^2 = x^2$$
$$12x^2 + 12x = x^2$$
$$11x^2 + 12x = 0$$
$$x(11x + 12) = 0$$
$$x = 0 \quad \textbf{or} \quad 11x + 12 = 0$$
$$x = -\tfrac{12}{11}$$

solution   not a solution

59.
$$\sqrt{\sqrt{a} + \sqrt{a+8}} = 2$$
$$\left(\sqrt{\sqrt{a} + \sqrt{a+8}}\right)^2 = 2^2$$
$$\sqrt{a} + \sqrt{a+8} = 4$$
$$\left(\sqrt{a} + \sqrt{a+8}\right)^2 = 4^2$$
$$a + 2\sqrt{a(a+8)} + a + 8 = 16$$
$$2\sqrt{a^2 + 8a} = -2a + 8$$
$$\left(2\sqrt{a^2 + 8a}\right)^2 = (-2a + 8)^2$$
$$4a^2 + 32a = 4a^2 - 32a + 64$$
$$64a = 64$$
$$a = 1: \text{ The answer checks.}$$

61.
$$\frac{6}{\sqrt{x+5}} = \sqrt{x}$$
$$\left(\frac{6}{\sqrt{x+5}}\right)^2 = \left(\sqrt{x}\right)^2$$
$$\frac{36}{x+5} = x$$
$$36 = x^2 + 5x$$
$$0 = x^2 + 5x - 36$$
$$0 = (x+9)(x-4)$$
$$x + 9 = 0 \quad \textbf{or} \quad x - 4 = 0$$
$$x = -9 \qquad\qquad x = 4$$

not a solution   solution

63.
$$\sqrt{x+2} + \sqrt{2x-3} = \sqrt{11-x}$$
$$\left(\sqrt{x+2} + \sqrt{2x-3}\right)^2 = \left(\sqrt{11-x}\right)^2$$
$$x+2+2\sqrt{(x+2)(2x-3)} + 2x - 3 = 11 - x$$
$$2\sqrt{2x^2 + x - 6} = -4x + 12$$
$$\left(2\sqrt{2x^2 + x - 6}\right)^2 = (-4x+12)^2$$
$$8x^2 + 4x - 24 = 16x^2 - 96x + 144$$
$$0 = 8x^2 - 100x + 168$$
$$0 = 4(2x-21)(x-2)$$

$$
\begin{array}{ll}
2x - 21 = 0 & \textbf{or} \quad x - 2 = 0 \\
x = \frac{21}{2} & \qquad x = 2 \\
\text{not a solution} & \quad \text{solution}
\end{array}
$$

65.
$$s = 1.45\sqrt{r}$$
$$65 = 1.45\sqrt{r}$$
$$65^2 = \left(1.45\sqrt{r}\right)^2$$
$$4225 = 2.1025r$$
$$2010 \text{ ft} \approx r$$

67.
$$v = \sqrt[3]{\frac{P}{0.02}}$$
$$v = \sqrt[3]{\frac{500}{0.02}}$$
$$v = \sqrt[3]{25000}$$
$$v \approx 29 \text{ mph}$$

69. $r = 1 - \sqrt[n]{\dfrac{T}{C}} = 1 - \sqrt[5]{\dfrac{9000}{22000}} \approx 1 - \sqrt[5]{0.40909} \approx 1 - 0.836 \approx 0.164 \approx 16\%$

71. Graph $y = \sqrt{5x}$ and $y = \sqrt{100 - 3x^2}$:

equilibrium price: $x = \$5$

73.
$$r = \sqrt[4]{\frac{8kl}{\pi R}}$$
$$r^4 = \left(\sqrt[4]{\frac{8kl}{\pi R}}\right)^4$$
$$r^4 = \frac{8kl}{\pi R}$$
$$\pi R r^4 = 8kl$$
$$R = \frac{8kl}{\pi r^4}$$

75. Answers may vary.

77.
$$\sqrt[3]{2x} = \sqrt{x}$$
$$\left(\sqrt[3]{2x}\right)^2 = \left(\sqrt{x}\right)^2$$
$$\left[\left(\sqrt[3]{2x}\right)^2\right]^3 = x^3$$
$$4x^2 = x^3$$
$$0 = x^3 - 4x^2$$
$$0 = x^2(x - 4)$$
$$x = 0 \text{ or } x = 4$$

Exercise 7.7 (page 488)

1. $\dfrac{x^2 - x - 6}{9 - x^2} \cdot \dfrac{x^2 + x - 6}{x^2 - 4} = \dfrac{(x-3)(x+2)}{(3+x)(3-x)} \cdot \dfrac{(x+3)(x-2)}{(x+2)(x-2)} = -1$

3. Let $w =$ the speed of the wind.

	Rate	Time	Dist.
Downwind	$200 + w$	$\frac{330}{200+w}$	330
Upwind	$200 - w$	$\frac{330}{200-w}$	330

$$\frac{330}{200 + w} + \frac{330}{200 - w} = \frac{10}{3}$$
$$\left(\frac{330}{200 + w} + \frac{330}{200 - w}\right) 3(200 + w)(200 - w) = \frac{10}{3} \cdot 3(200 + w)(200 - w)$$
$$330(3)(200 - w) + 330(3)(200 + w) = 10(40{,}000 - w^2)$$
$$198{,}000 - 990w + 198{,}000 + 990w = 400{,}000 - 10w^2$$
$$10w^2 = 4000$$
$$w^2 = 400$$
$$w = 20 \text{ mph}$$

5. imaginary

7. -1

9. 1

11. $\dfrac{\sqrt{a}}{\sqrt{b}}$

13. $5; 7$

15. conjugates

17. $\sqrt{-9} = \sqrt{-1 \cdot 9} = \sqrt{i^2 \cdot 3^2} = 3i$

19. $\sqrt{-36} = \sqrt{-1 \cdot 36} = \sqrt{i^2 \cdot 6^2} = 6i$

21. $\sqrt{-7} = \sqrt{-1 \cdot 7} = \sqrt{i^2 \cdot 7} = i\sqrt{7}$

23. $3 + 7i \overset{?}{=} \sqrt{9} + (5 + 2)i$
$3 + 7i \overset{?}{=} 3 + 7i$
They are equal.

25. $8 + 5i \overset{?}{=} 2^3 + \sqrt{25}i^3$

$8 + 5i \overset{?}{=} 8 + 5(-i)$

$8 + 5i \overset{?}{=} 8 - 5i$

They are not equal.

27. $\sqrt{4} + \sqrt{-4} \overset{?}{=} 2 - 2i$

$2 + \sqrt{-1 \cdot 4} \overset{?}{=} 2 - 2i$

$2 + 2i = 2 - 2i$

They are not equal.

29. $(3 + 4i) + (5 - 6i) = 3 + 4i + 5 - 6i$
$= 8 - 2i$

31. $(7 - 3i) - (4 + 2i) = 7 - 3i - 4 - 2i$
$= 3 - 5i$

33. $(8 + 5i) + (7 + 2i) = 8 + 5i + 7 + 2i$
$= 15 + 7i$

35. $(1 + i) - 2i + (5 - 7i) = 1 + i - 2i + 5 - 7i$
$= 6 - 8i$

37. $(5 + 3i) - (3 - 5i) + \sqrt{-1} = 5 + 3i - 3 + 5i + i = 2 + 9i$

39. $(-8 - \sqrt{3}i) - (7 - 3\sqrt{3}i) = -8 - \sqrt{3}i - 7 + 3\sqrt{3}i = -15 + 2\sqrt{3}i$

41. $3i(2 - i) = 6i - 3i^2 = 6i - 3(-1) = 6i + 3 = 3 + 6i$

43. $-5i(5 - 5i) = -25i + 25i^2 = -25i + 25(-1) = -25i - 25 = -25 - 25i$

45. $(2 + i)(3 - i) = 6 - 2i + 3i - i^2 = 6 + i - (-1) = 6 + i + 1 = 7 + i$

47. $(2 - 4i)(3 + 2i) = 6 + 4i - 12i - 8i^2 = 6 - 8i - 8(-1) = 6 - 8i + 8 = 14 - 8i$

49. $(2 + \sqrt{2}i)(3 - \sqrt{2}i) = 6 - 2\sqrt{2}i + 3\sqrt{2}i - 2i^2 = 6 + \sqrt{2}i - 2(-1) = 8 + \sqrt{2}i$

51. $(8 - \sqrt{-1})(-2 - \sqrt{-16}) = (8 - i)(-2 - 4i) = -16 - 32i + 2i + 4i^2 = -16 - 30i - 4$
$= -20 - 30i$

53. $(2 + i)^2 = (2 + i)(2 + i) = 4 + 2i + 2i + i^2 = 4 + 4i - 1 = 3 + 4i$

55. $(2 + 3i)^2 = (2 + 3i)(2 + 3i) = 4 + 6i + 6i + 9i^2 = 4 + 12i - 9 = -5 + 12i$

57. $i(5 + i)(3 - 2i) = i(15 - 10i + 3i - 2i^2) = i(15 - 7i + 2) = i(17 - 7i) = 17i - 7i^2 = 7 + 17i$

59. $(2 + i)(2 - i)(1 + i) = (4 - 2i + 2i - i^2)(1 + i) = 5(1 + i) = 5 + 5i$

61. $(3 + i)[(3 - 2i) + (2 + i)] = (3 + i)(5 - i) = 15 - 3i + 5i - i^2 = 16 + 2i$

63. $\dfrac{1}{i} = \dfrac{1i^3}{ii^3} = \dfrac{i^3}{i^4} = \dfrac{i^3}{1} = i^3 = -i = 0 - i$

65. $\dfrac{4}{5i^3} = \dfrac{4i}{5i^3 i} = \dfrac{4i}{5i^4} = \dfrac{4i}{5(1)} = \dfrac{4}{5}i = 0 + \dfrac{4}{5}i$

67. $\dfrac{3i}{8\sqrt{-9}} = \dfrac{3i}{8(3i)} = \dfrac{1}{8} = \dfrac{1}{8} + 0i$

69. $\dfrac{-3}{5i^5} = \dfrac{-3i^3}{5i^5 i^3} = \dfrac{-3(-i)}{5i^8} = \dfrac{3i}{5} = 0 + \dfrac{3}{5}i$

71. $\dfrac{5}{2-i} = \dfrac{5(2+i)}{(2-i)(2+i)} = \dfrac{5(2+i)}{4-i^2} = \dfrac{5(2+i)}{5} = 2+i$

73. $\dfrac{13i}{5+i} = \dfrac{13i(5-i)}{(5+i)(5-i)} = \dfrac{65i-13i^2}{25-i^2} = \dfrac{13+65i}{26} = \dfrac{13}{26} + \dfrac{65}{26}i = \dfrac{1}{2} + \dfrac{5}{2}i$

75. $\dfrac{-12}{7-\sqrt{-1}} = \dfrac{-12}{7-i} = \dfrac{-12(7+i)}{(7-i)(7+i)} = \dfrac{-84-12i}{49-i^2} = \dfrac{-84-12i}{50} = \dfrac{-84}{50} - \dfrac{12}{50}i = -\dfrac{42}{25} - \dfrac{6}{25}i$

77. $\dfrac{5i}{6+2i} = \dfrac{5i(6-2i)}{(6+2i)(6-2i)} = \dfrac{30i-10i^2}{36-4i^2} = \dfrac{10+30i}{40} = \dfrac{10}{40} + \dfrac{30}{40}i = \dfrac{1}{4} + \dfrac{3}{4}i$

79. $\dfrac{3-2i}{3+2i} = \dfrac{(3-2i)(3-2i)}{(3+2i)(3-2i)} = \dfrac{9-6i-6i+4i^2}{9-4i^2} = \dfrac{5-12i}{13} = \dfrac{5}{13} - \dfrac{12}{13}i$

81. $\dfrac{3+2i}{3+i} = \dfrac{(3+2i)(3-i)}{(3+i)(3-i)} = \dfrac{9-3i+6i-2i^2}{9-i^2} = \dfrac{11+3i}{10} = \dfrac{11}{10} + \dfrac{3}{10}i$

83. $\dfrac{\sqrt{5}-\sqrt{3}i}{\sqrt{5}+\sqrt{3}i} = \dfrac{\left(\sqrt{5}-\sqrt{3}i\right)\left(\sqrt{5}-\sqrt{3}i\right)}{\left(\sqrt{5}+\sqrt{3}i\right)\left(\sqrt{5}-\sqrt{3}i\right)} = \dfrac{5-\sqrt{15}i-\sqrt{15}i+3i^2}{5-3i^2} = \dfrac{2-2\sqrt{15}i}{8}$

$$= \dfrac{2}{8} - \dfrac{2\sqrt{15}}{8}i$$
$$= \dfrac{1}{4} - \dfrac{\sqrt{15}}{4}i$$

85. $\left(\dfrac{i}{3+2i}\right)^2 = \dfrac{i^2}{(3+2i)^2} = \dfrac{-1}{(3+2i)(3+2i)} = \dfrac{-1}{9+12i+4i^2} = \dfrac{-1}{5+12i} = \dfrac{-1(5-12i)}{(5+12i)(5-12i)}$

$$= \dfrac{-5+12i}{25-144i^2}$$
$$= -\dfrac{5}{169} + \dfrac{12}{169}i$$

87. $\dfrac{i(3-i)}{3+i} = \dfrac{(3i-i^2)(3-i)}{(3+i)(3-i)} = \dfrac{(1+3i)(3-i)}{9-i^2} = \dfrac{3-i+9i-3i^2}{10} = \dfrac{6+8i}{10} = \dfrac{3}{5} + \dfrac{4}{5}i$

89. $\dfrac{(2-5i)-(5-2i)}{5-i} = \dfrac{2-5i-5+2i}{5-i} = \dfrac{-3-3i}{5-i} = \dfrac{(-3-3i)(5+i)}{(5-i)(5+i)} = \dfrac{-15-3i-15i-3i^2}{25-i^2}$

$$= \dfrac{-12-18i}{26}$$
$$= \dfrac{-12}{26} - \dfrac{18}{26}i$$
$$= -\dfrac{6}{13} - \dfrac{9}{13}i$$

91. $i^{21} = i^{20}i^1 = (i^4)^5 i = 1^5 i = i$

93. $i^{27} = i^{24}i^3 = (i^4)^6 i^3 = 1^6 i^3 = i^3 = -i$

95. $i^{100} = (i^4)^{25} = 1^{25} = 1$

97. $i^{97} = i^{96}i^1 = (i^4)^{24} i = 1^{24} i = i$

99. $|6 + 8i| = \sqrt{6^2 + 8^2} = \sqrt{36 + 64} = \sqrt{100} = 10$

101. $|12 - 5i| = \sqrt{12^2 + (-5)^2} = \sqrt{144 + 25} = \sqrt{169} = 13$

103. $|5 + 7i| = \sqrt{5^2 + 7^2} = \sqrt{25 + 49} = \sqrt{74}$

105. $\left|\dfrac{3}{5} - \dfrac{4}{5}i\right| = \sqrt{\left(\dfrac{3}{5}\right)^2 + \left(\dfrac{4}{5}\right)^2} = \sqrt{\dfrac{9}{25} + \dfrac{16}{25}} = \sqrt{\dfrac{25}{25}} = \sqrt{1} = 1$

107.
$$x^2 - 2x + 26 = 0$$
$$(1 - 5i)^2 - 2(1 - 5i) + 26 = 0$$
$$(1 - 5i)(1 - 5i) - 2 + 10i + 26 = 0$$
$$1 - 10i + 25i^2 + 24 + 10i = 0$$
$$1 - 10i - 25 + 24 + 10i = 0$$
$$0 = 0$$

109.
$$x^4 - 3x^2 - 4 = 0$$
$$i^4 - 3i^2 - 4 = 0$$
$$1 - 3(-1) - 4 = 0$$
$$1 + 3 - 4 = 0$$
$$0 = 0$$

111. $V = IR = (2 - 3i)(2 + i) = 4 + 2i - 6i - 3i^2 = 4 - 4i - 3(-1) = 4 - 4i + 3 = 7 - 4i$ volts

113. $Z = \dfrac{V}{I} = \dfrac{1.7 + 0.5i}{0.5i} = \dfrac{(1.7 + 0.5i)i}{(0.5i)i} = \dfrac{1.7i + 0.5i^2}{0.5i^2} = \dfrac{-0.5 + 1.7i}{-0.5} = 1 - 3.4i$

115. Answers may vary.

117. $\dfrac{3 - i}{2} = \dfrac{(3 - i)(3 + i)}{2(3 + i)} = \dfrac{9 - i^2}{2(3 + i)} = \dfrac{10}{2(3 + i)} = \dfrac{5}{3 + i}$

Chapter 7 Summary (page 491)

1. $\sqrt{49} = \sqrt{7^2} = 7$

2. $-\sqrt{121} = -\sqrt{11^2} = -11$

3. $-\sqrt{36} = -\sqrt{6^2} = -6$

4. $\sqrt{225} = \sqrt{15^2} = 15$

5. $\sqrt[3]{-27} = \sqrt[3]{(-3)^3} = -3$

6. $-\sqrt[3]{216} = -\sqrt[3]{6^3} = -6$

7. $\sqrt[4]{625} = \sqrt[4]{5^4} = 5$

8. $\sqrt[5]{-32} = \sqrt[5]{(-2)^5} = -2$

9. $\sqrt{25x^2} = \sqrt{5^2 x^2} = |5x| = 5|x|$

10. $\sqrt{x^2 + 4x + 4} = \sqrt{(x + 2)^2} = |x + 2|$

11. $\sqrt[3]{27a^6 b^3} = 3a^2 b$

12. $\sqrt[4]{256x^8 y^4} = |4x^2 y| = 4x^2|y|$

13. $y = f(x) = \sqrt{x+2}$; Shift $y = \sqrt{x}$ left 2.

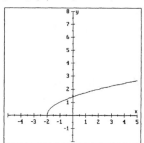

14. $y = f(x) = -\sqrt{x-1}$; Reflect $y = \sqrt{x}$ about the x-axis and shift right 1.

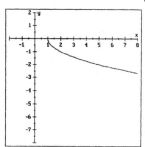

15. $y = f(x) = -\sqrt{x}+2$; Reflect $y = \sqrt{x}$ about the x-axis and shift up 2.

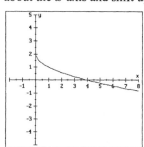

16. $y = f(x) = -\sqrt[3]{x}+3$; Reflect $y = \sqrt[3]{x}$ about the x-axis and shift up 3.

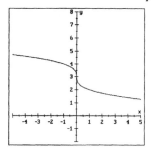

17. mean $= \dfrac{4+8+12+16+20}{5} = \dfrac{60}{5} = 12$

18.

Original term	Mean	Difference (term−mean)	Square of difference
4	12	−8	64
8	12	−4	16
12	12	0	0
16	12	4	16
20	12	8	64

$$s = \sqrt{\dfrac{64+16+0+16+64}{5}}$$
$$= \sqrt{\dfrac{160}{5}}$$
$$= \sqrt{32}$$
$$\approx 5.7$$

19. $d = 1.4\sqrt{h}$
$d = 1.4\sqrt{4.7} \approx 3.0$ miles

20. $d = 1.4\sqrt{h}$
$4 = 1.4\sqrt{h}$
$16 = 1.96h$
$8.2 \text{ ft} = h$

21. Let $x =$ one-half of d.

$$125^2 = x^2 + 117^2$$
$$15625 = x^2 + 13689$$
$$1936 = x^2$$
$$44 = x$$
$$d = 2x = 2(44) = 88 \text{ yd}$$

22. Let $x =$ one-half of d.

$$8900^2 = x^2 + 3900^2$$
$$79{,}210{,}000 = x^2 + 15{,}210{,}000$$
$$64{,}000{,}000 = x^2$$
$$8000 = x$$
$$d = 2x = 2(8000) = 16{,}000 \text{ yd}$$

23. $d = \sqrt{(x_2 - x_1)^2 + (y_2 - y_1)^2} = \sqrt{(0-5)^2 + [0-(-12)]^2} = \sqrt{(-5)^2 + 12^2} = \sqrt{169} = 13$

24. $d = \sqrt{(x_2 - x_1)^2 + (y_2 - y_1)^2} = \sqrt{(-4-(-2))^2 + (6-8)^2} = \sqrt{(-2)^2 + (-2)^2} = \sqrt{8} \approx 2.83$

25. $25^{1/2} = 5$

26. $-36^{1/2} = -6$

27. $9^{3/2} = \left(9^{1/2}\right)^3 = 3^3 = 27$

28. $16^{3/2} = \left(16^{1/2}\right)^3 = 4^3$
$\quad = 64$

29. $(-8)^{1/3} = -2$

30. $-8^{2/3} = -(8^{1/3})^2 = -2^2$
$\quad = -4$

31. $8^{-2/3} = \dfrac{1}{8^{2/3}} = \dfrac{1}{(8^{1/3})^2} = \dfrac{1}{2^2} = \dfrac{1}{4}$

32. $8^{-1/3} = \dfrac{1}{8^{1/3}} = \dfrac{1}{2}$

33. $-49^{5/2} = -(49^{1/2})^5 = -7^5 = -16{,}807$

34. $\dfrac{1}{25^{5/2}} = \dfrac{1}{(25^{1/2})^5} = \dfrac{1}{5^5} = \dfrac{1}{3125}$

35. $\left(\dfrac{1}{4}\right)^{-3/2} = 4^{3/2} = (4^{1/2})^3 = 2^3 = 8$

36. $\left(\dfrac{4}{9}\right)^{-3/2} = \left(\dfrac{9}{4}\right)^{3/2} = \left[\left(\dfrac{9}{4}\right)^{1/2}\right]^3$
$\quad = \left(\dfrac{3}{2}\right)^3 = \dfrac{27}{8}$

37. $(27x^3y)^{1/3} = 3xy^{1/3}$

38. $(81x^4y^2)^{1/4} = 3xy^{1/2}$

39. $(25x^3y^4)^{3/2} = 125x^{9/2}y^6$

40. $(8u^2v^3)^{-2/3} = \dfrac{1}{4u^{4/3}v^2}$

41. $5^{1/4}5^{1/2} = 5^{1/4+1/2} = 5^{3/4}$

42. $a^{3/7}a^{2/7} = a^{3/7+2/7} = a^{5/7}$

43. $u^{1/2}\left(u^{1/2} - u^{-1/2}\right) = u^{2/2} - u^0 = u - 1$

44. $v^{2/3}\left(v^{1/3} + v^{4/3}\right) = v^{3/3} + v^{6/3} = v + v^2$

45. $\left(x^{1/2} + y^{1/2}\right)^2 = \left(x^{1/2} + y^{1/2}\right)\left(x^{1/2} + y^{1/2}\right) = x^{2/2} + x^{1/2}y^{1/2} + x^{1/2}y^{1/2} + y^{2/2}$
$\quad = x + 2x^{1/2}y^{1/2} + y$

46. $\left(a^{2/3} + b^{2/3}\right)\left(a^{2/3} - b^{2/3}\right) = a^{4/3} - a^{2/3}b^{2/3} + a^{2/3}b^{2/3} - b^{4/3} = a^{4/3} - b^{4/3}$

47. $\sqrt[6]{5^2} = 5^{2/6} = 5^{1/3} = \sqrt[3]{5}$

48. $\sqrt[8]{x^4} = x^{4/8} = x^{1/2} = \sqrt{x}$

49. $\sqrt[9]{27a^3b^6} = (3^3a^3b^6)^{1/9} = 3^{3/9}a^{3/9}b^{6/9} = 3^{1/3}a^{1/3}b^{2/3} = \sqrt[3]{3ab^2}$

50. $\sqrt[4]{25a^2b^2} = 5^{2/4}a^{2/4}b^{2/4} = 5^{1/2}a^{1/2}b^{1/2} = \sqrt{5ab}$

51. $\sqrt{240} = \sqrt{16 \cdot 15} = \sqrt{16}\sqrt{15} = 4\sqrt{15}$ **52.** $\sqrt[3]{54} = \sqrt[3]{27 \cdot 2} = \sqrt[3]{27}\sqrt[3]{2} = 3\sqrt[3]{2}$

53. $\sqrt[4]{32} = \sqrt[4]{16 \cdot 2} = \sqrt[4]{16}\sqrt[4]{2} = 2\sqrt[4]{2}$ **54.** $\sqrt[5]{96} = \sqrt[5]{32 \cdot 3} = \sqrt[5]{32}\sqrt[5]{3} = 2\sqrt[5]{3}$

55. $\sqrt{8x^3} = \sqrt{4x^2}\sqrt{2x} = 2x\sqrt{2x}$ **56.** $\sqrt{18x^4y^3} = \sqrt{9x^4y^2}\sqrt{2y} = 3x^2y\sqrt{2y}$

57. $\sqrt[3]{16x^5y^4} = \sqrt[3]{8x^3y^3}\sqrt[3]{2x^2y} = 2xy\sqrt[3]{2x^2y}$ **58.** $\sqrt[3]{54x^7y^3} = \sqrt[3]{27x^6y^3}\sqrt[3]{2x} = 3x^2y\sqrt[3]{2x}$

59. $\dfrac{\sqrt{32x^3}}{\sqrt{2x}} = \sqrt{\dfrac{32x^3}{2x}} = \sqrt{16x^2} = 4x$ **60.** $\dfrac{\sqrt[3]{16x^5}}{\sqrt[3]{2x^2}} = \sqrt[3]{\dfrac{16x^5}{2x^2}} = \sqrt[3]{8x^3} = 2x$

61. $\sqrt[3]{\dfrac{2a^2b}{27x^3}} = \dfrac{\sqrt[3]{2a^2b}}{3x}$ **62.** $\sqrt{\dfrac{17xy}{64a^4}} = \dfrac{\sqrt{17xy}}{8a^2}$

63. $\sqrt{2} + \sqrt{8} = \sqrt{2} + \sqrt{4}\sqrt{2} = \sqrt{2} + 2\sqrt{2}$ **64.** $\sqrt{20} - \sqrt{5} = \sqrt{4}\sqrt{5} - \sqrt{5} = 2\sqrt{5} - \sqrt{5}$
$\qquad\qquad\qquad\qquad = 3\sqrt{2}$ $\qquad\qquad\qquad\qquad\qquad = \sqrt{5}$

65. $2\sqrt[3]{3} - \sqrt[3]{24} = 2\sqrt[3]{3} - \sqrt[3]{8}\sqrt[3]{3} = 2\sqrt[3]{3} - 2\sqrt[3]{3} = 0$

66. $\sqrt[4]{32} + 2\sqrt[4]{162} = \sqrt[4]{16}\sqrt[4]{2} + 2\sqrt[4]{81}\sqrt[4]{2} = 2\sqrt[4]{2} + 2(3)\sqrt[4]{2} = 2\sqrt[4]{2} + 6\sqrt[4]{2} = 8\sqrt[4]{2}$

67. $2x\sqrt{8} + 2\sqrt{200x^2} + \sqrt{50x^2} = 2x\sqrt{4}\sqrt{2} + 2\sqrt{100x^2}\sqrt{2} + \sqrt{25x^2}\sqrt{2}$
$\qquad\qquad\qquad\qquad\qquad = 2x(2)\sqrt{2} + 2(10x)\sqrt{2} + 5x\sqrt{2}$
$\qquad\qquad\qquad\qquad\qquad = 4x\sqrt{2} + 20x\sqrt{2} + 5x\sqrt{2} = 29x\sqrt{2}$

68. $3\sqrt{27a^3} - 2a\sqrt{3a} + 5\sqrt{75a^3} = 3\sqrt{9a^2}\sqrt{3a} - 2a\sqrt{3a} + 5\sqrt{25a^2}\sqrt{3a}$
$\qquad\qquad\qquad\qquad\qquad = 3(3a)\sqrt{3a} - 2a\sqrt{3a} + 5(5a)\sqrt{3a}$
$\qquad\qquad\qquad\qquad\qquad = 9a\sqrt{3a} - 2a\sqrt{3a} + 25a\sqrt{3a} = 32a\sqrt{3a}$

69. $\sqrt[3]{54} - 3\sqrt[3]{16} + 4\sqrt[3]{128} = \sqrt[3]{27}\sqrt[3]{2} - 3\sqrt[3]{8}\sqrt[3]{2} + 4\sqrt[3]{64}\sqrt[3]{2}$
$\qquad\qquad\qquad\qquad\qquad = 3\sqrt[3]{2} - 3(2)\sqrt[3]{2} + 4(4)\sqrt[3]{2} = 3\sqrt[3]{2} - 6\sqrt[3]{2} + 16\sqrt[3]{2} = 13\sqrt[3]{2}$

70. $2\sqrt[4]{32x^5} + 4\sqrt[4]{162x^5} - 5x\sqrt[4]{512x} = 2\sqrt[4]{16x^4}\sqrt[4]{2x} + 4\sqrt[4]{81x^4}\sqrt[4]{2x} - 5x\sqrt[4]{256}\sqrt[4]{2x}$
$\qquad\qquad\qquad\qquad\qquad = 2(2x)\sqrt[4]{2x} + 4(3x)\sqrt[4]{2x} - 5x(4)\sqrt[4]{2x}$
$\qquad\qquad\qquad\qquad\qquad = 4x\sqrt[4]{2x} + 12x\sqrt[4]{2x} - 20x\sqrt[4]{2x} = -4x\sqrt[4]{2x}$

71. hypotenuse $= 7\sqrt{2}$ m   **72.** shorter leg $= \dfrac{1}{2}\left(12\sqrt{3}\right) = 6\sqrt{3}$ cm
$\qquad\qquad\qquad\qquad\qquad$ longer leg $= \sqrt{3}\left(6\sqrt{3}\right) = 6(3) = 18$ cm

73. $x = 5\sqrt{2} \approx 7.07$ in.

74. $x = \sqrt{3}\left(\frac{1}{2} \cdot 10\right) = 5\sqrt{3} \approx 8.66$ cm

75. $\left(2\sqrt{5}\right)\left(3\sqrt{2}\right) = 6\sqrt{10}$

76. $2\sqrt{6}\sqrt{216} = 2\sqrt{6}\sqrt{36}\sqrt{6} = 2(6)(6) = 72$

77. $\sqrt{9x}\sqrt{x} = \sqrt{9x^2} = 3x$

78. $\sqrt[3]{3}\sqrt[3]{9} = \sqrt[3]{27} = 3$

79. $-\sqrt[3]{2x^2}\sqrt[3]{4x} = -\sqrt[3]{8x^3}$
$$= -2x$$

80. $-\sqrt[4]{256x^5y^{11}}\sqrt[4]{625x^9y^3} = -\sqrt[4]{256x^4y^8}\sqrt[4]{xy^3}\sqrt[4]{625x^8}\sqrt[4]{xy^3} = -4xy^2(5x^2)\sqrt[4]{x^2y^6}$
$$= -20x^3y^2\sqrt[4]{y^4}\sqrt[4]{x^2y^2}$$
$$= -20x^3y^3\sqrt[4]{x^2y^2} = -20x^3y^3\sqrt{xy}$$

81. $\sqrt{2}\left(\sqrt{8} - 3\right) = \sqrt{16} - 3\sqrt{2} = 4 - 3\sqrt{2}$

82. $\sqrt{2}\left(\sqrt{2} + 3\right) = \sqrt{4} + 3\sqrt{2} = 2 + 3\sqrt{2}$

83. $\sqrt{5}\left(\sqrt{2} - 1\right) = \sqrt{10} - \sqrt{5}$

84. $\sqrt{3}\left(\sqrt{3} + \sqrt{2}\right) = \sqrt{9} + \sqrt{6} = 3 + \sqrt{6}$

85. $\left(\sqrt{2} + 1\right)\left(\sqrt{2} - 1\right) = \sqrt{4} - \sqrt{2} + \sqrt{2} - 1 = 1$

86. $\left(\sqrt{3} + \sqrt{2}\right)\left(\sqrt{3} + \sqrt{2}\right) = \sqrt{9} + \sqrt{6} + \sqrt{6} + \sqrt{4} = 5 + 2\sqrt{6}$

87. $\left(\sqrt{x} + \sqrt{y}\right)\left(\sqrt{x} - \sqrt{y}\right) = \sqrt{x^2} - \sqrt{xy} + \sqrt{xy} - \sqrt{y^2} = x - y$

88. $\left(2\sqrt{u} + 3\right)\left(3\sqrt{u} - 4\right) = 6\sqrt{u^2} - 8\sqrt{u} + 9\sqrt{u} - 12 = 6u + \sqrt{u} - 12$

89. $\dfrac{1}{\sqrt{3}} = \dfrac{1\sqrt{3}}{\sqrt{3}\sqrt{3}} = \dfrac{\sqrt{3}}{3}$

90. $\dfrac{\sqrt{3}}{\sqrt{5}} = \dfrac{\sqrt{3}\sqrt{5}}{\sqrt{5}\sqrt{5}} = \dfrac{\sqrt{15}}{5}$

91. $\dfrac{x}{\sqrt{xy}} = \dfrac{x\sqrt{xy}}{\sqrt{xy}\sqrt{xy}} = \dfrac{x\sqrt{xy}}{xy} = \dfrac{\sqrt{xy}}{y}$

92. $\dfrac{\sqrt[3]{uv}}{\sqrt[3]{u^5v^7}} = \dfrac{\sqrt[3]{uv}\sqrt[3]{uv^2}}{\sqrt[3]{u^5v^7}\sqrt[3]{uv^2}} = \dfrac{\sqrt[3]{u^2v^3}}{\sqrt[3]{u^6v^9}} = \dfrac{v\sqrt[3]{u^2}}{u^2v^3} = \dfrac{\sqrt[3]{u^2}}{u^2v^2}$

93. $\dfrac{2}{\sqrt{2} - 1} = \dfrac{2\left(\sqrt{2} + 1\right)}{\left(\sqrt{2} - 1\right)\left(\sqrt{2} + 1\right)} = \dfrac{2\left(\sqrt{2} + 1\right)}{\sqrt{4} - 1} = \dfrac{2\left(\sqrt{2} + 1\right)}{2 - 1} = \dfrac{2\left(\sqrt{2} + 1\right)}{1} = 2\left(\sqrt{2} + 1\right)$

94. $\dfrac{\sqrt{2}}{\sqrt{3}-1} = \dfrac{\sqrt{2}\left(\sqrt{3}+1\right)}{\left(\sqrt{3}-1\right)\left(\sqrt{3}+1\right)} = \dfrac{\sqrt{2}\left(\sqrt{3}+1\right)}{\sqrt{9}-1} = \dfrac{\sqrt{2}\left(\sqrt{3}+1\right)}{3-1} = \dfrac{\sqrt{2}\left(\sqrt{3}+1\right)}{2}$

$$= \dfrac{\sqrt{6}+\sqrt{2}}{2}$$

95. $\dfrac{2x-32}{\sqrt{x}+4} = \dfrac{(2x-32)\left(\sqrt{x}-4\right)}{\left(\sqrt{x}+4\right)\left(\sqrt{x}-4\right)} = \dfrac{2(x-16)\left(\sqrt{x}-4\right)}{\sqrt{x^2}-16} = \dfrac{2(x-16)\left(\sqrt{x}-4\right)}{x-16} = 2\left(\sqrt{x}-4\right)$

96. $\dfrac{\sqrt{a}+1}{\sqrt{a}-1} = \dfrac{\left(\sqrt{a}+1\right)\left(\sqrt{a}+1\right)}{\left(\sqrt{a}-1\right)\left(\sqrt{a}+1\right)} = \dfrac{\sqrt{a^2}+\sqrt{a}+\sqrt{a}+1}{\sqrt{a^2}-1} = \dfrac{a+2\sqrt{a}+1}{a-1}$

97. $\dfrac{\sqrt{3}}{5} = \dfrac{\sqrt{3}\sqrt{3}}{5\sqrt{3}} = \dfrac{3}{5\sqrt{3}}$

98. $\dfrac{\sqrt[3]{9}}{3} = \dfrac{\sqrt[3]{9}\sqrt[3]{3}}{3\sqrt[3]{3}} = \dfrac{\sqrt[3]{27}}{3\sqrt[3]{3}} = \dfrac{3}{3\sqrt[3]{3}} = \dfrac{1}{\sqrt[3]{3}}$

99. $\dfrac{3-\sqrt{x}}{2} = \dfrac{\left(3-\sqrt{x}\right)\left(3+\sqrt{x}\right)}{2\left(3+\sqrt{x}\right)} = \dfrac{9-\sqrt{x^2}}{2\left(3+\sqrt{x}\right)} = \dfrac{9-x}{2\left(3+\sqrt{x}\right)}$

100. $\dfrac{\sqrt{a}-\sqrt{b}}{\sqrt{a}} = \dfrac{\left(\sqrt{a}-\sqrt{b}\right)\left(\sqrt{a}+\sqrt{b}\right)}{\sqrt{a}\left(\sqrt{a}+\sqrt{b}\right)} = \dfrac{\sqrt{a^2}-\sqrt{b^2}}{\sqrt{a^2}+\sqrt{ab}} = \dfrac{a-b}{a+\sqrt{ab}}$

101.
$$\sqrt{y+3} = \sqrt{2y-19}$$
$$\left(\sqrt{y+3}\right)^2 = \left(\sqrt{2y-19}\right)^2$$
$$y+3 = 2y-19$$
$$-y = -22$$
$$y = 22$$
The answer checks.

102.
$$u = \sqrt{25u-144}$$
$$u^2 = \left(\sqrt{25u-144}\right)^2$$
$$u^2 = 25u-144$$
$$u^2 - 25u + 144 = 0$$
$$(u-9)(u-16) = 0$$
$$u = 9 \text{ or } u = 16; \text{ Both answers check.}$$

103.
$$r = \sqrt{12r-27}$$
$$r^2 = \left(\sqrt{12r-27}\right)^2$$
$$r^2 = 12r-27$$
$$r^2 - 12r + 27 = 0$$
$$(r-9)(r-3) = 0$$
$$r = 9 \text{ or } r = 3; \text{ Both answers check.}$$

104.
$$\sqrt{z+1} + \sqrt{z} = 2$$
$$\sqrt{z+1} = 2 - \sqrt{z}$$
$$\left(\sqrt{z+1}\right)^2 = \left(2-\sqrt{z}\right)^2$$
$$z+1 = 4 - 4\sqrt{z} + z$$
$$4\sqrt{z} = 3$$
$$\left(4\sqrt{z}\right)^2 = 3^2$$
$$16z = 9$$
$$z = \tfrac{9}{16}; \text{ The answer checks.}$$

105. $\sqrt{2x+5} - \sqrt{2x} = 1$

$\qquad \sqrt{2x+5} = 1 + \sqrt{2x}$

$\qquad \left(\sqrt{2x+5}\right)^2 = \left(1+\sqrt{2x}\right)^2$

$\qquad\quad 2x+5 = 1 + 2\sqrt{2x} + 2x$

$\qquad\qquad 4 = 2\sqrt{2x}$

$\qquad\qquad 4^2 = \left(2\sqrt{2x}\right)^2$

$\qquad\qquad 16 = 8x$

$\qquad\qquad 2 = x;$ The answer checks.

106. $\sqrt[3]{x^3+8} = x+2$

$\qquad \left(\sqrt[3]{x^3+8}\right)^3 = (x+2)^3$

$\qquad\quad x^3+8 = x^3 + 6x^2 + 12x + 8$

$\qquad\qquad 0 = 6x^2 + 12x$

$\qquad\qquad 0 = 6x(x+2)$

$x = 0$ or $x = -2$; Both answers check.

107. $(5+4i) + (7-12i) = 5 + 4i + 7 - 12i = 12 - 8i$

108. $(-6-40i) - (-8+28i) = -6 - 40i + 8 - 28i = 2 - 68i$

109. $(-32+\sqrt{-144}) - (64+\sqrt{-81}) = -32 + \sqrt{144i^2} - 64 - \sqrt{81i^2} = -32 + 12i - 64 - 9i$
$$= -96 + 3i$$

110. $(-8+\sqrt{-8}) + (6-\sqrt{-32}) = -8 + \sqrt{4i^2}\sqrt{2} + 6 - \sqrt{16i^2}\sqrt{2}$
$$= -8 + 2i\sqrt{2} + 6 - 4i\sqrt{2} = -2 - 2\sqrt{2}i$$

111. $(2-7i)(-3+4i) = -6 + 8i + 21i - 28i^2 = -6 + 29i + 28 = 22 + 29i$

112. $(-5+6i)(2+i) = -10 - 5i + 12i + 6i^2 = -10 + 7i - 6 = -16 + 7i$

113. $(5-\sqrt{-27})(-6+\sqrt{-12}) = (5-3i\sqrt{3})(-6+2i\sqrt{3}) = -30 + 10i\sqrt{3} + 18i\sqrt{3} - 6i^2(3)$
$$= -30 + 28i\sqrt{3} + 18 = -12 + 28\sqrt{3}i$$

114. $(2+\sqrt{-128})(3-\sqrt{-98}) = (2+8i\sqrt{2})(3-7i\sqrt{2}) = 6 - 14i\sqrt{2} + 24i\sqrt{2} - 56i^2(2)$
$$= 6 + 10i\sqrt{2} + 112 = 118 + 10\sqrt{2}i$$

115. $\dfrac{3}{4i} = \dfrac{3}{4i} \cdot \dfrac{i}{i} = \dfrac{3i}{4i^2} = \dfrac{3i}{-4} = -\dfrac{3}{4}i = 0 - \dfrac{3}{4}i$

116. $\dfrac{-2}{5i^3} = \dfrac{-2}{5i^3} \cdot \dfrac{i}{i} = \dfrac{-2i}{5i^4} = \dfrac{-2i}{5} = 0 - \dfrac{2}{5}i$

117. $\dfrac{6}{2+i} = \dfrac{6}{2+i} \cdot \dfrac{2-i}{2-i} = \dfrac{6(2-i)}{4-i^2} = \dfrac{6(2-i)}{5} = \dfrac{12-6i}{5} = \dfrac{12}{5} - \dfrac{6}{5}i$

118. $\dfrac{7}{3-i} = \dfrac{7}{3-i} \cdot \dfrac{3+i}{3+i} = \dfrac{7(3+i)}{9-i^2} = \dfrac{7(3+i)}{10} = \dfrac{21+7i}{10} = \dfrac{21}{10} + \dfrac{7}{10}i$

119. $\dfrac{4+i}{4-i} = \dfrac{4+i}{4-i} \cdot \dfrac{4+i}{4+i} = \dfrac{16+8i+i^2}{16-i^2} = \dfrac{15+8i}{17} = \dfrac{15}{17} + \dfrac{8}{17}i$

120. $\dfrac{3-i}{3+i} = \dfrac{3-i}{3+i} \cdot \dfrac{3-i}{3-i} = \dfrac{9-6i+i^2}{9-i^2} = \dfrac{8-6i}{10} = \dfrac{8}{10} - \dfrac{6}{10}i = \dfrac{4}{5} - \dfrac{3}{5}i$

121. $\dfrac{3}{5+\sqrt{-4}} = \dfrac{3}{5+2i} = \dfrac{3}{5+2i} \cdot \dfrac{5-2i}{5-2i} = \dfrac{3(5-2i)}{25-4i^2} = \dfrac{3(5-2i)}{29} = \dfrac{15-6i}{29} = \dfrac{15}{29} - \dfrac{6}{29}i$

122. $\dfrac{2}{3-\sqrt{-9}} = \dfrac{2}{3-3i} = \dfrac{2}{3-3i} \cdot \dfrac{3+3i}{3+3i} = \dfrac{2(3+3i)}{9-9i^2} = \dfrac{2(3+3i)}{18} = \dfrac{3+3i}{9} = \dfrac{3}{9} + \dfrac{3}{9}i = \dfrac{1}{3} + \dfrac{1}{3}i$

123. $|9+12i| = \sqrt{9^2 + 12^2} = \sqrt{81+144} = \sqrt{225} = 15 = 15 + 0i$

124. $|24-10i| = \sqrt{24^2 + (-10)^2} = \sqrt{576+100} = \sqrt{676} = 26 = 26 + 0i$

125. $i^{12} = \left(i^4\right)^3 = 1^3 = 1$

126. $i^{583} = i^{580}i^3 = \left(i^4\right)^{145}i^3 = 1^{145}(-i) = -i$

Chapter 7 Test   (page 497)

1. $\sqrt{49} = \sqrt{7^2} = 7$

2. $\sqrt[3]{64} = \sqrt[3]{4^3} = 4$

3. $\sqrt{4x^2} = \sqrt{(2x)^2} = |2x| = 2|x|$

4. $\sqrt[3]{8x^3} = \sqrt[3]{(2x)^3} = 2x$

5. $f(x) = \sqrt{x-2}$; Shift $y = \sqrt{x}$ right 2.

6. $f(x) = \sqrt[3]{x} + 3$; Shift $y = \sqrt[3]{x}$ up 3.

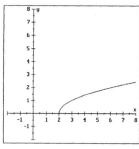

domain $= [2, \infty)$, range $= [0, \infty)$

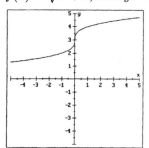

domain $= (-\infty, \infty)$, range $= (-\infty, \infty)$

7.
$$53^2 = 45^2 + h^2$$
$$2809 = 2025 + h^2$$
$$784 = h^2$$
$$28 \text{ in.} = h$$

8.
$$2^2 = \left(\dfrac{w}{2}\right)^2 + (1.9)^2$$
$$4 = \dfrac{w^2}{4} + 3.61$$
$$0.39 = \dfrac{w^2}{4}$$
$$1.56 = w^2$$
$$1.25 \text{ meters} = w$$

9. $d = \sqrt{(x_2 - x_1)^2 + (y_2 - y_1)^2} = \sqrt{(6-0)^2 + (8-0)^2} = \sqrt{6^2 + 8^2} = \sqrt{100} = 10$

10. $d = \sqrt{(x_2 - x_1)^2 + (y_2 - y_1)^2} = \sqrt{(-2 - 22)^2 + (5 - 12)^2} = \sqrt{(-24)^2 + (-7)^2} = \sqrt{625} = 25$

11. $16^{1/4} = 2$

12. $27^{2/3} = \left(27^{1/3}\right)^2 = 3^2 = 9$

13. $36^{-3/2} = \dfrac{1}{36^{3/2}} = \dfrac{1}{\left(36^{1/2}\right)^3} = \dfrac{1}{6^3} = \dfrac{1}{216}$

14. $\left(-\dfrac{8}{27}\right)^{-2/3} = \left(-\dfrac{27}{8}\right)^{2/3} = \left[\left(-\dfrac{27}{8}\right)^{1/3}\right]^2 = \left(-\dfrac{3}{2}\right)^2 = \dfrac{9}{4}$

15. $\dfrac{2^{5/3}2^{1/6}}{2^{1/2}} = \dfrac{2^{10/6}2^{1/6}}{2^{3/6}} = 2^{10/6 + 1/6 - 3/6} = 2^{8/6} = 2^{4/3}$

16. $\dfrac{\left(8x^3y\right)^{1/2}\left(8xy^5\right)^{1/2}}{\left(x^3y^6\right)^{1/3}} = \dfrac{8^{1/2}x^{3/2}y^{1/2}8^{1/2}x^{1/2}y^{5/2}}{x^{3/3}y^{6/3}} = \dfrac{8^{2/2}x^{4/2}y^{6/2}}{xy^2} = \dfrac{8x^2y^3}{xy^2} = 8xy$

17. $\sqrt{48} = \sqrt{16}\sqrt{3} = 4\sqrt{3}$

18. $\sqrt{250x^3y^5} = \sqrt{25x^2y^4}\sqrt{10xy} = 5xy^2\sqrt{10xy}$

19. $\dfrac{\sqrt[3]{24x^{15}y^4}}{\sqrt[3]{y}} = \sqrt[3]{\dfrac{24x^{15}y^4}{y}} = \sqrt[3]{24x^{15}y^3} = \sqrt[3]{8x^{15}y^3}\sqrt[3]{3} = 2x^5y\sqrt[3]{3}$

20. $\sqrt{\dfrac{3a^5}{48a^7}} = \sqrt{\dfrac{1}{16a^2}} = \dfrac{1}{4a}$

21. $\sqrt{12x^2} = \sqrt{4x^2}\sqrt{3} = 2|x|\sqrt{3}$

22. $\sqrt{8x^6} = \sqrt{4x^6}\sqrt{2} = 2|x^3|\sqrt{2}$

23. $\sqrt[3]{81x^3} = \sqrt[3]{27x^3}\sqrt[3]{3} = 3x\sqrt[3]{3}$

24. $\sqrt{18x^4y^9} = \sqrt{9x^4y^8}\sqrt{2y} = |3x^2y^4|\sqrt{2y} = 3x^2y^4\sqrt{2y}$

25. $\sqrt{12} - \sqrt{27} = \sqrt{4}\sqrt{3} - \sqrt{9}\sqrt{3} = 2\sqrt{3} - 3\sqrt{3} = -\sqrt{3}$

26. $2\sqrt[3]{40} - \sqrt[3]{5000} + 4\sqrt[3]{625} = 2\sqrt[3]{8}\sqrt[3]{5} - \sqrt[3]{1000}\sqrt[3]{5} + 4\sqrt[3]{125}\sqrt[3]{5} = 2(2)\sqrt[3]{5} - 10\sqrt[3]{5} + 4(5)\sqrt[3]{5}$
$$= 4\sqrt[3]{5} - 10\sqrt[3]{5} + 20\sqrt[3]{5}$$
$$= 14\sqrt[3]{5}$$

27. $2\sqrt{48y^5} - 3y\sqrt{12y^3} = 2\sqrt{16y^4}\sqrt{3y} - 3y\sqrt{4y^2}\sqrt{3y} = 2(4y^2)\sqrt{3y} - 3y(2y)\sqrt{3y}$
$$= 8y^2\sqrt{3y} - 6y^2\sqrt{3y} = 2y^2\sqrt{3y}$$

28. $\sqrt[4]{768z^5} + z\sqrt[4]{48z} = \sqrt[4]{256z^4}\sqrt[4]{3z} + z\sqrt[4]{16}\sqrt[4]{3z} = 4z\sqrt[4]{3z} + 2z\sqrt[4]{3z} = 6z\sqrt[4]{3z}$

29. $-2\sqrt{xy}\left(3\sqrt{x} + \sqrt{xy^3}\right) = -6\sqrt{x^2y} - 2\sqrt{x^2y^4} = -6x\sqrt{y} - 2xy^2$

30. $\left(3\sqrt{2}+\sqrt{3}\right)\left(2\sqrt{2}-3\sqrt{3}\right)=6\sqrt{4}-9\sqrt{6}+2\sqrt{6}-3\sqrt{9}=12-7\sqrt{6}-9=3-7\sqrt{6}$

31. $\dfrac{1}{\sqrt{5}}=\dfrac{1\sqrt{5}}{\sqrt{5}\sqrt{5}}=\dfrac{\sqrt{5}}{5}$

32. $\dfrac{3t-1}{\sqrt{3t}-1}=\dfrac{(3t-1)\left(\sqrt{3t}+1\right)}{\left(\sqrt{3t}-1\right)\left(\sqrt{3t}+1\right)}=\dfrac{(3t-1)\left(\sqrt{3t}+1\right)}{\sqrt{9t^2}-1}=\dfrac{(3t-1)\left(\sqrt{3t}+1\right)}{3t-1}=\sqrt{3t}+1$

33. $\dfrac{\sqrt{3}}{\sqrt{7}}=\dfrac{\sqrt{3}\sqrt{3}}{\sqrt{7}\sqrt{3}}=\dfrac{3}{\sqrt{21}}$

34. $\dfrac{\sqrt{a}+\sqrt{b}}{\sqrt{a}-\sqrt{b}}=\dfrac{\left(\sqrt{a}+\sqrt{b}\right)\left(\sqrt{a}-\sqrt{b}\right)}{\left(\sqrt{a}-\sqrt{b}\right)\left(\sqrt{a}-\sqrt{b}\right)}=\dfrac{\sqrt{a^2}-\sqrt{b^2}}{\sqrt{a^2}-\sqrt{ab}-\sqrt{ab}+\sqrt{b^2}}=\dfrac{a-b}{a-2\sqrt{ab}+b}$

35.
$$\sqrt[3]{6n+4}-4=0$$
$$\sqrt[3]{6n+4}=4$$
$$\left(\sqrt[3]{6n+4}\right)^3=4^3$$
$$6n+4=64$$
$$6n=60$$
$$n=10$$
The answer checks.

36.
$$1-\sqrt{u}=\sqrt{u-3}$$
$$\left(1-\sqrt{u}\right)^2=\left(\sqrt{u-3}\right)^2$$
$$1-2\sqrt{u}+u=u-3$$
$$4=2\sqrt{u}$$
$$4^2=\left(2\sqrt{u}\right)^2$$
$$16=4u$$
$$4=u$$
The answer does not check. $\Rightarrow$ no solution

37. $(2+4i)+(-3+7i)=2+4i-3+7i=-1+11i$

38. $(3-\sqrt{-9})-(-1+\sqrt{-16})=3-3i+1-4i=4-7i$

39. $2i(3-4i)=6i-8i^2=6i+8=8+6i$

40. $(3+2i)(-4-i)=-12-3i-8i-2i^2=-12-11i+2=-10-11i$

41. $\dfrac{1}{i\sqrt{2}}=\dfrac{1}{i\sqrt{2}}\cdot\dfrac{i\sqrt{2}}{i\sqrt{2}}=\dfrac{i\sqrt{2}}{2i^2}=-\dfrac{\sqrt{2}i}{2}=0-\dfrac{\sqrt{2}}{2}i$

42. $\dfrac{2+i}{3-i}=\dfrac{2+i}{3-i}\cdot\dfrac{3+i}{3+i}=\dfrac{6+5i+i^2}{9-i^2}=\dfrac{5+5i}{10}=\dfrac{1}{2}+\dfrac{1}{2}$

Exercise 8.1 (page 508)

1.
$$\frac{t+9}{2} + \frac{t+2}{5} = \frac{8}{5} + 4t$$
$$5(t+9) + 2(t+2) = 2(8) + 40t$$
$$5t + 45 + 2t + 4 = 16 + 40t$$
$$33 = 33t$$
$$1 = t$$

3.
$$3(t-3) + 3t \leq 2(t+1) + t + 1$$
$$3t - 9 + 3t \leq 2t + 2 + t + 1$$
$$3t \leq 12$$
$$t \leq 4$$

5. $x = \sqrt{c}; x = -\sqrt{c}$

7. positive or negative

9.
$$6x^2 + 12x = 0$$
$$6x(x+2) = 0$$
$$6x = 0 \quad \text{or} \quad x + 2 = 0$$
$$x = 0 \qquad\qquad x = -2$$

11.
$$2y^2 - 50 = 0$$
$$2(y+5)(y-5) = 0$$
$$y + 5 = 0 \quad \text{or} \quad y - 5 = 0$$
$$y = -5 \qquad\qquad y = 5$$

13.
$$r^2 + 6r + 8 = 0$$
$$(r+2)(r+4) = 0$$
$$r + 2 = 0 \quad \text{or} \quad r + 4 = 0$$
$$r = -2 \qquad\qquad r = -4$$

15.
$$7x - 6 = x^2$$
$$0 = x^2 - 7x + 6$$
$$0 = (x-6)(x-1)$$
$$x - 6 = 0 \quad \text{or} \quad x - 1 = 0$$
$$x = 6 \qquad\qquad x = 1$$

17.
$$2z^2 - 5z + 2 = 0$$
$$(2z-1)(z-2) = 0$$
$$2z - 1 = 0 \quad \text{or} \quad z - 2 = 0$$
$$z = \tfrac{1}{2} \qquad\qquad z = 2$$

19.
$$6s^2 + 11s - 10 = 0$$
$$(2s+5)(3s-2) = 0$$
$$2s + 5 = 0 \quad \text{or} \quad 3s - 2 = 0$$
$$s = -\tfrac{5}{2} \qquad\qquad s = \tfrac{2}{3}$$

21.
$$x^2 = 36$$
$$x = \pm\sqrt{36} = \pm 6$$

23.
$$z^2 = 5$$
$$z = \pm\sqrt{5}$$

25.
$$3x^2 - 16 = 0$$
$$3x^2 = 16$$
$$x^2 = \frac{16}{3}$$
$$x = \pm\sqrt{\frac{16}{3}} = \pm\frac{4}{\sqrt{3}} = \pm\frac{4\sqrt{3}}{3}$$

27.
$$(y+1)^2 = 1$$
$$y + 1 = \pm\sqrt{1}$$
$$y + 1 = \pm 1$$
$$y = -1 \pm 1$$
$$y = 0 \text{ or } y = -2$$

29.
$$(s-7)^2 - 9 = 0$$
$$(s-7)^2 = 9$$
$$s - 7 = \pm\sqrt{9}$$
$$s - 7 = \pm 3$$
$$s = 7 \pm 3$$
$$s = 10 \text{ or } s = 4$$

31.
$$(x+5)^2 - 3 = 0$$
$$(x+5)^2 = 3$$
$$x + 5 = \pm\sqrt{3}$$
$$x = -5 \pm\sqrt{3}$$

33. $(x-2)^2 - 5 = 0$
$(x-2)^2 = 5$
$x - 2 = \pm\sqrt{5}$
$x = 2 \pm \sqrt{5}$

35. $p^2 + 16 = 0$
$p^2 = -16$
$p = \pm\sqrt{-16}$
$p = \pm 4i$

37. $4m^2 + 81 = 0$
$4m^2 = -81$
$m^2 = -\dfrac{81}{4}$
$m = \pm\sqrt{-\dfrac{81}{4}}$
$m = \pm\dfrac{9}{2}i$

39. $x^2 + 2x - 8 = 0$
$x^2 + 2x = 8$
$x^2 + 2x + 1 = 8 + 1$
$(x+1)^2 = 9$
$x + 1 = \pm 3$
$x = -1 \pm 3$
$x = 2 \text{ or } x = -4$

41. $x^2 - 6x + 8 = 0$
$x^2 - 6x = -8$
$x^2 - 6x + 9 = -8 + 9$
$(x-3)^2 = 1$
$x - 3 = \pm 1$
$x = 3 \pm 1$
$x = 4 \text{ or } x = 2$

43. $x^2 + 5x + 4 = 0$
$x^2 + 5x = -4$
$x^2 + 5x + \dfrac{25}{4} = -4 + \dfrac{25}{4}$
$\left(x + \dfrac{5}{2}\right)^2 = \dfrac{9}{4}$
$x + \dfrac{5}{2} = \pm\dfrac{3}{2}$
$x = -\dfrac{5}{2} \pm \dfrac{3}{2}$
$x = -\frac{2}{2} = -1 \text{ or } x = -\frac{8}{2} = -4$

45. $x + 1 = 2x^2$
$-2x^2 + x = -1$
$x^2 - \dfrac{1}{2}x = \dfrac{1}{2}$
$x^2 - \dfrac{1}{2}x + \dfrac{1}{16} = \dfrac{1}{2} + \dfrac{1}{16}$
$\left(x - \dfrac{1}{4}\right)^2 = \dfrac{9}{16}$
$x - \dfrac{1}{4} = \pm\dfrac{3}{4}$
$x = \dfrac{1}{4} \pm \dfrac{3}{4}$
$x = \frac{4}{4} = 1 \text{ or } x = -\frac{2}{4} = -\frac{1}{2}$

47. $6x^2 + 11x + 3 = 0$
$6x^2 + 11x = -3$
$x^2 + \dfrac{11}{6}x = -\dfrac{1}{2}$
$x^2 + \dfrac{11}{6}x + \dfrac{121}{144} = -\dfrac{1}{2} + \dfrac{121}{144}$
$\left(x + \dfrac{11}{12}\right)^2 = \dfrac{49}{144}$
$x + \dfrac{11}{12} = \pm\dfrac{7}{12}$
$x = -\dfrac{11}{12} \pm \dfrac{7}{12}$
$x = -\frac{4}{12} = -\frac{1}{3} \text{ or } x = -\frac{18}{12} = -\frac{3}{2}$

49. $9 - 6r = 8r^2$
$-8r^2 - 6r = -9$
$r^2 + \dfrac{3}{4}r = \dfrac{9}{8}$
$r^2 + \dfrac{3}{4}r + \dfrac{9}{64} = \dfrac{9}{8} + \dfrac{9}{64}$
$\left(r + \dfrac{3}{8}\right)^2 = \dfrac{81}{64}$
$r + \dfrac{3}{8} = \pm\dfrac{9}{8}$
$r = -\dfrac{3}{8} \pm \dfrac{9}{8}$
$r = \frac{6}{8} = \frac{3}{4} \text{ or } r = -\frac{12}{8} = -\frac{3}{2}$

51.
$$\frac{7x+1}{5} = -x^2$$
$$7x+1 = -5x^2$$
$$5x^2 + 7x = -1$$
$$x^2 + \frac{7}{5}x = -\frac{1}{5}$$
$$x^2 + \frac{7}{5}x + \frac{49}{100} = -\frac{1}{5} + \frac{49}{100}$$
$$\left(x + \frac{7}{10}\right)^2 = \frac{29}{100}$$
$$x + \frac{7}{10} = \pm\frac{\sqrt{29}}{10}$$
$$x = -\frac{7}{10} \pm \frac{\sqrt{29}}{10}$$

53.
$$p^2 + 2p + 2 = 0$$
$$p^2 + 2p = -2$$
$$p^2 + 2p + 1 = -2 + 1$$
$$(p+1)^2 = -1$$
$$p + 1 = \pm\sqrt{-1}$$
$$p + 1 = \pm i$$
$$p = -1 \pm i$$

55.
$$y^2 + 8y + 18 = 0$$
$$y^2 + 8y = -18$$
$$y^2 + 8y + 16 = -18 + 16$$
$$(y+4)^2 = -2$$
$$y + 4 = \pm\sqrt{-2}$$
$$y = -4 \pm i\sqrt{2}$$

57.
$$3m^2 - 2m + 3 = 0$$
$$m^2 - \frac{2}{3}m = -1$$
$$m^2 - \frac{2}{3}m + \frac{1}{9} = -1 + \frac{1}{9}$$
$$\left(m - \frac{1}{3}\right)^2 = -\frac{8}{9}$$
$$m - \frac{1}{3} = \pm\sqrt{-\frac{8}{9}}$$
$$m = \frac{1}{3} \pm \frac{\sqrt{8}}{3}i$$
$$m = \frac{1}{3} \pm \frac{2\sqrt{2}}{3}i$$

59.
$$f(x) = 0$$
$$2x^2 + x - 5 = 0$$
$$x^2 + \frac{1}{2}x = \frac{5}{2}$$
$$x^2 + \frac{1}{2}x + \frac{1}{16} = \frac{5}{2} + \frac{1}{16}$$
$$\left(x + \frac{1}{4}\right)^2 = \frac{41}{16}$$
$$x + \frac{1}{4} = \pm\sqrt{\frac{41}{16}}$$
$$x + \frac{1}{4} = \pm\frac{\sqrt{41}}{4}$$
$$x = -\frac{1}{4} \pm \frac{\sqrt{41}}{4}$$

61.
$$f(x) = 0$$
$$x^2 + x - 3 = 0$$
$$x^2 + x = 3$$
$$x^2 + x + \frac{1}{4} = 3 + \frac{1}{4}$$
$$\left(x + \frac{1}{2}\right)^2 = \frac{13}{4}$$
$$x + \frac{1}{2} = \pm\sqrt{\frac{13}{4}}$$
$$x = -\frac{1}{2} \pm \frac{\sqrt{13}}{2}$$

63.
$$s = 16t^2$$
$$256 = 16t^2$$
$$16 = t^2$$
$$\pm\sqrt{16} = t$$
$$\pm 4 = t$$
$t = 4$ is the only answer that makes sense, so it will take 4 seconds.

67.
$$A = P(1+r)^t$$
$$9193.60 = 8500(1+r)^2$$
$$\frac{9193.60}{8500} = (1+r)^2$$
$$1.0816 = (1+r)^2$$
$$\pm\sqrt{1.0816} = \sqrt{(1+r)^2}$$
$$\pm 1.04 = 1+r$$
$$-1 \pm 1.04 = r$$
$r = 0.04$ or $r = -2.04$; r must be positive, so $r = 0.04$, or 4%.

65.
$$s^2 = 10.5l$$
$$s^2 = 10.5(500)$$
$$s^2 = 5250$$
$$s = \pm\sqrt{5250}$$
$$s \approx \pm 72.5$$
s must be positive, so the speed was about 72.5 mph.

69. Answers may vary.

71. $\left(\frac{1}{2}\sqrt{3}\right)^2 = \left(\frac{\sqrt{3}}{2}\right)^2 = \frac{3}{4}$

Exercise 8.2 (page 514)

1.
$$Ax + By = C$$
$$By = -Ax + C$$
$$B = \frac{-Ax + C}{y}$$

3. $\sqrt{24} = \sqrt{4}\sqrt{6} = 2\sqrt{6}$

5. $\dfrac{3}{\sqrt{3}} = \dfrac{3}{\sqrt{3}} \cdot \dfrac{\sqrt{3}}{\sqrt{3}} = \dfrac{3\sqrt{3}}{3} = \sqrt{3}$

7. $3; -2; 6$

9.
$$x^2 + 3x + 2 = 0$$
$$a = 1, b = 3, c = 2$$
$$x = \frac{-b \pm \sqrt{b^2 - 4ac}}{2a}$$
$$= \frac{-3 \pm \sqrt{3^2 - 4(1)(2)}}{2(1)}$$
$$= \frac{-3 \pm \sqrt{9 - 8}}{2}$$
$$= \frac{-3 \pm \sqrt{1}}{2} = \frac{-3 \pm 1}{2}$$
$x = -\frac{2}{2} = -1$ or $x = -\frac{4}{2} = -2$

11.
$$a^2 - 2a - 15 = 0$$
$$a = 1, b = -2, c = -15$$
$$a = \frac{-b \pm \sqrt{b^2 - 4ac}}{2a}$$
$$= \frac{2 \pm \sqrt{(-2)^2 - 4(1)(-15)}}{2(1)}$$
$$= \frac{2 \pm \sqrt{4 + 60}}{2}$$
$$= \frac{2 \pm \sqrt{64}}{2} = \frac{2 \pm 8}{2}$$
$x = \frac{10}{2} = 5$ or $x = -\frac{6}{2} = -3$

13.
$$x^2 + 12x = -36$$
$$x^2 + 12x + 36 = 0$$
$$a = 1, b = 12, c = 36$$
$$x = \frac{-b \pm \sqrt{b^2 - 4ac}}{2a}$$
$$= \frac{-12 \pm \sqrt{12^2 - 4(1)(36)}}{2(1)}$$
$$= \frac{-12 \pm \sqrt{144 - 144}}{2}$$
$$= \frac{-12 \pm \sqrt{0}}{2} = \frac{-12 \pm 0}{2}$$
$$x = -\frac{12}{2} = -6 \text{ or } x = -\frac{12}{2} = -6$$

15.
$$2x^2 - x - 3 = 0$$
$$a = 2, b = -1, c = -3$$
$$x = \frac{-b \pm \sqrt{b^2 - 4ac}}{2a}$$
$$= \frac{1 \pm \sqrt{(-1)^2 - 4(2)(-3)}}{2(2)}$$
$$= \frac{1 \pm \sqrt{1 + 24}}{4}$$
$$= \frac{1 \pm \sqrt{25}}{4} = \frac{1 \pm 5}{4}$$
$$x = \frac{6}{4} = \frac{3}{2} \text{ or } x = -\frac{4}{4} = -1$$

17.
$$5x^2 + 5x + 1 = 0$$
$$a = 5, b = 5, c = 1$$
$$x = \frac{-b \pm \sqrt{b^2 - 4ac}}{2a}$$
$$= \frac{-5 \pm \sqrt{5^2 - 4(5)(1)}}{2(5)}$$
$$= \frac{-5 \pm \sqrt{25 - 20}}{10}$$
$$= \frac{-5 \pm \sqrt{5}}{10}$$
$$= \frac{-5}{10} \pm \frac{\sqrt{5}}{10} = -\frac{1}{2} \pm \frac{\sqrt{5}}{10}$$

19.
$$8u = -4u^2 - 3$$
$$4u^2 + 8u + 3 = 0$$
$$a = 4, b = 8, c = 3$$
$$u = \frac{-b \pm \sqrt{b^2 - 4ac}}{2a}$$
$$= \frac{-8 \pm \sqrt{8^2 - 4(4)(3)}}{2(4)}$$
$$= \frac{-8 \pm \sqrt{64 - 48}}{8}$$
$$= \frac{-8 \pm \sqrt{16}}{8} = \frac{-8 \pm 4}{8}$$
$$u = \frac{-4}{8} = -\frac{1}{2} \text{ or } u = \frac{-12}{8} = -\frac{3}{2}$$

21.
$$16y^2 + 8y - 3 = 0$$
$$a = 16, b = 8, c = -3$$
$$y = \frac{-b \pm \sqrt{b^2 - 4ac}}{2a}$$
$$= \frac{-8 \pm \sqrt{8^2 - 4(16)(-3)}}{2(16)}$$
$$= \frac{-8 \pm \sqrt{64 + 192}}{32}$$
$$= \frac{-8 \pm \sqrt{256}}{32} = \frac{-8 \pm 16}{32}$$
$$y = \frac{8}{32} = \frac{1}{4} \text{ or } y = \frac{-24}{32} = -\frac{3}{4}$$

23.
$$\frac{x^2}{2} + \frac{5}{2}x = -1$$
$$x^2 + 5x = -2$$
$$x^2 + 5x + 2 = 0$$
$$a = 1, b = 5, c = 2$$
$$x = \frac{-b \pm \sqrt{b^2 - 4ac}}{2a}$$
$$= \frac{-5 \pm \sqrt{5^2 - 4(1)(2)}}{2(1)}$$
$$= \frac{-5 \pm \sqrt{25 - 8}}{2}$$
$$= \frac{-5 \pm \sqrt{17}}{2} = -\frac{5}{2} \pm \frac{\sqrt{17}}{2}$$

25. $x^2 + 2x + 2 = 0$

$a = 1, b = 2, c = 2$

$x = \dfrac{-b \pm \sqrt{b^2 - 4ac}}{2a}$

$= \dfrac{-2 \pm \sqrt{2^2 - 4(1)(2)}}{2(1)}$

$= \dfrac{-2 \pm \sqrt{4 - 8}}{2}$

$= \dfrac{-2 \pm \sqrt{-4}}{2}$

$= \dfrac{-2 \pm \sqrt{-1 \cdot 4}}{2}$

$= \dfrac{-2 \pm 2i}{2} = \dfrac{-2}{2} \pm \dfrac{2i}{2} = -1 \pm i$

27. $2x^2 + x + 1 = 0$

$a = 2, b = 1, c = 1$

$x = \dfrac{-b \pm \sqrt{b^2 - 4ac}}{2a}$

$= \dfrac{-1 \pm \sqrt{1^2 - 4(2)(1)}}{2(2)}$

$= \dfrac{-1 \pm \sqrt{1 - 8}}{4}$

$= \dfrac{-1 \pm \sqrt{-7}}{4}$

$= \dfrac{-1 \pm \sqrt{-1 \cdot 7}}{4}$

$= \dfrac{-1 \pm i\sqrt{7}}{4} = -\dfrac{1}{4} \pm \dfrac{\sqrt{7}}{4}i$

29. $3x^2 - 4x = -2$

$3x^2 - 4x + 2 = 0$

$a = 3, b = -4, c = 2$

$x = \dfrac{-b \pm \sqrt{b^2 - 4ac}}{2a}$

$= \dfrac{4 \pm \sqrt{(-4)^2 - 4(3)(2)}}{2(3)}$

$= \dfrac{4 \pm \sqrt{16 - 24}}{6}$

$= \dfrac{4 \pm \sqrt{-8}}{6}$

$= \dfrac{4 \pm \sqrt{-1 \cdot 4 \cdot 2}}{6}$

$= \dfrac{4 \pm 2i\sqrt{2}}{6}$

$= \dfrac{4}{6} \pm \dfrac{2\sqrt{2}}{6}i = \dfrac{2}{3} \pm \dfrac{\sqrt{2}}{3}i$

31. $3x^2 - 2x = -3$

$3x^2 - 2x + 3 = 0$

$a = 3, b = -2, c = 3$

$x = \dfrac{-b \pm \sqrt{b^2 - 4ac}}{2a}$

$= \dfrac{2 \pm \sqrt{(-2)^2 - 4(3)(3)}}{2(3)}$

$= \dfrac{2 \pm \sqrt{4 - 36}}{6}$

$= \dfrac{2 \pm \sqrt{-32}}{6}$

$= \dfrac{2 \pm \sqrt{-1 \cdot 16 \cdot 2}}{6}$

$= \dfrac{2 \pm 4i\sqrt{2}}{6}$

$= \dfrac{2}{6} \pm \dfrac{4\sqrt{2}}{6}i = \dfrac{1}{3} \pm \dfrac{2\sqrt{2}}{3}i$

33.
$$f(x) = 0$$
$$4x^2 + 4x - 19 = 0$$
$$a = 4, b = 4, c = -19$$
$$x = \frac{-b \pm \sqrt{b^2 - 4ac}}{2a}$$
$$= \frac{-4 \pm \sqrt{4^2 - 4(4)(-19)}}{2(4)}$$
$$= \frac{-4 \pm \sqrt{16 + 304}}{8}$$
$$= \frac{-4 \pm \sqrt{320}}{8}$$
$$= -\frac{4}{8} \pm \frac{8\sqrt{5}}{8} = -\frac{1}{2} \pm \sqrt{5}$$

35.
$$f(x) = 0$$
$$3x^2 + 2x + 2 = 0$$
$$a = 3, b = 2, c = 2$$
$$x = \frac{-b \pm \sqrt{b^2 - 4ac}}{2a}$$
$$= \frac{-2 \pm \sqrt{2^2 - 4(3)(2)}}{2(3)}$$
$$= \frac{-2 \pm \sqrt{4 - 24}}{6}$$
$$= \frac{-2 \pm \sqrt{-20}}{6}$$
$$= -\frac{2}{6} \pm \frac{2i\sqrt{5}}{6} = -\frac{1}{3} \pm \frac{\sqrt{5}}{3}i$$

37.
$$0.7x^2 - 3.5x - 25 = 0$$
$$a = 0.7, b = -3.5, c = -25$$
$$x = \frac{-b \pm \sqrt{b^2 - 4ac}}{2a}$$
$$= \frac{3.5 \pm \sqrt{(-3.5)^2 - 4(0.7)(-25)}}{2(0.7)}$$
$$= \frac{3.5 \pm \sqrt{12.25 + 70}}{1.4}$$
$$= \frac{3.5 \pm \sqrt{82.25}}{1.4}$$
$$= \frac{3.5 \pm 9.069}{1.4}$$
$$x = 8.98 \text{ or } x = -3.98$$

39.
$$C = \frac{N^2 - N}{2}$$
$$2C = N^2 - N$$
$$N^2 - N - 2C = 0$$
$$a = 1, b = -1, c = -2C$$
$$N = \frac{-b \pm \sqrt{b^2 - 4ac}}{2a}$$
$$= \frac{1 \pm \sqrt{(-1)^2 - 4(1)(-2C)}}{2(1)}$$
$$= \frac{1 \pm \sqrt{1 + 8C}}{2}$$

41. Let x and $x + 2$ represent the integers.
$$x(x + 2) = 288$$
$$x^2 + 2x - 288 = 0$$
$$(x + 18)(x - 16) = 0$$
$$x = -18 \text{ or } x = 16$$
Since the integers are positive, they must be 16 and 18.

43. Let x and $x + 1$ represent the integers.
$$x^2 + (x + 1)^2 = 85$$
$$x^2 + x^2 + 2x + 1 = 85$$
$$2x^2 + 2x - 84 = 0$$
$$2(x + 7)(x - 6) = 0$$
$$x = -7 \text{ or } x = 6$$
Since the integers are positive, they must be 6 and 7.

45.
$$(x - 3)(x - 5) = 0$$
$$x^2 - 8x + 15 = 0$$

47.
$$(x - 2)(x - 3)(x + 4) = 0$$
$$(x^2 - 5x + 6)(x + 4) = 0$$
$$x^3 - x^2 - 14x + 24 = 0$$

49.
$$\text{length} \cdot \text{width} = \text{Area}$$
$$(x + 4)x = 96$$
$$x^2 + 4x - 96 = 0$$
$$(x + 12)(x - 8) = 0$$
$$x = -12 \text{ or } x = 8$$
Since the width is positive, the dimensions are 8 ft by 12 ft.

51. Let s = the length of a side.
$$\text{Area} = \text{perimeter}$$
$$s^2 = 4s$$
$$s^2 - 4s = 0$$
$$s(s - 4) = 0$$
$$s = 0 \text{ or } s = 4$$
Since the length cannot be 0, the length of a side is 4 units.

53. Let b represent the base. Then $3b + 5$ represents the height.
$$\tfrac{1}{2}\text{base} \cdot \text{height} = \text{Area}$$
$$\tfrac{1}{2}b(3b + 5) = 6$$
$$b(3b + 5) = 12$$
$$3b^2 + 5b - 12 = 0$$
$$(3b - 4)(b + 3) = 0$$
$$b = \tfrac{4}{3} \text{ or } b = -3 \quad \text{Since the base is positive, it must be } \tfrac{4}{3} \text{ cm.}$$

55. Let r = the slower rate. Then $r + 20$ = the faster rate.

	Rate	Time	Dist.
Slower	r	$\frac{150}{r}$	150
Faster	$r + 20$	$\frac{150}{r+20}$	150

$$\boxed{\text{Faster time}} + 2 = \boxed{\text{Slower time}}$$
$$\frac{150}{r + 20} + 2 = \frac{150}{r}$$
$$\left(\frac{150}{r + 20} + 2\right)(r)(r + 20) = \frac{150}{r} \cdot r(r + 20)$$
$$150r + 2r(r + 20) = 150(r + 20)$$
$$2r^2 + 40r - 3000 = 0$$
$$2(r + 50)(r - 30) = 0$$
$$r = -50 \quad \text{or} \quad r = 30 \quad r = 30 \text{ is the only answer that makes sense.}$$
$$\text{Her original speed was 30 mph.}$$

57. Let $x =$ the number of 10¢ increases. Then the ticket price will be $4 + 0.10x$, while the projected attendance will be $300 - 5x$, for total receipts of $(4 + 0.10x)(300 - 5x)$.

$$\text{Total} = 1248$$
$$(4 + 0.10x)(300 - 5x) = 1248$$
$$1200 + 10x - 0.5x^2 = 1248$$
$$-0.5x^2 + 10x - 48 = 0$$
$$x^2 - 20x + 96 = 0$$
$$(x - 12)(x - 8) = 0$$
$$x = 12 \text{ or } x = 8 \Rightarrow 4 + 0.10(12) = 5.20; \; 4 + 0.10(8) = 4.80$$

The ticket price would be either $5.20 or $4.80.

59. Let $x =$ the number of additional subscribers. Then the profit per subscriber will be $20 + 0.01x$, for a total profit of $(20 + 0.01x)(3000 + x)$.

$$\text{Total profit} = 120000$$
$$(20 + 0.01x)(3000 + x) = 120000$$
$$60,000 + 50x + 0.01x^2 = 120000$$
$$0.01x^2 + 50x - 60,000 = 0$$
$$x^2 + 5000x - 6,000,000 = 0$$
$$(x + 6000)(x - 1000) = 0$$
$$x = -6000 \text{ (impossible) or } x = 1000 \Rightarrow \text{The total number of subscribers would be 4000.}$$

61. Let $w =$ the constant width.

$$\text{Frame area} = \text{Picture area}$$
$$(12 + 2w)(10 + 2w) - 12(10) = 12(10)$$
$$120 + 44w + 4w^2 - 240 = 0$$
$$4w^2 + 44w - 120 = 0$$
$$4(w^2 + 11w - 30) = 0$$
$$w^2 + 11w - 30 = 0 \Rightarrow a = 1, b = 11, c = -30$$
$$w = \frac{-11 \pm \sqrt{11^2 - 4(1)(-30)}}{2(1)}$$
$$= \frac{-11 \pm \sqrt{121 + 120}}{2} = \frac{-11 \pm \sqrt{241}}{2} = \frac{-11 \pm 15.52}{2} = \frac{4.52}{2} \text{ or } \frac{-26.52}{2} \text{ (impossible)}$$
$$w = 2.26 \text{ in.}$$

63. Let $[H^+]$ (and then $[A^-]$) $= x$ and $[HA] = 0.1 - x$.

$$\frac{[H^+][A^-]}{[HA]} = 4 \times 10^{-4}$$

$$\frac{x^2}{0.1 - x} = 4 \times 10^{-4}$$

$$x^2 = 4 \times 10^{-5} - \left(4 \times 10^{-4}\right)x$$

$$x^2 + \left(4 \times 10^{-4}\right)x - 4 \times 10^{-5} = 0$$

$$x = \frac{-b \pm \sqrt{b^2 - 4ac}}{2a} = \frac{-4 \times 10^{-4} \pm \sqrt{(4 \times 10^{-4})^2 - 4(1)(-4 \times 10^{-5})}}{2(1)}$$

$$\approx \frac{-4 \times 10^{-4} \pm 0.012655}{2} \approx \frac{0.012255}{2} \text{ or } -\frac{0.013055}{2} \text{ (impossible)}$$

The concentration is about $0.00613 \text{ M} = 6.13 \times 10^{-3} \text{ M}$.

65. Answers may vary.

67. $x^2 + 2\sqrt{2}x - 6 = 0$

$a = 1, b = 2\sqrt{2}, c = -6$

$$x = \frac{-b \pm \sqrt{b^2 - 4ac}}{2a}$$

$$= \frac{-2\sqrt{2} \pm \sqrt{\left(2\sqrt{2}\right)^2 - 4(1)(-6)}}{2(1)}$$

$$= \frac{-2\sqrt{2} \pm \sqrt{8 + 24}}{2}$$

$$= \frac{-2\sqrt{2} \pm \sqrt{32}}{2} = \frac{-2\sqrt{2} \pm 4\sqrt{2}}{2}$$

$x = \frac{2\sqrt{2}}{2} = \sqrt{2}$ or $x = \frac{-6\sqrt{2}}{2} = -3\sqrt{2}$

69. $x^2 - 3ix - 2 = 0$

$a = 1, b = -3i, c = -2$

$$x = \frac{-b \pm \sqrt{b^2 - 4ac}}{2a}$$

$$= \frac{3i \pm \sqrt{(-3i)^2 - 4(1)(-2)}}{2(1)}$$

$$= \frac{3i \pm \sqrt{9i^2 + 8}}{2}$$

$$= \frac{3i \pm \sqrt{-1}}{2} = \frac{3i \pm i}{2}$$

$x = \frac{4i}{2} = 2i$ or $x = \frac{2i}{2} = i$

Exercise 8.3 (page 522)

1. $\dfrac{1}{4} + \dfrac{1}{t} = \dfrac{1}{2t}$

$\left(\dfrac{1}{4} + \dfrac{1}{t}\right)4t = \dfrac{1}{2t} \cdot 4t$

$t + 4 = 2$

$t = -2$

3. $m = \dfrac{\Delta y}{\Delta x} = \dfrac{-4 - 5}{-2 - 3} = \dfrac{-9}{-5} = \dfrac{9}{5}$

5. $b^2 - 4ac$

7. rational; unequal

9. $4x^2 - 4x + 1 = 0; a = 4, b = -4, c = 1$
$b^2 - 4ac = (-4)^2 - 4(4)(1)$
$\qquad = 16 - 16 = 0$
The solutions are rational and equal.

11. $5x^2 + x + 2 = 0; a = 5, b = 1, c = 2$
$b^2 - 4ac = 1^2 - 4(5)(2)$
$\qquad = 1 - 40 = -39$
The solutions are complex conjugates.

13. $\qquad 2x^2 = 4x - 1$
$2x^2 - 4x + 1 = 0; a = 2, b = -4, c = 1$
$b^2 - 4ac = (-4)^2 - 4(2)(1)$
$\qquad = 16 - 8 = 8$
The solutions are irrational and unequal.

15. $\qquad x(2x - 3) = 20$
$2x^2 - 3x - 20 = 0; a = 2, b = -3, c = -20$
$b^2 - 4ac = (-3)^2 - 4(2)(-20)$
$\qquad = 9 + 160 = 169$
The solutions are rational and unequal.

17. $x^2 + kx + 9 = 0; a = 1, b = k, c = 9$
Set the discriminant equal to 0:
$$b^2 - 4ac = 0$$
$$k^2 - 4(1)(9) = 0$$
$$k^2 - 36 = 0$$
$$k^2 = 36$$
$$k = \pm 6$$

19. $\qquad 9x^2 + 4 = -kx$
$9x^2 + kx + 4 = 0; a = 9, b = k, c = 4$
Set the discriminant equal to 0:
$$b^2 - 4ac = 0$$
$$k^2 - 4(9)(4) = 0$$
$$k^2 - 144 = 0$$
$$k^2 = 144$$
$$k = \pm 12$$

21. $(k - 1)x^2 + (k - 1)x + 1 = 0$
$a = k - 1, b = k - 1, c = 1$
Set the discriminant equal to 0:
$$b^2 - 4ac = 0$$
$$(k - 1)^2 - 4(k - 1)(1) = 0$$
$$k^2 - 2k + 1 - 4k + 4 = 0$$
$$k^2 - 6k + 5 = 0$$
$$(k - 5)(k - 1) = 0$$
$$k - 5 = 0 \quad \text{or} \quad k - 1 = 0$$
$$k = 5 \qquad\qquad k = 1$$
$$\qquad\qquad\qquad\qquad \text{doesn't work}$$

23. $(k + 4)x^2 + 2kx + 9 = 0$
$a = k + 4, b = 2k, c = 9$
Set the discriminant equal to 0:
$$b^2 - 4ac = 0$$
$$(2k)^2 - 4(k + 4)(9) = 0$$
$$4k^2 - 36(k + 4) = 0$$
$$4k^2 - 36k - 144 = 0$$
$$4(k^2 - 9k - 36) = 0$$
$$4(k - 12)(k + 3) = 0$$
$$k - 12 = 0 \quad \text{or} \quad k + 3 = 0$$
$$k = 12 \qquad\qquad k = -3$$

25. $1492x^2 + 1776x - 1984 = 0$
$a = 1492, b = 1776, c = -1984$
$b^2 - 4ac = (1776)^2 - 4(1492)(-1984)$
$\qquad = 3,154,176 + 11,840,512$
$\qquad = 14,994,688$
The solutions are real numbers.

27. $\qquad 3x^2 + 4x = k$
$3x^2 + 4x - k = 0$
$a = 3, b = 4, c = -k$
Set the discriminant less than 0:
$$b^2 - 4ac < 0$$
$$4^2 - 4(3)(-k) < 0$$
$$16 + 12k < 0$$
$$12k < -16$$
$$k < -\frac{16}{12}, \text{ or } k < -\frac{4}{3}$$

29.
$$x^4 - 17x^2 + 16 = 0$$
$$(x^2 - 16)(x^2 - 1) = 0$$

$x^2 - 16 = 0$	**or**	$x^2 - 1 = 0$
$x^2 = 16$		$x^2 = 1$
$x = \pm 4$		$x = \pm 1$

31.
$$x^4 - 3x^2 = -2$$
$$x^4 - 3x^2 + 2 = 0$$
$$(x^2 - 2)(x^2 - 1) = -0$$

$x^2 - 2 = 0$	**or**	$x^2 - 1 = 0$
$x^2 = 2$		$x^2 = 1$
$x = \pm \sqrt{2}$		$x = \pm 1$

33.
$$x^4 = 6x^2 - 5$$
$$x^4 - 6x^2 + 5 = 0$$
$$(x^2 - 5)(x^2 - 1) = -0$$

$x^2 - 5 = 0$	**or**	$x^2 - 1 = 0$
$x^2 = 5$		$x^2 = 1$
$x = \pm \sqrt{5}$		$x = \pm 1$

35.
$$2x^4 - 10x^2 = -8$$
$$2x^4 - 10x^2 + 8 = 0$$
$$2(x^2 - 4)(x^2 - 1) = -0$$

$x^2 - 4 = 0$	**or**	$x^2 - 1 = 0$
$x^2 = 4$		$x^2 = 1$
$x = \pm 2$		$x = \pm 1$

37.
$$2x^4 + 24 = 26x^2$$
$$2x^4 - 26x^2 + 24 = 0$$
$$2(x^2 - 12)(x^2 - 1) = -0$$

$x^2 - 12 = 0$	**or**	$x^2 - 1 = 0$
$x^2 = 12$		$x^2 = 1$
$x = \pm \sqrt{12}$		$x = \pm 1$
$= \pm 2\sqrt{3}$		

39.
$$2x + x^{1/2} - 3 = 0$$
$$\left(2x^{1/2} + 3\right)\left(x^{1/2} - 1\right) = 0$$

$2x^{1/2} + 3 = 0$	**or**	$x^{1/2} - 1 = 0$
$2x^{1/2} = -3$		$x^{1/2} = 1$
$x^{1/2} = -\frac{3}{2}$		$\left(x^{1/2}\right)^2 = 1^2$
$\left(x^{1/2}\right)^2 = \left(-\frac{3}{2}\right)^2$		$x = 1$
$x = \frac{9}{4}$		Solution
Does not check		

41.
$$3x + 5x^{1/2} + 2 = 0$$
$$\left(3x^{1/2} + 2\right)\left(x^{1/2} + 1\right) = 0$$

$3x^{1/2} + 2 = 0$	**or**	$x^{1/2} + 1 = 0$
$3x^{1/2} = -2$		$x^{1/2} = -1$
$x^{1/2} = -\frac{2}{3}$		$\left(x^{1/2}\right)^2 = (-1)^2$
$\left(x^{1/2}\right)^2 = \left(-\frac{2}{3}\right)^2$		$x = 1$
$x = \frac{4}{9}$		Does not check
Does not check		

43.
$$x^{2/3} + 5x^{1/3} + 6 = 0$$
$$\left(x^{1/3} + 2\right)\left(x^{1/3} + 3\right) = 0$$

$x^{1/3} + 2 = 0$	**or**	$x^{1/3} + 3 = 0$
$x^{1/3} = -2$		$x^{1/3} = -3$
$\left(x^{1/3}\right)^3 = (-2)^3$		$\left(x^{1/3}\right)^3 = (-3)^3$
$x = -8$		$x = -27$
Solution		Solution

45.
$$x^{2/3} - 2x^{1/3} - 3 = 0$$
$$\left(x^{1/3} - 3\right)\left(x^{1/3} + 1\right) = 0$$

$x^{1/3} - 3 = 0$	**or**	$x^{1/3} + 1 = 0$
$x^{1/3} = 3$		$x^{1/3} = -1$
$\left(x^{1/3}\right)^3 = (3)^3$		$\left(x^{1/3}\right)^3 = (-1)^3$
$x = 27$		$x = -1$
Solution		Solution

47.
$$x + 5 + \frac{4}{x} = 0$$
$$x\left(x + 5 + \frac{4}{x}\right) = x(0)$$
$$x^2 + 5x + 4 = 0$$
$$(x + 4)(x + 1) = 0$$

$x + 4 = 0$	**or**	$x + 1 = 0$
$x = -4$		$x = -1$

49.
$$x + 1 = \frac{20}{x}$$
$$x + 1 - \frac{20}{x} = 0$$
$$x\left(x + 1 - \frac{20}{x}\right) = x(0)$$
$$x^2 + x - 20 = 0$$
$$(x + 5)(x - 4) = 0$$
$$x + 5 = 0 \quad \textbf{or} \quad x - 4 = 0$$
$$x = -5 \qquad\qquad x = 4$$

51.
$$\frac{1}{x - 1} + \frac{3}{x + 1} = 2$$
$$\left(\frac{1}{x - 1} + \frac{3}{x + 1}\right)(x - 1)(x + 1) = 2(x + 1)(x - 1)$$
$$1(x + 1) + 3(x - 1) = 2(x^2 - 1)$$
$$x + 1 + 3x - 3 = 2x^2 - 2$$
$$0 = 2x^2 - 4x$$
$$0 = 2x(x - 2) \qquad 2x = 0 \quad \textbf{or} \quad x - 2 = 0$$
$$x = 0 \qquad\qquad x = 2$$

53.
$$\frac{1}{x + 2} + \frac{24}{x + 3} = 13$$
$$\left(\frac{1}{x + 2} + \frac{24}{x + 3}\right)(x + 2)(x + 3) = 13(x + 2)(x + 3)$$
$$1(x + 3) + 24(x + 2) = 13(x^2 + 5x + 6)$$
$$x + 3 + 24x + 48 = 13x^2 + 65x + 78$$
$$0 = 13x^2 + 40x + 27$$
$$0 = (13x + 27)(x + 1) \qquad 13x + 27 = 0 \quad \textbf{or} \quad x + 1 = 0$$
$$13x = -27 \qquad\qquad x = -1$$
$$x = -\frac{27}{13}$$

55.
$$x^{-4} - 2x^{-2} + 1 = 0$$
$$(x^{-2} - 1)(x^{-2} - 1) = 0$$
$$x^{-2} - 1 = 0 \quad \textbf{or} \quad x^{-2} - 1 = 0$$
$$x^{-2} = 1 \qquad\qquad x^{-2} = 1$$
$$\frac{1}{x^2} = 1 \qquad\qquad \frac{1}{x^2} = 1$$
$$1 = x^2 \qquad\qquad 1 = x^2$$
$$\pm 1 = x \qquad\qquad \pm 1 = x$$

57.
$$x + \frac{2}{x - 2} = 0$$
$$\left(x + \frac{2}{x - 2}\right)(x - 2) = 0(x - 2)$$
$$x(x - 2) + 2 = 0$$
$$x^2 - 2x + 2 = 0$$
$$x^2 - 2x = -2$$
$$x^2 - 2x + 1 = -2 + 1$$
$$(x - 1)^2 = -1$$
$$x - 1 = \pm\sqrt{-1}$$
$$x = 1 \pm i$$

59. $x^2 + y^2 = r^2$
$$x^2 = r^2 - y^2$$
$$x = \pm \sqrt{r^2 - y^2}$$

61. $I = \dfrac{k}{d^2}$
$$Id^2 = k$$
$$d^2 = \frac{k}{I}$$
$$d = \pm \sqrt{\frac{k}{I}} = \pm \frac{\sqrt{kI}}{I}$$

63. $xy^2 + 3xy + 7 = 0; a = x, b = 3x, c = 7$
$$y = \frac{-b \pm \sqrt{b^2 - 4ac}}{2a}$$
$$= \frac{-3x \pm \sqrt{(3x)^2 - 4(x)(7)}}{2x}$$
$$= \frac{-3x \pm \sqrt{9x^2 - 28x}}{2x}$$

65. $\sigma = \sqrt{\dfrac{\Sigma x^2}{N} - \mu^2}$
$$\sigma^2 = \frac{\Sigma x^2}{N} - \mu^2$$
$$\mu^2 = \frac{\Sigma x^2}{N} - \sigma^2$$

67. $12x^2 - 5x - 2 = 0; a = 12, b = -5, c = -2$
$(4x + 1)(3x - 2) = 0$

$4x + 1 = 0 \quad$ **or** $\quad 3x - 2 = 0 \qquad -\dfrac{b}{a} = -\dfrac{-5}{12} = \dfrac{5}{12} \qquad\qquad \dfrac{c}{a} = \dfrac{-2}{12} = -\dfrac{1}{6}$

$\qquad 4x = -1 \qquad\qquad 3x = 2 \qquad -\dfrac{1}{4} + \dfrac{2}{3} = -\dfrac{3}{12} + \dfrac{8}{12} = \dfrac{5}{12} \quad \left(-\dfrac{1}{4}\right)\left(\dfrac{2}{3}\right) = -\dfrac{1}{6}$

$\qquad x = -\dfrac{1}{4} \qquad\qquad x = \dfrac{2}{3}$

69. $2x^2 + 5x + 1 = 0; a = 2, b = 5, c = 1; -\dfrac{b}{a} = -\dfrac{5}{2}; \dfrac{c}{a} = \dfrac{1}{2}$
$$x = \frac{-b \pm \sqrt{b^2 - 4ac}}{2a} = \frac{-5 \pm \sqrt{5^2 - 4(2)(1)}}{2(2)} = \frac{-5 \pm \sqrt{17}}{4} = -\frac{5}{4} \pm \frac{\sqrt{17}}{4}$$
$$\frac{-5 + \sqrt{17}}{4} + \frac{-5 - \sqrt{17}}{4} = \frac{-10}{4} = -\frac{5}{2}$$
$$\left(\frac{-5 + \sqrt{17}}{4}\right)\left(\frac{-5 - \sqrt{17}}{4}\right) = \frac{25 + 5\sqrt{17} - 5\sqrt{17} - 17}{16} = \frac{8}{16} = \frac{1}{2}$$

71. $3x^2 - 2x + 4 = 0; a = 3, b = -2, c = 4; -\dfrac{b}{a} = -\dfrac{-2}{3} = \dfrac{2}{3}; \dfrac{c}{a} = \dfrac{4}{3}$
$$x = \frac{-b \pm \sqrt{b^2 - 4ac}}{2a} = \frac{-(-2) \pm \sqrt{(-2)^2 - 4(3)(4)}}{2(3)} = \frac{2 \pm \sqrt{-44}}{6} = \frac{1}{3} \pm \frac{i\sqrt{11}}{3}$$
$$\frac{1 + i\sqrt{11}}{3} + \frac{1 - i\sqrt{11}}{3} = \frac{2}{3}$$
$$\left(\frac{1 + i\sqrt{11}}{3}\right)\left(\frac{1 - i\sqrt{11}}{3}\right) = \frac{1 - i\sqrt{11} + i\sqrt{11} - 11i^2}{9} = \frac{1 + 11}{9} = \frac{12}{9} = \frac{4}{3}$$

73. $x^2 + 2x + 5 = 0$; $a = 1, b = 2, c = 5$; $-\dfrac{b}{a} = -\dfrac{2}{1} = -2$; $\dfrac{c}{a} = \dfrac{5}{1} = 5$

$$x = \frac{-b \pm \sqrt{b^2 - 4ac}}{2a} = \frac{-2 \pm \sqrt{2^2 - 4(1)(5)}}{2(1)} = \frac{-2 \pm \sqrt{-16}}{2} = \frac{-2 \pm 4i}{2} = -1 \pm 2i$$

$(-1 + 2i) + (-1 - 2i) = -2$; $(-1 + 2i)(-1 - 2i) = 1 + 2i - 2i - 4i^2 = 1 + 4 = 5$

75. Answers may vary.   **77.** No

Exercise 8.4   (page 532)

1. $3x + 5 = 5x - 15$
   $-2x = -20$
   $x = 10$

3. Let $t =$ the time of the second train.
   Then $t + 3 =$ the time of the first train.

	Rate	Time	Dist.
First	30	$t+3$	$30(t+3)$
Second	55	t	$55t$

1st distance = 2nd distance
$$30(t+3) = 55t$$
$$30t + 90 = 55t$$
$$-25t = -90$$
$$t = \frac{-90}{-25} = \frac{18}{5} = 3\tfrac{3}{5} \text{ hours}$$

5. $f(x) = ax^2 + bx + c$;
   $a \neq 0$

7. vertex

9. upward

11. to the right

13. upward

15. $f(x) = x^2$
   vertex: $(0, 0)$; opens U

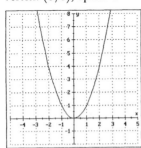

17. $f(x) = x^2 + 2$
   vertex: $(0, 2)$; opens U

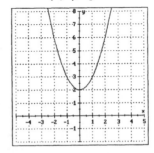

19. $f(x) = -(x - 2)^2$
   vertex: $(2, 0)$; opens D

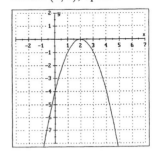

21. $f(x) = (x-3)^2 + 2$
vertex: $(3,2)$; opens U

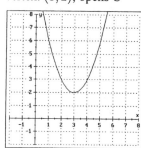

23. $f(x) = x^2 + x - 6$
$f(x) = \left(x + \frac{1}{2}\right)^2 - \frac{25}{4}$
vertex: $\left(-\frac{1}{2}, -\frac{25}{4}\right)$; opens U

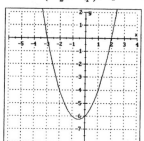

25. $f(x) = -2x^2 + 4x + 1$
$f(x) = -2(x-1)^2 + 3$
vertex: $(1,3)$; opens D

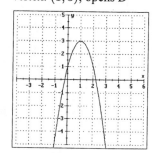

27. $y = (x-1)^2 + 2$; $V(1,2)$; axis: $x = 1$

29. $y = 2(x+3)^2 - 4$;
$V(-3,-4)$; axis: $x = -3$

31. $y = -3x^2 \Rightarrow y = -3(x-0)^2 + 0$
$V(0,0)$; axis: $x = 0$

33. $y = 2x^2 - 4x$
$y = 2(x^2 - 2x)$
$y = 2(x^2 - 2x + 1) - 2$
$y = 2(x-1)^2 - 2$
$V(1,-2)$; axis: $x = 1$

35. $y = -4x^2 + 16x + 5$
$y = -4(x^2 - 4x) + 5$
$y = -4(x^2 - 4x + 4) + 5 + 16$
$y = -4(x-2)^2 + 21$
$V(2,21)$; axis: $x = 2$

37. $y - 7 = 6x^2 - 5x$
$y = 6\left(x^2 - \frac{5}{6}x\right) + 7$
$y = 6\left(x^2 - \frac{5}{6}x + \frac{25}{144}\right) + 7 - \frac{25}{24}$
$y = 6\left(x - \frac{5}{12}\right)^2 + \frac{143}{24}$
$V\left(\frac{5}{12}, \frac{143}{24}\right)$; axis: $x = \frac{5}{12}$

39. $y - 2 = (x-5)^2 \Rightarrow y = (x-5)^2 + 2 \Rightarrow V(5,2)$

41. $y = 2x^2 - x + 1$

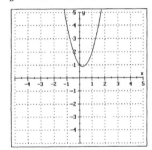

$V(0.25, 0.88)$

43. $y = 7 + x - x^2$

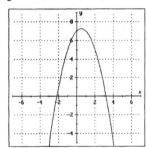

$V(0.5, 7.25)$

45. $y = x^2 + x - 6$

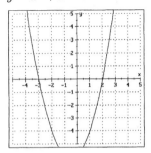

solution set: $\{2, -3\}$

47. $y = 0.5x^2 - 0.7x - 3$

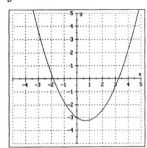

solution set: $\{-1.85, 3.25\}$

49. Since the graph of the height equation is a parabola, the max. height occurs at the vertex.

$s = 48t - 16t^2$

$s = -16(t^2 - 3t)$

$s = -16\left(t^2 - 3t + \frac{9}{4}\right) + 36$

$s = -16\left(t - \frac{3}{2}\right)^2 + 36$

$V\left(\frac{3}{2}, 36\right) \Rightarrow$ max. height $= 36$ ft

To find the time it takes for the ball to return to earth, set $s = 0$ and solve for t.

$s = 48t - 16t^2$

$0 = 48t - 16t^2$

$0 = 16t(3 - t)$

$t = 0$ or $t = 3$

The ball returns to earth after 3 seconds.

(0 seconds is when the ball is originally thrown.)

51. Let $w =$ the width of the rectangle.
Then $100 - w =$ the length.

$A = w(100 - w)$

$A = -w^2 + 100w$

$A = -(w^2 - 100w + 2500) + 2500$

$A = -(w - 50)^2 + 2500$

dim: 50 ft by 50 ft; area $= 2500$ ft^2

53. Let $w =$ the width of the rectangle.
Then $150 - w =$ the length.

$A = w(150 - w)$

$A = -w^2 + 150w$

$A = -(w^2 - 150w + 5625) + 5625$

$A = -(w - 75)^2 + 5625$

dim: 75 ft by 75 ft; area $= 5625$ ft^2

55. Graph $y = R = -\frac{x^2}{1000} + 10x$ and find the x-coordinate of the vertex:

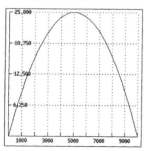

5000 stereos should be sold.

57. Graph $y = R = -\frac{x^2}{728} + 9x$ and find the x- and y- coordinates of the vertex:

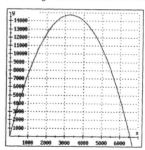

Max. revenue = \$14,742; # radios = 3276

59. Let $x =$ the number of \$1 increases to the price. Then the sales will be $4000 - 100x$, and the revenue will be $(30 + x)(4000 - 100x)$. Find the vertex of the parabola $y = (30 + x)(4000 - 100x)$.

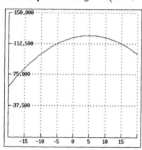

The price should increase \$5, to a total of \$35.

61. Graph $y = 50x(1 - x)$ and find the x-coordinate(s) when $y = 9.375$.

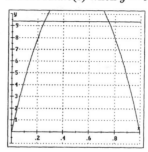

$p = 0.25$ or $p = 0.75$

63. **Answers may vary.**

65. Graph $y = x^2 + x + 1$ to find x-intercept(s):

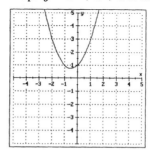

There are no x-intercepts, which means there is no solution to the equation.

Exercise 8.5 (page 542)

1. $y = kx$

3. $t = kxy$

5. $y = 3x - 4$
$m = 3$

7. greater

9. undefined

11. $x^2 - 5x + 4 < 0$
$(x - 4)(x - 1) < 0$

```
x - 4   - - - - - - - 0 ++++
x - 1   - - - 0++++++++++
        ←——(———)——→
            1       4
```

solution set: $(1, 4)$

13. $x^2 - 8x + 15 > 0$
$(x - 5)(x - 3) > 0$

```
x - 5   - - - - - - - -0++++
x - 3   - - - 0++++++++++
        ←——)————(——→
            3       5
```

solution set: $(-\infty, 3) \cup (5, \infty)$

15. $x^2 + x - 12 \leq 0$
$(x - 3)(x + 4) \leq 0$

```
x - 3   - - - - - - - - 0++++
x + 4   - - -   0++++++++++
        ←—[———————]—→
           -4       3
```

solution set: $[-4, 3]$

17. $x^2 + 2x \geq 15$
$x^2 + 2x - 15 \geq 0$
$(x - 3)(x + 5) \geq 0$

```
x - 3   - - - - - - - -0++++
x + 5   - - - 0++++++++++
        ←——]————[——→
           -5      3
```

solution set: $(-\infty, -5] \cup [3, \infty)$

19. $x^2 + 8x < -16$
$x^2 + 8x + 16 < 0$
$(x + 4)(x + 4) < 0$

```
x + 4   - - - - 0+++++++
x + 4   - - - - 0+++++++
        ←————————————→
              -4
```

Since the product is never negative, there is no solution.

21. $x^2 \geq 9$
$x^2 - 9 \geq 0$
$(x - 3)(x + 3) \geq 0$

```
x - 3   - - - - - - - - -0++++
x + 3   - - - 0++++++++++
        ←——]————[——→
           -3      3
```

solution set: $(-\infty, -3] \cup [3, \infty)$

23.
$$2x^2 - 50 < 0$$
$$2(x - 5)(x + 5) < 0$$

$$x - 5 \quad -------- \; 0++++$$
$$x + 5 \quad --- \; 0++++++ \; ++++$$

$$\longleftarrow \; (\!\!-\!\!-\!\!-\!\!-\!\!) \!\!\longrightarrow$$
$$\qquad -5 \qquad 5$$

solution set: $(-5, 5)$

25.
$$\frac{1}{x} < 2$$
$$\frac{1}{x} - 2 < 0$$
$$\frac{1}{x} - \frac{2x}{x} < 0$$
$$\frac{1 - 2x}{x} < 0$$

$$1 - 2x \quad +++++++++ \; 0---$$
$$x \qquad\qquad ---0++++++++++$$

$$\longleftarrow \;)\!\!-\!\!-\!\!-\!\!(\!\!\longrightarrow$$
$$\qquad 0 \qquad \frac{1}{2}$$

solution set: $(-\infty, 0) \cup \left(\frac{1}{2}, \infty\right)$

27.
$$\frac{4}{x} \geq 2$$
$$\frac{4}{x} - 2 \geq 0$$
$$\frac{4}{x} - \frac{2x}{x} \geq 0$$
$$\frac{4 - 2x}{x} \geq 0$$

$$4 - 2x \quad +++++++++++0---$$
$$x \qquad\qquad ---0+++++++++++$$

$$\longleftarrow \; (\!\!-\!\!-\!\!-\!\!] \!\!\longrightarrow$$
$$\qquad 0 \qquad 2$$

solution set: $(0, 2]$

29.
$$-\frac{5}{x} < 3$$
$$-\frac{5}{x} - 3 < 0$$
$$\frac{-5}{x} - \frac{3x}{x} < 0$$
$$\frac{-5 - 3x}{x} < 0$$

$$-5 - 3x \quad +++ \; 0--------$$
$$x \qquad\qquad ---------0++++$$

$$\longleftarrow \;)\!\!-\!\!-\!\!-\!\!(\!\!\longrightarrow$$
$$\quad -\frac{5}{3} \qquad 0$$

solution set: $\left(-\infty, -\frac{5}{3}\right) \cup (0, \infty)$

31.
$$\frac{x^2 - x - 12}{x - 1} < 0$$
$$\frac{(x - 4)(x + 3)}{x - 1} < 0$$

$$x - 4 \quad -------- \quad --- \; 0++++$$
$$x - 1 \quad -------- \quad 0+++ \; +++++$$
$$x + 3 \quad --- \; 0++++++ \; ++++ \; +++++$$

$$\longleftarrow \;)\!\!-\!\!-\!\!-\!\!(\!\!-\!\!-\!\!-\!\!) \!\!\longrightarrow$$
$$\quad -3 \qquad 1 \qquad 4$$

solution set: $(-\infty, -3) \cup (1, 4)$

33.
$$\frac{x^2 + x - 20}{x + 2} \geq 0$$
$$\frac{(x - 4)(x + 5)}{x + 2} \geq 0$$

$$x - 4 \quad -------- \quad ---- \; 0++++$$
$$x + 2 \quad -------- \quad 0++++ \; +++++$$
$$x + 5 \quad --- \; 0++++++++++++ \; +++++$$

$$\longleftarrow \; [\!\!-\!\!-\!\!-\!\!) \!\!-\!\![\!\!\longrightarrow$$
$$\quad -5 \qquad -2 \qquad 4$$

solution set: $[-5, -2) \cup [4, \infty)$

35.

$$\frac{x^2 - 4x + 4}{x + 4} < 0$$

$$\frac{(x-2)(x-2)}{x+4} < 0$$

$x - 2$ $\quad$ – – – – – – – – $\;$ 0++++
$x - 2$ $\quad$ – – – – – – – – $\;$ 0+++++
$x + 4$ $\quad$ – – – $\;$ 0++++++ +++++

solution set: $(-\infty, -4)$

37.

$$\frac{6x^2 - 5x + 1}{2x + 1} > 0$$

$$\frac{(2x-1)(3x-1)}{2x+1} > 0$$

$2x - 1$ $\quad$ – – – – – – – – $\;$ 0++++
$3x - 1$ $\quad$ – – – – – – – – $\;$ 0+++++++++
$2x + 1$ $\quad$ – – – $\;$ 0++++++++++++++++

solution set: $\left(-\dfrac{1}{2}, \dfrac{1}{3}\right) \cup \left(\dfrac{1}{2}, \infty\right)$

39.

$$\frac{3}{x - 2} < \frac{4}{x}$$

$$\frac{3}{x - 2} - \frac{4}{x} < 0$$

$$\frac{3x}{x(x-2)} - \frac{4(x-2)}{x(x-2)} < 0$$

$$\frac{-x + 8}{x(x-2)} < 0$$

$-x + 8$ $\quad$ ++++++++++++++0– – –
$x - 2$ $\quad$ – – – – – – – $\;$ 0+++++++++
x $\quad$ – – – $\;$ 0++++ +++++++++

solution set: $(0, 2) \cup (8, \infty)$

41.

$$\frac{-5}{x + 2} \geq \frac{4}{2 - x}$$

$$\frac{-5}{x + 2} - \frac{4}{2 - x} \geq 0$$

$$\frac{-5(2 - x)}{(x+2)(2-x)} - \frac{4(x+2)}{(x+2)(2-x)} \geq 0$$

$$\frac{x - 18}{(x+2)(2-x)} \geq 0$$

$x - 18$ $\quad$ – – – – – – – – – – – $\;$ 0++++
$2 - x$ $\quad$ ++++++++++++ 0– – – – – –
$x + 2$ $\quad$ – – – $\;$ 0++++++++++++++++++

solution set: $(-\infty, -2) \cup (2, 18]$

43.

$$\frac{7}{x - 3} \geq \frac{2}{x + 4}$$

$$\frac{7}{x - 3} - \frac{2}{x + 4} \geq 0$$

$$\frac{7(x+4)}{(x-3)(x+4)} - \frac{2(x-3)}{(x-3)(x+4)} \geq 0$$

$$\frac{5x + 34}{(x-3)(x+4)} \geq 0$$

$x - 3$ $\quad$ – – – – – – – – $\;$ – – – – $\;$ 0++++
$x + 4$ $\quad$ – – – – – – – – $\;$ 0++++ +++++
$5x + 34$ $\quad$ – – – $\;$ 0++++++ +++++ +++++

solution set: $\left[-\dfrac{34}{5}, -4\right) \cup (3, \infty)$

45.

$$\frac{x}{x + 4} \leq \frac{1}{x + 1}$$

$$\frac{x}{x + 4} - \frac{1}{x + 1} \leq 0$$

$$\frac{x(x+1)}{(x+4)(x+1)} - \frac{1(x+4)}{(x+4)(x+1)} \leq 0$$

$$\frac{x^2 - 4}{(x+4)(x+1)} \leq 0$$

$$\frac{(x+2)(x-2)}{(x+4)(x+1)} \leq 0$$

$x - 2$ $\quad$ – – – – – – – – – – – $\;$ – – – –0+++
$x + 1$ $\quad$ – – – – – – – – – – $\;$ 0++++ ++++
$x + 2$ $\quad$ – – – – – – $\;$ 0++++ +++++ ++++
$x + 4$ $\quad$ – – $\;$ 0+++++++++ +++++ ++++

solution set: $(-4, -2] \cup (-1, 2]$

47.

$$\frac{x}{x+16} > \frac{1}{x+1}$$

$$\frac{x}{x+16} - \frac{1}{x+1} > 0$$

$$\frac{x(x+1)}{(x+16)(x+1)} - \frac{1(x+16)}{(x+16)(x+1)} > 0$$

$$\frac{x^2-16}{(x+16)(x+1)} > 0$$

$$\frac{(x+4)(x-4)}{(x+16)(x+1)} > 0$$

$x-4$ — — — — — — — — — — — — — —0+++
$x+1$ — — — — — — — — — — 0++++ ++++
$x+4$ — — — — — — — 0++++ +++++ ++++
$x+16$ — — — 0++++ +++++ +++++ ++++

⟵ —) — (—) — (⟶
      −16   −4   −1   4

solution set: $(-\infty, -16) \cup (-4, -1) \cup (4, \infty)$

49.

$$(x+2)^2 > 0$$
$$(x+2)(x+2) > 0$$

$x+2$ — — — — —0+++++++
$x+2$ — — — — —0+++++++

⟵ — — —) (— — —⟶
          −2

solution set: $(-\infty, -2) \cup (-2, \infty)$

51. $x^2 - 2x - 3 < 0$

Graph $y = x^2 - 2x - 3$
and find the x-coordinates
of points below the x-axis.

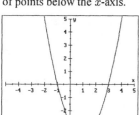

$(-1, 3)$

53. $\frac{x+3}{x-2} > 0$

Graph $y = (x+3)/(x-2)$
and find the x-coordinates
of points above the x-axis.

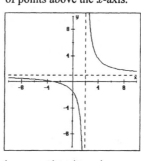

$(-\infty, -3) \cup (2, \infty)$

55. $y < x^2 + 1$

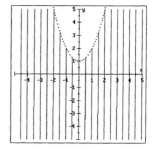

57. $y \le x^2 + 5x + 6$

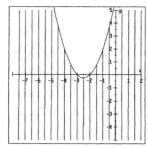

59. $y \ge (x-1)^2$

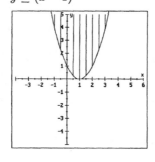

61. $-x^2 - y + 6 > -x$
$-x^2 + x + 6 > y$

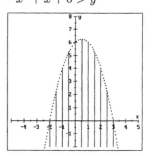

63. $y < |x + 4|$

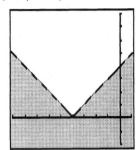

65. $y \leq -|x| + 2$

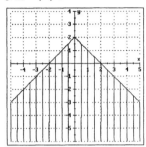

67. Answers may vary.

69. It will be positive if 4, 2 or 0 factors are negative.

Exercise 8.6 (page 550)

1. $\dfrac{3x^2 + x - 14}{4 - x^2} = \dfrac{(3x + 7)(x - 2)}{(2 + x)(2 - x)} = -\dfrac{3x + 7}{x + 2}$

3. $\dfrac{8 + 2x - x^2}{12 + x - 3x^2} \div \dfrac{3x^2 + 5x - 2}{3x - 1} = \dfrac{x^2 - 2x - 8}{3x^2 - x - 12} \cdot \dfrac{3x - 1}{3x^2 + 5x - 2}$

$= \dfrac{(x - 4)(x + 2)}{3x^2 - x - 12} \cdot \dfrac{3x - 1}{(3x - 1)(x + 2)} = \dfrac{x - 4}{3x^2 - x - 12}$

5. $f(x) + g(x)$ **7.** $f(x)g(x)$ **9.** domain   **11.** $f(x)$

13. $f + g = f(x) + g(x) = 3x + 4x = 7x$
domain $= (-\infty, \infty)$

15. $f \cdot g = f(x) \cdot g(x) = 3x \cdot 4x = 12x^2$
domain $= (-\infty, \infty)$

17. $g - f = g(x) - f(x) = 4x - 3x = x$
domain $= (-\infty, \infty)$

19. $g/f = \dfrac{g(x)}{f(x)} = \dfrac{4x}{3x} = \dfrac{4}{3}$ (for $x \neq 0$)
domain $= (-\infty, 0) \cup (0, \infty)$

21. $f + g = f(x) + g(x) = 2x + 1 + x - 3 = 3x - 2$; domain $= (-\infty, \infty)$

23. $f \cdot g = f(x) \cdot g(x) = (2x + 1)(x - 3) = 2x^2 - 5x - 3$; domain $= (-\infty, \infty)$

25. $g - f = g(x) - f(x) = (x - 3) - (2x + 1) = x - 3 - 2x - 1 = -x - 4$; domain $= (-\infty, \infty)$

27. $g/f = \dfrac{g(x)}{f(x)} = \dfrac{x - 3}{2x + 1}$; domain $= \left(-\infty, -\dfrac{1}{2}\right) \cup \left(-\dfrac{1}{2}, \infty\right)$

29. $f - g = f(x) - g(x) = (3x - 2) - (2x^2 + 1) = 3x - 2 - 2x^2 - 1 = -2x^2 + 3x - 3$
domain $= (-\infty, \infty)$

31. $f/g = \dfrac{f(x)}{g(x)} = \dfrac{3x-2}{2x^2+1}$; domain $= (-\infty, \infty)$

33. $f - g = f(x) - g(x) = (x^2 - 1) - (x^2 - 4) = x^2 - 1 - x^2 + 4 = 3$; domain $= (-\infty, \infty)$

35. $g/f = \dfrac{g(x)}{f(x)} = \dfrac{x^2-4}{x^2-1} = \dfrac{(x+2)(x-2)}{(x+1)(x-1)}$; domain $= (-\infty, -1) \cup (-1, 1) \cup (1, \infty)$

37. $(f \circ g)(2) = f(g(2)) = f(2^2 - 1) = f(3) = 2(3) + 1 = 7$

39. $(g \circ f)(-3) = g(f(-3)) = g(2(-3) + 1) = g(-5) = (-5)^2 - 1 = 24$

41. $(f \circ g)(0) = f(g(0)) = f(0^2 - 1) = f(-1) = 2(-1) + 1 = -1$

43. $(f \circ g)\left(\dfrac{1}{2}\right) = f\left(g\left(\dfrac{1}{2}\right)\right) = f\left(\left(\dfrac{1}{2}\right)^2 - 1\right) = f\left(-\dfrac{3}{4}\right) = 2\left(-\dfrac{3}{4}\right) + 1 = -\dfrac{3}{2} + 1 = -\dfrac{1}{2}$

45. $(f \circ g)(x) = f(g(x)) = f(x^2 - 1) = 2(x^2 - 1) + 1 = 2x^2 - 2 + 1 = 2x^2 - 1$

47. $(g \circ f)(2x) = g(f(2x)) = g(2(2x) + 1) = g(4x + 1) = (4x + 1)^2 - 1 = 16x^2 + 8x + 1 - 1$
$$= 16x^2 + 8x$$

49. $(f \circ g)(4) = f(g(4)) = f(4^2 + 4) = f(20) = 3(20) - 2 = 58$

51. $(g \circ f)(-3) = g(f(-3)) = g(3(-3) - 2) = g(-11) = (-11)^2 + (-11) = 110$

53. $(g \circ f)(0) = g(f(0)) = g(3(0) - 2) = g(-2) = (-2)^2 + (-2) = 2$

55. $(g \circ f)(x) = g(f(x)) = g(3x - 2) = (3x - 2)^2 + 3x - 2 = 9x^2 - 12x + 4 + 3x - 2$
$$= 9x^2 - 9x + 2$$

57. $\dfrac{f(x+h) - f(x)}{h} = \dfrac{2(x+h) + 3 - (2x+3)}{h} = \dfrac{2x + 2h + 3 - 2x - 3}{h} = \dfrac{2h}{h} = 2$

59. $\dfrac{f(x+h) - f(x)}{h} = \dfrac{(x+h)^2 - x^2}{h} = \dfrac{x^2 + 2xh + h^2 - x^2}{h} = \dfrac{2xh + h^2}{h} = 2x + h$

61. $\dfrac{f(x+h) - f(x)}{h} = \dfrac{2(x+h)^2 - 1 - (2x^2 - 1)}{h} = \dfrac{2x^2 + 4xh + 2h^2 - 1 - 2x^2 + 1}{h}$
$$= \dfrac{4xh + 2h^2}{h} = 4x + 2h$$

63. $\dfrac{f(x+h) - f(x)}{h} = \dfrac{(x+h)^2 + (x+h) - (x^2 + x)}{h} = \dfrac{x^2 + 2xh + h^2 + x + h - x^2 - x}{h}$
$$= \dfrac{2xh + h^2 + h}{h} = 2x + h + 1$$

65.
$$\frac{f(x+h) - f(x)}{h} = \frac{(x+h)^2 + 3(x+h) - 4 - (x^2 + 3x - 4)}{h}$$
$$= \frac{x^2 + 2xh + h^2 + 3x + 3h - 4 - x^2 - 3x + 4}{h}$$
$$= \frac{2xh + h^2 + 3h}{h} = 2x + h + 3$$

67.
$$\frac{f(x+h) - f(x)}{h} = \frac{2(x+h)^2 + 3(x+h) - 7 - (2x^2 + 3x - 7)}{h}$$
$$= \frac{2x^2 + 4xh + 2h^2 + 3x + 3h - 7 - 2x^2 - 3x + 7}{h}$$
$$= \frac{4xh + 2h^2 + 3h}{h} = 4x + 2h + 3$$

69.
$$\frac{f(x) - f(a)}{x - a} = \frac{(2x+3) - (2a+3)}{x - a} = \frac{2x + 3 - 2a - 3}{x - a} = \frac{2x - 2a}{x - a} = \frac{2(x-a)}{x - a} = 2$$

71.
$$\frac{f(x) - f(a)}{x - a} = \frac{x^2 - a^2}{x - a} = \frac{(x+a)(x-a)}{x - a} = x + a$$

73.
$$\frac{f(x) - f(a)}{x - a} = \frac{(2x^2 - 1) - (2a^2 - 1)}{x - a} = \frac{2x^2 - 1 - 2a^2 + 1}{x - a} = \frac{2x^2 - 2a^2}{x - a}$$
$$= \frac{2(x+a)(x-a)}{x - a}$$
$$= 2(x+a) = 2x + 2a$$

75.
$$\frac{f(x) - f(a)}{x - a} = \frac{(x^2 + x) - (a^2 + a)}{x - a} = \frac{x^2 + x - a^2 - a}{x - a} = \frac{x^2 - a^2 + x - a}{x - a}$$
$$= \frac{(x+a)(x-a) + 1(x-a)}{x - a}$$
$$= \frac{(x-a)(x+a+1)}{x - a} = x + a + 1$$

77.
$$\frac{f(x) - f(a)}{x - a} = \frac{(x^2 + 3x - 4) - (a^2 + 3a - 4)}{x - a} = \frac{x^2 + 3x - 4 - a^2 - 3a + 4}{x - a}$$
$$= \frac{x^2 - a^2 + 3x - 3a}{x - a}$$
$$= \frac{(x+a)(x-a) + 3(x-a)}{x - a}$$
$$= \frac{(x-a)(x+a+3)}{x - a} = x + a + 3$$

79. $\dfrac{f(x) - f(a)}{x - a} = \dfrac{(2x^2 + 3x - 7) - (2a^2 + 3a - 7)}{x - a} = \dfrac{2x^2 + 3x - 7 - 2a^2 - 3a + 7}{x - a}$

$$= \dfrac{2x^2 - 2a^2 + 3x - 3a}{x - a}$$

$$= \dfrac{2(x + a)(x - a) + 3(x - a)}{x - a}$$

$$= \dfrac{(x - a)(2(x + a) + 3)}{x - a}$$

$$= 2(x + a) + 3 = 2x + 2a + 3$$

81. $(f \circ g)(x) = f(g(x)) = f(2x - 5) = (2x - 5) + 1 = 2x - 4$
$(g \circ f)(x) = g(f(x)) = g(x + 1) = 2(x + 1) - 5 = 2x + 2 - 5 = 2x - 3$

83. $f(a) = a^2 + 2a - 3;\ f(h) = h^2 + 2h - 3 \Rightarrow f(a) + f(h) = a^2 + h^2 + 2a + 2h - 6$
$f(a + h) = (a + h)^2 + 2(a + h) - 3 = a^2 + 2ah + h^2 + 2a + 2h - 3$
$$= a^2 + h^2 + 2ah + 2a + 2h - 3$$

85. $\dfrac{f(x + h) - f(x)}{h} = \dfrac{(x + h)^3 - 1 - (x^3 - 1)}{h} = \dfrac{x^3 + 3x^2h + 3xh^2 + h^3 - 1 - x^3 + 1}{h}$

$$= \dfrac{3x^2h + 3xh^2 + h^3}{h}$$

$$= \dfrac{h(3x^2 + 3xh + h^2)}{h} = 3x^2 + 3xh + h^2$$

87. $F(t) = 2700 - 200t;\ C(F) = \frac{5}{9}(F - 32)$
$C(F(t)) = C(2700 - 200t) = \frac{5}{9}(2700 - 200t - 32) = \frac{5}{9}(2668 - 200t)$

89. Answers may vary.   **91.** It is associative. Examples will vary.

Exercise 8.7 (page 559)

1. $3 - \sqrt{-64} = 3 - \sqrt{64i^2} = 3 - 8i$

3. $(3 + 4i)(2 - 3i) = 6 - 9i + 8i - 12i^2 = 6 - i - 12(-1) = 6 - i + 12 = 18 - i$

5. $|6 - 8i| = \sqrt{6^2 + (-8)^2} = \sqrt{36 + 64} = \sqrt{100} = 10$

7. one-to-one   **9.** 2   **11.** x

13. Each input has a different output.   **15.** The inputs $x = 2$ and $x = -2$ have the same
one-to-one   output. not one-to-one

17. one-to-one   **19.** one-to-one   **21.** not one-to-one   **23.** one-to-one

25. inverse $= \{(2,3),(1,2),(0,1)\}$. Since each x-coordinate is paired with only one y-coordinate, the inverse relation **is a function**.

27. inverse $= \{(2,1),(3,2),(3,1),(5,1)\}$. Since $x = 3$ is paired with more than one y-coordinate, the inverse relation **is not a function**.

29. inverse $= \{(1,1),(4,2),(9,3),(16,4)\}$. Since each x-coordinate is paired with only one y-coordinate, the inverse relation **is a function**.

31.

$$f(x) = 3x + 1$$
$$y = 3x + 1$$
$$x = 3y + 1$$
$$x - 1 = 3y$$
$$\frac{x-1}{3} = y$$
$$\frac{1}{3}x - \frac{1}{3} = y$$
$$f^{-1}(x) = \tfrac{1}{3}x - \tfrac{1}{3}$$

$$\underline{f \circ f^{-1}}$$
$$f\left[f^{-1}(x)\right] = f\left(\tfrac{1}{3}x - \tfrac{1}{3}\right)$$
$$= 3 \cdot \left(\tfrac{1}{3}x - \tfrac{1}{3}\right) + 1$$
$$= x - 1 + 1$$
$$= x$$

$$\underline{f^{-1} \circ f}$$
$$f^{-1}[f(x)] = f^{-1}(3x + 1)$$
$$= \tfrac{1}{3}(3x + 1) - \tfrac{1}{3}$$
$$= x + \tfrac{1}{3} - \tfrac{1}{3}$$
$$= x$$

33.

$$x + 4 = 5y$$
$$y = f(x) = \frac{x+4}{5}$$
$$x = \frac{y+4}{5}$$
$$5x = y + 4$$
$$5x - 4 = y$$
$$f^{-1}(x) = 5x - 4$$

$$\underline{f \circ f^{-1}}$$
$$f\left[f^{-1}(x)\right] = f(5x - 4)$$
$$= \frac{(5x - 4) + 4}{5}$$
$$= \frac{5x}{5}$$
$$= x$$

$$\underline{f^{-1} \circ f}$$
$$f^{-1}[f(x)] = f^{-1}\left(\frac{x+4}{5}\right)$$
$$= 5 \cdot \left(\frac{x+4}{5}\right) - 4$$
$$= x + 4 - 4$$
$$= x$$

35.

$$f(x) = \frac{x-4}{5}$$
$$y = \frac{x-4}{5}$$
$$x = \frac{y-4}{5}$$
$$5x = y - 4$$
$$5x + 4 = y$$
$$f^{-1}(x) = 5x + 4$$

$$\underline{f \circ f^{-1}}$$
$$f\left[f^{-1}(x)\right] = f(5x + 4)$$
$$= \frac{(5x + 4) - 4}{5}$$
$$= \frac{5x}{5}$$
$$= x$$

$$\underline{f^{-1} \circ f}$$
$$f^{-1}[f(x)] = f^{-1}\left(\frac{x-4}{5}\right)$$
$$= 5 \cdot \left(\frac{x-4}{5}\right) + 4$$
$$= x - 4 + 4$$
$$= x$$

37.

$$4x - 5y = 20$$
$$5y = 4x - 20$$
$$y = f(x) = \tfrac{4}{5}x - 4$$
$$x = \tfrac{4}{5}y - 4$$
$$x + 4 = \tfrac{4}{5}y$$
$$\tfrac{5}{4}(x + 4) = y$$
$$f^{-1}(x) = \tfrac{5}{4}x + 5$$

$f \circ f^{-1}$

$$f\left[f^{-1}(x)\right] = f\left(\tfrac{5}{4}x + 5\right)$$
$$= \tfrac{4}{5}\left(\tfrac{5}{4}x + 5\right) - 4$$
$$= x + 4 - 4$$
$$= x$$

$f^{-1} \circ f$

$$f^{-1}[f(x)] = f^{-1}\left(\tfrac{4}{5}x - 4\right)$$
$$= \tfrac{5}{4}\left(\tfrac{4}{5}x - 4\right) + 5$$
$$= x - 5 + 5$$
$$= x$$

39.

$$y = 4x + 3$$
$$x = 4y + 3$$
$$x - 3 = 4y$$
$$\frac{x - 3}{4} = y$$

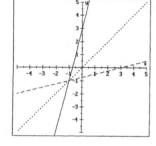

41.

$$x = \frac{y - 2}{3}$$
$$y = \frac{x - 2}{3}$$

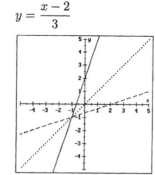

43.

$$3x - y = 5$$
$$3y - x = 5$$
$$3y = x + 5$$
$$y = \frac{x + 5}{3}$$

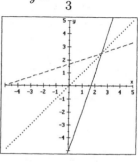

45.

$$3(x + y) = 2x + 4$$
$$3(y + x) = 2y + 4$$
$$3y + 3x = 2y + 4$$
$$y = 4 - 3x$$

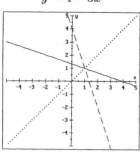

47.

$$y = x^2 + 4$$
$$x = y^2 + 4$$
$$x - 4 = y^2$$
$$\pm\sqrt{x - 4} = y$$

The relation **is not** a function.

49.

$$y = x^3$$
$$x = y^3$$
$$\sqrt[3]{x} = y$$

The relation **is** a function.

51.

$$y = |x|$$
$$x = |y|$$

The relation **is not** a function.

53.
$$y = 2x^3 - 3$$
$$x = 2y^3 - 3$$
$$x + 3 = 2y^3$$
$$\frac{x+3}{2} = y^3$$
$$\sqrt[3]{\frac{x+3}{2}} = \sqrt[3]{y^3}, \text{ or } y = f^{-1}(x) = \sqrt[3]{\frac{x+3}{2}}$$

55. $y = x^2 + 1$
inverse: $x = y^2 + 1$

57. $y = \sqrt{x}$
inverse: $x = \sqrt{y}$

59. **Answers may vary.**

61.
$$y = \frac{x+1}{x-1}$$
$$x = \frac{y+1}{y-1}$$
$$x(y-1) = y+1$$
$$xy - x = y + 1$$
$$xy - y = x + 1$$
$$y(x-1) = x + 1$$
$$y = \frac{x+1}{x-1}$$

Chapter 8 Summary (page 563)

1.
$$12x^2 + x - 6 = 0$$
$$(4x+3)(3x-2) = 0$$
$$4x+3 = 0 \quad \text{or} \quad 3x-2 = 0$$
$$x = -\tfrac{3}{4} \qquad\qquad x = \tfrac{2}{3}$$

2.
$$6x^2 + 17x + 5 = 0$$
$$(2x+5)(3x+1) = 0$$
$$2x+5 = 0 \quad \text{or} \quad 3x+1 = 0$$
$$x = -\tfrac{5}{2} \qquad\qquad x = -\tfrac{1}{3}$$

3.
$$15x^2 + 2x - 8 = 0$$
$$(3x-2)(5x+4) = 0$$
$$3x-2 = 0 \quad \text{or} \quad 5x+4 = 0$$
$$x = \tfrac{2}{3} \qquad\qquad x = -\tfrac{4}{5}$$

4.
$$(x+2)^2 = 36$$
$$x+2 = \pm\sqrt{36}$$
$$x+2 = \pm 6$$
$$x = -2 \pm 6$$
$$x = 4 \quad \text{or} \quad x = -8$$

5.
$$x^2 + 6x + 8 = 0$$
$$x^2 + 6x = -8$$
$$x^2 + 6x + 9 = -8 + 9$$
$$(x+3)^2 = 1$$
$$x + 3 = \pm 1$$
$$x = -3 \pm 1$$
$$x = -2 \quad \text{or} \quad x = -4$$

6.
$$2x^2 - 9x + 7 = 0$$
$$x^2 - \frac{9}{2}x + \frac{7}{2} = 0$$
$$x^2 - \frac{9}{2}x = -\frac{7}{2}$$
$$x^2 - \frac{9}{2}x + \frac{81}{16} = -\frac{56}{16} + \frac{81}{16}$$
$$\left(x - \frac{9}{4}\right)^2 = \frac{25}{16}$$
$$x - \frac{9}{4} = \pm \frac{5}{4}$$
$$x = \frac{9}{4} \pm \frac{5}{4}$$
$$x = \frac{7}{2} \quad \text{or} \quad x = 1$$

7.
$$2x^2 - x - 5 = 0$$
$$x^2 - \frac{1}{2}x - \frac{5}{2} = 0$$
$$x^2 - \frac{1}{2}x = \frac{5}{2}$$
$$x^2 - \frac{1}{2}x + \frac{1}{16} = \frac{5}{2} + \frac{1}{16}$$
$$\left(x - \frac{1}{4}\right)^2 = \frac{41}{16}$$
$$x - \frac{1}{4} = \pm \frac{\sqrt{41}}{4}$$
$$x = \frac{1}{4} \pm \frac{\sqrt{41}}{4}$$

8.
$$x^2 - 8x - 9 = 0$$
$$a = 1, b = -8, c = -9$$
$$x = \frac{-b \pm \sqrt{b^2 - 4ac}}{2a}$$
$$= \frac{-(-8) \pm \sqrt{(-8)^2 - 4(1)(-9)}}{2(1)}$$
$$= \frac{8 \pm \sqrt{64 + 36}}{2}$$
$$= \frac{8 \pm \sqrt{100}}{2} = \frac{8 \pm 10}{2}$$
$$x = \frac{18}{2} = 9 \text{ or } x = \frac{-2}{2} = -1$$

9.
$$x^2 - 10x = 0$$
$$a = 1, b = -10, c = 0$$
$$x = \frac{-b \pm \sqrt{b^2 - 4ac}}{2a}$$
$$= \frac{-(-10) \pm \sqrt{(-10)^2 - 4(1)(0)}}{2(1)}$$
$$= \frac{10 \pm \sqrt{100 + 0}}{2}$$
$$= \frac{10 \pm \sqrt{100}}{2} = \frac{10 \pm 10}{2}$$
$$x = \frac{20}{2} = 10 \text{ or } x = \frac{0}{2} = 0$$

10.
$$2x^2 + 13x - 7 = 0$$
$$a = 2, b = 13, c = -7$$
$$x = \frac{-b \pm \sqrt{b^2 - 4ac}}{2a}$$
$$= \frac{-(13) \pm \sqrt{13^2 - 4(2)(-7)}}{2(2)}$$
$$= \frac{-13 \pm \sqrt{169 + 56}}{4}$$
$$= \frac{-13 \pm \sqrt{225}}{4} = \frac{-13 \pm 15}{4}$$
$$x = \frac{2}{4} = \frac{1}{2} \text{ or } x = \frac{-28}{4} = -7$$

11. $3x^2 + 20x - 7 = 0$

$a = 3, b = 20, c = -7$

$$x = \frac{-b \pm \sqrt{b^2 - 4ac}}{2a}$$

$$= \frac{-20 \pm \sqrt{(20)^2 - 4(3)(-7)}}{2(3)}$$

$$= \frac{-20 \pm \sqrt{400 + 84}}{6}$$

$$= \frac{-20 \pm \sqrt{484}}{6} = \frac{-20 \pm 22}{6}$$

$$x = \frac{2}{6} = \frac{1}{3} \text{ or } x = \frac{-42}{6} = -7$$

12. $2x^2 - x - 2 = 0$

$a = 2, b = -1, c = -2$

$$x = \frac{-b \pm \sqrt{b^2 - 4ac}}{2a}$$

$$= \frac{-(-1) \pm \sqrt{(-1)^2 - 4(2)(-2)}}{2(2)}$$

$$= \frac{1 \pm \sqrt{1 + 16}}{4}$$

$$= \frac{1 \pm \sqrt{17}}{4} = \frac{1}{4} \pm \frac{\sqrt{17}}{4}$$

13. $x^2 + x + 2 = 0$

$a = 1, b = 1, c = 2$

$$x = \frac{-b \pm \sqrt{b^2 - 4ac}}{2a}$$

$$= \frac{-1 \pm \sqrt{1^2 - 4(1)(2)}}{2(1)}$$

$$= \frac{-1 \pm \sqrt{1 - 8}}{2}$$

$$= \frac{-1 \pm \sqrt{-7}}{2} = -\frac{1}{2} \pm \frac{\sqrt{7}}{2}i$$

14. Let w represent the original width.
Then $w + 2$ represents the original length.
The new dimensions are then $2w$ and
$2(w + 2) = 2w + 4$.
Old Area $+ 72 = $ New Area
$w(w + 2) + 72 = 2w(2w + 4)$
$w^2 + 2w + 72 = 4w^2 + 8w$
$0 = 3w^2 + 6w - 72$
$0 = 3(w + 6)(w - 4)$
$w = -6$ or $w = 4$
Since the width is positive, the
dimensions are 4 cm by 6 cm.

15. Let w represent the original width.
Then $w + 1$ represents the original length.
The new dimensions are then $2w$ and
$3(w + 1) = 3w + 3$.
Old Area $+ 30 = $ New Area
$w(w + 1) + 30 = 2w(3w + 3)$
$w^2 + w + 30 = 6w^2 + 6w$
$0 = 5w^2 + 5w - 30$
$0 = 5(w + 3)(w - 2)$
$w = -3$ or $w = 2$
Since the width is positive, the
dimensions are 2 ft by 3 ft.

16. When the rocket hits the ground, $h = 0$:
$h = 112t - 16t^2$
$0 = 112t - 16t^2$
$0 = 16t(7 - t)$
$t = 0$ or $t = 7$
It hits the ground after 7 seconds.

17. The maximum height occurs at the vertex:
$$h = 112t - 16t^2$$
$$h = -16t^2 + 112t$$
$$h = -16(t^2 - 7t)$$
$$h = -16\left(t^2 - 7t + \frac{49}{4}\right) + 196$$

$$h = -16\left(t^2 - 7t + \frac{49}{4}\right) + 196$$
$$h = -16\left(t - \frac{7}{2}\right)^2 + 196$$
Vertex: $\left(\frac{7}{2}, 196\right) \Rightarrow$ max. height $= 196$ ft

18. $3x^2 + 4x - 3 = 0$
$a = 3, b = 4, c = -3$
$b^2 - 4ac = 4^2 - 4(3)(-3)$
$\qquad = 16 + 36 = 52$
irrational unequal solutions

19. $4x^2 - 5x + 7 = 0$
$a = 4, b = -5, c = 7$
$b^2 - 4ac = (-5)^2 - 4(4)(7)$
$\qquad = 25 - 112 = -87$
complex conjugate solutions

20. $(k-8)x^2 + (k+16)x = -49$
$(k-8)x^2 + (k+16)x + 49 = 0$
$a = k - 8, b = k + 16, c = 49$
Set the discriminant equal to 0:
$$b^2 - 4ac = 0$$
$$(k+16)^2 - 4(k-8)(49) = 0$$
$$k^2 + 32k + 256 - 196k + 1568 = 0$$
$$k^2 - 164k + 1824 = 0$$
$$(k-12)(k-152) = 0$$
$k - 12 = 0 \quad$ **or** $\quad k - 152 = 0$
$\qquad k = 12 \qquad\qquad k = 152$

21. $3x^2 + 4x = k + 1$
$3x^2 + 4x - k - 1 = 0$
$a = 3, b = 4, c = -k - 1$
Set the discriminant ≥ 0:
$$b^2 - 4ac \geq 0$$
$$4^2 - 4(3)(-k-1) \geq 0$$
$$16 + 12k + 12 \geq 0$$
$$12k \geq -28$$
$$k \geq -\frac{28}{12}$$
$$k \geq -\frac{7}{3}$$

22.
$$x - 13x^{1/2} + 12 = 0$$
$$\left(x^{1/2} - 12\right)\left(x^{1/2} - 1\right) = 0$$
$x^{1/2} - 12 = 0 \quad$ **or** $\quad x^{1/2} - 1 = 0$
$\qquad x^{1/2} = 12 \qquad\qquad x^{1/2} = 1$
$\qquad \left(x^{1/2}\right)^2 = (12)^2 \qquad \left(x^{1/2}\right)^2 = 1^2$
$\qquad x = 144 \qquad\qquad\quad x = 1$
$\qquad$ Solution. $\qquad\qquad$ Solution.

23.
$$a^{2/3} + a^{1/3} - 6 = 0$$
$$\left(a^{1/3} - 2\right)\left(a^{1/3} + 3\right) = 0$$
$a^{1/3} - 2 = 0 \quad$ **or** $\quad a^{1/3} + 3 = 0$
$\qquad a^{1/3} = 2 \qquad\qquad a^{1/3} = -3$
$\qquad \left(a^{1/3}\right)^3 = (2)^3 \qquad \left(a^{1/3}\right)^3 = (-3)^3$
$\qquad a = 8 \qquad\qquad\quad a = -27$
$\qquad$ Solution. $\qquad\qquad$ Solution.

24.
$$\frac{1}{x+1} - \frac{1}{x} = -\frac{1}{x+1}$$
$$\left(\frac{1}{x+1} - \frac{1}{x}\right)(x)(x+1) = -\frac{1}{x+1}(x)(x+1)$$
$$1(x) - 1(x+1) = -x$$
$$x - x - 1 = -x$$
$$-1 = -x$$
$$1 = x$$

25.
$$\frac{6}{x+2} + \frac{6}{x+1} = 5$$

$$\left(\frac{6}{x+2} + \frac{6}{x+1}\right)(x+2)(x+1) = 5(x+2)(x+1)$$

$$6(x+1) + 6(x+2) = 5(x^2+3x+2)$$

$$6x + 6 + 6x + 12 = 5x^2 + 15x + 10$$

$$0 = 5x^2 + 3x - 8$$

$$0 = (5x+8)(x-1) \qquad 5x+8=0 \quad \textbf{or} \quad x-1=0$$
$$x = -\tfrac{8}{5} \qquad\qquad x = 1$$

26. $3x^2 - 14x + 3 = 0$

$$\text{sum} = -\frac{b}{a} = -\frac{-14}{3} = \frac{14}{3}$$

27. $3x^2 - 14x + 3 = 0$

$$\text{product} = \frac{c}{a} = \frac{3}{3} = 1$$

28. $y = 2x^2 - 3$

$y = 2(x-0)^2 - 3$

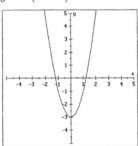

vertex: $(0, -3)$

29. $y = -2x^2 - 1$

$y = -2(x-0)^2 - 1$

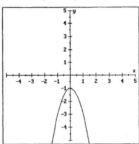

vertex: $(0, -1)$

30. $y = -4(x-2)^2 + 1$

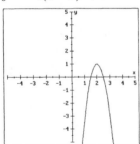

vertex: $(2, 1)$

31. $y = 5x^2 + 10x - 1$

$$= 5(x^2 + 2x) - 1$$
$$= 5(x^2 + 2x + 1) - 1 - 5$$
$$= 5(x+1)^2 - 6$$

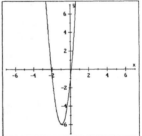

vertex: $(-1, -6)$

32. $x^2 + 2x - 35 > 0$

$(x + 7)(x - 5) > 0$

$x - 5 \quad -\,-\,-\,-\,-\,-\,-\,-\,0\,+\,+\,+\,+$

$x + 7 \quad -\,-\,-\ 0\,+\,+\,+\,+\,+\,+\ +\,+\,+\,+$

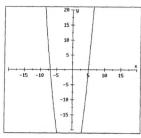

$\qquad\quad -7 \qquad\quad 5$

solution set: $(-\infty, -7) \cup (5, \infty)$

33. $x^2 + 7x - 18 < 0$

$(x - 2)(x + 9) < 0$

$x - 2 \quad -\,-\,-\,-\,-\,-\,-\,-\,-\ 0\,+\,+\,+\,+$

$x + 9 \quad -\,-\,-\ 0\,+\,+\,+\,+\,+\,+\,+\,+\,+\,+$

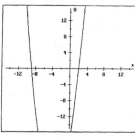

$\qquad\quad -9 \qquad 2$

solution set: $(-9, 2)$

34. $\dfrac{3}{x} \leq 5$

$\dfrac{3}{x} - 5 \leq 0$

$\dfrac{3}{x} - \dfrac{5x}{x} \leq 0$

$\dfrac{3 - 5x}{x} \leq 0$

$3 - 5x \quad +\,+\,+\,+\,+\,+\,+\,+\,+\ 0\,-\,-\,-$

$x \qquad\quad\ -\,-\,-\ 0\,+\,+\,+\,+\,+\,+\,+\,+\,+\,+$

$\qquad\qquad\qquad 0 \qquad\quad \dfrac{3}{5}$

solution set: $(-\infty, 0) \cup \left[\dfrac{3}{5}, \infty\right)$

35. $\dfrac{2x^2 - x - 28}{x - 1} > 0$

$\dfrac{(2x + 7)(x - 4)}{x - 1} > 0$

$x - 4 \quad -\,-\,-\,-\,-\,-\,-\,-\ -\,-\,-\,-\ 0\,+\,+\,+\,+$

$x - 1 \quad -\,-\,-\,-\,-\,-\,-\,-\ 0\,+\,+\,+\,+\,+\,+\,+\,+\,+$

$2x + 7 \quad -\,-\,-\ 0\,+\,+\,+\,+\,+\,+\,+\,+\,+\,+\,+\,+\,+\,+\,+$

$\qquad\quad -\dfrac{7}{2} \qquad\quad 1 \qquad\quad 4$

solution set: $\left(-\dfrac{7}{2}, 1\right) \cup (4, \infty)$

36. $x^2 + 2x - 35 > 0$

Graph $y = x^2 + 2x - 35$ and find the x-coordinates of points above the x-axis.

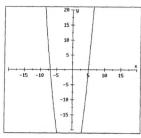

$(-\infty, -7) \cup (5, \infty)$

37. $x^2 + 7x - 18 < 0$

Graph $y = x^2 + 7x - 18$ and find the x-coordinates of points below the x-axis.

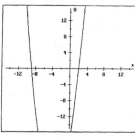

$(-9, 2)$

38. $\frac{3}{x} \le 5 \Rightarrow \frac{3}{x} - 5 \le 0$
Graph $y = (3/x) - 5$
and find the x-coordinates
of points below or on the x-axis.

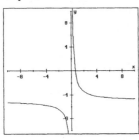

$$(-\infty, 0) \cup \left[\frac{3}{5}, \infty\right)$$

39. $\frac{2x^2 - x - 28}{x - 1} > 0$
Graph $y = (2x^2 - x - 28)/(x - 1)$
and find the x-coordinates
of points above the x-axis.

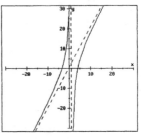

$$\left(-\frac{7}{2}, 1\right) \cup (4, \infty)$$

40. $y < \frac{1}{2}x^2 - 1$

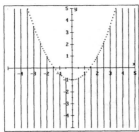

41. $y \ge -|x|$

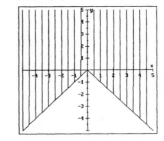

42. $f + g = f(x) + g(x) = 2x + x + 1$
$ = 3x + 1$

43. $f - g = f(x) - g(x) = 2x - (x + 1)$
$ = x - 1$

44. $f \cdot g = f(x)g(x) = 2x(x + 1) = 2x^2 + 2x$

45. $f/g = \frac{f(x)}{g(x)} = \frac{2x}{x+1}$

46. $(f \circ g)(2) = f(g(2)) = f(2 + 1)$
$ = f(3) = 2(3) = 6$

47. $(g \circ f)(-1) = g(f(-1)) = g(2(-1))$
$ = g(-2)$
$ = -2 + 1 = -1$

48. $(f \circ g)(x) = f(g(x)) = f(x + 1) = 2(x + 1)$

49. $(g \circ f)(x) = g(f(x)) = g(2x) = 2x + 1$

50. $f(x) = 2(x - 3)$

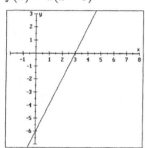

one-to-one

51. $f(x) = x(2x - 3)$

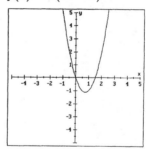

not one-to-one

52. $f(x) = -3(x - 2)^2 + 5$

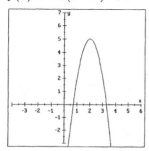

not one-to-one

53. $f(x) = |x|$

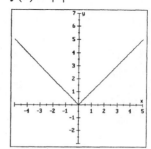

not one-to-one

54.
$$y = 6x - 3$$
$$x = 6y - 3$$
$$x + 3 = 6y$$
$$\tfrac{x+3}{6} = y, \text{ or } y = f^{-1}(x) = \tfrac{x+3}{6}$$

55.
$$y = 4x + 5$$
$$x = 4y + 5$$
$$x - 5 = 4y$$
$$\tfrac{x-5}{4} = y, \text{ or } y = f^{-1}(x) = \tfrac{x-5}{4}$$

56.
$$y = 2x^2 - 1$$
$$x = 2y^2 - 1$$
$$x + 1 = 2y^2$$
$$\tfrac{x+1}{2} = y^2$$
$$\sqrt{\tfrac{x+1}{2}} = y, \text{ or } y = f^{-1}(x) = \sqrt{\tfrac{x+1}{2}}$$

57.
$$y = |x|$$
$$x = |y|$$

Chapter 8 Test (page 567)

1.
$$x^2 + 3x - 18 = 0$$
$$(x + 6)(x - 3) = 0$$
$$x + 6 = 0 \quad \text{or} \quad x - 3 = 0$$
$$x = -6 \qquad\qquad x = 3$$

2.
$$x(6x + 19) = -15$$
$$6x^2 + 19x + 15 = 0$$
$$(2x + 3)(3x + 5) = 0$$
$$2x + 3 = 0 \quad \text{or} \quad 3x + 5 = 0$$
$$x = -\tfrac{3}{2} \qquad\qquad x = -\tfrac{5}{3}$$

3. $\left(\frac{1}{2} \cdot 24\right)^2 = 12^2 = 144$

4. $\left(\frac{1}{2} \cdot (-50)\right)^2 = (-25)^2 = 625$

5.
$$x^2 + 4x + 1 = 0$$
$$x^2 + 4x = -1$$
$$x^2 + 4x + 4 = -1 + 4$$
$$(x+2)^2 = 3$$
$$x + 2 = \pm \sqrt{3}$$
$$x = -2 \pm \sqrt{3}$$

6.
$$x^2 - 5x - 3 = 0$$
$$x^2 - 5x = 3$$
$$x^2 - 5x + \frac{25}{4} = 3 + \frac{25}{4}$$
$$\left(x - \frac{5}{2}\right)^2 = \frac{37}{4}$$
$$x - \frac{5}{2} = \pm \sqrt{\frac{37}{4}}$$
$$x = \frac{5}{2} \pm \frac{\sqrt{37}}{2}$$

7.
$$2x^2 + 5x + 1 = 0$$
$$a = 2, b = 5, c = 1$$
$$x = \frac{-b \pm \sqrt{b^2 - 4ac}}{2a}$$
$$= \frac{-5 \pm \sqrt{5^2 - 4(2)(1)}}{2(2)}$$
$$= \frac{-5 \pm \sqrt{25 - 8}}{4}$$
$$= \frac{-5 \pm \sqrt{17}}{4} = -\frac{5}{4} \pm \frac{\sqrt{17}}{4}$$

8.
$$x^2 - x + 3 = 0$$
$$a = 1, b = -1, c = 3$$
$$x = \frac{-b \pm \sqrt{b^2 - 4ac}}{2a}$$
$$= \frac{-(-1) \pm \sqrt{(-1)^2 - 4(1)(3)}}{2(1)}$$
$$= \frac{1 \pm \sqrt{1 - 12}}{2}$$
$$= \frac{1 \pm \sqrt{-11}}{2} = \frac{1}{2} \pm \frac{\sqrt{11}}{2}i$$

9.
$$3x^2 + 5x + 17 = 0$$
$$a = 3, b = 5, c = 17$$
$$b^2 - 4ac = 5^2 - 4(3)(17)$$
$$= 25 - 208 = -183$$
nonreal solutions

10.
$$4x^2 - 2kx + k - 1 = 0$$
$$a = 4, b = -2k, c = k - 1$$
Set the discriminant equal to 0:
$$b^2 - 4ac = 0$$
$$(-2k)^2 - 4(4)(k-1) = 0$$
$$4k^2 - 16k + 16 = 0$$
$$4(k-2)(k-2) = 0$$
$$k - 2 = 0 \quad \text{or} \quad k - 2 = 0$$
$$k = 2 \qquad\qquad k = 2$$

11. Let x = the length of the shorter leg.
Then $x + 14$ = the other length.
$$x^2 + (x+14)^2 = 26^2$$
$$x^2 + x^2 + 28x + 196 = 676$$
$$2x^2 + 28x - 480 = 0$$
$$2(x + 24)(x - 10) = 0$$
$$x = -24 \quad \text{or} \quad x = 10$$
The shorter leg is 10 inches long.

12.
$$2y - 3y^{1/2} + 1 = 0$$
$$\left(2y^{1/2} - 1\right)\left(y^{1/2} - 1\right) = 0$$
$$2y^{1/2} - 1 = 0 \quad \text{or} \quad y^{1/2} - 1 = 0$$
$$2y^{1/2} = 1 \qquad\qquad y^{1/2} = 1$$
$$\left(y^{1/2}\right)^2 = \left(\tfrac{1}{2}\right)^2 \qquad \left(y^{1/2}\right)^2 = 1^2$$
$$y = \tfrac{1}{4} \qquad\qquad y = 1$$
Solution $\qquad\qquad$ Solution

13. $y = \dfrac{1}{2}x^2 - 4 = \dfrac{1}{2}(x-0)^2 - 4$

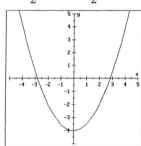

vertex: $(0, -4)$

14. $y \le -x^2 + 3$

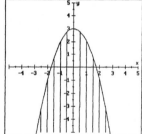

15. $\quad x^2 - 2x - 8 > 0$

$\quad (x+2)(x-4) > 0$

$\quad x-4 \quad --------0\ ++++$

$\quad x+2 \quad ---\ 0++++++\ ++++$

$\qquad \overset{\longleftarrow}{}\ )\ \overline{}\ (\overset{\longrightarrow}{}$

$\qquad\qquad -2 \qquad 4$

solution set: $(-\infty, -2) \cup (4, \infty)$

16. $\quad \dfrac{x-2}{x+3} \le 0$

$\quad x-2 \quad --------0\ ++++$

$\quad x+3 \quad ---\ 0++++++\ ++++$

$\qquad \overset{\longleftarrow}{}\ (\overline{}]\overset{\longrightarrow}{}$

$\qquad\qquad -3 \qquad 2$

solution set: $(-3, 2]$

17. $\quad g + f = g(x) + f(x) = x - 1 + 4x$

$\qquad\qquad\qquad = 5x - 1$

18. $\quad f - g = f(x) - g(x) = 4x - (x-1)$

$\qquad\qquad\qquad = 3x + 1$

19. $\quad g \cdot f = g(x)f(x) = (x-1)4x = 4x^2 - 4x$

20. $\quad g/f = \dfrac{g(x)}{f(x)} = \dfrac{x-1}{4x}$

21. $\quad (g \circ f)(1) = g(f(1)) = g(4(1))$

$\qquad\qquad\qquad = g(4) = 4 - 1 = 3$

22. $\quad (f \circ g)(0) = f(g(0)) = f(0-1)$

$\qquad\qquad\qquad = f(-1) = 4(-1) = -4$

23. $\quad (f \circ g)(-1) = f(g(-1)) = f(-1-1) = f(-2) = 4(-2) = -8$

24. $\quad (g \circ f)(-2) = g(f(-2)) = g(4(-2)) = g(-8) = -8 - 1 = -9$

25. $\quad (f \circ g)(x) = f(g(x)) = f(x-1) = 4(x-1)$

26. $\quad (g \circ f)(x) = g(f(x)) = g(4x) = 4x - 1$

27. $\quad 3x + 2y = 12$

$\qquad 3y + 2x = 12$

$\qquad\quad 3y = -2x + 12$

$\qquad\quad\ y = \dfrac{-2x + 12}{3}$

28. $\qquad\qquad y = 3x^2 + 4$

$\qquad\qquad\quad x = 3y^2 + 4$

$\qquad\qquad x - 4 = 3y^2$

$\qquad\qquad \dfrac{x-4}{3} = y^2$

$\qquad\qquad -\sqrt{\dfrac{x-4}{3}} = y$

Cumulative Review Exercises (page 568)

1. $y = f(x) = 2x^2 - 3$
domain $= (-\infty, \infty)$
range $= [-3, \infty)$

2. $y = f(x) = -|x - 4|$
domain $= (-\infty, \infty)$
range $= (-\infty, 0]$

3. $y - y_1 = m(x - x_1)$
$y + 4 = 3(x + 2)$
$y = 3x + 2$

4. $2x + 3y = 6$
$3y = -2x + 6$
$y = -\dfrac{2}{3}x + 2$
$y - y_1 = m(x - x_1)$
$y + 2 = -\dfrac{2}{3}(x - 0)$
$y = -\dfrac{2}{3}x - 2$

5. $(2a^2 + 4a - 7) - 2(3a^2 - 4a) = 2a^2 + 4a - 7 - 6a^2 + 8a = -4a^2 + 12a - 7$

6. $(3x + 2)(2x - 3) = 6x^2 - 9x + 4x - 6 = 6x^2 - 5x - 6$

7. $x^4 - 16y^4 = (x^2 + 4y^2)(x^2 - 4y^2) = (x^2 + 4y^2)(x + 2y)(x - 2y)$

8. $15x^2 - 2x - 8 = (5x - 4)(3x + 2)$

9. $x^2 - 5x - 6 = 0$
$(x - 6)(x + 1) = 0$
$x - 6 = 0 \quad \textbf{or} \quad x + 1 = 0$
$x = 6 \qquad\qquad x = -1$

10. $6a^3 - 2a = a^2$
$6a^3 - a^2 - 2a = 0$
$a(6a^2 - a - 2) = 0$
$a(3a - 2)(2a + 1) = 0$
$a = 0 \quad \textbf{or} \quad 3a - 2 = 0 \quad \textbf{or} \quad 2a + 1 = 0$
$a = \frac{2}{3} \qquad\qquad a = -\frac{1}{2}$

11. $\sqrt{25x^4} = 5x^2$

12. $\sqrt{48t^3} = \sqrt{16t^2}\sqrt{3t} = 4t\sqrt{3t}$

13. $\sqrt[3]{-27x^3} = -3x$

14. $\sqrt[3]{\dfrac{128x^4}{2x}} = \sqrt[3]{64x^3} = 4x$

15. $8^{-1/3} = \dfrac{1}{8^{1/3}} = \dfrac{1}{2}$

16. $64^{2/3} = \left(64^{1/3}\right)^2 = 4^2 = 16$

17. $\dfrac{y^{2/3}y^{5/3}}{y^{1/3}} = \dfrac{y^{7/3}}{y^{1/3}} = y^{6/3} = y^2$

18. $\dfrac{x^{5/3}x^{1/2}}{x^{3/4}} = \dfrac{x^{13/6}}{x^{3/4}} = x^{17/12}$

CUMULATIVE REVIEW EXERCISES

19. $f(x) = \sqrt{x-2}$; Shift $y = \sqrt{x}$ right 2.

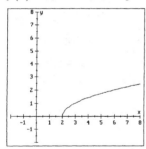

$D = [2, \infty), R = [0, \infty)$

20. $f(x) = -\sqrt{x+2}$; Reflect $y = \sqrt{x}$ about the x-axis and shift left 2.

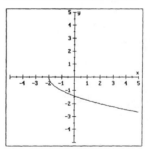

$D = [-2, \infty), R = (-\infty, 0]$

21. $\left(x^{2/3} - x^{1/3}\right)\left(x^{2/3} + x^{1/3}\right) = x^{4/3} + x^{3/3} - x^{3/3} - x^{2/3} = x^{4/3} - x^{2/3}$

22. $\left(x^{-1/2} + x^{1/2}\right)^2 = \left(x^{-1/2} + x^{1/2}\right)\left(x^{-1/2} + x^{1/2}\right) = x^{-2/2} + x^0 + x^0 + x^{2/2} = x + 2 + \frac{1}{x}$

23. $\sqrt{50} - \sqrt{8} + \sqrt{32} = \sqrt{25}\sqrt{2} - \sqrt{4}\sqrt{2} + \sqrt{16}\sqrt{2} = 5\sqrt{2} - 2\sqrt{2} + 4\sqrt{2} = 7\sqrt{2}$

24. $-3\sqrt[4]{32} - 2\sqrt[4]{162} + 5\sqrt[4]{48} = -3\sqrt[4]{16}\sqrt[4]{2} - 2\sqrt[4]{81}\sqrt[4]{2} + 5\sqrt[4]{16}\sqrt[4]{3}$
$$= -3(2)\sqrt[4]{2} - 2(3)\sqrt[4]{2} + 5(2)\sqrt[4]{3}$$
$$= -6\sqrt[4]{2} - 6\sqrt[4]{2} + 10\sqrt[4]{3} = -12\sqrt[4]{2} + 10\sqrt[4]{3}$$

25. $3\sqrt{2}(2\sqrt{3} - 4\sqrt{12}) = 6\sqrt{6} - 12\sqrt{24} = 6\sqrt{6} - 12\sqrt{4}\sqrt{6} = 6\sqrt{6} - 24\sqrt{6} = -18\sqrt{6}$

26. $\dfrac{5}{\sqrt[3]{x}} = \dfrac{5}{\sqrt[3]{x}} \cdot \dfrac{\sqrt[3]{x^2}}{\sqrt[3]{x^2}} = \dfrac{5\sqrt[3]{x^2}}{\sqrt[3]{x^3}} = \dfrac{5\sqrt[3]{x^2}}{x}$

27. $\dfrac{\sqrt{x}+2}{\sqrt{x}-1} = \dfrac{\sqrt{x}+2}{\sqrt{x}-1} \cdot \dfrac{\sqrt{x}+1}{\sqrt{x}+1} = \dfrac{x + 3\sqrt{x} + 2}{x-1}$

28. $\sqrt[6]{x^3 y^3} = (x^3 y^3)^{1/6} = x^{3/6} y^{3/6} = x^{1/2} y^{1/2} = \sqrt{xy}$

CUMULATIVE REVIEW EXERCISES

29.
$$5\sqrt{x+2} = x+8$$
$$\left(5\sqrt{x+2}\right)^2 = (x+8)^2$$
$$25(x+2) = x^2 + 16x + 64$$
$$25x + 50 = x^2 + 16x + 64$$
$$0 = x^2 - 9x + 14$$
$$0 = (x-7)(x-2)$$
$$x = 7 \quad \text{or} \quad x = 2 \quad \text{(Both check.)}$$

30.
$$\sqrt{x} + \sqrt{x+2} = 2$$
$$\sqrt{x} = 2 - \sqrt{x+2}$$
$$\left(\sqrt{x}\right)^2 = \left(2 - \sqrt{x+2}\right)^2$$
$$x = 4 - 4\sqrt{x+2} + x + 2$$
$$4\sqrt{x+2} = 6$$
$$\left(4\sqrt{x+2}\right)^2 = 6^2$$
$$16(x+2) = 36$$
$$16x + 32 = 36$$
$$16x = 4$$
$$x = \frac{4}{16} = \frac{1}{4}$$

31. hypotenuse $= 3\sqrt{2}$ in.

32. hypotenuse $= 2 \cdot \dfrac{3}{\sqrt{3}} = \dfrac{6\sqrt{3}}{3} = 2\sqrt{3}$ in.

33. $d = \sqrt{(-2-4)^2 + (6-14)^2} = \sqrt{(-6)^2 + (-8)^2} = \sqrt{36 + 64} = \sqrt{100} = 10$

34. $\left(\frac{1}{2} \cdot 6\right)^2 = 3^2 = 9$

35.
$$2x^2 + x - 3 = 0$$
$$x^2 + \frac{1}{2}x - \frac{3}{2} = 0$$
$$x^2 + \frac{1}{2}x = \frac{3}{2}$$
$$x^2 + \frac{1}{2}x + \frac{1}{16} = \frac{3}{2} + \frac{1}{16}$$
$$\left(x + \frac{1}{4}\right)^2 = \frac{25}{16}$$
$$x + \frac{1}{4} = \pm\frac{5}{4}$$
$$x = -\frac{1}{4} \pm \frac{5}{4}$$
$$x = \frac{4}{4} = 1 \quad \text{or} \quad x = -\frac{6}{4} = -\frac{3}{2}$$

36.
$$3x^2 + 4x - 1 = 0$$
$$a = 3, b = 4, c = -1$$
$$x = \frac{-b \pm \sqrt{b^2 - 4ac}}{2a}$$
$$= \frac{-4 \pm \sqrt{4^2 - 4(3)(-1)}}{2(3)}$$
$$= \frac{-4 \pm \sqrt{16 + 12}}{6}$$
$$= \frac{-4 \pm \sqrt{28}}{6}$$
$$= \frac{-4 \pm 2\sqrt{7}}{6} = -\frac{2}{3} \pm \frac{\sqrt{7}}{3}$$

CUMULATIVE REVIEW EXERCISES

37. $y = \dfrac{1}{2}x^2 + 5 = \dfrac{1}{2}(x-0)^2 + 5$

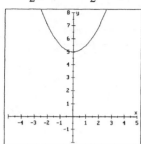

vertex: $(0, 5)$

38. $y \le -x^2 + 3$

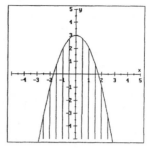

vertex: $(0, 3)$

39. $(3 + 5i) + (4 - 3i) = 3 + 5i + 4 - 3i = 7 + 2i$

40. $(7 - 4i) - (12 + 3i) = 7 - 4i - 12 - 3i = -5 - 7i$

41. $(2 - 3i)(2 + 3i) = 4 + 6i - 6i - 9i^2 = 4 + 9 = 13 = 13 + 0i$

42. $(3 + i)(3 - 3i) = 9 - 9i + 3i - 3i^2 = 9 - 6i + 3 = 12 - 6i$

43. $(3 - 2i) - (4 + i)^2 = 3 - 2i - (16 + 8i + i^2) = 3 - 2i - (15 + 8i) = 3 - 2i - 15 - 8i$
$$= -12 - 10i$$

44. $\dfrac{5}{3-i} = \dfrac{5}{3-i} \cdot \dfrac{3+i}{3+i} = \dfrac{5(3+i)}{9-i^2} = \dfrac{5(3+i)}{10} = \dfrac{3+i}{2} = \dfrac{3}{2} + \dfrac{1}{2}i$

45. $|3 + 2i| = \sqrt{3^2 + 2^2} = \sqrt{9 + 4} = \sqrt{13}$

46. $|5 - 6i| = \sqrt{5^2 + (-6)^2} = \sqrt{25 + 36} = \sqrt{61}$

47.
$$2x^2 + 4x = k$$
$$2x^2 + 4x - k = 0$$
$$a = 2, b = 4, c = -k$$
Set the discriminant equal to 0:
$$b^2 - 4ac = 0$$
$$4^2 - 4(2)(-k) = 0$$
$$16 + 8k = 0$$
$$8k = -16$$
$$k = -2$$

48.
$$a - 7a^{1/2} + 12 = 0$$
$$\left(a^{1/2} - 3\right)\left(a^{1/2} - 4\right) = 0$$
$$a^{1/2} - 3 = 0 \quad \text{or} \quad a^{1/2} - 4 = 0$$
$$a^{1/2} = 3 \qquad\qquad a^{1/2} = 4$$
$$\left(a^{1/2}\right)^2 = 3^2 \qquad \left(a^{1/2}\right)^2 = 4^2$$
$$a = 9 \qquad\qquad a = 16$$
$$\text{Solution} \qquad\qquad \text{Solution}$$

49.
$$x^2 - x - 6 > 0$$
$$(x+2)(x-3) > 0$$

$x - 3$ --------0 ++++
$x + 2$ --- 0+++++++ ++++

⟵——)———(——⟶
    −2   3

solution set: $(-\infty, -2) \cup (3, \infty)$

50.
$$x^2 - x - 6 \le 0$$
$$(x+2)(x-3) \le 0$$

$x - 3$ -------0 ++++
$x + 2$ --- 0+++++++ ++++

⟵——[———]——⟶
    −2   3

solution set: $[-2, 3]$

51. $f(-1) = 3(-1)^2 + 2 = 3(1) + 2 = 3 + 2 = 5$

52. $(g \circ f)(2) = g(f(2)) = g(3(2)^2 + 2) = g(14) = 2(14) - 1 = 27$

53. $(f \circ g)(x) = f(g(x)) = f(2x - 1) = 3(2x - 1)^2 + 2 = 3(4x^2 - 4x + 1) + 2 = 12x^2 - 12x + 5$

54. $(g \circ f)(x) = g(f(x)) = g(3x^2 + 2) = 2(3x^2 + 2) - 1 = 6x^2 + 3$

55.
$$y = 3x + 2$$
$$x = 3y + 2$$
$$x - 2 = 3y$$
$$\frac{x-2}{3} = y, \text{ or } y = f^{-1}(x) = \frac{x-2}{3}$$

56.
$$y = x^3 + 4$$
$$x = y^3 + 4$$
$$x - 4 = y^3$$
$$\sqrt[3]{x-4} = y, \text{ or } y = f^{-1}(x) = \sqrt[3]{x-4}$$

Exercise 9.1 (page 580)

1.
$$3x + 2x - 20 = 180$$
$$5x = 200$$
$$x = 40$$

3. $m(\angle 2) = 3x = 3(40) = 120°$

5. exponential

7. $(0, \infty)$

9. increasing

11. $P\left(1 + \frac{r}{k}\right)^{kt}$

Problems 13-15 are to be solved using a calculator. The keystrokes needed to solve each problem using a TI-83 graphing calculator appear in each solution. There may be other solutions. Keystrokes for other calculators may be slightly different.

13. $2^{\sqrt{2}} \Rightarrow$ ⌷2⌷ ⌷^⌷ ⌷$\sqrt{\ }$⌷ ⌷2⌷ ⌷ENTER⌷
$\{2.6651\}$

15. $5^{\sqrt{5}} \Rightarrow$ ⌷5⌷ ⌷^⌷ ⌷$\sqrt{\ }$⌷ ⌷5⌷ ⌷ENTER⌷
$\{36.5548\}$

17. $\left(2^{\sqrt{3}}\right)^{\sqrt{3}} = 2^{(\sqrt{3})(\sqrt{3})} = 2^3 = 8$

19. $7^{\sqrt{3}}7^{\sqrt{12}} = 7^{\sqrt{3}+\sqrt{12}} = 7^{\sqrt{3}+2\sqrt{3}} = 7^{3\sqrt{3}}$

21. $y = f(x) = 3^x$
through $(0, 1)$ and $(1, 3)$

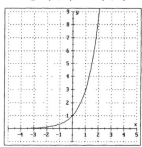

23. $y = f(x) = \left(\frac{1}{3}\right)^x$
through $(0, 1)$ and $\left(1, \frac{1}{3}\right)$

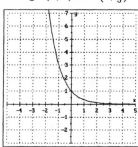

25. $f(x) = 3^x - 2$
Shift $y = 3^x$ down 2.

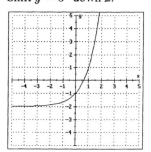

27. $f(x) = 3^{x-1}$
Shift $y = 3^x$ right 1.

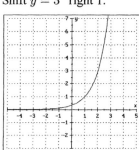

29. $y = b^x$
$\dfrac{1}{2} = b^1$
$\dfrac{1}{2} = b$

31. $y = b^x$
$2 = b^0$
no such value of b

33. $y = b^x$
$2 = b^1$
$2 = b$

35. $y = b^x$
$9 = b^2$
$3 = b$

37. $y = f(x) = \dfrac{1}{2}\left(3^{x/2}\right)$

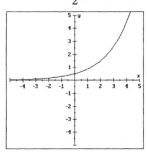

increasing

39. $y = f(x) = 2\left(3^{-x/2}\right)$

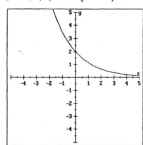

decreasing

41. $A = P\left(1 + \dfrac{r}{k}\right)^{kt}$

$\qquad = 10{,}000\left(1 + \dfrac{0.08}{4}\right)^{4(10)}$

$\qquad = 10{,}000(1.02)^{40} \approx \$22{,}080.40$

43. $A = P\left(1 + \dfrac{r}{k}\right)^{kt}$ $\qquad A = P\left(1 + \dfrac{r}{k}\right)^{kt}$ $\qquad$ difference $= \$1314.07 - \$1282.04$

$\qquad = 1000\left(1 + \dfrac{0.05}{4}\right)^{4(5)}$ $\qquad = 1000\left(1 + \dfrac{0.055}{4}\right)^{4(5)}$ $\qquad\qquad = \$32.03$

$\qquad = 1000(1.0125)^{20}$ $\qquad\quad = 1000(1.01375)^{20}$

$\qquad \approx \$1282.040$ $\qquad\qquad \approx \$1314.07$

45. $A = P\left(1 + \dfrac{r}{k}\right)^{kt}$ $\qquad\qquad$ **47.** $A = A_0\left(\dfrac{2}{3}\right)^t$

$\qquad = 1\left(1 + \dfrac{0.05}{1}\right)^{1(300)}$ $\qquad\qquad\qquad A = A_0\left(\dfrac{2}{3}\right)^5$

$\qquad = 1(1.05)^{300}$ $\qquad\qquad\qquad\qquad A = \dfrac{32}{243}A_0$

$\qquad \approx \$2{,}273{,}996.13$

49. $C = (3 \times 10^{-4})(0.7)^t = (3 \times 10^{-4})(0.7)^5 \approx 5.0421 \times 10^{-5}$ coulombs

51. $A = P\left(1 + \dfrac{r}{k}\right)^{kt}$ $\qquad\qquad\qquad$ **53.** **Answers may vary.**

$\qquad = 4700\left(1 + \dfrac{-0.25}{1}\right)^{1(5)}$

$\qquad = 4700(0.75)^5$

$\qquad \approx \$1115.33$

55. If the base were 0, then the function would not be defined for $x = 0 \Rightarrow y = 0^0$.

Exercise 9.2 (page 587)

1. $\sqrt{240x^5} = \sqrt{16x^4}\sqrt{15x} = 4x^2\sqrt{15x}$

3. $4\sqrt{48y^3} - 3y\sqrt{12y} = 4\sqrt{16y^2}\sqrt{3y} - 3y\sqrt{4}\sqrt{3y} = 4(4y)\sqrt{3y} - 3y(2)\sqrt{3y}$
$\qquad\qquad\qquad\qquad\qquad\qquad\qquad\qquad = 16y\sqrt{3y} - 6y\sqrt{3y} = 10y\sqrt{3y}$

5. 2.72 $\qquad\qquad\qquad\qquad$ **7.** increasing $\qquad\qquad\qquad$ **9.** $A = Pe^{rt}$

11. $y = f(x) = e^x + 1$
Shift $y = e^x$ up 1.

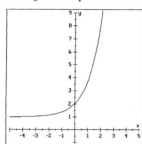

13. $y = f(x) = e^{(x+3)}$
Shift $y = e^x$ left 3.

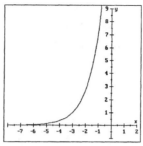

15. $y = f(x) = -e^x$; Reflect
$y = e^x$ about the x-axis.

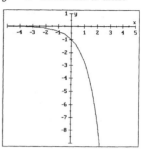

17. $y = f(x) = 2e^x$; Stretch
$y = e^x$ vertically by a
factor of 2.

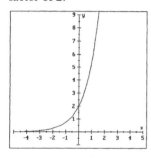

19. The graph should be
increasing. The graph could
not look like this.

21. The graph should go
through the point $(0, 1)$.
The graph could not look
like this.

23. $A = Pe^{rt}$
$= 5000e^{0.06(12)}$
$= 5000e^{0.72}$
$\approx \$10{,}272.17$

25. $A = Pe^{rt}$
$12000 = Pe^{0.07(9)}$
$12000 = Pe^{0.63}$
$12000 \approx P(1.8776106)$
$P \approx \dfrac{12000}{1.8776106}$
$P \approx \$6{,}391.10$

27. $A = Pe^{rt}$
$= 5000e^{0.085(5)}$
$= 5000e^{0.425}$
$\approx \$7{,}647.95$ (continuous)

$A = P\left(1 + \dfrac{r}{k}\right)^{kt}$
$= 5000\left(1 + \dfrac{0.085}{1}\right)^{1(5)}$
$= 5000(1.085)^5$
$\approx \$7{,}518.28$ (annual)

29. $A = Pe^{rt}$
$= 6e^{0.019(30)}$
$= 6e^{0.57}$
≈ 10.6 billion people

31. $A = Pe^{rt}$
$= 6e^{0.019(50)}$
$= 6e^{0.95}$
$\approx 6(2.6)$
It will increase by a factor of about 2.6.

33. $A = 8000e^{-0.008t}$
$= 8000e^{-0.008(20)}$
$= 8000e^{-0.16}$
≈ 6817

35. $P = 0.3\left(1 - e^{-0.05t}\right)$
$= 0.3\left(1 - e^{-0.05(15)}\right)$
$= 0.3\left(1 - e^{-0.75}\right)$
$\approx 0.3(1 - 0.47237)$
≈ 0.16

37. $x = 0.08\left(1 - e^{-0.1t}\right)$
$= 0.08\left(1 - e^{-0.1(0)}\right)$
$= 0.08\left(1 - e^{0}\right)$
$= 0.08(1 - 1)$
$= 0.08(0) = 0$

39. $v = 50\left(1 - e^{-0.2t}\right)$
$= 50\left(1 - e^{-0.2(20)}\right)$
$= 50\left(1 - e^{-4}\right)$
$\approx 50(1 - 0.01832)$
$\approx 50(0.98168) \approx 49$ mps

41. $A = Pe^{rt}$
$= 4570e^{-0.06(6.5)}$
$\approx \$3094.15$

43. $y = 1000e^{0.02x}$
$y = 31x + 2000$

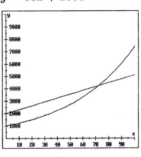

about 72 years

45. **Answers may vary.**

47. $e \approx 2.7182$; $1 + 1 + \frac{1}{2} + \frac{1}{2\cdot3} + \frac{1}{2\cdot3\cdot4} + \frac{1}{2\cdot3\cdot4\cdot5} \approx 2.7167$

49. $e^{t+5} = ke^t$
$e^t \cdot e^5 = ke^t$
$e^5 e^t = ke^t$
$k = e^5$

Exercise 9.3 (page 597)

1. $\sqrt[3]{6x + 4} = 4$
$6x + 4 = 4^3$
$6x + 4 = 64$
$6x = 60$
$x = 10$

3. $\sqrt{a+1} - 1 = 3a$
$\sqrt{a+1} = 3a + 1$
$a + 1 = (3a + 1)^2$
$a + 1 = 9a^2 + 6a + 1$
$0 = 9a^2 + 5a$
$0 = a(9a + 5)$
$a = 0$ or $a = -\frac{5}{9}$ $\left(-\frac{5}{9}$ does not check.$\right)$

5. $x = b^y$

7. range

9. inverse

11. exponent

13. $(b, 1); (1, 0)$

15. $20 \log \dfrac{E_O}{E_I}$

17. $\log_3 27 = 3 \Rightarrow 3^3 = 27$

19. $\log_{1/2} \dfrac{1}{4} = 2 \Rightarrow \left(\dfrac{1}{2}\right)^2 = \dfrac{1}{4}$

21. $\log_4 \dfrac{1}{64} = -3 \Rightarrow 4^{-3} = \dfrac{1}{64}$

23. $\log_{1/2} \dfrac{1}{8} = 3 \Rightarrow \left(\dfrac{1}{2}\right)^3 = \dfrac{1}{8}$

25. $6^2 = 36 \Rightarrow \log_6 36 = 2$

27. $5^{-2} = \dfrac{1}{25} \Rightarrow \log_5 \dfrac{1}{25} = -2$

29. $\left(\dfrac{1}{2}\right)^{-5} = 32 \Rightarrow \log_{1/2} 32 = -5$

31. $x^y = z \Rightarrow \log_x z = y$

33. $\log_2 16 = x \Rightarrow 2^x = 16 \Rightarrow x = 4$

35. $\log_4 16 = x \Rightarrow 4^x = 16 \Rightarrow x = 2$

37. $\log_{1/2} \dfrac{1}{8} = x \Rightarrow \left(\dfrac{1}{2}\right)^x = \dfrac{1}{8} \Rightarrow x = 3$

39. $\log_9 3 = x \Rightarrow 9^x = 3 \Rightarrow x = \dfrac{1}{2}$

41. $\log_{1/2} 8 = x \Rightarrow \left(\dfrac{1}{2}\right)^x = 8 \Rightarrow x = -3$

43. $\log_7 x = 2 \Rightarrow 7^2 = x \Rightarrow x = 49$

45. $\log_6 x = 1 \Rightarrow 6^1 = x \Rightarrow x = 6$

47. $\log_{25} x = \dfrac{1}{2} \Rightarrow 25^{1/2} = x \Rightarrow x = 5$

49. $\log_5 x = -2 \Rightarrow 5^{-2} = x \Rightarrow x = \dfrac{1}{25}$

51. $\log_{36} x = -\dfrac{1}{2} \Rightarrow 36^{-1/2} = x \Rightarrow x = \dfrac{1}{6}$

53. $\log_{100} \dfrac{1}{1000} = x \Rightarrow 100^x = \dfrac{1}{1000} \Rightarrow x = -\dfrac{3}{2}$

55. $\log_{27} 9 = x \Rightarrow 27^x = 9 \Rightarrow x = \dfrac{2}{3}$

57. $\log_x 5^3 = 3 \Rightarrow x^3 = 5^3 \Rightarrow x = 5$

59. $\log_x \dfrac{9}{4} = 2 \Rightarrow x^2 = \dfrac{9}{4} \Rightarrow x = \dfrac{3}{2}$

61. $\log_x \dfrac{1}{64} = -3 \Rightarrow x^{-3} = \dfrac{1}{64} \Rightarrow x = 4$

63. $\log_{2\sqrt{2}} x = 2 \Rightarrow (2\sqrt{2})^2 = x \Rightarrow x = 8$

65. $2^{\log_2 4} = x \Rightarrow x = 4$

67. $x^{\log_4 6} = 6 \Rightarrow x = 4$

69. $\log 10^3 = x \Rightarrow 10^x = 10^3 \Rightarrow x = 3$

71. $10^{\log x} = 100 \Rightarrow \log x = 2 \Rightarrow x = 100$

73. $\log 8.25 \approx 0.9165$

75. $\log 0.00867 \approx -2.0620$

77. $\log y = 1.4023 \Rightarrow y = 25.25$

79. $\log y = 4.24 \Rightarrow y = 17{,}378.01$

81. $\log y = -3.71 \Rightarrow y = 0.00$ **83.** $\log y = \log 8 \Rightarrow \log y = 0.9030 \Rightarrow y = 8$

85. $y = f(x) = \log_3 x$

through $(1, 0)$ and $(3, 1)$

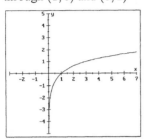

increasing

87. $y = f(x) = \log_{1/2} x$

through $(1, 0)$ and $\left(\frac{1}{2}, 1\right)$

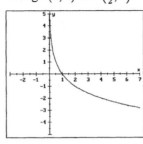

decreasing

89. $y = f(x) = 3 + \log_3 x$

Shift $y = \log_3 x$ up 3.

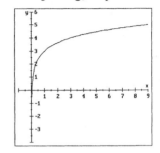

91. $y = f(x) = \log_{1/2}(x - 2)$

Shift $y = \log_{1/2} x$ right 2.

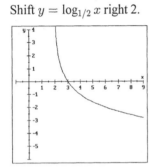

93. $y = f(x) = 2^x$

$y = g(x) = \log_2 x$

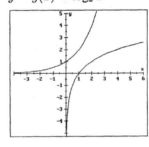

95. $y = f(x) = \left(\dfrac{1}{4}\right)^x$

$y = g(x) = \log_{1/4} x$

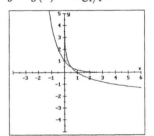

97. $\log_b 2 = 0 \Rightarrow b^0 = 2$; No such value exists.   **99.** $\log_b 9 = 2 \Rightarrow b^2 = 9 \Rightarrow b = 3$

101. dB gain $= 20 \log \dfrac{E_O}{E_I} = 20 \log \dfrac{20}{0.71} = 20 \log 28.169 \approx 29.0$ dB

103. dB gain $= 20 \log \dfrac{E_O}{E_I} = 20 \log \dfrac{30}{0.1} = 20 \log 300 \approx 49.5$ dB

105. $R = \log \dfrac{A}{P} = \log \dfrac{5000}{0.2} = \log 25,000 \approx 4.4$ **107.** $R = \log \dfrac{A}{P} = \log \dfrac{2500}{0.25} = \log 10000 = 4$

109. $n = \dfrac{\log V - \log C}{\log \left(1 - \frac{2}{N}\right)} = \dfrac{\log 2000 - \log 17000}{\log \left(1 - \frac{2}{5}\right)} \approx \dfrac{-0.929419}{-0.221849} \approx 4.2$ years old

111. $n = \dfrac{\log \left[\frac{Ar}{P} + 1\right]}{\log (1 + r)} = \dfrac{\log \left[\frac{20,000(0.12)}{1000} + 1\right]}{\log (1 + 0.12)} = \dfrac{\log 3.4}{\log 1.12} \approx 10.8$ years

113. Answers may vary.   **115. Answers may vary.**   **117. Answers may vary.**

Exercise 9.4 (page 604)

1. $y = 5x + 8 \Rightarrow m = 5$
Use the parallel slope:
$y - y_1 = m(x - x_1)$
$y - 0 = 5(x - 0)$
$y = 5x$

3. $y = \dfrac{2}{3}x - 12 \Rightarrow m = \dfrac{2}{3}$
Use the perpendicular slope:
$y - y_1 = m(x - x_1)$
$y - 2 = -\dfrac{3}{2}(x - 3)$
$y - 2 = -\dfrac{3}{2}x + \dfrac{9}{2}$
$y = -\dfrac{3}{2}x + \dfrac{13}{2}$

5. $x = 5$

7. $\dfrac{2x + 3}{4x^2 - 9} = \dfrac{2x + 3}{(2x + 3)(2x - 3)} = \dfrac{1}{2x - 3}$

9. $\dfrac{x^2 + 3x + 2}{3x + 12} \cdot \dfrac{x + 4}{x^2 - 4} = \dfrac{(x + 2)(x + 1)}{3(x + 4)} \cdot \dfrac{x + 4}{(x + 2)(x - 2)} = \dfrac{x + 1}{3(x - 2)}$

11. $\log_e x$

13. $(-\infty, \infty)$

15. 10

17. $\dfrac{\ln 2}{r}$

19. $\ln 25.25 \approx 3.2288$

21. $\ln 9.89 \approx 2.2915$

23. $\log(\ln 2) \approx \log(0.6931) \approx -0.1592$

25. $\ln(\log 0.5) = \ln(-0.3010) \Rightarrow$ impossible

27. $\ln y = 2.3015 \Rightarrow y = 9.9892$

29. $\ln y = 3.17 \Rightarrow y = 23.8075$

31. $\ln y = -4.72 \Rightarrow y = 0.0089$

33. $\log y = \ln 6 \Rightarrow \log y \approx 1.7918 \Rightarrow y \approx 61.9098$
(The answer will vary if rounding is used on the calculator.)

35. The graph must be increasing. The graph could not look like this.

37. The graph must go through $(1, 0)$. The graph could not look like this.

39. $y = -\ln x$

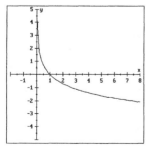

41. $y = \ln(-x)$

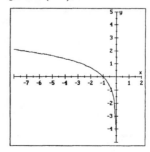

43. $t = \dfrac{\ln 2}{r} = \dfrac{\ln 2}{0.12} \approx 5.8$ years

45. $t = \dfrac{\ln 3}{r} = \dfrac{\ln 3}{0.12} \approx 9.2$ years

47. $t = -\dfrac{1}{0.9}\ln\dfrac{50 - T_r}{200 - T_r} = -\dfrac{1}{0.9}\ln\dfrac{50 - 38}{200 - 38} = -\dfrac{1}{0.9}\ln\dfrac{12}{162} \approx -\dfrac{1}{0.9}(-2.6027) \approx 2.9$ hours

49. Answers may vary.

51. $P = P_0 e^{rt} = P_0 e^{r\frac{\ln 2}{r}} = P_0 e^{\ln 2} = 2P_0$

53. Let $t = \dfrac{\ln 4}{r} \Rightarrow P = P_0 e^{rt} = P_0 e^{r\frac{\ln 4}{r}} = P_0 e^{\ln 4} = 4P_0$

Exercise 9.5 (page 613)

1. $m = \dfrac{\Delta y}{\Delta x} = \dfrac{3 - (-4)}{-2 - 4} = \dfrac{7}{-6} = -\dfrac{7}{6}$

3. $x = \dfrac{-2+4}{2} = \dfrac{2}{2} = 1$
$y = \dfrac{3+(-4)}{2} = \dfrac{-1}{2} = -\dfrac{1}{2}$
midpoint: $\left(1, -\dfrac{1}{2}\right)$

5. 0

7. $M; N$

9. $x; y$

11. x

13. $\neq$

15. 0

17. 7

19. 10

21. 1

23. 0

25. 7

27. 10

29. 1

Problems 31-35 are to be solved using a calculator. The keystrokes needed to solve each problem using a TI-83 graphing calculator appear in each solution. There may be other solutions. Keystrokes for other calculators may be slightly different.

31. `log` `2` `.` `5` `×` `3` `.` `7` `ENTER` {0.96614}
`log` `2` `.` `5` `)` `+` `log` `3` `.` `7` `)` `ENTER` {0.96614}

33. `ln` `2` `.` `2` `5` `^` `4` `ENTER` {3.24372}
`4` `ln` `2` `.` `2` `5` `ENTER` {3.24372}

35. `log` `√` `2` `4` `.` `3` `ENTER` {0.69280}
`.` `5` `log` `2` `4` `.` `3` `ENTER` {0.69280}

37. $\log_b xyz = \log_b x + \log_b y + \log_b z$

39. $\log_b \dfrac{2x}{y} = \log_b 2x - \log_b y$
$= \log_b 2 + \log_b x - \log_b y$

41. $\log_b x^3 y^2 = \log_b x^3 + \log_b y^2 = 3\log_b x + 2\log_b y$

43. $\log_b (xy)^{1/2} = \dfrac{1}{2}\log_b xy = \dfrac{1}{2}(\log_b x + \log_b y) = \dfrac{1}{2}\log_b x + \dfrac{1}{2}\log_b y$

45. $\log_b x\sqrt{z} = \log_b xz^{1/2} = \log_b x + \log_b z^{1/2} = \log_b x + \dfrac{1}{2}\log_b z$

47. $\log_b \dfrac{\sqrt[3]{x}}{\sqrt[4]{yz}} = \log_b \dfrac{x^{1/3}}{(yz)^{1/4}} = \log_b x^{1/3} - \log_b (yz)^{1/4} = \dfrac{1}{3}\log_b x - \dfrac{1}{4}\log_b yz$

$$= \dfrac{1}{3}\log_b x - \dfrac{1}{4}(\log_b y + \log_b z)$$

$$= \dfrac{1}{3}\log_b x - \dfrac{1}{4}\log_b y - \dfrac{1}{4}\log_b z$$

49. $\log_b (x+1) - \log_b x = \log_b \dfrac{x+1}{x}$

51. $2\log_b x + \dfrac{1}{2}\log_b y = \log_b x^2 + \log_b y^{1/2} = \log_b x^2 y^{1/2}$

53. $-3\log_b x - 2\log_b y + \dfrac{1}{2}\log_b z = \log_b x^{-3} + \log_b y^{-2} + \log_b z^{1/2} = \log_b x^{-3}y^{-2}z^{1/2} = \log_b \dfrac{z^{1/2}}{x^3 y^2}$

55. $\log_b \left(\dfrac{x}{z} + x\right) - \log_b \left(\dfrac{y}{z} + y\right) = \log_b \dfrac{\frac{x}{z} + x}{\frac{y}{z} + y} = \log_b \dfrac{x + xz}{y + yz} = \log_b \dfrac{x(1+z)}{y(1+z)} = \log_b \dfrac{x}{y}$

57. $\log_b 0 = 1 \Rightarrow b^1 = 0 \Rightarrow b = 0 \Rightarrow$ FALSE $(b \neq 0)$

59. $\log_b xy = (\log_b x)(\log_b y) \Rightarrow$ FALSE $(\log_b xy = \log_b x + \log_b y)$

61. $\log_7 7^7 = 7 \Rightarrow 7^7 = 7^7 \Rightarrow$ TRUE

63. $\dfrac{\log_b A}{\log_b B} = \log_b A - \log_b B \Rightarrow$ FALSE $\left(\log_b \dfrac{A}{B} = \log_b A - \log_b B\right)$

65. $3\log_b \sqrt[3]{a} = 3\log_b a^{1/3} = \dfrac{1}{3} \cdot 3\log_b a = \log_b a \Rightarrow$ TRUE

67. $\log_b \dfrac{1}{a} = \log_b 1 - \log_b a = 0 - \log_b a = -\log_b a \Rightarrow$ TRUE

69. $\log 28 = \log 4 \cdot 7 = \log 4 + \log 7 = 0.6021 + 0.8451 = 1.4472$

71. $\log 2.25 = \log \dfrac{9}{4} = \log 9 - \log 4 = 0.9542 - 0.6021 = 0.3521$

73. $\log \dfrac{63}{4} = \log \dfrac{7 \cdot 9}{4} = \log 7 + \log 9 - \log 4 = 0.8451 + 0.9542 - 0.6021 = 1.1972$

75. $\log 252 = \log 4 \cdot 7 \cdot 9 = \log 4 + \log 7 + \log 9 = 0.6021 + 0.8451 + 0.9542 = 2.4014$

77. $\log 112 = \log 4^2 \cdot 7 = \log 4^2 + \log 7 = 2\log 4 + \log 7 = 2(0.6021) + 0.8451 = 2.0493$

79. $\log \dfrac{144}{49} = \log \dfrac{16 \cdot 9}{7^2} = \log \dfrac{4^2 \cdot 9}{7^2} = \log 4^2 + \log 9 - \log 7^2 = 2\log 4 + \log 9 - 2\log 7$

$$= 2(0.6021) + 0.9542 - 2(0.8451)$$
$$= 0.4682$$

81. $\log_3 7 = \dfrac{\log 7}{\log 3} \approx 1.7712$ **83.** $\log_{1/3} 3 = \dfrac{\log 3}{\log \frac{1}{3}} \approx -1.0000$

85. $\log_3 8 = \dfrac{\log 8}{\log 3} \approx 1.8928$ **87.** $\log_{\sqrt{2}} \sqrt{5} = \dfrac{\log \sqrt{5}}{\log \sqrt{2}} \approx 2.3219$

89. $\text{pH} = -\log [\text{H}^+] = -\log (1.7 \times 10^{-5}) \approx 4.77$

91. low pH: high pH:

$\text{pH} = -\log [\text{H}^+]$ $\quad\quad\quad$ $\text{pH} = -\log [\text{H}^+]$

$6.8 = -\log [\text{H}^+]$ $\quad\quad\quad$ $7.6 = -\log [\text{H}^+]$

$-6.8 = \log [\text{H}^+]$ $\quad\quad\quad$ $-7.6 = \log [\text{H}^+]$

$[\text{H}^+] = 1.5849 \times 10^{-7}$ $\quad\quad$ $[\text{H}^+] = 2.5119 \times 10^{-8}$

93. $k \ln 2I = k(\ln 2 + \ln I)$ **95.** $L = 3k \ln I = k \cdot 3 \ln I$

$\quad\quad\quad = k \ln 2 + k \ln I$ $\quad\quad\quad\quad\quad = k \ln I^3$

$\quad\quad\quad = k \ln 2 + L$ $\quad\quad\quad$ The intensity must be cubed.

The loudness increases by $k \ln 2$.

97. Answers may vary. **99.** $\ln(e^x) = \log_e(e^x) = x$

101. Let $\log_{b^2} x = y$. Then

$$(b^2)^y = x$$
$$b^{2y} = x$$
$$(b^{2y})^{1/2} = x^{1/2}$$
$$b^y = x^{1/2}$$
$$\log_b x^{1/2} = y$$
$$\frac{1}{2}\log_b x = y$$

Exercise 9.6 (page 622)

1. $5x^2 - 25x = 0$ **3.** $\quad\quad 3p^2 + 10p = 8$

$\quad 5x(x - 5) = 0$ $\quad\quad\quad\quad 3p^2 + 10p - 8 = 0$

$\quad 5x = 0 \quad \text{or} \quad x - 5 = 0$ $\quad\quad (3p - 2)(p + 4) = 0$

$\quad\quad x = 0 \quad\quad\quad\quad x = 5$ $\quad\quad 3p - 2 = 0 \quad \text{or} \quad p + 4 = 0$

$\quad\quad\quad\quad\quad\quad\quad\quad\quad\quad\quad\quad p = \frac{2}{3} \quad\quad\quad\quad p = -4$

5. exponential

7. $A_0 2^{-t/h}$

9.
$$4^x = 5$$
$$\log 4^x = \log 5$$
$$x \log 4 = \log 5$$
$$x = \frac{\log 5}{\log 4}$$
$$x \approx 1.1610$$

11.
$$e^t = 50$$
$$\ln e^t = \ln 50$$
$$t \ln e = \ln 50$$
$$t = \ln 50$$
$$t \approx 3.9120$$

13.
$$5 = 2.1(1.04)^t$$
$$\frac{5}{2.1} = (1.04)^t$$
$$\log \frac{5}{2.1} = \log (1.04)^t$$
$$\log \frac{5}{2.1} = t \log 1.04$$
$$\frac{\log \frac{5}{2.1}}{\log 1.04} = t$$
$$22.1184 \approx t$$

15.
$$13^{x-1} = 2$$
$$\log 13^{x-1} = \log 2$$
$$(x - 1) \log 13 = \log 2$$
$$x - 1 = \frac{\log 2}{\log 13}$$
$$x = \frac{\log 2}{\log 13} + 1$$
$$x \approx 1.2702$$

17.
$$2^{x+1} = 3^x$$
$$\log 2^{x+1} = \log 3^x$$
$$(x + 1) \log 2 = x \log 3$$
$$x \log 2 + \log 2 = x \log 3$$
$$\log 2 = x \log 3 - x \log 2$$
$$\log 2 = x(\log 3 - \log 2)$$
$$\frac{\log 2}{\log 3 - \log 2} = x$$
$$1.7095 \approx x$$

19.
$$2^x = 3^x$$
$$\log 2^x = \log 3^x$$
$$x \log 2 = x \log 3$$
$$0 = x \log 3 - x \log 2$$
$$0 = x(\log 3 - \log 2)$$
$$\frac{0}{\log 3 - \log 2} = x$$
$$0 = x$$

21.
$$7^{x^2} = 10$$
$$\log 7^{x^2} = \log 10$$
$$x^2 \log 7 = \log 10$$
$$x^2 = \frac{\log 10}{\log 7}$$
$$x^2 \approx 1.1833$$
$$x \approx \pm 1.0878$$

23.
$$8^{x^2} = 9^x$$
$$\log 8^{x^2} = \log 9^x$$
$$x^2 \log 8 = x \log 9$$
$$x^2 \log 8 - x \log 9 = 0$$
$$x(x \log 8 - \log 9) = 0$$
$$x = 0 \quad \text{or} \quad x \log 8 - \log 9 = 0$$
$$x \log 8 = \log 9$$
$$x = \frac{\log 9}{\log 8}$$
$$x \approx 1.0566$$

25.
$$2^{x^2-2x} = 8$$
$$2^{x^2-2x} = 2^3$$
$$x^2 - 2x = 3$$
$$x^2 - 2x - 3 = 0$$
$$(x+1)(x-3) = 0$$
$$x + 1 = 0 \quad \textbf{or} \quad x - 3 = 0$$
$$x = -1 \qquad\qquad x = 3$$

27.
$$3^{x^2+4x} = \frac{1}{81}$$
$$3^{x^2+4x} = 3^{-4}$$
$$x^2 + 4x = -4$$
$$x^2 + 4x + 4 = 0$$
$$(x+2)(x+2) = 0$$
$$x + 2 = 0 \quad \textbf{or} \quad x + 2 = 0$$
$$x = -2 \qquad\qquad x = -2$$

29.
$$4^{x+2} - 4^x = 15$$
$$4^x 4^2 - 4^x = 15$$
$$16 \cdot 4^x - 4^x = 15$$
$$15 \cdot 4^x = 15$$
$$4^x = 1$$
$$x = 0$$

31.
$$2(3^x) = 6^{2x}$$
$$\log 2(3^x) = \log 6^{2x}$$
$$\log 2 + \log 3^x = 2x \log 6$$
$$\log 2 + x \log 3 = 2x \log 6$$
$$\log 2 = 2x \log 6 - x \log 3$$
$$\log 2 = x(2 \log 6 - \log 3)$$
$$\frac{\log 2}{2 \log 6 - \log 3} = x$$
$$0.2789 \approx x$$

33. $2^{x+1} = 7 \Rightarrow$ Graph $y = 2^{x+1}$ and $y = 7$.

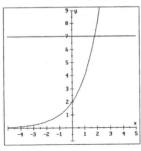

$x \approx 1.8$

35. $2^{x^2-2x} - 8 = 0 \Rightarrow$ Graph $y = 2^{x^2-2x} - 8$ and find any x-intercept(s).

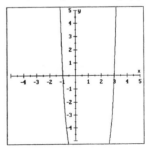

$x = 3$ or $x = -1$

37.
$$\log 2x = \log 4$$
$$2x = 4$$
$$x = 2$$

39.
$$\log (3x + 1) = \log (x + 7)$$
$$3x + 1 = x + 7$$
$$2x = 6$$
$$x = 3$$

41.
$$\log (3 - 2x) - \log (x + 24) = 0$$
$$\log (3 - 2x) = \log (x + 24)$$
$$3 - 2x = x + 24$$
$$-21 = 3x$$
$$-7 = x$$

43. $\log \dfrac{4x+1}{2x+9} = 0$

$$10^0 = \dfrac{4x+1}{2x+9}$$
$$1 = \dfrac{4x+1}{2x+9}$$
$$2x+9 = 4x+1$$
$$8 = 2x$$
$$4 = x$$

45. $\log x^2 = 2$

$$10^2 = x^2$$
$$100 = x^2$$
$$\pm 10 = x$$

47. $\log x + \log (x-48) = 2$

$$\log x(x-48) = 2$$
$$10^2 = x(x-48)$$
$$0 = x^2 - 48x - 100$$
$$0 = (x-50)(x+2)$$
$$x - 50 = 0 \quad \textbf{or} \quad x+2 = 0$$
$$x = 50 \qquad\qquad x = -2$$
Does not check.

49. $\log x + \log (x-15) = 2$

$$\log x(x-15) = 2$$
$$10^2 = x(x-15)$$
$$0 = x^2 - 15x - 100$$
$$0 = (x-20)(x+5)$$
$$x - 20 = 0 \quad \textbf{or} \quad x+5 = 0$$
$$x = 20 \qquad\qquad x = -5$$
Does not check.

51.
$$\log (x+90) = 3 - \log x$$
$$\log x + \log (x+90) = 3$$
$$\log x(x+90) = 3$$
$$10^3 = x(x+90)$$
$$0 = x^2 + 90x - 1000$$
$$0 = (x-10)(x+100)$$
$$x - 10 = 0 \quad \textbf{or} \quad x+100 = 0$$
$$x = 10 \qquad\qquad x = -100$$
Does not check.

53. $\log (x-6) - \log (x-2) = \log \dfrac{5}{x}$

$$\log \dfrac{x-6}{x-2} = \log \dfrac{5}{x}$$
$$\dfrac{x-6}{x-2} = \dfrac{5}{x}$$
$$x(x-6) = 5(x-2)$$
$$x^2 - 6x = 5x - 10$$
$$x^2 - 11x + 10 = 0$$
$$(x-10)(x-1) = 0$$
$$x - 10 = 0 \quad \textbf{or} \quad x-1 = 0$$
$$x = 10 \qquad\qquad x = 1$$
Does not check.

55. $\log x^2 = (\log x)^2$

$$2 \log x = (\log x)^2$$
$$0 = (\log x)^2 - 2 \log x$$
$$0 = \log x (\log x - 2)$$
$$\log x = 0 \quad \textbf{or} \quad \log x - 2 = 0$$
$$x = 1 \qquad\qquad \log x = 2$$
$$\qquad\qquad\qquad x = 100$$

57. $\dfrac{\log (3x-4)}{\log x} = 2$

$$\log (3x-4) = 2 \log x$$
$$\log (3x-4) = \log x^2$$
$$3x - 4 = x^2$$
$$0 = x^2 - 3x + 4$$
$$b^2 - 4ac = (-3)^2 - 4(1)(4) = -7 \Rightarrow$$
solutions are nonreal $\Rightarrow$ no solution

59.
$$\frac{\log (5x + 6)}{2} = \log x$$
$$\log (5x + 6) = 2 \log x$$
$$\log (5x + 6) = \log x^2$$
$$5x + 6 = x^2$$
$$0 = x^2 - 5x - 6$$
$$0 = (x - 6)(x + 1)$$
$$x - 6 = 0 \quad \text{or} \quad x + 1 = 0$$
$$x = 6 \qquad\qquad x = -1$$
$$\text{Does not check.}$$

61.
$$\log_3 x = \log_3 \left(\frac{1}{x}\right) + 4$$
$$\log_3 x = \log_3 \left(\frac{1}{x}\right) + \log_3 81$$
$$\log_3 x = \log_3 \left(\frac{81}{x}\right)$$
$$x = \frac{81}{x}$$
$$x^2 = 81$$
$$x = 9 \ (-9 \text{ does not check.})$$

63.
$$2 \log_2 x = 3 + \log_2 (x - 2)$$
$$\log_2 x^2 - \log_2 (x - 2) = 3$$
$$\log_2 \frac{x^2}{x - 2} = 3$$
$$\frac{x^2}{x - 2} = 8$$
$$x^2 = 8(x - 2)$$
$$x^2 = 8x - 16$$
$$x^2 - 8x + 16 = 0$$
$$(x - 4)(x - 4) = 0$$
$$x - 4 = 0 \quad \text{or} \quad x - 4 = 0$$
$$x = 4 \qquad\qquad x = 4$$

65.
$$\log (7y + 1) = 2 \log (y + 3) - \log 2$$
$$\log (7y + 1) = \log (y + 3)^2 - \log 2$$
$$\log (7y + 1) = \log \frac{y^2 + 6y + 9}{2}$$
$$7y + 1 = \frac{y^2 + 6y + 9}{2}$$
$$2(7y + 1) = y^2 + 6y + 9$$
$$14y + 2 = y^2 + 6y + 9$$
$$0 = y^2 - 8y + 7$$
$$0 = (y - 7)(y - 1)$$
$$y - 7 = 0 \quad \text{or} \quad y - 1 = 0$$
$$y = 7 \qquad\qquad y = 1$$

67. $\log x + \log (x - 15) = 2$
Graph $y = \log x + \log (x - 15)$ and $y = 2$.

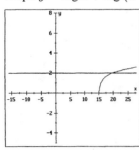

$x = 20$

69. $\ln (2x + 5) - \ln 3 = \ln (x - 1) \Rightarrow$ Graph
$y = \ln (2x + 5) - \ln 3$ and $y = \ln (x - 1)$.

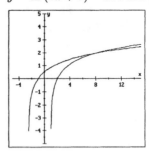

$x = 8$

71.
$$A = A_0 2^{-t/h}$$
$$0.75 A_0 = A_0 2^{-t/12.4}$$
$$0.75 = 2^{-t/12.4}$$
$$\log 0.75 = \log 2^{-t/12.4}$$
$$\log 0.75 = -\frac{t}{12.4} \log 2$$
$$\frac{\log 0.75}{-\log 2} = \frac{t}{12.4}$$
$$-12.4 \frac{\log 0.75}{\log 2} = t$$
$$5.1 \text{ years} \approx t$$

73.
$$A = A_0 2^{-t/h}$$
$$0.20 A_0 = A_0 2^{-t/18.4}$$
$$0.20 = 2^{-t/18.4}$$
$$\log 0.20 = \log 2^{-t/18.4}$$
$$\log 0.20 = -\frac{t}{18.4} \log 2$$
$$\frac{\log 0.20}{-\log 2} = \frac{t}{18.4}$$
$$-18.4 \frac{\log 0.20}{\log 2} = t$$
$$42.7 \text{ days} \approx t$$

75.
$$A = A_0 2^{-t/h}$$
$$0.60 A_0 = A_0 2^{-t/5700}$$
$$0.60 = 2^{-t/5700}$$
$$\log 0.60 = \log 2^{-t/5700}$$
$$\log 0.60 = -\frac{t}{5700} \log 2$$
$$\frac{\log 0.60}{-\log 2} = \frac{t}{5700}$$
$$-5700 \frac{\log 0.60}{\log 2} = t$$
$$4200 \text{ years} \approx t$$

77.
$$A = P\left(1 + \frac{r}{k}\right)^{kt}$$
$$800 = 500\left(1 + \frac{0.085}{2}\right)^{2t}$$
$$1.6 = (1.0425)^{2t}$$
$$\log 1.6 = \log (1.0425)^{2t}$$
$$\log 1.6 = 2t \log (1.0425)$$
$$\frac{\log 1.6}{2 \log 1.0425} = t$$
$$5.6 \text{ years} \approx t$$

79.
$$A = P\left(1 + \frac{r}{k}\right)^{kt}$$
$$2100 = 1300\left(1 + \frac{0.09}{4}\right)^{4t}$$
$$\frac{2100}{1300} = (1.0225)^{4t}$$
$$\log \frac{2100}{1300} = \log (1.0225)^{4t}$$
$$\log \frac{21}{13} = 4t \log (1.0225)$$
$$\frac{\log \frac{21}{13}}{4 \log 1.0225} = t$$
$$5.4 \text{ years} \approx t$$

81. doubling time $= t = \dfrac{\ln 2}{r}$
$$= \frac{100 \ln 2}{100r}$$
$$\approx \frac{70}{100r}$$
$$= \frac{70}{r, \text{ written as a \%}}$$

83.
$$P = P_0 e^{kt}$$
$$2P_0 = P_0 e^{5k}$$
$$2 = e^{5k}$$
$$\ln 2 = \ln e^{5k}$$
$$\ln 2 = 5k$$
$$\frac{\ln 2}{5} = k$$

$$P = P_0 e^{kt}$$
$$1{,}000{,}000 = 30{,}000 e^{\frac{\ln 2}{5}t}$$
$$33.333 \approx e^{\frac{\ln 2}{5}t}$$
$$\ln 33.333 \approx \ln e^{\frac{\ln 2}{5}t}$$
$$\ln 33.333 \approx \frac{\ln 2}{5}t$$
$$\frac{5 \ln 33.333}{\ln 2} \approx t$$
$$25.3 \text{ years} \approx t$$

85
$$P = P_0 e^{kt}$$
$$2P_0 = P_0 e^{24k}$$
$$2 = e^{24k}$$
$$\ln 2 = \ln e^{24k}$$
$$\ln 2 = 24k$$
$$\frac{\ln 2}{24} = k$$

$$P = P_0 e^{kt}$$
$$P = P_0 e^{\frac{\ln 2}{24}(36)}$$
$$P = P_0 e^{\frac{3\ln 2}{2}}$$
$$P = P_0(2.828)$$

It will be about 2.828 times larger.

87. $n = \dfrac{1}{\log 2}\left(\log\dfrac{B}{b}\right) = \dfrac{1}{\log 2}\log\dfrac{5 \times 10^6}{500} = \dfrac{1}{\log 2}\log 10{,}000 \approx 13.3$ generations

89. Answers may vary.

91. Since the logarithm of a negative number is not defined (as a real number), the values $x - 3$ and $x^2 + 2$ must be nonnegative. Since $x^2 + 2$ is always greater than 0, the only restriction is that $x - 3 > 0$, or $x > 3$. Thus, x cannot be a solution if $x \le 3$.

Chapter 9 Summary (page 626)

1. $5^{\sqrt{2}} \cdot 5^{\sqrt{2}} = 5^{\sqrt{2}+\sqrt{2}} = 5^{2\sqrt{2}}$

2. $\left(2^{\sqrt{5}}\right)^{\sqrt{2}} = 2^{(\sqrt{5})(\sqrt{2})} = 2^{\sqrt{10}}$

3. $y = 3^x$; through $(0,1)$ and $(1,3)$

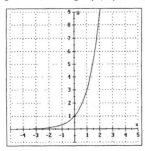

4. $y = \left(\frac{1}{3}\right)^x$; through $(0,1)$ and $\left(1,\frac{1}{3}\right)$

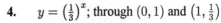
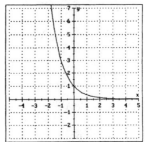

5. The graph will go through $(0,1)$ and $(1,6)$, so $x = 1$ and $y = 6$.

6. domain $= (-\infty, \infty)$; range $= (0, \infty)$

7. $y = f(x) = \left(\frac{1}{2}\right)^x - 2$

Shift $y = \left(\frac{1}{2}\right)^x$ down 2.

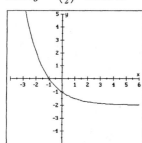

8. $y = f(x) = \left(\frac{1}{2}\right)^{x+2}$

Shift $y = \left(\frac{1}{2}\right)^x$ left 2.

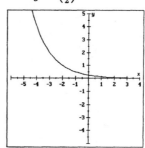

9. $A = P\left(1 + \frac{r}{k}\right)^{kt} = 10500\left(1 + \frac{0.09}{4}\right)^{4\cdot60} = 10500(1.0225)^{240} \approx \$2,189,703.45$

10. $A = Pe^{rt} = 10500e^{0.09(60)} = 10500e^{5.4} \approx \$2,324,767.37$

11. $y = f(x) = e^x + 1$; Shift $y = e^x$ up 1.

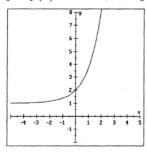

12. $y = f(x) = e^{x-3}$; Shift $y = e^x$ right 3.

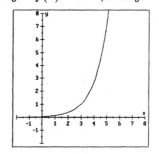

13. $P = P_0 e^{kt} = 275,000,000e^{0.015(50)}$
$\approx 582,000,000$

14. domain $= (0, \infty)$; range $= (-\infty, \infty)$

15. $\log_3 9 = 2$

16. $\log_9 \frac{1}{3} = -\frac{1}{2}$

17. $\log_\pi 1 = 0$

18. $\log_5 0.04 = \log_5 \frac{1}{25} = -2$

19. $\log_a \sqrt{a} = \log_a a^{1/2} = \frac{1}{2}$

20. $\log_a \sqrt[3]{a} = \log_a a^{1/3} = \frac{1}{3}$

21. $\log_2 x = 5 \Rightarrow 2^5 = x \Rightarrow x = 32$

22. $\log_{\sqrt{3}} x = 4 \Rightarrow \left(\sqrt{3}\right)^4 = x \Rightarrow x = 9$

23. $\log_{\sqrt{3}} x = 6 \Rightarrow \left(\sqrt{3}\right)^6 = x \Rightarrow x = 27$

24. $\log_{0.1} 10 = x \Rightarrow (0.1)^x = 10$
$\left(\frac{1}{10}\right)^x = 10 \Rightarrow x = -1$

25. $\log_x 2 = -\dfrac{1}{3} \Rightarrow x^{-1/3} = 2$

$\left(x^{-1/3}\right)^{-3} = 2^{-3} \Rightarrow x = \dfrac{1}{8}$

26. $\log_x 32 = 5 \Rightarrow x^5 = 32 \Rightarrow x = 2$

27. $\log_{0.25} x = -1 \Rightarrow (0.25)^{-1} = x$

$\left(\dfrac{1}{4}\right)^{-1} = x \Rightarrow x = 4$

28. $\log_{0.125} x = -\dfrac{1}{3} \Rightarrow (0.125)^{-1/3} = x$

$\left(\dfrac{1}{8}\right)^{-1/3} = x \Rightarrow x = 2$

29. $\log_{\sqrt{2}} 32 = x \Rightarrow \left(\sqrt{2}\right)^x = 32$

$\left(2^{1/2}\right)^x = 2^5 \Rightarrow \dfrac{1}{2}x = 5 \Rightarrow x = 10$

30. $\log_{\sqrt{5}} x = -4 \Rightarrow \left(\sqrt{5}\right)^{-4} = x$

$\left(5^{1/2}\right)^{-4} = x \Rightarrow 5^{-2} = x \Rightarrow x = \dfrac{1}{25}$

31. $\log_{\sqrt{3}} 9\sqrt{3} = x \Rightarrow \left(\sqrt{3}\right)^x = 9\sqrt{3}$

$\left(3^{1/2}\right)^x = 3^{5/2} \Rightarrow x = 5$

32. $\log_{\sqrt{5}} 5\sqrt{5} = x \Rightarrow \left(\sqrt{5}\right)^x = 5\sqrt{5}$

$\left(5^{1/2}\right)^x = 5^{3/2} \Rightarrow x = 3$

33. $y = f(x) = \log(x - 2)$
Shift $y = \log x$ right 2.

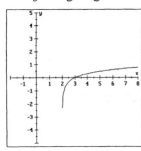

34. $y = f(x) = 3 + \log x$
Shift $y = \log x$ up 3.

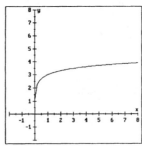

35. $y = 4^x$
$y = \log_4 x$

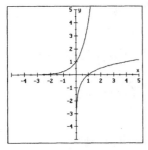

36. $y = \left(\frac{1}{3}\right)^x$
$y = \log_{1/3} x$

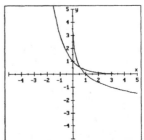

37. dB gain $= 20 \log \dfrac{E_O}{E_I} = 20 \log \dfrac{18}{0.04} = 20 \log 450 \approx 53$ dB

38. $R = \log \frac{A}{P} = \log \frac{7500}{0.3} = \log 25{,}000 \approx 4.4$

39. $\ln 452 \approx 6.1137$

40. $\ln\left(\log 7.85\right) \approx \ln 0.8949 \approx -0.1111$

41. $\ln x = 2.336 \Rightarrow x = 10.3398$

42. $\ln x = \log 8.8 \Rightarrow x = 2.5715$

43. $y = f(x) = 1 + \ln x$
Shift $y = \ln x$ up 1.

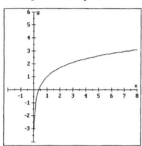

44. $y = f(x) = \ln\left(x + 1\right)$
Shift $y = \ln x$ left 1.

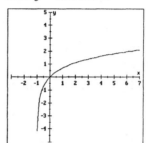

45. $t = \dfrac{\ln 2}{r} = \dfrac{\ln 2}{0.03} \approx 23$ years

46. $\log_7 1 = 0$

47. $\log_7 7 = 1$

48. $\log_7 7^3 = 3$

49. $7^{\log_7 4} = 4$

50. $\ln e^4 = 4$

51. $\ln 1 = 0$

52. $10^{\log_{10} 7} = 7$

53. $e^{\ln 3} = 3$

54. $\log_b b^4 = 4$

55. $\ln e^9 = 9$

56. $\log_b \dfrac{x^2 y^3}{z^4} = \log_b x^2 + \log_b y^3 - \log_b z^4 = 2\log_b x + 3\log_b y - 4\log_b z$

57. $\log_b \sqrt{\dfrac{x}{yz^2}} = \log_b \left(\dfrac{x}{yz^2}\right)^{1/2} = \dfrac{1}{2}\log_b \dfrac{x}{yz^2} = \dfrac{1}{2}\left(\log_b x - \log_b y - \log_b z^2\right)$

$= \dfrac{1}{2}\left(\log_b x - \log_b y - 2\log_b z\right)$

58. $3\log_b x - 5\log_b y + 7\log_b z = \log_b x^3 - \log_b y^5 + \log_b z^7 = \log_b \dfrac{x^3 z^7}{y^5}$

59. $\dfrac{1}{2}\log_b x + 3\log_b y - 7\log_b z = \log_b x^{1/2} + \log_b y^3 - \log_b z^7 = \log_b \dfrac{y^3 \sqrt{x}}{z^7}$

60. $\log abc = \log a + \log b + \log c = 0.6 + 0.36 + 2.4 = 3.36$

61. $\log a^2 b = \log a^2 + \log b = 2\log a + \log b = 2(0.6) + 0.36 = 1.56$

62. $\log \dfrac{ac}{b} = \log a + \log c - \log b = 0.6 + 2.4 - 0.36 = 2.64$

63. $\log \dfrac{a^2}{c^3 b^2} = \log a^2 - \log c^3 - \log b^2 = 2\log a - 3\log c - 2\log b$

$$= 2(0.6) - 3(2.4) - 2(0.36) = -6.72$$

64. $\log_5 17 = \dfrac{\log 17}{\log 5} \approx 1.7604$

65.
$$\text{pH} = -\log [\text{H}^+]$$
$$3.1 = -\log [\text{H}^+]$$
$$-3.1 = \log [\text{H}^+]$$
$$[\text{H}^+] = 7.94 \times 10^{-4} \text{ gram-ions per liter}$$

66.
$$k \ln \left(\tfrac{1}{2} I\right) = k\left(\ln \tfrac{1}{2} + \ln I\right)$$
$$= k \ln \tfrac{1}{2} + k \ln I$$
$$= k \ln 2^{-1} + L$$
$$= -k \ln 2 + L$$
The loudness decreases by $k \ln 2$.

67.
$$3^x = 7$$
$$\log 3^x = \log 7$$
$$x \log 3 = \log 7$$
$$x = \frac{\log 7}{\log 3} \approx 1.7712$$

68.
$$5^{x+2} = 625$$
$$5^{x+2} = 5^4$$
$$x + 2 = 4$$
$$x = 2$$

69.
$$25 = 5.5(1.05)^t$$
$$\frac{25}{5.5} = (1.05)^t$$
$$\log \frac{25}{5.5} = \log (1.05)^t$$
$$\log \frac{25}{5.5} = t \log 1.05$$
$$\frac{\log \frac{25}{5.5}}{\log 1.05} = t$$
$$31.0335 \approx t$$

70.
$$4^{2t-1} = 64$$
$$4^{2t-1} = 4^3$$
$$2t - 1 = 3$$
$$2t = 4$$
$$t = 2$$

71.
$$2^x = 3^{x-1}$$
$$\log 2^x = \log 3^{x-1}$$
$$x \log 2 = (x-1) \log 3$$
$$x \log 2 = x \log 3 - \log 3$$
$$\log 3 = x \log 3 - x \log 2$$
$$\log 3 = x(\log 3 - \log 2)$$
$$\frac{\log 3}{\log 3 - \log 2} = x$$
$$2.7095 \approx x$$

72.
$$2^{x^2+4x} = \frac{1}{8}$$
$$2^{x^2+4x} = 2^{-3}$$
$$x^2 + 4x = -3$$
$$x^2 + 4x + 3 = 0$$
$$(x+3)(x+1) = 0$$
$$x + 3 = 0 \quad \text{or} \quad x + 1 = 0$$
$$x = -3 \qquad\qquad x = -1$$

73.
$$\log x + \log (29 - x) = 2$$
$$\log x(29 - x) = 2$$
$$10^2 = x(29 - x)$$
$$100 = 29x - x^2$$
$$x^2 - 29x + 100 = 0$$
$$(x - 25)(x - 4) = 0$$
$$x - 25 = 0 \quad \text{or} \quad x - 4 = 0$$
$$x = 25 \qquad\qquad x = 4$$

74.
$$\log_2 x + \log_2 (x - 2) = 3$$
$$\log_2 x(x - 2) = 3$$
$$x(x - 2) = 2^3$$
$$x^2 - 2x = 8$$
$$x^2 - 2x - 8 = 0$$
$$(x - 4)(x + 2) = 0$$
$$x - 4 = 0 \quad \text{or} \quad x + 2 = 0$$
$$x = 4 \qquad\qquad x = -2$$
$$\text{Does not check.}$$

75.
$$\log_2 (x + 2) + \log_2 (x - 1) = 2$$
$$\log_2 (x + 2)(x - 1) = 2$$
$$(x + 2)(x - 1) = 2^2$$
$$x^2 + x - 2 = 4$$
$$x^2 + x - 6 = 0$$
$$(x - 2)(x + 3) = 0$$
$$x - 2 = 0 \quad \text{or} \quad x + 3 = 0$$
$$x = 2 \qquad\qquad x = -3$$
$$\text{Does not check.}$$

76.
$$\frac{\log (7x - 12)}{\log x} = 2$$
$$\log (7x - 12) = 2 \log x$$
$$\log (7x - 12) = \log x^2$$
$$7x - 12 = x^2$$
$$0 = x^2 - 7x + 12$$
$$0 = (x - 4)(x - 3)$$
$$x - 4 = 0 \quad \text{or} \quad x - 3 = 0$$
$$x = 4 \qquad\qquad x = 3$$

77.
$$\log x + \log (x - 5) = \log 6$$
$$\log x(x - 5) = \log 6$$
$$x(x - 5) = 6$$
$$x^2 - 5x - 6 = 0$$
$$(x - 6)(x + 1) = 0$$
$$x - 6 = 0 \quad \text{or} \quad x + 1 = 0$$
$$x = 6 \qquad\qquad x = -1$$
$$\text{Does not check.}$$

78.
$$\log 3 - \log (x - 1) = -1$$
$$\log \frac{3}{x - 1} = -1$$
$$\frac{3}{x - 1} = 10^{-1}$$
$$\frac{3}{x - 1} = \frac{1}{10}$$
$$30 = x - 1$$
$$31 = x$$

79.
$$e^{x \ln 2} = 9$$
$$e^{\ln 2^x} = 9$$
$$2^x = 9$$
$$\ln 2^x = \ln 9$$
$$x \ln 2 = \ln 9$$
$$x = \frac{\ln 9}{\ln 2} \approx 3.1699$$

80.
$$\ln x = \ln (x - 1)$$
$$x = x - 1$$
$$0 = -1$$
$$\text{There is no solution.}$$

81.
$$\ln x = \ln (x - 1) + 1$$
$$\ln x - \ln (x - 1) = 1$$
$$\ln \frac{x}{x - 1} = 1$$
$$\frac{x}{x - 1} = e^1$$
$$x = e(x - 1)$$
$$x = ex - e$$
$$e = ex - x$$
$$e = x(e - 1)$$
$$\frac{e}{e - 1} = x$$
$$1.5820 \approx x$$

82.
$$\ln x = \log_{10} x$$
$$\ln x = \frac{\ln x}{\ln 10}$$
$$\ln x \ln 10 = \ln x$$
$$\ln x \ln 10 - \ln x = 0$$
$$\ln x \left(\ln 10 - 1\right) = 0$$
$$\ln x = 0$$
$$x = 1$$

83.
$$A = A_0 2^{-t/h}$$
$$\frac{2}{3} A_0 = A_0 2^{-t/5700}$$
$$\frac{2}{3} = 2^{-t/5700}$$
$$\log \frac{2}{3} = \log 2^{-t/5700}$$
$$\log \frac{2}{3} = -\frac{t}{5700} \log 2$$
$$\frac{\log \frac{2}{3}}{-\log 2} = \frac{t}{5700}$$
$$-5700 \frac{\log \frac{2}{3}}{\log 2} = t$$
$$3300 \text{ years} \approx t$$

Chapter 9 Test (page 631)

1. $f(x) = 2^x + 1$; Shift $y = 2^x$ up 1.

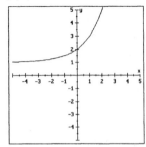

2. $f(x) = 2^{-x}$; Reflect $y = 2^x$ about y-axis.

3. $A = A_0(2)^{-t} = 3(2)^{-6} = \frac{3}{2^6} = \frac{3}{64}$ gram

CHAPTER 9 TEST

4. $A = A_0\left(1 + \dfrac{r}{k}\right)^{kt} = 1000\left(1 + \dfrac{0.06}{2}\right)^{2(1)} = 1000(1.03)^2 \approx \$1060.90$

5. $f(x) = e^x$

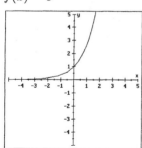

6. $A = A_0 e^{rt} = 2000e^{(0.08)10}$
$= 2000e^{0.8}$
$\approx \$4451.08$

7. $\log_4 16 = x \Rightarrow 4^x = 16 \Rightarrow x = 2$

8. $\log_x 81 = 4 \Rightarrow x^4 = 81 \Rightarrow x = 3$

9. $\log_3 x = -3 \Rightarrow 3^{-3} = x \Rightarrow x = \dfrac{1}{27}$

10. $\log_x 100 = 2 \Rightarrow x^2 = 100 \Rightarrow x = 10$

11. $\log_{3/2} \dfrac{9}{4} = x \Rightarrow \left(\dfrac{3}{2}\right)^x = \dfrac{9}{4} \Rightarrow x = 2$

12. $\log_{2/3} x = -3 \Rightarrow \left(\dfrac{2}{3}\right)^{-3} = x \Rightarrow x = \dfrac{27}{8}$

13. $f(x) = -\log_3 x$

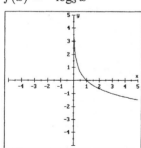

14. $f(x) = \ln x$

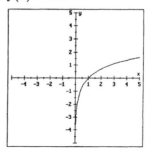

15. $\log a^2 bc^3 = \log a^2 + \log b + \log c^3 = 2\log a + \log b + 3\log c$

16. $\ln \sqrt{\dfrac{a}{b^2 c}} = \ln \left(\dfrac{a}{b^2 c}\right)^{1/2} = \dfrac{1}{2}\ln \dfrac{a}{b^2 c} = \dfrac{1}{2}(\ln a - \ln b^2 - \ln c) = \dfrac{1}{2}\ln a - \ln b - \dfrac{1}{2}\ln c$

17. $\dfrac{1}{2}\log (a+2) + \log b - 3\log c = \log (a+2)^{1/2} + \log b - \log c^3 = \log \dfrac{b\sqrt{a+2}}{c^3}$

18. $\dfrac{1}{3}(\log a - 2\log b) - \log c = \dfrac{1}{3}(\log a - \log b^2) - \log c = \dfrac{1}{3}\log \dfrac{a}{b^2} - \log c = \log \sqrt[3]{\dfrac{a}{b^2}} - \log c$
$= \log \dfrac{\sqrt[3]{a}}{c\sqrt[3]{b^2}}$

19. $\log 24 = \log 8 \cdot 3 = \log 2^3 \cdot 3 = \log 2^3 + \log 3 = 3 \log 2 + \log 3 = 3(0.3010) + 0.4771 = 1.3801$

20. $\log \dfrac{8}{3} = \log \dfrac{2^3}{3} = \log 2^3 - \log 3 = 3 \log 2 - \log 3 = 3(0.3010) - 0.4771 = 0.4259$

21. $\log_7 3 = \dfrac{\log 3}{\log 7} \text{ or } \dfrac{\ln 3}{\ln 7}$

22. $\log_\pi e = \dfrac{\log e}{\log \pi} \text{ or } \dfrac{\ln e}{\ln \pi}$

23. $\log_a ab = \log_a a + \log_a b = 1 + \log_a b \Rightarrow$ TRUE

24. $\dfrac{\log a}{\log b} = \log a - \log b \Rightarrow$ FALSE $\left(\log \dfrac{a}{b} = \log a - \log b \right)$

25. $\log a^{-3} = -3 \log a \neq \dfrac{1}{3 \log a} \Rightarrow$ FALSE

26. $\ln(-x) = -\ln x \Rightarrow$ FALSE (This implies one of the logarithms is negative, which is impossible.)

27. $\text{pH} = -\log [\text{H}^+] = -\log (3.7 \times 10^{-7}) \approx 6.4$

28. $\text{dB gain} = 20 \log \dfrac{E_O}{E_I} = 20 \log \dfrac{60}{0.3} = 20 \log 200 \approx 46$

29.
$$5^x = 3$$
$$\log 5^x = \log 3$$
$$x \log 5 = \log 3$$
$$x = \dfrac{\log 3}{\log 5}$$

30.
$$3^{x-1} = 100^x$$
$$\log 3^{x-1} = \log 100^x$$
$$(x-1)\log 3 = x \log 100$$
$$x \log 3 - \log 3 = 2x$$
$$x \log 3 - 2x = \log 3$$
$$x (\log 3 - 2) = \log 3$$
$$x = \dfrac{\log 3}{\log 3 - 2}$$

31.
$$\log (5x + 2) = \log (2x + 5)$$
$$5x + 2 = 2x + 5$$
$$3x = 3$$
$$x = 1$$

32.
$$\log x + \log (x - 9) = 1$$
$$\log x(x - 9) = 1$$
$$x(x - 9) = 10$$
$$x^2 - 9x - 10 = 0$$
$$(x - 10)(x + 1) = 0$$
$$x - 10 = 0 \quad \textbf{or} \quad x + 1 = 0$$
$$x = 10 \qquad\qquad x = -1$$
$$\qquad\qquad\qquad \text{Does not check.}$$

Exercise 10.1 (page 643)

1. $|3x - 4| = 11$
$3x - 4 = 11$ **or** $3x - 4 = -11$
$3x = 15 \qquad\qquad 3x = -7$
$x = 5 \qquad\qquad x = -\frac{7}{3}$

3. $|3x + 4| = |5x - 2|$
$3x + 4 = 5x - 2$ **or** $3x + 4 = -(5x - 2)$
$-2x = -6 \qquad\qquad 3x + 4 = -5x + 2$
$x = 3 \qquad\qquad\qquad 8x = -2$
$x = -\frac{1}{4}$

5. circle; plane

7. $r^2 < 0$

9. parabola; $(3, 2)$; right

11. $x^2 + y^2 = 9$
C $(0, 0)$; $r = \sqrt{9} = 3$

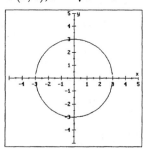

13. $(x - 2)^2 + y^2 = 9$
C $(2, 0)$; $r = \sqrt{9} = 3$

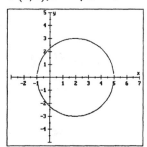

15. $(x - 2)^2 + (y - 4)^2 = 4$
C $(2, 4)$; $r = \sqrt{4} = 2$

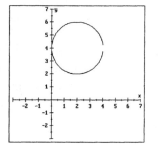

17. $(x + 3)^2 + (y - 1)^2 = 16$
C $(-3, 1)$; $r = \sqrt{16} = 4$

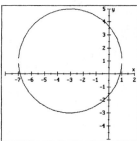

19. $x^2 + (y + 3)^2 = 1$
C $(0, -3)$; $r = \sqrt{1} = 1$

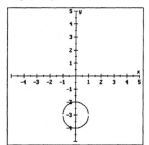

21. $3x^2 + 3y^2 = 16$
$3y^2 = 16 - 3x^2$
$y^2 = \dfrac{16 - 3x^2}{3}$
$y = \pm\sqrt{\dfrac{16 - 3x^2}{3}}$

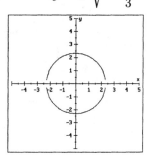

23. $(x+1)^2 + y^2 = 16$
$$y^2 = 16 - (x+1)^2$$
$$y = \pm\sqrt{16 - (x+1)^2}$$

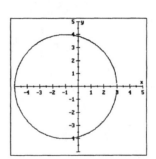

25. $(x-h)^2 + (y-k)^2 = r^2$
$$(x-0)^2 + (y-0)^2 = 1^2$$
$$x^2 + y^2 = 1$$

27. $(x-h)^2 + (y-k)^2 = r^2$
$$(x-6)^2 + (y-8)^2 = 5^2$$
$$(x-6)^2 + (y-8)^2 = 25$$

29. $(x-h)^2 + (y-k)^2 = r^2$
$$(x-(-2))^2 + (y-6)^2 = 12^2$$
$$(x+2)^2 + (y-6)^2 = 144$$

31. $(x-h)^2 + (y-k)^2 = r^2$
$$(x-0)^2 + (y-0)^2 = \left(\sqrt{2}\right)^2$$
$$x^2 + y^2 = 2$$

33. $x^2 + y^2 + 2x - 8 = 0$
$$x^2 + 2x + y^2 = 8$$
$$x^2 + 2x + 1 + y^2 = 8 + 1$$
$$(x+1)^2 + y^2 = 9$$
$$C\,(-1,0);\, r = \sqrt{9} = 3$$

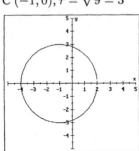

35. $9x^2 + 9y^2 - 12y = 5$
$$x^2 + y^2 - \frac{4}{3}y = \frac{5}{9}$$
$$x^2 + y^2 - \frac{4}{3}y + \frac{4}{9} = \frac{5}{9} + \frac{4}{9}$$
$$x^2 + \left(y - \frac{2}{3}\right)^2 = 1$$
$$C\left(0, \frac{2}{3}\right);\, r = \sqrt{1} = 1$$

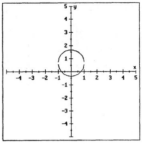

SECTION 10.1

37.
$$x^2 + y^2 - 2x + 4y = -1$$
$$x^2 - 2x + y^2 + 4y = -1$$
$$x^2 - 2x + 1 + y^2 + 4y + 4 = -1 + 1 + 4$$
$$(x-1)^2 + (y+2)^2 = 4$$
C $(1, -2); r = \sqrt{4} = 2$

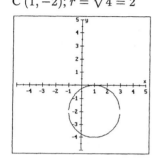

39.
$$x^2 + y^2 + 6x - 4y = -12$$
$$x^2 + 6x + y^2 - 4y = -12$$
$$x^2 + 6x + 9 + y^2 - 4y + 4 = -12 + 9 + 4$$
$$(x+3)^2 + (y-2)^2 = 1$$
C $(-3, 2); r = \sqrt{1} = 1$

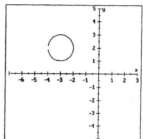

41. $x = y^2$
$x = (y-0)^2 + 0$
V $(0,0)$; opens R

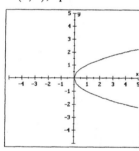

43. $x = -\dfrac{1}{4}y^2$

$x = -\dfrac{1}{4}(y-0)^2 + 0$

V $(0,0)$; opens L

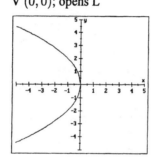

45. $y = x^2 + 4x + 5$
$y = x^2 + 4x + 4 + 5 - 4$
$y = (x+2)^2 + 1$
V $(-2, 1)$; opens U

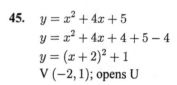
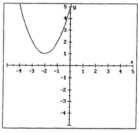

47. $y = -x^2 - x + 1$
$y = -(x^2 + x) + 1$
$y = -\left(x^2 + x + \dfrac{1}{4}\right) + 1 + \dfrac{1}{4}$

$y = -\left(x + \dfrac{1}{2}\right)^2 + \dfrac{5}{4}$

V $\left(-\frac{1}{2}, \frac{5}{4}\right)$; opens D

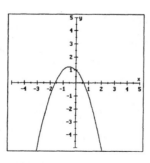

49. $y^2 + 4x - 6y = -1$ V $(2, 3)$; opens L

$$4x = -y^2 + 6y - 1$$

$$x = -\frac{1}{4}y^2 + \frac{3}{2}y - \frac{1}{4}$$

$$x = -\frac{1}{4}(y^2 - 6y) - \frac{1}{4}$$

$$x = -\frac{1}{4}(y^2 - 6y + 9) - \frac{1}{4} + \frac{9}{4}$$

$$x = -\frac{1}{4}(y - 3)^2 + 2$$

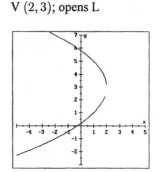

51. $y = 2(x - 1)^2 + 3$ **53.** $x = 2y^2$ **55.** $x^2 - 2x + y = 6$

V $(1, 3)$; opens U $y^2 = \dfrac{x}{2}$ $y = 6 - x^2 + 2x$

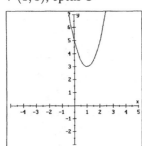

$$y = \pm\sqrt{\frac{x}{2}}$$

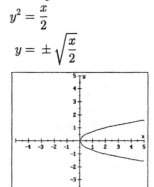

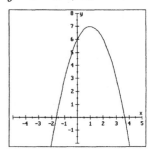

57. The radius of the larger gear is $\sqrt{16} = 4$. Centers: **7 units apart** $\Rightarrow$ smaller gear $r = 3$.

$$(x - h)^2 + (y - k)^2 = r^2$$

$$(x - 7)^2 + (y - 0)^2 = 3^2 \Rightarrow (x - 7)^2 + y^2 = 9$$

59.

$$x^2 + y^2 - 8x - 20y + 16 = 0$$

$$x^2 - 8x + y^2 - 20y = -16$$

$$x^2 - 8x + 16 + y^2 - 20y + 100 = -16 + 16 + 100$$

$$(x - 4)^2 + (y - 10)^2 = 100$$

center: $(4, 10)$; radius $= 10$

$$x^2 + y^2 + 2x + 4y - 11 = 0$$

$$x^2 + 2x + 1 + y^2 + 4y + 4 = 11 + 1 + 4$$

$$(x + 1)^2 + (y + 2)^2 = 16$$

center: $(-1, -2)$; radius $= 4$

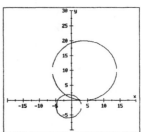

Since the ranges overlap (see graph), they can not be licensed for the same frequency.

61. Set $y = 0$:
$$y = 30x - x^2$$
$$0 = 30x - x^2$$
$$0 = x(30 - x)$$
$$x = 0 \text{ or } x = 30$$
It lands 30 feet away.

63. Find the vertex:
$$2y^2 - 9x = 18$$
$$-9x = -2y^2 + 18$$
$$x = \frac{2}{9}y^2 - 2$$
$$x = \frac{2}{9}(y - 0)^2 - 2$$
vertex: $(-2, 0) \Rightarrow$ distance $= 2$ AU

65. Answers may vary.

67. Answers may vary.

Exercise 10.2 (page 654)

1. $3x^{-2}y^2(4x^2 + 3y^{-2}) = 12x^0y^2 + 9x^{-2}y^0 = 12y^2 + \dfrac{9}{x^2}$

3. $\dfrac{x^{-2} + y^{-2}}{x^{-2} - y^{-2}} = \dfrac{x^{-2} + y^{-2}}{x^{-2} - y^{-2}} \cdot \dfrac{x^2y^2}{x^2y^2} = \dfrac{y^2 + x^2}{y^2 - x^2}$

5. ellipse; sum

7. center

9. $(0, 0)$

11. $\dfrac{x^2}{4} + \dfrac{y^2}{9} = 1$
C $(0, 0)$; move 2 horiz. and 3 vert.

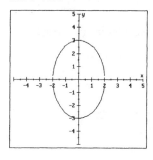

13. $x^2 + 9y^2 = 9$
$$\frac{x^2}{9} + \frac{9y^2}{9} = \frac{9}{9}$$
$$\frac{x^2}{9} + \frac{y^2}{1} = 1$$
C $(0, 0)$; move 3 horiz. and 1 vert.

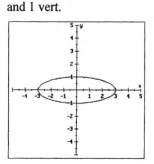

15. $16x^2 + 4y^2 = 64$
$$\frac{16x^2}{64} + \frac{4y^2}{64} = \frac{64}{64}$$
$$\frac{x^2}{4} + \frac{y^2}{16} = 1$$
C $(0, 0)$; move 2 horiz. and 4 vert.

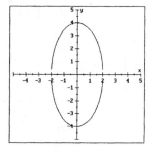

17. $\dfrac{(x-2)^2}{9} + \dfrac{(y-1)^2}{4} = 1$

C $(2,1)$; move 3 horiz. and 2 vert.

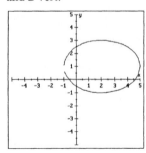

19. $(x+1)^2 + 4(y+2)^2 = 4$

$\dfrac{(x+1)^2}{4} + \dfrac{4(y+2)^2}{4} = \dfrac{4}{4}$

$\dfrac{(x+1)^2}{4} + \dfrac{(y+2)^2}{1} = 1$

C $(-1,-2)$; move 2 horiz. and 1 vert.

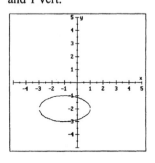

21. $\dfrac{x^2}{9} + \dfrac{y^2}{4} = 1$

$\dfrac{y^2}{4} = 1 - \dfrac{x^2}{9}$

$y^2 = 4\left(1 - \dfrac{x^2}{9}\right)$

$y = \pm\sqrt{4\left(1 - \dfrac{x^2}{9}\right)}$

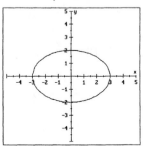

23. $\dfrac{x^2}{4} + \dfrac{(y-1)^2}{9} = 1$

$\dfrac{(y-1)^2}{9} = 1 - \dfrac{x^2}{4}$

$(y-1)^2 = 9\left(1 - \dfrac{x^2}{4}\right)$

$y - 1 = \pm\sqrt{9\left(1 - \dfrac{x^2}{4}\right)}$

$y = 1 \pm\sqrt{9\left(1 - \dfrac{x^2}{4}\right)}$

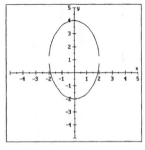

25. $x^2 + 4y^2 - 4x + 8y + 4 = 0$

$x^2 - 4x + 4\left(y^2 + 2y\right) = -4$

$x^2 - 4x + 4 + 4\left(y^2 + 2y + 1\right) = -4 + 4 + 4$

$(x-2)^2 + 4(y+1)^2 = 4$

$\dfrac{(x-2)^2}{4} + \dfrac{(y+1)^2}{1} = 1$

C $(2,-1)$; move 2 horiz. and 1 vert.

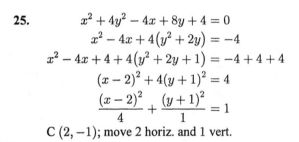

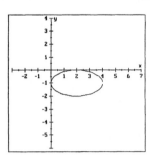

27.
$$9x^2 + 4y^2 - 18x + 16y = 11$$
$$9(x^2 - 2x) + 4(y^2 + 4y) = 11$$
$$9(x^2 - 2x + 1) + 4(y^2 + 4y + 4) = 11 + 9 + 16$$
$$9(x - 1)^2 + 4(y + 2)^2 = 36$$
$$\frac{(x-1)^2}{4} + \frac{(y+2)^2}{9} = 1$$
C $(1, -2)$; move 2 horiz. and 3 vert.

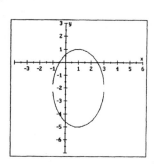

29. Note: $a = 40/2 = 20$, $b = 10$
$$\frac{x^2}{a^2} + \frac{y^2}{b^2} = 1$$
$$\frac{x^2}{400} + \frac{y^2}{100} = 1$$
$$x^2 + 4y^2 = 400$$
$$4y^2 = 400 - x^2$$
$$y^2 = \tfrac{1}{4}(400 - x^2)$$
$$y = \tfrac{1}{2}\sqrt{400 - x^2}$$

31.
$$9x^2 + 16y^2 = 144$$
$$\frac{9x^2}{144} + \frac{16y^2}{144} = \frac{144}{144}$$
$$\frac{x^2}{16} + \frac{y^2}{9} = 1$$
area $= \pi ab$
$$= \pi(4)(3)$$
$$= 12\pi \text{ square units}$$

33. Answers may vary.

35. It is a circle.

Exercise 10.3 (page 663)

1. $-6x^4 + 9x^3 - 6x^2 = -3x^2(2x^2 - 3x + 2)$

3. $15a^2 - 4ab - 4b^2 = (3a - 2b)(5a + 2b)$

5. hyperbola; difference

7. center

9. $(0,0)$

11. $\dfrac{x^2}{9} - \dfrac{y^2}{4} = 1$
C $(0,0)$; open horiz.;
move 3 horiz. and 2 vert.

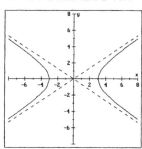

13. $\dfrac{y^2}{4} - \dfrac{x^2}{9} = 1$
C $(0,0)$; open vert.;
move 3 horiz. and 2 vert.

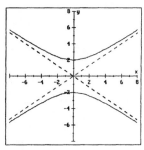

15. $25x^2 - y^2 = 25$

$$\frac{25x^2}{25} - \frac{y^2}{25} = 1$$

$$\frac{x^2}{1} - \frac{y^2}{25} = 1$$

C $(0,0)$; open horiz.;
move 1 horiz. and 5 vert.

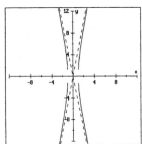

17. $\frac{(x-2)^2}{9} - \frac{y^2}{16} = 1$

C $(2,0)$; open horiz.;
move 3 horiz. and 4 vert.

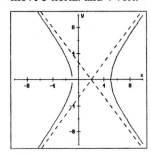

19. $4(x+3)^2 - (y-1)^2 = 4$

$$\frac{4(x+3)^2}{4} - \frac{(y-1)^2}{4} = \frac{4}{4}$$

$$\frac{(x+3)^2}{1} - \frac{(y-1)^2}{4} = 1$$

C $(-3,1)$; open horiz.;
move 1 horiz. and 2 vert.

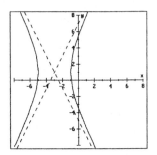

21. $xy = 8$

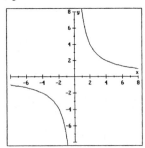

23. $\frac{x^2}{9} - \frac{y^2}{4} = 1$

$$\frac{y^2}{4} = \frac{x^2}{9} - 1$$

$$y^2 = 4\left(\frac{x^2}{9} - 1\right)$$

$$y = \pm\sqrt{4\left(\frac{x^2}{9} - 1\right)}$$

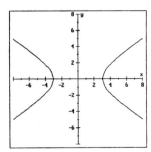

25. $\dfrac{x^2}{4} - \dfrac{(y-1)^2}{9} = 1$

$\dfrac{(y-1)^2}{9} = \dfrac{x^2}{4} - 1$

$(y-1)^2 = 9\left(\dfrac{x^2}{4} - 1\right)$

$y - 1 = \pm\sqrt{9\left(\dfrac{x^2}{4} - 1\right)}$

$y = 1 \pm\sqrt{9\left(\dfrac{x^2}{4} - 1\right)}$

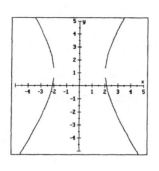

27.

$4x^2 - y^2 + 8x - 4y = 4$

$4x^2 + 8x - y^2 - 4y = 4$

$4\left(x^2 + 2x\right) - \left(y^2 + 4y\right) = 4$

$4\left(x^2 + 2x + 1\right) - \left(y^2 + 4y + 4\right) = 4 + 4 - 4$

$4(x+1)^2 - (y+2)^2 = 4$

$\dfrac{(x+1)^2}{1} - \dfrac{(y+2)^2}{4} = 1$

C $(-1, -2)$; opens horiz.; move 1 horiz. and 2 vert.

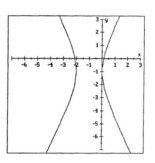

29.

$4y^2 - x^2 + 8y + 4x = 4$

$4y^2 + 8y - x^2 + 4x = 4$

$4\left(y^2 + 2y\right) - \left(x^2 - 4x\right) = 4$

$4\left(y^2 + 2y + 1\right) - \left(x^2 - 4x + 4\right) = 4 + 4 - 4$

$4(y+1)^2 - (x-2)^2 = 4$

$\dfrac{(y+1)^2}{1} - \dfrac{(x-2)^2}{4} = 1$

C $(2, -1)$; opens vert.; move 2 horiz. and 1 vert.

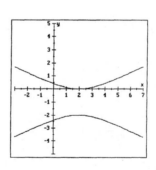

31. $9y^2 - x^2 = 81$

$\dfrac{9y^2}{81} - \dfrac{x^2}{81} = \dfrac{81}{81}$

$\dfrac{y^2}{9} - \dfrac{x^2}{81} = 1$

distance $= \sqrt{9} = 3$ units

33. $y^2 - x^2 = 25$

$\dfrac{y^2}{25} - \dfrac{x^2}{25} = 1$

vertex: $(0, 5)$

Let $y = 10$:

$10^2 - x^2 = 25$

$-x^2 = -75$

$x = \sqrt{75} = 5\sqrt{3}$

width $= 2(5\sqrt{3}) = 10\sqrt{3}$ miles

35. Answers may vary.

37. If $a = b$, the rectangle is a square.

Exercise 10.4 (page 669)

1. $\sqrt{200x^2} - 3\sqrt{98x^2} = \sqrt{100x^2}\sqrt{2} - 3\sqrt{49x^2}\sqrt{2} = 10x\sqrt{2} - 3(7x)\sqrt{2} = -11x\sqrt{2}$

3. $\dfrac{3t\sqrt{2t} - 2\sqrt{2t^3}}{\sqrt{18t} - \sqrt{2t}} = \dfrac{3t\sqrt{2t} - 2\sqrt{t^2}\sqrt{2t}}{\sqrt{9}\sqrt{2t} - \sqrt{2t}} = \dfrac{3t\sqrt{2t} - 2t\sqrt{2t}}{3\sqrt{2t} - \sqrt{2t}} = \dfrac{t\sqrt{2t}}{2\sqrt{2t}} = \dfrac{t}{2}$

5. graphing; substitution

7. $\begin{cases} 8x^2 + 32y^2 = 256 \\ x = 2y \end{cases}$

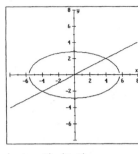

$(-4, -2), (4, 2)$

9. $\begin{cases} x^2 + y^2 = 10 \\ y = 3x^2 \end{cases}$

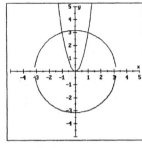

$(-1, 3), (1, 3)$

11. $\begin{cases} x^2 + y^2 = 25 \\ 12x^2 + 64y^2 = 768 \end{cases}$

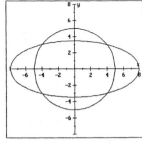

$(-4, 3), (4, 3), (4, -3)$
$(-4, -3)$

13. $\begin{cases} x^2 - 13 = -y^2 \\ y = 2x - 4 \end{cases}$

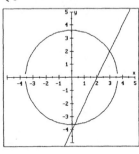

$\left(\frac{1}{5}, -\frac{18}{5}\right), (3, 2)$

15. $\begin{cases} x^2 - 6x - y = -5 \\ x^2 - 6x + y = -5 \end{cases}$
$\begin{cases} y = x^2 - 6x + 5 \\ y = -x^2 + 6x - 5 \end{cases}$

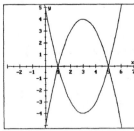

$(1, 0), (5, 0)$

17. $\begin{cases} (1) & 25x^2 + 9y^2 = 225 \\ (2) & 5x + 3y = 15 \end{cases}$

Substitute $x = -\frac{3}{5}y + 3$ from (2) into (1):

$$25x^2 + 9y^2 = 225$$
$$25\left(-\frac{3}{5}y + 3\right)^2 + 9y^2 = 225$$
$$25\left(\frac{9}{25}y^2 - \frac{18}{5}y + 9\right) + 9y^2 = 225$$
$$9y^2 - 90y + 225 + 9y^2 = 225$$
$$18y^2 - 90y = 0$$
$$18y(y - 5) = 0$$

$18y = 0$ **or** $y - 5 = 0$
$\quad y = 0 \qquad\qquad y = 5$

Substitute these and solve for x:

$\begin{array}{ll} 5x + 3y = 15 & 5x + 3y = 15 \\ 5x + 3(0) = 15 & 5x + 3(5) = 15 \\ \quad 5x = 15 & \qquad 5x = 0 \\ \qquad x = 3 & \qquad\quad x = 0 \end{array}$

Solutions: $(3, 0), (0, 5)$

19. $\begin{cases} (1) & x^2 + y^2 = 2 \\ (2) & x + y = 2 \end{cases}$

Substitute $x = 2 - y$ from (2) into (1):

$$x^2 + y^2 = 2$$
$$(2 - y)^2 + y^2 = 2$$
$$4 - 4y + y^2 + y^2 = 2$$
$$2y^2 - 4y + 2 = 0$$
$$2(y - 1)(y - 1) = 0$$

$y - 1 = 0$ **or** $y - 1 = 0$
$\quad y = 1 \qquad\qquad y = 1$

Substitute this and solve for x:

$x = 2 - y$
$x = 2 - 1$
$x = 1$

Solution: $(1, 1)$

21. $\begin{cases} (1) & x^2 + y^2 = 5 \\ (2) & x + y = 3 \end{cases}$

Substitute $x = 3 - y$ from (2) into (1):

$$x^2 + y^2 = 5$$
$$(3 - y)^2 + y^2 = 5$$
$$9 - 6y + y^2 + y^2 = 5$$
$$2y^2 - 6y + 4 = 0$$
$$2(y - 2)(y - 1) = 0$$

$y - 2 = 0$ **or** $y - 1 = 0$
$\quad y = 2 \qquad\qquad y = 1$

Substitute these and solve for x:

$\begin{array}{ll} x = 3 - y & x = 3 - y \\ x = 3 - 2 & x = 3 - 1 \\ x = 1 & x = 2 \end{array}$

Solutions: $(1, 2), (2, 1)$

23. $\begin{cases} (1) & x^2 + y^2 = 13 \\ (2) & y = x^2 - 1 \end{cases}$

Substitute $x^2 = 13 - y^2$ from (1) into (2):

$$y = x^2 - 1$$
$$y = 13 - y^2 - 1$$
$$y^2 + y - 12 = 0$$
$$(y + 4)(y - 3) = 0$$

$y + 4 = 0$ **or** $y - 3 = 0$
$\quad y = -4 \qquad\qquad y = 3$

Substitute these and solve for x:

$\begin{array}{ll} x^2 = 13 - y^2 & x^2 = 13 - y^2 \\ x^2 = 13 - 16 & x^2 = 13 - 9 \\ x^2 = -4 & x^2 = 4 \\ \text{complex} & x = \pm 2 \end{array}$

Solutions: $(2, 3), (-2, 3)$

25. $\begin{cases} (1) & x^2 + y^2 = 30 \\ (2) & y = x^2 \end{cases}$

Substitute $x^2 = y$ from (2) into (1):

$$x^2 + y^2 = 30$$
$$y + y^2 = 30$$
$$y^2 + y - 30 = 0$$
$$(y - 5)(y + 6) = 0$$

$y - 5 = 0$ **or** $y + 6 = 0$
$\qquad y = 5 \qquad\qquad y = -6$

Substitute these and solve for x:

$\begin{array}{ll} x^2 = y & x^2 = y \\ x^2 = 5 & x^2 = -6 \\ x = \pm\sqrt{5} & \text{complex} \end{array}$

Solutions: $\left(\sqrt{5}, 5\right), \left(-\sqrt{5}, 5\right)$

27. $\begin{array}{ll} x^2 + y^2 = & 13 \\ x^2 - y^2 = & 5 \\ \hline 2x^2 \quad = & 18 \\ x^2 \quad = & 9 \\ x \quad = & \pm 3 \end{array}$

Substitute and solve for y:

$\begin{array}{ll} x^2 + y^2 = 13 & x^2 + y^2 = 13 \\ 3^2 + y^2 = 13 & (-3)^2 + y^2 = 13 \\ y^2 = 4 & y^2 = 4 \\ y = \pm 2 & y = \pm 2 \end{array}$

Solutions: $(3, 2), (3, -2), (-3, 2), (-3, -2)$

29. $\begin{array}{ll} x^2 + y^2 = & 20 \\ x^2 - y^2 = & -12 \\ \hline 2x^2 \quad = & 8 \\ x^2 \quad = & 4 \\ x \quad = & \pm 2 \end{array}$

Substitute and solve for y:

$\begin{array}{ll} x^2 + y^2 = 20 & x^2 + y^2 = 20 \\ 2^2 + y^2 = 20 & (-2)^2 + y^2 = 20 \\ y^2 = 16 & y^2 = 16 \\ y = \pm 4 & y = \pm 4 \end{array}$

Solutions: $(2, 4), (2, -4), (-2, 4), (-2, -4)$

31. $\begin{cases} (1) & y^2 = 40 - x^2 \\ (2) & y = x^2 - 10 \end{cases}$

Substitute $x^2 = 40 - y^2$ from (1) into (2):

$$y = x^2 - 10$$
$$y = 40 - y^2 - 10$$
$$y^2 + y - 30 = 0$$
$$(y + 6)(y - 5) = 0$$

$y + 6 = 0$ **or** $y - 5 = 0$
$\qquad y = -6 \qquad\qquad y = 5$

Substitute these and solve for x:

$\begin{array}{ll} x^2 = 40 - y^2 & x^2 = 40 - y^2 \\ x^2 = 40 - 36 & x^2 = 40 - 25 \\ x^2 = 4 & x^2 = 15 \\ x = \pm 2 & x = \pm\sqrt{15} \end{array}$

$(2, -6), (-2, -6), \left(\sqrt{15}, 5\right), \left(-\sqrt{15}, 5\right)$

33. $\begin{cases} (1) & y = x^2 - 4 \\ (2) & x^2 - y^2 = -16 \end{cases}$

Substitute $x^2 = y + 4$ from (1) into (2):

$$x^2 - y^2 = -16$$
$$y + 4 - y^2 = -16$$
$$y^2 - y - 20 = 0$$
$$(y + 4)(y - 5) = 0$$

$y + 4 = 0$ **or** $y - 5 = 0$
$\qquad y = -4 \qquad\qquad y = 5$

Substitute these and solve for x:

$\begin{array}{ll} x^2 = y + 4 & x^2 = y + 4 \\ x^2 = -4 + 4 & x^2 = 5 + 4 \\ x^2 = 0 & x^2 = 9 \\ x = 0 & x = \pm 3 \end{array}$

Solutions: $(0, -4), (3, 5), (-3, 5)$

35. $\begin{aligned} x^2 - y^2 &= -5 \Rightarrow (\times 2) \\ 3x^2 + 2y^2 &= 30 \Rightarrow \end{aligned}$ $\begin{aligned} 2x^2 - 2y^2 &= -10 \\ 3x^2 + 2y^2 &= 30 \\ \hline 5x^2 &= 20 \\ x^2 &= 4 \\ x &= \pm 2 \end{aligned}$

Substitute and solve for y:

$\begin{aligned} x^2 - y^2 &= -5 \\ 2^2 - y^2 &= -5 \\ -y^2 &= -9 \\ y^2 &= 9 \\ y &= \pm 3 \end{aligned}$ $\qquad$ $\begin{aligned} x^2 - y^2 &= -5 \\ (-2)^2 - y^2 &= -5 \\ -y^2 &= -9 \\ y^2 &= 9 \\ y &= \pm 3 \end{aligned}$

Solutions: $(2,3),(2,-3),(-2,3),(-2,-3)$

37. $\begin{aligned} \frac{1}{x} + \frac{2}{y} &= 1 \Rightarrow \\ \frac{2}{x} - \frac{1}{y} &= \frac{1}{3} \Rightarrow (\times 2) \end{aligned}$ $\begin{aligned} \frac{1}{x} + \frac{2}{y} &= 1 \\ \frac{4}{x} - \frac{2}{y} &= \frac{2}{3} \\ \hline \frac{5}{x} &= \frac{5}{3} \\ 15 &= 5x \\ 3 &= x \end{aligned}$

Substitute and solve for y:

$\begin{aligned} \frac{1}{x} + \frac{2}{y} &= 1 \\ \frac{1}{3} + \frac{2}{y} &= 1 \\ \frac{2}{y} &= \frac{2}{3} \\ 6 &= 2y \\ 3 &= y \end{aligned}$

Solution: $(3,3)$

39. $\begin{cases} (1) & 3y^2 = xy \\ (2) & 2x^2 + xy - 84 = 0 \end{cases}$

From (1): $3y^2 - xy = 0$

$y(3y - x) = 0$

$y = 0 \quad \text{or} \quad y = \frac{1}{3}x$

Substitute these into (2):

$\begin{aligned} 2x^2 + xy - 84 &= 0 \\ 2x^2 + x(0) - 84 &= 0 \\ 2x^2 &= 84 \\ x^2 &= 42 \\ x &= \pm\sqrt{42} \end{aligned}$

$\begin{aligned} 2x^2 + xy - 84 &= 0 \\ 2x^2 + x\left(\tfrac{1}{3}x\right) - 84 &= 0 \\ 2x^2 + \tfrac{1}{3}x^2 &= 84 \\ 6x^2 + x^2 &= 252 \\ 7x^2 &= 252 \\ x^2 &= 36 \\ x &= \pm 6 \end{aligned}$

(substitute and solve for y)

Solutions: $\left(\sqrt{42},0\right), \left(-\sqrt{42},0\right), (6,2), (-6,-2)$

41. $\begin{cases} (1) & xy = \frac{1}{6} \\ (2) & y + x = 5xy \end{cases}$

Substitute $x = \frac{1}{6y}$ from (1) into (2):

$\begin{aligned} y + \frac{1}{6y} &= \frac{5y}{6y} \\ 6y^2 + 1 &= 5y \\ 6y^2 - 5y + 1 &= 0 \\ (2y - 1)(3y - 1) &= 0 \end{aligned}$

$2y - 1 = 0 \quad \text{or} \quad 3y - 1 = 0$

$y = \frac{1}{2} \qquad\qquad y = \frac{1}{3}$

Substitute these and solve for x:

$x = \frac{1}{6y} \qquad\qquad x = \frac{1}{6y}$

$x = \dfrac{1}{6\left(\frac{1}{2}\right)} \qquad x = \dfrac{1}{6\left(\frac{1}{3}\right)}$

$x = \frac{1}{3} \qquad\qquad x = \frac{1}{2}$

Solutions: $\left(\frac{1}{3},\frac{1}{2}\right), \left(\frac{1}{2},\frac{1}{3}\right)$

43. Let the integers be x and y. Then the equations are

$$\begin{cases} (1) & xy = 32 \\ (2) & x + y = 12 \end{cases}$$

Substitute $x = \frac{32}{y}$ from (1) into (2):

$$\frac{32}{y} + y = 12$$
$$32 + y^2 = 12y$$
$$y^2 - 12y + 32 = 0$$
$$(y - 4)(y - 8) = 0$$

$y - 4 = 0$ **or** $y - 8 = 0$
$y = 4 \qquad\qquad y = 8$

Substitute these and solve for x:

$$x = \frac{32}{y} = \frac{32}{4} = 8 \qquad x = \frac{32}{y} = \frac{32}{8} = 4$$

The integers are 8 and 4.

45. Let l = the length of the rectangle, and w = the width of the rectangle. Then the equations are:

$$\begin{cases} (1) & lw = 63 \\ (2) & 2l + 2w = 32 \end{cases}$$

Substitute $l = \frac{63}{w}$ from (1) into (2):

$$2\left(\frac{63}{w}\right) + 2w = 32$$
$$\frac{126}{w} + 2w = 32$$
$$126 + 2w^2 = 32w$$
$$2w^2 - 32y + 126 = 0$$
$$2(w - 7)(w - 9) = 0$$

$w - 7 = 0$ **or** $w - 9 = 0$
$w = 7 \qquad\qquad w = 9$

Substitute these and solve for l:

$$l = \frac{63}{w} = \frac{63}{7} = 9 \qquad l = \frac{63}{w} = \frac{63}{9} = 7$$

The dimensions are 7 cm by 9 cm.

47. Let r = Rania's rate, and let p = the amount Rania invested.
Then Jerome invested $p + 150$ at a rate of $r + 0.015$. The equations are

$$\begin{cases} (1) & pr = 67.50 \Rightarrow p = \frac{67.5}{r} \\ (2) & (p + 150)(r + 0.015) = 94.5 \end{cases}$$

Substitute $p = \frac{67.5}{r}$ from (1) into (2):

$$\left(\frac{67.5}{r} + 150\right)(r + 0.015) = 94.5$$
$$67.5 + \frac{1.0125}{r} + 150r + 2.25 = 94.5$$
$$67.5r + 1.0125 + 150r^2 + 2.25r = 94.5r$$
$$150r^2 - 24.75r - 1.0125 = 0$$
$$12{,}000r^2 - 1980r + 81 = 0$$
$$(100r - 9)(120r - 9) = 0$$

$100r - 9 = 0$ **or** $120r - 9 = 0$
$r = 0.09 \qquad\qquad r = 0.075$

Substitute and solve for p:

$$p = \frac{67.5}{r} = \frac{67.5}{0.09} = 750 \text{ or } p = \frac{67.5}{r} = \frac{67.5}{0.075} = 900$$

Rania invested \$750 at 9% or \$900 at 7.5%.

49. Let r = Jim's rate and t = Jim's time. Then his brother's rate was $r - 17$ and his time was $t + 1.5$.

$$\begin{cases} (1) \ rt = 306 \Rightarrow t = \frac{306}{r} \\ (2) \ (r-17)(t+1.5) = 306 \end{cases}$$

Substitute $t = \frac{306}{r}$ from (1) into (2):

$$(r-17)\left(\frac{306}{r} + 1.5\right) = 306$$

$$306 + 1.5r - \frac{5202}{r} - 25.5 = 306$$

$$306r + 1.5r^2 - 5202 - 25.5r = 306r$$

$$1.5r^2 - 25.5r - 5202 = 0$$

$$3r^2 - 51r - 10{,}404 = 0$$

$$(3r + 153)(r - 68) = 0$$

$3r + 153 = 0$ **or** $r - 68 = 0$

$r = -153/3$ $r = 68$

Substitute and solve for t:

$$t = \frac{306}{r} = \frac{306}{68} = 4.5$$

Jim drove for 4.5 hours at 68 miles per hour.

51. Answers may vary.

53. $0, 1, 2, 3, 4$

Exercise 10.5 (page 676)

1. $(6x - 10) + (3x + 10) = 180$

$$9x = 180$$

$$x = 20$$

3. domains

5. constant; $f(x)$

7. step

9. increasing on $(-\infty, 0)$, decreasing on $(0, \infty)$

11. decreasing on $(-\infty, 0)$, constant on $(0, 2)$ increasing on $(2, \infty)$

13. $f(x) = \begin{cases} -1 & \text{if } x \le 0 \\ x & \text{if } x > 0 \end{cases}$

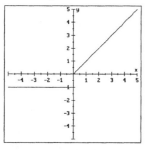

constant on $(-\infty, 0)$
increasing on $(0, \infty)$

15. $f(x) = \begin{cases} -x & \text{if } x \le 0 \\ x & \text{if } 0 < x < 2 \\ -x & \text{if } x \ge 2 \end{cases}$

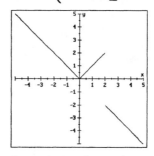

decreasing on $(-\infty, 0)$
increasing on $(0, 2)$
decreasing on $(2, \infty)$

17. $f(x) = -[[x]]$

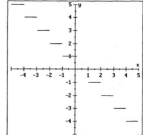

SECTION 10.5

19. $f(x) = 2[[x]]$

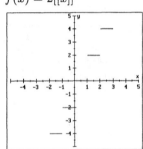

21. $f(x) = \begin{cases} -1 & \text{if } x < 0 \\ 0 & \text{if } x = 0 \\ 1 & \text{if } x > 0 \end{cases}$

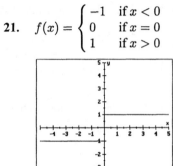

23. Find y when $x = 2.5$.
cost = \$30

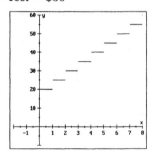

25. After 2 hours, B is cheaper.

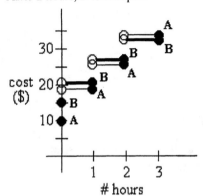

27. Answers may vary.

29. $f(x) = \begin{cases} x & \text{if } x < -2 \\ -x & \text{if } x > -2 \end{cases}$

Chapter 10 Summary (page 679)

1. $(x-1)^2 + (y+2)^2 = 9$
$C(1,-2); r = \sqrt{9} = 3$

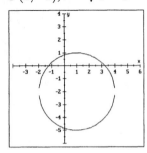

2. $x^2 + y^2 = 16$
$C(0,0); r = \sqrt{16} = 4$

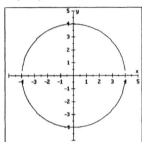

3.

$$x^2 + y^2 + 4x - 2y = 4$$
$$x^2 + 4x + y^2 - 2y = 4$$
$$x^2 + 4x + 4 + y^2 - 2y + 1 = 4 + 4 + 1$$
$$(x + 2)^2 + (y - 1)^2 = 9$$
$$C\,(-2, 1);\, r = \sqrt{9} = 3$$

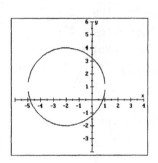

4. $x = -3(y - 2)^2 + 5$
V $(5, 2)$; opens L

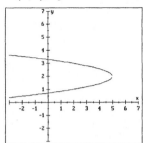

5. $x = 2(y + 1)^2 - 2$
V $(-2, -1)$; opens R

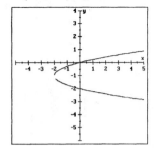

6. $9x^2 + 16y^2 = 144$
$$\frac{9x^2}{144} + \frac{16y^2}{144} = \frac{144}{144}$$
$$\frac{x^2}{16} + \frac{y^2}{9} = 1$$
C $(0, 0)$; move 4 horiz. and 3 vert.

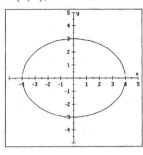

7. $\dfrac{(x - 2)^2}{4} + \dfrac{(y - 1)^2}{9} = 1$
C $(2, 1)$; move 2 horiz. and 3 vert.

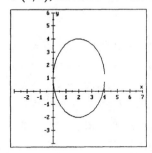

8.

$$4x^2 + 9y^2 + 8x - 18y = 23$$
$$4x^2 + 8x + 9y^2 - 18y = 23$$
$$4(x^2 + 2x) + 9(y^2 - 2y) = 23$$
$$4(x^2 + 2x + 1) + 9(y^2 - 2y + 1) = 23 + 4 + 9$$
$$4(x + 1)^2 + 9(y - 1)^2 = 36$$
$$\frac{(x + 1)^2}{9} + \frac{(y - 1)^2}{4} = 1$$

C $(-1, 1)$; move 3 horiz. and 2 vert.

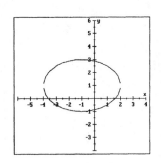

9.

$$9x^2 - y^2 = -9$$
$$\frac{9x^2}{-9} - \frac{y^2}{-9} = \frac{-9}{-9}$$
$$\frac{y^2}{9} - \frac{x^2}{1} = 1$$

C $(0, 0)$; opens vert.; move 1 horiz. and 3 vert.

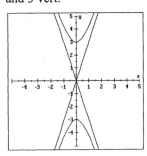

10. $xy = 9$

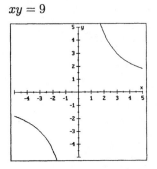

11.

$$4x^2 - 2y^2 + 8x - 8y = 8$$
$$4x^2 + 8x - 2y^2 - 8y = 8$$
$$4(x^2 + 2x) - 2(y^2 + 4y) = 8$$
$$4(x^2 + 2x + 1) - 2(y^2 + 4y + 4) = 8$$
$$4(x + 1)^2 - 2(y + 2)^2 = 8$$
$$\frac{(x + 1)^2}{2} - \frac{(y + 2)^2}{4} = 1 \Rightarrow \text{hyperbola}$$

12.

$$9x^2 - 4y^2 - 18x - 8y = 31$$
$$9x^2 - 18x - 4y^2 - 8y = 31$$
$$9(x^2 - 2x) - 4(y^2 + 2y) = 31$$
$$9(x^2 - 2x + 1) - 4(y^2 + 2y + 1) = 31 + 9 - 4$$
$$9(x - 1)^2 - 4(y + 1)^2 = 36$$
$$\frac{(x-1)^2}{4} - \frac{(y+1)^2}{9} = 1$$

C $(1, -1)$; opens horiz.; move 2 horiz. and 3 vert.

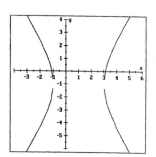

13. $3x^2 + y^2 = 52$ Substitute and solve for y:

$$\underline{x^2 - y^2 = 12}$$
$$4x^2 = 64$$
$$x^2 = 16$$
$$x = \pm 4$$

$x^2 - y^2 = 12$	$x^2 - y^2 = 12$
$4^2 - y^2 = 12$	$(-4)^2 - y^2 = 12$
$y^2 = 4$	$y^2 = 4$
$y = \pm 2$	$y = \pm 2$

Solutions: $(4, 2), (4, -2), (-4, 2), (-4, -2)$.

14. $\frac{x^2}{16} + \frac{y^2}{12} = 1 \Rightarrow \times 48$ $3x^2 + 4y^2 = 48$ Substitute and solve for x:

$$\underline{x^2 - \frac{y^2}{3} = 1 \Rightarrow \times (-3) \quad -3x^2 + y^2 = -3}$$
$$5y^2 = 45$$
$$y^2 = 9$$
$$y = \pm 3$$

$$3x^2 + 4y^2 = 48$$
$$3x^2 + 4(9) = 48$$
$$3x^2 = 12$$
$$x^2 = 4 \Rightarrow x = \pm 2$$

Solutions: $(2, 3), (2, -3), (-2, 3), (-2, -3)$

15. increasing on $(-\infty, -2)$; constant on $(-2, 1)$; decreasing on $(1, \infty)$

16. $f(x) = \begin{cases} x & \text{if } x \le 1 \\ -x^2 & \text{if } x > 1 \end{cases}$

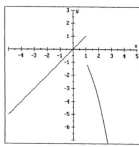

17. $f(x) = 3[[x]]$

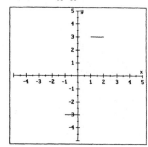

Chapter 10 Test (page 681)

1. $(x-2)^2 + (y+3)^2 = 4$
Center: $(2, -3)$; radius $= 2$

2.
$$x^2 + y^2 + 4x - 6y = 3$$
$$x^2 + 4x + y^2 - 6y = 3$$
$$x^2 + 4x + 4 + y^2 - 6y + 9 = 3 + 4 + 9$$
$$(x+2)^2 + (y-3)^2 = 16$$
Center: $(-2, 3)$; radius $= 4$

3. $(x+1)^2 + (y-2)^2 = 9$
$C(-1, 2); r = \sqrt{9} = 3$

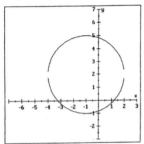

4. $x = (y-2)^2 - 1$
$V(-1, 2)$; opens R

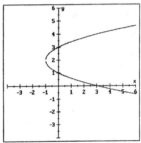

5.
$$9x^2 + 4y^2 = 36$$
$$\frac{9x^2}{36} + \frac{4y^2}{36} = \frac{36}{36}$$
$$\frac{x^2}{4} + \frac{y^2}{9} = 1$$
$C(0,0)$; move 2 horiz. and 3 vert.

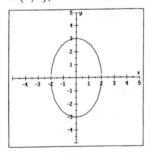

6. $\dfrac{(x-2)^2}{9} - y^2 = 1$
$C(2,0)$; opens horiz; move 3 horiz and 1 vert

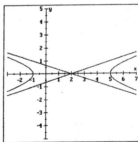

7.
$$4x^2 + y^2 - 24x + 2y = -33$$
$$4(x^2 - 6x) + (y^2 + 2y) = -33$$
$$4(x^2 - 6x + 9) + (y^2 + 2y + 1) = -33 + 36 + 1$$
$$4(x - 3)^2 + (y + 1)^2 = 4$$
$$\frac{(x - 3)^2}{1} + \frac{(y + 1)^2}{4} = 1$$
C $(3, -1)$; move 1 horiz. and 2 vert.

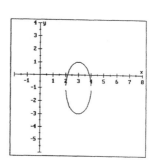

8.
$$x^2 - 9y^2 + 2x + 36y = 44$$
$$x^2 + 2x - 9y^2 + 36y = 44$$
$$(x^2 + 2x) - 9(y^2 - 4y) = 44$$
$$(x^2 + 2x + 1) - 9(y^2 - 4y + 4) = 44 + 1 - 36$$
$$(x + 1)^2 - 9(y - 2)^2 = 9$$
$$\frac{(x + 1)^2}{9} - \frac{(y - 2)^2}{1} = 1$$
C $(-1, 2)$; opens horiz.; move 3 horiz. and 1 vert.

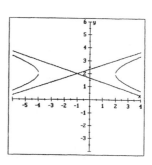

9.
$$\begin{cases} (1) & 2x - y = -2 \\ (2) & x^2 + y^2 = 16 + 4y \end{cases}$$
Substitute $y = 2x + 2$ from (1) into (2):
$$x^2 + y^2 = 16 + 4y$$
$$x^2 + (2x + 2)^2 = 16 + 4(2x + 2)$$
$$x^2 + 4x^2 + 8x + 4 = 16 + 8x + 8$$
$$5x^2 - 20 = 0$$
$$5(x + 2)(x - 2) = 0$$

$x + 2 = 0$ **or** $x - 2 = 0$
$\qquad x = -2 \qquad\qquad x = 2$
Substitute these and solve for y:
$y = 2x + 2 \qquad\qquad y = 2x + 2$
$y = 2(-2) + 2 \qquad\quad y = 2(2) + 2$
$y = -4 + 2 = -2 \qquad y = 4 + 2 = 6$
Solutions: $(-2, -2), (2, 6)$

10.
$$\begin{cases} (1) & x^2 + y^2 = 25 \\ (2) & 4x^2 - 9y = 0 \end{cases}$$
Substitute $x^2 = 25 - y^2$ from (1) into (2):
$$4x^2 - 9y = 0$$
$$4(25 - y^2) - 9y = 0$$
$$100 - 4y^2 - 9y = 0$$
$$-4y^2 - 9y + 100 = 0$$
$$4y^2 + 9y - 100 = 0$$
$$(y - 4)(4y + 25) = 0$$

$y - 4 = 0$ **or** $4y + 25 = 0$
$\qquad y = 4 \qquad\qquad y = -\frac{25}{4}$
Substitute these and solve for x:
$x^2 = 25 - y^2 \qquad\qquad x^2 = 25 - y^2$
$x^2 = 25 - 4^2 \qquad\qquad x^2 = 25 - \left(-\frac{25}{4}\right)^2$
$x^2 = 25 - 16 = 9 \qquad x^2 = 25 - \frac{625}{16} = -\frac{225}{16}$
$x = \pm 3 \qquad\qquad\qquad x$ is nonreal.
Solutions: $(3, 4), (-3, 4)$

11. increasing: $(-3,0)$; decreasing: $(0,3)$

12. $f(x) = \begin{cases} -x^2 & \text{when } x < 0 \\ -x & \text{when } x \geq 0 \end{cases}$

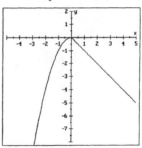

Cumulative Review Exercises (page 683)

1. $(4x - 3y)(3x + y) = 12x^2 + 4xy - 9xy - 3y^2 = 12x^2 - 5xy - 3y^2$

2. $(a^n + 1)(a^n - 3) = a^n a^n - 3a^n + a^n - 3 = a^{2n} - 2a^n - 3$

3. $\dfrac{5a - 10}{a^2 - 4a + 4} = \dfrac{5(a - 2)}{(a - 2)(a - 2)} = \dfrac{5}{a - 2}$

4. $\dfrac{a^4 - 5a^2 + 4}{a^2 + 3a + 2} = \dfrac{(a^2 - 4)(a^2 - 1)}{(a + 2)(a + 1)} = \dfrac{(a + 2)(a - 2)(a + 1)(a - 1)}{(a + 2)(a + 1)} = (a - 2)(a - 1)$
$$= a^2 - 3a + 2$$

5. $\dfrac{a^2 - a - 6}{a^2 - 4} \div \dfrac{a^2 - 9}{a^2 + a - 6} = \dfrac{a^2 - a - 6}{a^2 - 4} \cdot \dfrac{a^2 + a - 6}{a^2 - 9} = \dfrac{(a - 3)(a + 2)}{(a + 2)(a - 2)} \cdot \dfrac{(a + 3)(a - 2)}{(a + 3)(a - 3)} = 1$

6. $\dfrac{2}{a - 2} + \dfrac{3}{a + 2} - \dfrac{a - 1}{a^2 - 4} = \dfrac{2}{a - 2} + \dfrac{3}{a + 2} - \dfrac{a - 1}{(a + 2)(a - 2)}$
$$= \dfrac{2(a + 2)}{(a - 2)(a + 2)} + \dfrac{3(a - 2)}{(a + 2)(a - 2)} - \dfrac{a - 1}{(a + 2)(a - 2)}$$
$$= \dfrac{2(a + 2) + 3(a - 2) - (a - 1)}{(a + 2)(a - 2)}$$
$$= \dfrac{2a + 4 + 3a - 6 - a + 1}{(a + 2)(a - 2)} = \dfrac{4a - 1}{(a + 2)(a - 2)}$$

7. $3x - 4y = 12$ $\qquad y = \dfrac{3}{4}x - 5$
$\qquad -4y = -3x + 12$
$\qquad\qquad y = \dfrac{3}{4}x - 3 \qquad m = \dfrac{3}{4}$
$\qquad\qquad m = \dfrac{3}{4}$
Parallel

8. $y = 3x + 4 \qquad x = -3y + 4$
$\qquad m = 3 \qquad 3y = -x + 4$
$\qquad\qquad y = -\dfrac{1}{3}x + \dfrac{4}{3}$
$\qquad\qquad m = -\dfrac{1}{3}$
Perpendicular

9. $\quad y - y_1 = m(x - x_1)$
$\quad\quad y - 5 = -2(x - 0)$
$\quad\quad y - 5 = -2x$
$\quad\quad\quad\quad y = -2x + 5$

10. $\quad m = \dfrac{y_2 - y_1}{x_2 - x_1} = \dfrac{4 - (-5)}{-5 - 8} = \dfrac{9}{-13} = -\dfrac{9}{13}$
$\quad\quad y - y_1 = m(x - x_1)$
$\quad\quad y - (-5) = -\dfrac{9}{13}(x - 8)$
$\quad\quad\quad y + 5 = -\dfrac{9}{13}x + \dfrac{72}{13}$
$\quad\quad\quad\quad\quad y = -\dfrac{9}{13}x + \dfrac{7}{13}$

11. $\quad 2x - 3y < 6$

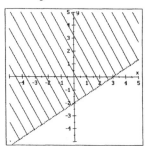

12. $\quad y \geq x^2 - 4$

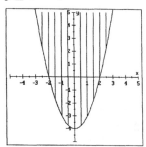

13. $\quad \sqrt{98} + \sqrt{8} - \sqrt{32} = \sqrt{49}\sqrt{2} + \sqrt{4}\sqrt{2} - \sqrt{16}\sqrt{2} = 7\sqrt{2} + 2\sqrt{2} - 4\sqrt{2} = 5\sqrt{2}$

14. $\quad 12\sqrt[3]{648x^4} + 3\sqrt[3]{81x^4} = 12\sqrt[3]{216x^3}\sqrt[3]{3x} + 3\sqrt[3]{27x^3}\sqrt[3]{3x} = 12(6x)\sqrt[3]{3x} + 3(3x)\sqrt[3]{3x}$
$$= 72x\sqrt[3]{3x} + 9x\sqrt[3]{3x} = 81x\sqrt[3]{3x}$$

15. $\quad \sqrt{3a + 1} = a - 1$
$\quad\quad \left(\sqrt{3a + 1}\right)^2 = (a - 1)^2$
$\quad\quad\quad 3a + 1 = a^2 - 2a + 1$
$\quad\quad\quad\quad\quad 0 = a^2 - 5a$
$\quad\quad\quad\quad\quad 0 = a(a - 5)$
$\quad\quad a = 0 \quad$ **or** $\quad a - 5 = 0$
$\quad\quad$ doesn't check $\quad\quad\quad a = 5$

16. $\quad \sqrt{x + 3} - \sqrt{3} = \sqrt{x}$
$\quad\quad\quad \sqrt{x + 3} = \sqrt{x} + \sqrt{3}$
$\quad\quad \left(\sqrt{x + 3}\right)^2 = \left(\sqrt{x} + \sqrt{3}\right)^2$
$\quad\quad\quad x + 3 = x + 2\sqrt{3x} + 3$
$\quad\quad\quad\quad\quad 0 = 2\sqrt{3x}$
$\quad\quad\quad\quad\quad 0^2 = \left(2\sqrt{3x}\right)^2$
$\quad\quad\quad\quad\quad 0 = 4(3x)$
$\quad\quad\quad\quad\quad 0 = 12x$
$\quad\quad\quad\quad\quad 0 = x$

17. $\quad 6a^2 + 5a - 6 = 0$
$\quad\quad (2a + 3)(3a - 2) = 0$
$\quad\quad 2a + 3 = 0 \quad$ **or** $\quad 3a - 2 = 0$
$\quad\quad\quad a = -\frac{3}{2} \quad\quad\quad\quad a = \frac{2}{3}$

18. $3x^2 + 8x - 1 = 0; a = 3, b = 8, c = -1$

$$x = \frac{-b \pm \sqrt{b^2 - 4ac}}{2a} = \frac{-8 \pm \sqrt{8^2 - 4(3)(-1)}}{2(3)} = \frac{-8 \pm \sqrt{64 + 12}}{6} = \frac{-8 \pm \sqrt{76}}{6}$$

$$= -\frac{8}{6} \pm \frac{2\sqrt{19}}{6}$$

$$= -\frac{4}{3} \pm \frac{\sqrt{19}}{3}$$

19. $(f \circ g)(x) = f(g(x)) = f(2x + 1) = (2x + 1)^2 - 2 = 4x^2 + 4x + 1 - 2 = 4x^2 + 4x - 1$

20.
$$y = 2x^3 - 1$$
$$x = 2y^3 - 1$$
$$x + 1 = 2y^3$$
$$\frac{x+1}{2} = y^3$$
$$\sqrt[3]{\frac{x+1}{2}} = y$$
$$y = f^{-1}(x) = \sqrt[3]{\frac{x+1}{2}}$$

21. $y = \left(\frac{1}{2}\right)^x$

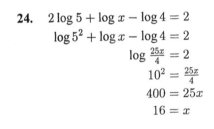

22. $y = \log_2 x \Rightarrow 2^y = x$

23.
$$2^{x+2} = 3^x$$
$$\log 2^{x+2} = \log 3^x$$
$$(x + 2) \log 2 = x \log 3$$
$$x \log 2 + 2 \log 2 = x \log 3$$
$$2 \log 2 = x \log 3 - x \log 2$$
$$2 \log 2 = x(\log 3 - \log 2)$$
$$\frac{2 \log 2}{\log 3 - \log 2} = x$$

24.
$$2 \log 5 + \log x - \log 4 = 2$$
$$\log 5^2 + \log x - \log 4 = 2$$
$$\log \frac{25x}{4} = 2$$
$$10^2 = \frac{25x}{4}$$
$$400 = 25x$$
$$16 = x$$

25. $x^2 + (y + 1)^2 = 9$

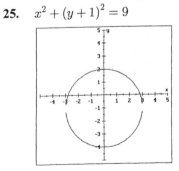

26. $x^2 - 9(y + 1)^2 = 9$

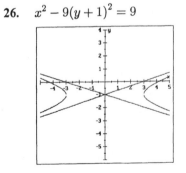

Exercise 11.1 (page 689)

1. $\log_4 16 = x \Rightarrow 4^x = 16 \Rightarrow x = 2$

3. $\log_{25} x = \frac{1}{2} \Rightarrow 25^{1/2} = x \Rightarrow x = 5$

5. one

7. Pascal's

9. 6!

11. 1

13. $3! = 3 \cdot 2 \cdot 1 = 6$

15. $-5! = -1 \cdot 5! = -1 \cdot 5 \cdot 4 \cdot 3 \cdot 2 \cdot 1 = -120$

17. $3! + 4! = 3 \cdot 2 \cdot 1 + 4 \cdot 3 \cdot 2 \cdot 1$
$\qquad = 6 + 24 = 30$

19. $3!(4!) = 3 \cdot 2 \cdot 1 \cdot 4 \cdot 3 \cdot 2 \cdot 1 = 144$

21. $8(7!) = 8 \cdot 7 \cdot 6 \cdot 5 \cdot 4 \cdot 3 \cdot 2 \cdot 1 = 40{,}320$

23. $\dfrac{9!}{11!} = \dfrac{9!}{11 \cdot 10 \cdot 9!} = \dfrac{1}{11 \cdot 10} = \dfrac{1}{110}$

25. $\dfrac{49!}{47!} = \dfrac{49 \cdot 48 \cdot 47!}{47!} = 49 \cdot 48 = 2352$

27. $\dfrac{5!}{3!(5-3)!} = \dfrac{5!}{3!2!} = \dfrac{5 \cdot 4 \cdot 3!}{3! \cdot 2 \cdot 1} = \dfrac{5 \cdot 4}{2 \cdot 1} = 10$

29. $\dfrac{7!}{5!(7-5)!} = \dfrac{7!}{5!2!} = \dfrac{7 \cdot 6 \cdot 5!}{5! \cdot 2 \cdot 1} = \dfrac{7 \cdot 6}{2 \cdot 1} = 21$

31. $\dfrac{5!(8-5)!}{4!7!} = \dfrac{5!3!}{4!7!} = \dfrac{5!}{7!} \cdot \dfrac{3!}{4!} = \dfrac{5!}{7 \cdot 6 \cdot 5!} \cdot \dfrac{3!}{4 \cdot 3!} = \dfrac{1}{7 \cdot 6} \cdot \dfrac{1}{4} = \dfrac{1}{168}$

33. $(x+y)^3 = x^3 + \dfrac{3!}{1!(3-1)!}x^2 y + \dfrac{3!}{2!(3-2)!}xy^2 + y^3 = x^3 + \dfrac{3!}{1!2!}x^2 y + \dfrac{3!}{2!1!}xy^2 + y^3$

$\qquad\qquad = x^3 + \dfrac{3 \cdot 2!}{1!2!}x^2 y + \dfrac{3 \cdot 2!}{2!1!}xy^2 + y^3$

$\qquad\qquad = x^3 + \dfrac{3}{1}x^2 y + \dfrac{3}{1}xy^2 + y^3$

$\qquad\qquad = x^3 + 3x^2 y + 3xy^2 + y^3$

35. $(x-y)^4 = x^4 + \dfrac{4!}{1!(4-1)!}x^3(-y) + \dfrac{4!}{2!(4-2)!}x^2(-y)^2 + \dfrac{4!}{3!(4-3)!}x(-y)^3 + (-y)^4$

$\qquad = x^4 + \dfrac{4!}{1!3!}(-x^3 y) + \dfrac{4!}{2!2!}x^2 y^2 + \dfrac{4!}{3!1!}(-xy^3) + y^4$

$\qquad = x^4 - \dfrac{4 \cdot 3!}{1!3!}x^3 y + \dfrac{4 \cdot 3 \cdot 2!}{2! \cdot 2 \cdot 1}x^2 y^2 - \dfrac{4 \cdot 3!}{3!1!}xy^3 + y^4$

$\qquad = x^4 - \dfrac{4}{1}x^3 y + \dfrac{12}{2}x^2 y^2 - \dfrac{4}{1}xy^3 + y^4$

$\qquad = x^4 - 4x^3 y + 6x^2 y^2 - 4xy^3 + y^4$

37. $(2x + y)^3 = (2x)^3 + \dfrac{3!}{1!(3-1)!}(2x)^2 y + \dfrac{3!}{2!(3-2)!}2xy^2 + y^3$

$$= 8x^3 + \dfrac{3!}{1!2!} \cdot 4x^2 y + \dfrac{3!}{2!1!} \cdot 2xy^2 + y^3$$

$$= 8x^3 + \dfrac{3 \cdot 2!}{1!2!} \cdot 4x^2 y + \dfrac{3 \cdot 2!}{2!1!} \cdot 2xy^2 + y^3$$

$$= 8x^3 + \dfrac{3}{1} \cdot 4x^2 y + \dfrac{3}{1} \cdot 2xy^2 + y^3$$

$$= 8x^3 + 12x^2 y + 6xy^2 + y^3$$

39. $(x - 2y)^3 = x^3 + \dfrac{3!}{1!(3-1)!}x^2(-2y) + \dfrac{3!}{2!(3-2)!}x(-2y)^2 + (-2y)^3$

$$= x^3 + \dfrac{3!}{1!2!} \cdot (-2x^2 y) + \dfrac{3!}{2!1!} \cdot 4xy^2 - 8y^3$$

$$= x^3 - \dfrac{3 \cdot 2!}{1!2!} \cdot 2x^2 y + \dfrac{3 \cdot 2!}{2!1!} \cdot 4xy^2 - 8y^3$$

$$= x^3 - \dfrac{3}{1} \cdot 2x^2 y + \dfrac{3}{1} \cdot 4xy^2 - 8y^3$$

$$= x^3 - 6x^2 y + 12xy^2 - 8y^3$$

41. $(2x + 3y)^3 = (2x)^3 + \dfrac{3!}{1!(3-1)!}(2x)^2(3y) + \dfrac{3!}{2!(3-2)!}2x(3y)^2 + (3y)^3$

$$= 8x^3 + \dfrac{3!}{1!2!} \cdot 4x^2(3y) + \dfrac{3!}{2!1!} \cdot 2x(9y^2) + 27y^3$$

$$= 8x^3 + \dfrac{3 \cdot 2!}{1!2!} \cdot 12x^2 y + \dfrac{3 \cdot 2!}{2!1!} \cdot 18xy^2 + 27y^3$$

$$= 8x^3 + \dfrac{3}{1} \cdot 12x^2 y + \dfrac{3}{1} \cdot 18xy^2 + 27y^3$$

$$= 8x^3 + 36x^2 y + 54xy^2 + 27y^3$$

43. $\left(\dfrac{x}{2} - \dfrac{y}{3}\right)^3 = \left(\dfrac{x}{2}\right)^3 + \dfrac{3!}{1!(3-1)!}\left(\dfrac{x}{2}\right)^2\left(-\dfrac{y}{3}\right) + \dfrac{3!}{2!(3-2)!}\left(\dfrac{x}{2}\right)\left(-\dfrac{y}{3}\right)^2 + \left(-\dfrac{y}{3}\right)^3$

$$= \dfrac{x^3}{8} - \dfrac{3!}{1!2!} \cdot \dfrac{x^2}{4} \cdot \dfrac{y}{3} + \dfrac{3!}{2!1!} \cdot \dfrac{x}{2} \cdot \dfrac{y^2}{9} - \dfrac{y^3}{27}$$

$$= \dfrac{x^3}{8} - \dfrac{3 \cdot 2!}{1!2!} \cdot \dfrac{x^2 y}{12} + \dfrac{3 \cdot 2!}{2!1!} \cdot \dfrac{xy^2}{18} - \dfrac{y^3}{27}$$

$$= \dfrac{x^3}{8} - \dfrac{3}{1} \cdot \dfrac{x^2 y}{12} + \dfrac{3}{1} \cdot \dfrac{xy^2}{18} - \dfrac{y^3}{27}$$

$$= \dfrac{x^3}{8} - \dfrac{x^2 y}{4} + \dfrac{xy^2}{6} - \dfrac{y^3}{27}$$

45. $(3 + 2y)^4 = 3^4 + \dfrac{4!}{1!(4-1)!}3^3(2y) + \dfrac{4!}{2!(4-2)!}3^2(2y)^2 + \dfrac{4!}{3!(4-3)!}3(2y)^3 + (2y)^4$

$= 81 + \dfrac{4!}{1!3!} \cdot 27(2y) + \dfrac{4!}{2!2!} \cdot 9(4y^2) + \dfrac{4!}{3!1!} \cdot 3(8y^3) + 16y^4$

$= 81 + \dfrac{4 \cdot 3!}{1!3!} \cdot 54y + \dfrac{4 \cdot 3 \cdot 2!}{2! \cdot 2 \cdot 1} \cdot 36y^2 + \dfrac{4 \cdot 3!}{3!1!} \cdot 24y^3 + 16y^4$

$= 81 + \dfrac{4}{1} \cdot 54y + \dfrac{12}{2} \cdot 36y^2 + \dfrac{4}{1} \cdot 24y^3 + 16y^4$

$= 81 + 216y + 216y^2 + 96y^3 + 16y^4$

47. $\left(\dfrac{x}{3} - \dfrac{y}{2}\right)^4$

$= \left(\dfrac{x}{3}\right)^4 + \dfrac{4!}{1!(4-1)!}\left(\dfrac{x}{3}\right)^3\left(-\dfrac{y}{2}\right) + \dfrac{4!}{2!(4-2)!}\left(\dfrac{x}{3}\right)^2\left(-\dfrac{y}{2}\right)^2 + \dfrac{4!}{3!(4-3)!}\left(\dfrac{x}{3}\right)\left(-\dfrac{y}{2}\right)^3$
$\qquad\qquad + \left(-\dfrac{y}{2}\right)^4$

$= \dfrac{x^4}{81} - \dfrac{4!}{1!3!} \cdot \dfrac{x^3}{27} \cdot \dfrac{y}{2} + \dfrac{4!}{2!2!} \cdot \dfrac{x^2}{9} \cdot \dfrac{y^2}{4} - \dfrac{4!}{3!1!} \cdot \dfrac{x}{3} \cdot \dfrac{y^3}{8} + \dfrac{y^4}{16}$

$= \dfrac{x^4}{81} - \dfrac{4 \cdot 3!}{1!3!} \cdot \dfrac{x^3 y}{54} + \dfrac{4 \cdot 3 \cdot 2!}{2!2!} \cdot \dfrac{x^2 y^2}{36} - \dfrac{4 \cdot 3!}{3!1!} \cdot \dfrac{xy^3}{24} + \dfrac{y^4}{16}$

$= \dfrac{x^4}{81} - \dfrac{4}{1} \cdot \dfrac{x^3 y}{54} + \dfrac{12}{2} \cdot \dfrac{x^2 y^2}{36} - \dfrac{4}{1} \cdot \dfrac{xy^3}{24} + \dfrac{y^4}{16}$

$= \dfrac{x^4}{81} - \dfrac{2x^3 y}{27} + \dfrac{x^2 y^2}{6} - \dfrac{xy^3}{6} + \dfrac{y^4}{16}$

49.

```
                          1
                       1     1
                    1     2     1
                 1     3     3     1
              1     4     6     4     1
           1     5    10    10     5     1
        1     6    15    20    15     6     1
     1     7    21    35    35    21     7     1
  1     8    28    56    70    56    28     8     1
1     9    36    84   126   126    84    36     9     1
```

51. $1, 1, 2, 3, 5, 8, 13, \ldots$; Beginning with 2, each number is the sum of the previous two numbers.

53. Answers may vary.

55. $\dfrac{n!}{n!(n-n)!} = \dfrac{n!}{n!(0)!} = \dfrac{n!}{n! \cdot 1} = 1$

Exercise 11.2 (page 692)

1. $3x + 2y = 12 \Rightarrow \quad 3x + 2y = 12$ Substitute and solve for y:

$\underline{2x - y = 1 \Rightarrow \times 2 \quad 4x - 2y = 2}$

$\qquad\qquad\qquad\qquad 7x \qquad = 14$

$\qquad\qquad\qquad\qquad\quad x = 2$

$2x - y = 1$

$2(2) - y = 1$

$4 - y = 1$

$-y = -3$

$y = 3$ Solution: $(2, 3)$

3. $\begin{vmatrix} 2 & -3 \\ 4 & -2 \end{vmatrix} = 2(-2) - (-3)(4) = -4 - (-12) = -4 + 12 = 8$

5. 3

7. 7

9. In the 2nd term, the exponent on b is 1.

Variables: $a^2 b^1 = a^2 b$

$\text{Coef.} = \dfrac{n!}{r!(n-r)!} = \dfrac{3!}{1!2!} = 3$

Term $= 3a^2 b$

11. In the 4th term, the exponent on $-y$ is 3.

Variables: $x^1(-y)^3 = -xy^3$

$\text{Coef.} = \dfrac{n!}{r!(n-r)!} = \dfrac{4!}{3!1!} = 4$

Term $= 4(-xy^3) = -4xy^3$

13. In the 5th term, the exponent on y is 4.

Variables: $x^2 y^4$

$\text{Coef.} = \dfrac{n!}{r!(n-r)!} = \dfrac{6!}{4!2!} = 15$

Term $= 15x^2 y^4$

15. In the 3rd term, the exponent on $-y$ is 2.

Variables: $x^6(-y)^2 = x^6 y^2$

$\text{Coef.} = \dfrac{n!}{r!(n-r)!} = \dfrac{8!}{2!6!} = 28$

Term $= 28x^6 y^2$

17. In the 3rd term, the exponent on 3 is 2.

Variables: $x^3(3)^2 = 9x^3$

$\text{Coef.} = \dfrac{n!}{r!(n-r)!} = \dfrac{5!}{2!3!} = 10$

Term $= 10(9x^3) = 90x^3$

19. In the 3rd term, the exponent on y is 2.

Variables: $(4x)^3 y^2 = 64x^3 y^2$

$\text{Coef.} = \dfrac{n!}{r!(n-r)!} = \dfrac{5!}{2!3!} = 10$

Term $= 10(64x^3 y^2) = 640x^3 y^2$

21. In the 2nd term, the exponent on $-3y$ is 1.

Variables: $x^3(-3y)^1 = -3x^3 y$

$\text{Coef.} = \dfrac{n!}{r!(n-r)!} = \dfrac{4!}{1!3!} = 4$

Term $= 4(-3x^3 y) = -12x^3 y$

23. In the 4th term, the exponent on -5 is 3. Variables: $(2x)^4(-5)^3 = (16x^4)(-125) = -2000x^4$

$\text{Coef.} = \dfrac{n!}{r!(n-r)!} = \dfrac{7!}{3!4!} = 35;$ Term $= 35(-2000x^4) = -70{,}000x^4$

SECTION 11.2

25. In the 5th term, the exponent on $-3y$ is 4. Variables: $(2x)^1(-3y)^4 = 2x(81y^4) = 162xy^4$

Coef. $= \dfrac{n!}{r!(n-r)!} = \dfrac{5!}{4!1!} = 5$; Term $= 5(162xy^4) = 810xy^4$

27. In the 3rd term, the exponent on $\sqrt{3}y$ is 2. Variables: $(\sqrt{2}x)^4(\sqrt{3}y)^2 = (4x^4)(3y^2) = 12x^4y^2$

Coef. $= \dfrac{n!}{r!(n-r)!} = \dfrac{6!}{2!4!} = 15$; Term $= 15(12x^4y^2) = 180x^4y^2$

29. In the 2nd term, the exponent on $-\dfrac{y}{3}$ is 1. Variables: $\left(\dfrac{x}{2}\right)^3\left(-\dfrac{y}{3}\right)^1 = \left(\dfrac{x^3}{8}\right)\left(-\dfrac{y}{3}\right) = -\dfrac{x^3y}{24}$

Coef. $= \dfrac{n!}{r!(n-r)!} = \dfrac{4!}{1!3!} = 4$; Term $= 4\left(-\dfrac{x^3y}{24}\right) = -\dfrac{x^3y}{6} = -\dfrac{1}{6}x^3y$

31. In the 4th term, the exponent on b is 3.
Variables: $a^{n-3}b^3$

Coef. $= \dfrac{n!}{r!(n-r)!} = \dfrac{n!}{3!(n-3)!}$

Term $= \dfrac{n!}{3!(n-3)!}a^{n-3}b^3$

33. In the 5th term, the exponent on $-b$ is 4.
Variables: $a^{n-4}(-b)^4 = a^{n-4}b^4$

Coef. $= \dfrac{n!}{r!(n-r)!} = \dfrac{n!}{4!(n-4)!}$

Term $= \dfrac{n!}{4!(n-4)!}a^{n-4}b^4$

35. In the rth term, the coefficient on b is $r-1$. Variables: $a^{n-(r-1)}b^{r-1} = a^{n-r+1}b^{r-1}$

Coef. $= \dfrac{n!}{r!(n-r)!} = \dfrac{n!}{(r-1)![n-(r-1)]!} = \dfrac{n!}{(r-1)!(n-r+1)!}$

Term $= \dfrac{n!}{(r-1)!(n-r+1)!}a^{n-r+1}b^{r-1}$

37. **Answers may vary.**

39. $\left(x+\dfrac{1}{x}\right)^{10} = (x+x^{-1})^{10}$. The constant term occurs when the exponent is 0.

The $(r+1)$th term of $(x+x^{-1})^{10}$ is $\dfrac{10!}{r!(10-r)!}x^{10-r}(x^{-1})^r = \dfrac{10!}{r!(10-r)!}x^{10-r}x^{-r}$.

But $\dfrac{10!}{r!(10-r)!}x^{10-r}x^{-r} = \dfrac{10!}{r!(10-r)!}x^{10-2r}$. If $10-2r=0$, then $r=5$.

The term is $\dfrac{10!}{5!(10-5)!}x^{10-5}x^{-5} = \dfrac{10!}{5!5!} = \dfrac{10\cdot9\cdot8\cdot7\cdot6\cdot5!}{5!\cdot5\cdot4\cdot3\cdot2\cdot1} = 252$

Exercise 11.3 (page 698)

1. $3(2x^2-4x+7)+4(3x^2+5x-6) = 6x^2-12x+21+12x^2+20x-24 = 18x^2+8x-3$

SECTION 11.3

3. $\dfrac{3a+4}{a-2} + \dfrac{3a-4}{a+2} = \dfrac{(3a+4)(a+2)}{(a-2)(a+2)} + \dfrac{(3a-4)(a-2)}{(a+2)(a-2)}$

$$= \dfrac{3a^2 + 10a + 8 + 3a^2 - 10a + 8}{(a+2)(a-2)} = \dfrac{6a^2 + 16}{(a+2)(a-2)}$$

5. sequence

7. arithmetic; difference

9. arithmetic mean

11. $1 + 2 + 3 + 4 + 5$

13. $3, 5, 7, 9, 11$

15. $-5, -8, -11, -14, -17$

17. nth term $= a + (n-1)d$
$29 = 5 + (5-1)d$
$29 = 5 + 4d$
$24 = 4d$
$6 = d$
$5, 11, 17, 23, 29$

19. nth term $= a + (n-1)d$
$-39 = -4 + (6-1)d$
$-39 = -4 + 5d$
$-35 = 5d$
$-7 = d$
$-4, -11, -18, -25, -32$

21. nth term $= a + (n-1)d$
$-83 = a + (6-1)7$
$-83 = a + 5(7)$
$-83 = a + 35$
$-118 = a$
$-118, -111, -104, -97, -90$

23. nth term $= a + (n-1)d$
$16 = a + (7-1)(-3)$
$16 = a + 6(-3)$
$16 = a - 18$
$34 = a$
$34, 31, 28, 25, 22$

25. nth term $= a + (n-1)d$
$131 = a + (19-1)d$
$131 = a + 18d$
nth term $= a + (n-1)d$
$138 = a + (20-1)d$
$138 = a + 19d$

$a + 18d = 131 \Rightarrow \times(-1) \quad -a - 18d = -131$
$a + 19d = 138 \Rightarrow \qquad \underline{a + 19d = 138}$
$ d = 7$

Substitute and solve for a:
$a + 18d = 131$
$a + 18(7) = 131$
$a + 126 = 131$
$a = 5 \Rightarrow 5, 12, 19, 26, 33$

27. nth term $= a + (n-1)d = 7 + (30-1)12 = 7 + 29(12) = 7 + 348 = 355$

29. nth term $= a + (n-1)d$
$-4 = a + (2-1)d$
$-4 = a + d$
nth term $= a + (n-1)d$
$-9 = a + (3-1)d$
$-9 = a + 2d$

$a + d = -4 \Rightarrow \times(-1) \quad -a - d = 4$
$a + 2d = -9 \Rightarrow \qquad \underline{a + 2d = -9}$
$ d = -5$

Substitute and solve for a:
$a + d = -4$
$a + (-5) = -4$
$a = 1$
nth term $= a + (n-1)d$
$= 1 + (37-1)(-5)$
$= 1 + 36(-5)$
$= 1 - 180 = -179$

31.
$$\text{nth term} = a + (n-1)d$$
$$263 = a + (27-1)(11)$$
$$263 = a + 26(11)$$
$$263 = a + 286$$
$$-23 = a$$

33.
$$\text{nth term} = a + (n-1)d$$
$$556 = 40 + (44-1)d$$
$$556 = 40 + 43d$$
$$516 = 43d$$
$$12 = d$$

35. Form an arithmetic sequence with a 1st term of 2 and a 5th term of 11:
$$\text{nth term} = a + (n-1)d$$
$$11 = 2 + (5-1)d$$
$$11 = 2 + 4d$$
$$9 = 4d$$
$$\frac{9}{4} = d$$
$$2, \boxed{\frac{17}{4}, \frac{13}{2}, \frac{35}{4}}, 11$$

37. Form an arithmetic sequence with a 1st term of 10 and a 6th term of 20:
$$\text{nth term} = a + (n-1)d$$
$$20 = 10 + (6-1)d$$
$$20 = 10 + 5d$$
$$10 = 5d$$
$$2 = d$$
$$10, \boxed{12, 14, 16, 18}, 20$$

39. Form an arithmetic sequence with a 1st term of 10 and a 3rd term of 19:
$$\text{nth term} = a + (n-1)d$$
$$19 = 10 + (3-1)d$$
$$19 = 10 + 2d$$
$$9 = 2d$$
$$\frac{9}{2} = d$$
$$10, \boxed{\frac{29}{2}}, 19$$

41. Form an arithmetic sequence with a 1st term of -4.5 and a 3rd term of 7:
$$\text{nth term} = a + (n-1)d$$
$$7 = -4.5 + (3-1)d$$
$$7 = -4.5 + 2d$$
$$11.5 = 2d$$
$$5.75 = d$$
$$-4.5, \boxed{1.25}, 7$$

43.
$$a = 1, d = 3, n = 30$$
$$l = a + (n-1)d = 1 + 29(3) = 88$$
$$S_n = \frac{n(a+l)}{2} = \frac{30(1+88)}{2} = 1335$$

45.
$$a = -5, d = 4, n = 17$$
$$l = a + (n-1)d = -5 + 16(4) = 59$$
$$S_n = \frac{n(a+l)}{2} = \frac{17(-5+59)}{2} = 459$$

47.
$$\text{nth term} = a + (n-1)d$$
$$7 = a + (2-1)d$$
$$7 = a + d$$
$$\text{nth term} = a + (n-1)d$$
$$12 = a + (3-1)d$$
$$12 = a + 2d$$

$$
\begin{aligned}
a + d &= 7 \Rightarrow \times(-1) & -a - d &= -7 \\
a + 2d &= 12 \Rightarrow & \underline{a + 2d} &= \underline{12} \\
& & d &= 5
\end{aligned}
$$

Substitute and solve for a:
$$a + d = 7$$
$$a + 5 = 7$$
$$a = 2, d = 5, n = 12$$
$$l = a + (n-1)d = 2 + 11(5) = 57$$
$$S_n = \frac{n(a+l)}{2} = \frac{12(2+57)}{2} = 354$$

SECTION 11.3

49. $f(n) = 2n + 1 \Rightarrow f(1) = 3$
$f(n) = 2n + 1 = 31$
$\quad\quad 2n = 30$
$\quad\quad\quad n = 15$
$S_n = \dfrac{n(a+l)}{2} = \dfrac{15(3+31)}{2} = 255$

51. $a = 1, d = 1, n = 50$
$l = a + (n-1)d = 1 + 49(1) = 50$
$S_n = \dfrac{n(a+l)}{2} = \dfrac{50(1+50)}{2} = 1275$

53. $a = 1, d = 2, n = 50; l = a + (n-1)d = 1 + 49(2) = 99$
$S_n = \dfrac{n(a+l)}{2} = \dfrac{50(1+99)}{2} = 2500$

55. $\displaystyle\sum_{k=1}^{4} 6k = 6(1) + 6(2) + 6(3) + 6(4)$
$\quad\quad = 6 + 12 + 18 + 24 = 60$

57. $\displaystyle\sum_{k=3}^{4} (k^2 + 3) = (3^2 + 3) + (4^2 + 3)$
$\quad\quad = 9 + 3 + 16 + 3 = 31$

59. $\displaystyle\sum_{k=4}^{4} (2k + 4) = 2(4) + 4 = 8 + 4 = 12$

61. $a = 60, d = 50 \Rightarrow 60, 110, 160, 210, 260, 310; n = 121$
$n\text{th term} = a + (n-1)d = 60 + (121-1)(50) = 60 + 120(50) = \$6060$

63. $a = 1, d = 1, n = 150, l = 150 \Rightarrow S_n = \dfrac{n(a+l)}{2} = \dfrac{150(1+150)}{2} = 11{,}325 \text{ bricks}$

65. After 1 sec.: $s = 16(1)^2 = 16$; After 2 sec.: $s = 16(2)^2 = 64$; After 3 sec.: $s = 16(3)^2 = 144$
During 2nd second $\Rightarrow$ falls $64 - 16 = 48$ ft; During 3rd second $\Rightarrow$ falls $144 - 64 = 80$ ft
The sequence of the amounts fallen during each second is $16, 48, 80 \Rightarrow a = 16, d = 32$
$n\text{th term} = a + (n-1)d = 16 + (12-1)(32) = 16 + 11(32) = 368$ ft

67. **Answers may vary.**

69. $\displaystyle\sum_{n=1}^{6} \left(\tfrac{1}{2}n + 1\right): \tfrac{3}{2}, 2, \tfrac{5}{2}, 3, \tfrac{7}{2}, 4$

71. Form an arithmetic sequence with a 1st term of a and a 3rd term of b:
$n\text{th term} = a + (n-1)d$
$\quad\quad b = a + (3-1)d$
$\quad\quad b = a + 2d$
$b - a = 2d$
$\dfrac{b-a}{2} = d \Rightarrow \text{mean} = a + \dfrac{b-a}{2} = \dfrac{2a}{2} + \dfrac{b-a}{2} = \dfrac{a+b}{2}$

73. $\displaystyle\sum_{k=1}^{5} 5k = 5(1) + 5(2) + 5(3) + 5(4) + 5(5) = 5(1 + 2 + 3 + 4 + 5) = 5 \sum_{k=1}^{5} k.$

75. $\displaystyle\sum_{k=1}^{n} 3 = \sum_{k=1}^{n} 3k^0 = 3(1)^0 + 3(2)^0 + \cdots + 3(n)^0 = 3 + 3 + \cdots + 3 = 3n$

Exercise 11.4 (page 706)

1. $x^2 - 5x - 6 \le 0$

$(x-6)(x+1) \le 0$

$x - 6 \quad --------0++++$
$x + 1 \quad ---\ 0++++++++++$

solution set: $[-1, 6]$

3. $\dfrac{x-4}{x+3} > 0$

$x - 4 \quad --------0++++$
$x + 3 \quad ---0++++++++++$

solution set: $(-\infty, -3) \cup (4, \infty)$

5. geometric

7. common ratio

9. $S_n = \dfrac{a - ar^n}{1 - r}$

11. $3, 6, 12, 24, 48$

13. $-5, -1, -\frac{1}{5}, -\frac{1}{25}, -\frac{1}{125}$

15. nth term $= ar^{n-1}$

$32 = 2r^{3-1}$

$32 = 2r^2$

$16 = r^2$

$\pm 4 = r$, so $r = 4$ $(r > 0)$

$2, 8, 32, 128, 512$

17. nth term $= ar^{n-1}$

$-192 = -3r^{4-1}$

$-192 = -3r^3$

$64 = r^3$

$4 = r$

$-3, -12, -48, -192, -768$

19. nth term $= ar^{n-1}$

$-4 = -64r^{5-1}$

$-4 = -64r^4$

$\frac{1}{16} = r^4$

$\pm \frac{1}{2} = r$, so $r = -\frac{1}{2}$ $(r < 0)$

$-64, 32, -16, 8, -4$

21. nth term $= ar^{n-1}$

$-2 = -64r^{6-1}$

$-2 = -64r^5$

$\frac{1}{32} = r^5$

$\frac{1}{2} = r$

$-64, -32, -16, -8, -4$

23. If the 3rd term is 50 and the 2nd term is 10, then the common ratio $r = 50 \div 10 = 5$.
1st term = 2nd term $\div r = 10 \div 5 = 2 \Rightarrow 2, 10, 50, 250, 1250$

25. nth term $= ar^{n-1} = 7 \cdot 2^{10-1} = 7 \cdot 2^9 = 7 \cdot 512 = 3584$

27. nth term $= ar^{n-1}$

$-81 = a(-3)^{8-1}$

$-81 = a(-3)^7$

$-81 = -2187a$

$\frac{1}{27} = a$

29. nth term $= ar^{n-1}$

$-1944 = -8r^{6-1}$

$-1944 = -8r^5$

$243 = r^5$

$3 = r$

31. 1st term $= 2$, 5th term $= 162$

nth term $= ar^{n-1}$

$162 = 2r^{5-1}$

$81 = r^4$

$3 = r$

$2, \boxed{6, 18, 54}, 162$

33. 1st term $= -4$, 6th term $= -12500$

nth term $= ar^{n-1}$

$-12500 = -4r^{6-1}$

$3125 = r^5$

$5 = r$

$-4, \boxed{-20, -100, -500, -2500}, -12500$

35. 1st term $= 2$, 3rd term $= 128$

nth term $= ar^{n-1}$

$128 = 2r^{3-1}$

$64 = r^2$

$-8 = r$

$2, \boxed{-16}, 128$

37. 1st term $= 10$, 3rd term $= 20$

nth term $= ar^{n-1}$

$20 = 10r^{3-1}$

$2 = r^2$

$\sqrt{2} = r$

$10, \boxed{10\sqrt{2}}, 20$

39. 1st term $= -50$, 3rd term $= 10$

nth term $= ar^{n-1}$

$10 = -50r^{3-1}$

$-\frac{1}{5} = r^2$

No such mean exists.

41. $a = 2, r = 3, n = 6$; $S_n = \dfrac{a - ar^n}{1 - r} = \dfrac{2 - 2(3)^6}{1 - 3} = \dfrac{2 - 2(729)}{-2} = \dfrac{-1456}{-2} = 728$

43. $a = 2, r = -3, n = 5$; $S_n = \dfrac{a - ar^n}{1 - r} = \dfrac{2 - 2(-3)^5}{1 - (-3)} = \dfrac{2 - 2(-243)}{4} = \dfrac{488}{4} = 122$

45. $a = 3, r = -2, n = 8$; $S_n = \dfrac{a - ar^n}{1 - r} = \dfrac{3 - 3(-2)^8}{1 - (-2)} = \dfrac{3 - 3(256)}{3} = \dfrac{-765}{3} = -255$

47. $a = 3, r = 2, n = 7$; $S_n = \dfrac{a - ar^n}{1 - r} = \dfrac{3 - 3(2)^7}{1 - 2} = \dfrac{3 - 3(128)}{-1} = \dfrac{-381}{-1} = 381$

49. If the 3rd term is $\frac{1}{5}$ and the 2nd term is 1, then the common ratio $r = \frac{1}{5} \div 1 = \frac{1}{5}$.

$a = 1 \div \frac{1}{5} = 5, r = \frac{1}{5}, n = 4$; $S_n = \dfrac{a - ar^n}{1 - r} = \dfrac{5 - 5\left(\frac{1}{5}\right)^4}{1 - \frac{1}{5}} = \dfrac{5 - \frac{1}{125}}{\frac{4}{5}} = \dfrac{\frac{624}{125}}{\frac{4}{5}} = \dfrac{156}{25}$

51. If the 4th term is 1 and the 3rd term is -2, then the common ratio $r = 1 \div (-2) = -\frac{1}{2}$.

2nd term $= -2 \div \left(-\frac{1}{2}\right) = 4$; 1st term $= 4 \div \left(-\frac{1}{2}\right) = -8$; $a = -8, r = -\frac{1}{2}, n = 6$

$S_n = \dfrac{a - ar^n}{1 - r} = \dfrac{-8 - (-8)\left(-\frac{1}{2}\right)^6}{1 - \left(-\frac{1}{2}\right)} = \dfrac{-8 + \frac{1}{8}}{\frac{3}{2}} = \dfrac{-\frac{63}{8}}{\frac{3}{2}} = -\dfrac{21}{4}$

53. Sequence of population: $500, 500(1.06), 500(1.06)^2, \ldots$
$a = 500, r = 1.06, n = 6 \Rightarrow n\text{th term} = ar^{n-1} = 500(1.06)^5 \approx 669$

55. Sequence of amounts: $10000, 10000(0.88), 10000(0.88)^2, \ldots$
$a = 10000, r = 0.88, n = 16 \Rightarrow n\text{th term} = ar^{n-1} = 10000(0.88)^{15} \approx \$1,469.74$

57. Sequence of values: $70000, 70000(1.06), 70000(1.06)^2, \ldots$
$a = 70000, r = 1.06, n = 13 \Rightarrow n\text{th term} = ar^{n-1} = 70000(1.06)^{12} \approx \$140,853.75$

59. Sequence of areas: $1, \frac{1}{2}, \frac{1}{4}, \ldots$
$a = 1, r = \frac{1}{2}, n = 12 \Rightarrow n\text{th term} = ar^{n-1}1 = 1\left(\frac{1}{2}\right)^{11} = \left(\frac{1}{2}\right)^{11} \approx 0.0005$

61. Sequence of amounts: $1000(1.03), 1000(1.03)^2, 1000(1.03)^3, \ldots$
$a = 1030, r = 1.03, n = 4 \Rightarrow S_n = \dfrac{a - ar^n}{1 - r} = \dfrac{1030 - 1030(1.03)^4}{1 - 1.03} = \dfrac{-129.2740743}{-0.03}$
$\approx \$4,309.14$

63. Answers may vary.   **65.** Answers may vary.

67. arithmetic mean   **69.** Answers may vary.

Exercise 11.5 (page 711)

1. $y = 3x^3 - 4$ **3.** $3x = y^2 + 4$ **5.** infinite   **7.** $S = \dfrac{a}{1 - r}$
function   not a function

9. $a = 8, r = \dfrac{1}{2}$

$S = \dfrac{a}{1 - r} = \dfrac{8}{1 - \frac{1}{2}} = \dfrac{8}{\frac{1}{2}} = 16$

11. $a = 54, r = \dfrac{1}{3}$

$S = \dfrac{a}{1 - r} = \dfrac{54}{1 - \frac{1}{3}} = \dfrac{54}{\frac{2}{3}} = 81$

13. $a = 12, r = -\dfrac{1}{2}$

$S = \dfrac{a}{1 - r} = \dfrac{12}{1 - \left(-\frac{1}{2}\right)} = \dfrac{12}{\frac{3}{2}} = 8$

15. $a = -45, r = -\dfrac{1}{3}$

$S = \dfrac{a}{1 - r} = \dfrac{-45}{1 - \left(-\frac{1}{3}\right)} = \dfrac{-45}{\frac{4}{3}} = -\dfrac{135}{4}$

17. $a = \dfrac{9}{2}, r = \dfrac{4}{3} \Rightarrow$ no sum $(|r| > 1)$

19. $a = -\dfrac{27}{2}, r = \dfrac{2}{3}$

$S = \dfrac{a}{1 - r} = \dfrac{-\frac{27}{2}}{1 - \frac{2}{3}} = \dfrac{-\frac{27}{2}}{\frac{1}{3}} = -\dfrac{81}{2}$

21. $0.\overline{1} = \dfrac{1}{10} + \dfrac{1}{100} + \dfrac{1}{1000} + \cdots \Rightarrow a = \dfrac{1}{10}, r = \dfrac{1}{10} \Rightarrow S = \dfrac{a}{1 - r} = \dfrac{\frac{1}{10}}{1 - \frac{1}{10}} = \dfrac{\frac{1}{10}}{\frac{9}{10}} = \dfrac{1}{9}$

23. $-0.\overline{3} = -\frac{3}{10} - \frac{3}{100} - \frac{3}{1000} + \cdots \Rightarrow a = -\frac{3}{10}, r = \frac{1}{10} \Rightarrow S = \frac{a}{1-r} = \frac{-\frac{3}{10}}{1-\frac{1}{10}} = \frac{-\frac{3}{10}}{\frac{9}{10}} = -\frac{1}{3}$

25. $0.\overline{12} = \frac{12}{100} + \frac{12}{10,000} + \frac{12}{1,000,000} + \cdots \Rightarrow a = \frac{12}{100}, r = \frac{1}{100}$

$S = \frac{a}{1-r} = \frac{\frac{12}{100}}{1-\frac{1}{100}} = \frac{\frac{12}{100}}{\frac{99}{100}} = \frac{12}{99} = \frac{4}{33}$

27. $0.\overline{75} = \frac{75}{100} + \frac{75}{10,000} + \frac{75}{1,000,000} + \cdots \Rightarrow a = \frac{75}{100}, r = \frac{1}{100}$

$S = \frac{a}{1-r} = \frac{\frac{75}{100}}{1-\frac{1}{100}} = \frac{\frac{75}{100}}{\frac{99}{100}} = \frac{75}{99} = \frac{25}{33}$

29. Distance ball travels down $= 10 + 5 + 2.5 + \cdots = \frac{a}{1-r} = \frac{10}{1-\frac{1}{2}} = \frac{10}{\frac{1}{2}} = 20$

Distance ball travels up $= 5 + 2.5 + 1.25 + \cdots = \frac{a}{1-r} = \frac{5}{1-\frac{1}{2}} = \frac{5}{\frac{1}{2}} = 10$

Total distance $= 20 + 10 = 30$ m

31. $S = \frac{a}{1-r} = \frac{1000}{1-0.8} = \frac{1000}{0.2}$
$= 5,000$ moths

33. **Answers may vary.**

35. $S = \frac{a}{1-r}$

$5 = \frac{1}{1-r}$

$5(1-r) = 1$

$5 - 5r = 1$

$4 = 5r$

$\frac{4}{5} = r$

37. $0.\overline{9} = \frac{9}{10} + \frac{9}{100} + \frac{9}{1000} + \cdots$

$a = \frac{9}{10}, r = \frac{1}{10} \Rightarrow S = \frac{a}{1-r}$

$= \frac{\frac{9}{10}}{1-\frac{1}{10}}$

$= \frac{\frac{9}{10}}{\frac{9}{10}} = \frac{9}{9} = 1$

39. No. $0.999999 = \frac{999,999}{1,000,000} < 1$

Exercise 11.6 (page 720)

1. $|2x - 3| = 9$

$2x - 3 = 9 \quad \text{or} \quad 2x - 3 = -9$

$2x = 12 \qquad\qquad 2x = -6$

$x = 6 \qquad\qquad\quad x = -3$

3. $\frac{3}{x-5} = \frac{8}{x}$

$3x = 8(x - 5)$

$3x = 8x - 40$

$-5x = -40$

$x = 8$

5. $p \cdot q$ **7.** $P(n, r)$ **9.** $n!$

11. $\binom{n}{r}$; combinations   **13.** 1

15. $P(3, 3) = \dfrac{3!}{(3-3)!} = \dfrac{3!}{0!} = \dfrac{6}{1} = 6$ **17.** $P(5, 3) = \dfrac{5!}{(5-3)!} = \dfrac{5!}{2!} = \dfrac{120}{2} = 60$

19. $P(2, 2) \cdot P(3, 3) = \dfrac{2!}{(2-2)!} \cdot \dfrac{3!}{(3-3)!} = \dfrac{2!}{0!} \cdot \dfrac{3!}{0!} = \dfrac{2}{1} \cdot \dfrac{6}{1} = 12$

21. $\dfrac{P(5, 3)}{P(4, 2)} = \dfrac{\frac{5!}{(5-3)!}}{\frac{4!}{(4-2)!}} = \dfrac{\frac{5!}{2!}}{\frac{4!}{2!}} = \dfrac{\frac{120}{2}}{\frac{24}{2}} = \dfrac{60}{12} = 5$

23. $\dfrac{P(6, 2) \cdot P(7, 3)}{P(5, 1)} = \dfrac{\frac{6!}{(6-2)!} \cdot \frac{7!}{(7-3)!}}{\frac{5!}{(5-1)!}} = \dfrac{\frac{6!}{4!} \cdot \frac{7!}{4!}}{\frac{5!}{4!}} = \dfrac{\frac{720}{24} \cdot \frac{5040}{24}}{\frac{120}{24}} = \dfrac{30 \cdot 210}{5} = 1{,}260$

25. $C(5, 3) = \dfrac{5!}{3!(5-3)!} = \dfrac{5!}{3!2!} = \dfrac{120}{6 \cdot 2} = \dfrac{120}{12} = 10$

27. $\binom{6}{3} = \dfrac{6!}{3!(6-3)!} = \dfrac{6!}{3!3!} = \dfrac{720}{6 \cdot 6} = \dfrac{720}{36} = 20$

29. $\binom{5}{4}\binom{5}{3} = \dfrac{5!}{4!(5-4)!} \cdot \dfrac{5!}{3!(5-3)!} = \dfrac{5!}{4!1!} \cdot \dfrac{5!}{3!2!} = \dfrac{120}{24 \cdot 1} \cdot \dfrac{120}{6 \cdot 2} = \dfrac{120}{24} \cdot \dfrac{120}{12} = 5 \cdot 10 = 50$

31. $\dfrac{C(38, 37)}{C(19, 18)} = \dfrac{\frac{38!}{37!(38-37)!}}{\frac{19!}{18!(19-18)!}} = \dfrac{\frac{38 \cdot 37!}{37!1!}}{\frac{19 \cdot 18!}{18!1!}} = \dfrac{\frac{38}{1}}{\frac{19}{1}} = \dfrac{38}{19} = 2$

33. $C(12, 0)C(12, 12) = \dfrac{12!}{0!(12-0)!} \cdot \dfrac{12!}{12!(12-12)!} = \dfrac{12!}{0!12!} \cdot \dfrac{12!}{12!0!} = \dfrac{12!}{12!} \cdot \dfrac{12!}{12!} = 1 \cdot 1 = 1$

35. $C(n, 2) = \dfrac{n!}{2!(n-2)!}$

37. $(x + y)^4 = \binom{4}{0}x^4y^0 + \binom{4}{1}x^3y^1 + \binom{4}{2}x^2y^2 + \binom{4}{3}x^1y^3 + \binom{4}{4}x^0y^4$

$= \dfrac{4!}{0!4!}x^4 + \dfrac{4!}{1!3!}x^3y + \dfrac{4!}{2!2!}x^2y^2 + \dfrac{4!}{3!1!}xy^3 + \dfrac{4!}{4!0!}y^4 = x^4 + 4x^3y + 6x^2y^2 + 4xy^3 + y^4$

39. $(2x + y)^3 = \binom{3}{0}(2x)^3y^0 + \binom{3}{1}(2x)^2y^1 + \binom{3}{2}(2x)^1y^2 + \binom{3}{3}(2x)^0y^3$

$= \dfrac{3!}{0!3!} \cdot 8x^3 + \dfrac{3!}{1!2!} \cdot 4x^2y + \dfrac{3!}{2!1!} \cdot 2xy^2 + \dfrac{3!}{3!0!}y^3$

$= 8x^3 + 12x^2y + 6xy^2 + y^3$

41. $(3x - 2)^4$

$$= \binom{4}{0}(3x)^4(-2)^0 + \binom{4}{1}(3x)^3(-2)^1 + \binom{4}{2}(3x)^2(-2)^2 + \binom{4}{3}(3x)^1(-2)^3$$
$$+ \binom{4}{4}(3x)^0(-2)^4$$

$$= \frac{4!}{0!4!} \cdot 81x^4(1) + \frac{4!}{1!3!} \cdot 27x^3(-2) + \frac{4!}{2!2!} \cdot 9x^2(4) + \frac{4!}{3!1!} \cdot 3x(-8) + \frac{4!}{4!0!} \cdot 1(16)$$
$$= 81x^4 - 216x^3 + 216x^2 - 96x + 16$$

43. $\binom{5}{3}x^2(-5y)^3 = \frac{5!}{3!2!}x^2(-125y^3) = 10x^2(-125y^3) = -1{,}250x^2y^3$

45. $\binom{4}{1}(x^2)^3(-y^3)^1 = \frac{4!}{1!3!}x^6(-y^3) = -4x^6y^3$

47. $7 \cdot 5 = 35$

49. $10 \cdot 10 \cdot 10 \cdot 10 \cdot 10 \cdot 10 = 1{,}000{,}000$

51. $9 \cdot 9 \cdot 8 \cdot 7 \cdot 6 \cdot 5 = 136{,}080$

53. $8 \cdot 10 \cdot 10 \cdot 10 \cdot 10 \cdot 10 = 8{,}000{,}000$

55. $6! = 720$

57. $4! \cdot 5! = 24 \cdot 120 = 2{,}880$

59. $25 \cdot 24 \cdot 23 = 13{,}800$

61. $P(10, 3) = \dfrac{10!}{(10 - 3)!} = \dfrac{10!}{7!} = 720$

63. $9 \cdot 10 \cdot 10 \cdot 1 \cdot 1 = 900$

65. $C(14, 3) = \dfrac{14!}{3!(14 - 3)!} = \dfrac{14!}{3!11!} = 364$

67. $C(5, 3) = 10 \Rightarrow 5$ persons

69. $C(100, 6) = \dfrac{100!}{6!(100 - 6)!} = \dfrac{100!}{6!94!} = \dfrac{100 \cdot 99 \cdot 98 \cdot 97 \cdot 96 \cdot 95 \cdot 94!}{6!94!} = 1{,}192{,}052{,}400$

71. $C(3, 2) \cdot C(4, 2) = \dfrac{3!}{2!1!} \cdot \dfrac{4!}{2!2!} = 3 \cdot 6 = 18$

73. $C(12, 2) \cdot C(10, 3) = \dfrac{12!}{2!10!} \cdot \dfrac{10!}{3!7!} = 66 \cdot 120 = 7{,}920$

75. Answers may vary.

77. Consider the two people who insist on standing together as one person. Then there are a total of 4 "persons" to be arranged. This can be done in $4! = 24$ ways. However, the two people who are standing together can be arranged in 2 different ways, so there are $24 \cdot 2 = 48$ arrangements.

Chapter 11 Summary (page 724)

1. $(4!)(3!) = 4 \cdot 3 \cdot 2 \cdot 1 \cdot 3 \cdot 2 \cdot 1 = 144$

2. $\dfrac{5!}{3!} = \dfrac{5 \cdot 4 \cdot 3!}{3!} = 5 \cdot 4 = 20$

3. $\dfrac{6!}{2!(6-2)!} = \dfrac{6!}{2!4!} = \dfrac{6 \cdot 5 \cdot 4!}{2 \cdot 1 \cdot 4!} = \dfrac{30}{2} = 15$

4. $\dfrac{12!}{3!(12-3)!} = \dfrac{12!}{3!9!} = \dfrac{12 \cdot 11 \cdot 10 \cdot 9!}{3 \cdot 2 \cdot 1 \cdot 9!}$
$= \dfrac{1320}{6} = 220$

5. $(n-n)! = 0! = 1$

6. $\dfrac{8!}{7!} = \dfrac{8 \cdot 7!}{7!} = 8$

7. $(x+y)^5 = x^5 + \dfrac{5!}{1!(5-1)!}x^4 y + \dfrac{5!}{2!(5-2)!}x^3 y^2 + \dfrac{5!}{3!(5-3)!}x^2 y^3 + \dfrac{5!}{4!(5-4)!}xy^4 + y^5$

$= x^5 + \dfrac{5!}{1!4!}x^4 y + \dfrac{5!}{2!3!}x^3 y^2 + \dfrac{5!}{3!2!}x^2 y^3 + \dfrac{5!}{4!1!}xy^4 + y^5$

$= x^5 + \dfrac{5 \cdot 4!}{1!4!}x^4 y + \dfrac{5 \cdot 4 \cdot 3!}{2 \cdot 1 \cdot 3!}x^3 y^2 + \dfrac{5 \cdot 4 \cdot 3!}{3! \cdot 2 \cdot 1}x^2 y^3 + \dfrac{5 \cdot 4!}{4! \cdot 1}xy^4 + y^5$

$= x^5 + \dfrac{5}{1}x^4 y + \dfrac{20}{2}x^3 y^2 + \dfrac{20}{2}x^2 y^3 + \dfrac{5}{1}xy^4 + y^5$

$= x^5 + 5x^4 y + 10x^3 y^2 + 10x^2 y^3 + 5xy^4 + y^5$

8. $(x-y)^4 = x^4 + \dfrac{4!}{1!(4-1)!}x^3(-y) + \dfrac{4!}{2!(4-2)!}x^2(-y)^2 + \dfrac{4!}{3!(4-3)!}x(-y)^3 + (-y)^4$

$= x^4 + \dfrac{4!}{1!3!}(-x^3 y) + \dfrac{4!}{2!2!}x^2 y^2 + \dfrac{4!}{3!1!}(-xy^3) + y^4$

$= x^4 - \dfrac{4 \cdot 3!}{1!3!}x^3 y + \dfrac{4 \cdot 3 \cdot 2!}{2! \cdot 2 \cdot 1}x^2 y^2 - \dfrac{4 \cdot 3!}{3!1!}xy^3 + y^4$

$= x^4 - \dfrac{4}{1}x^3 y + \dfrac{12}{2}x^2 y^2 - \dfrac{4}{1}xy^3 + y^4 = x^4 - 4x^3 y + 6x^2 y^2 - 4xy^3 + y^4$

9. $(4x-y)^3 = (4x)^3 + \dfrac{3!}{1!(3-1)!}(4x)^2(-y) + \dfrac{3!}{2!(3-2)!}(4x)(-y)^2 + (-y)^3$

$= 64x^3 + \dfrac{3!}{1!2!} \cdot (-16x^2 y) + \dfrac{3!}{2!1!} \cdot 4xy^2 - y^3$

$= 64x^3 - \dfrac{3 \cdot 2!}{1!2!} \cdot 16x^2 y + \dfrac{3 \cdot 2!}{2!1!} \cdot 4xy^2 - y^3$

$= 64x^3 - \dfrac{3}{1} \cdot 16x^2 y + \dfrac{3}{1} \cdot 4xy^2 - y^3 = 64x^3 - 48x^2 y + 12xy^2 - y^3$

10. $(x + 4y)^3 = x^3 + \dfrac{3!}{1!(3-1)!}x^2(4y) + \dfrac{3!}{2!(3-2)!}x(4y)^2 + (4y)^3$

$\qquad = x^3 + \dfrac{3!}{1!2!} \cdot 4x^2y + \dfrac{3!}{2!1!} \cdot 16xy^2 + 64y^3$

$\qquad = x^3 + \dfrac{3 \cdot 2!}{1!2!} \cdot 4x^2y + \dfrac{3 \cdot 2!}{2!1!} \cdot 16xy^2 + 64y^3$

$\qquad = x^3 + \dfrac{3}{1} \cdot 4x^2y + \dfrac{3}{1} \cdot 16xy^2 + 64y^3 = x^3 + 12x^2y + 48xy^2 + 64y^3$

11. In the 3rd term, the exponent on 3 is 2.

Variables: x^2y^2

Coef. $= \dfrac{n!}{r!(n-r)!} = \dfrac{4!}{2!2!} = 6$

Term $= 6x^2y^2$

12. In the 4th term, the exponent on $-y$ is 3.

Variables: $x^2(-y)^3 = -x^2y^3$

Coef. $= \dfrac{n!}{r!(n-r)!} = \dfrac{5!}{3!2!} = 10$

Term $= 10(-x^2y^3) = -10x^2y^3$

13. 2nd term: The exponent on $-4y$ is 1. Variables: $(3x)^2(-4y)^1 = (9x^2)(-4y) = -36x^2y$

Coef. $= \dfrac{n!}{r!(n-r)!} = \dfrac{3!}{2!1!} = 3$; Term $= 3(-36x^2y) = -108x^2y$

14. 3rd term: The exponent on $3y$ is 2. Variables: $(4x)^2(3y)^2 = (16x^2)(9y^2) = 144x^2y^2$

Coef. $= \dfrac{n!}{r!(n-r)!} = \dfrac{4!}{2!2!} = 6$; Term $= 6(144x^2y^2) = 864x^2y^2$

15. nth term $= a + (n-1)d = 7 + (8-1)5 = 7 + 7 \cdot 5 = 42$

16. nth term $= a + (n-1)d \qquad a + 8d = 242 \Rightarrow \times(-1) \; -a - 8d = -242$

$\qquad 242 = a + (9-1)d \qquad \underline{a + 6d = 212} \Rightarrow \qquad \underline{\quad a + 6d = \quad 212}$

$\qquad 242 = a + 8d \qquad\qquad\qquad\qquad\qquad\qquad\quad -2d = \quad -30$

$\qquad n$th term $= a + (n-1)d \qquad\qquad\qquad\qquad\qquad\quad d = \quad 15$

$\qquad 212 = a + (7-1)d \qquad$ Substitute and solve for a:

$\qquad 212 = a + 6d \qquad\qquad\qquad a + 6d = 212$

$\qquad\qquad\qquad\qquad\qquad\qquad a + 6(15) = 212$

$\qquad\qquad\qquad\qquad\qquad\qquad a + 90 = 212$

$\qquad\qquad\qquad\qquad\qquad\qquad a = 122 \Rightarrow 122, 137, 152, 167, 182$

17. Form an arithmetic sequence with a 1st term of 8 and a 4th term of 25:

nth term $= a + (n-1)d$

$\qquad 25 = 8 + (4-1)d$

$\qquad 17 = 3d$

$\qquad \frac{17}{3} = d \Rightarrow 8, \boxed{\frac{41}{3}, \frac{58}{3}}, 25$

18. $a = 11, d = 7, n = 20$

$l = a + (n-1)d = 11 + 19(7) = 144$

$S_n = \dfrac{n(a+l)}{2} = \dfrac{20(11+144)}{2} = 1{,}550$

19. $a = 9, d = -\frac{5}{2}, n = 10$

$l = a + (n-1)d = 9 + 9\left(-\frac{5}{2}\right) = -\frac{27}{2}$

$S_n = \dfrac{n(a+l)}{2} = \dfrac{10\left(9 - \frac{27}{2}\right)}{2} = -\dfrac{45}{2}$

20. $\displaystyle\sum_{k=4}^{6}\frac{1}{2}k = \frac{1}{2}(4) + \frac{1}{2}(5) + \frac{1}{2}(6) = 2 + \frac{5}{2} + 3 = \frac{15}{2}$

21. $\displaystyle\sum_{k=2}^{5}7k^2 = 7(2)^2 + 7(3)^2 + 7(4)^2 + 7(5)^2 = 28 + 63 + 112 + 175 = 378$

22. $\displaystyle\sum_{k=1}^{4}(3k-4) = (3(1)-4) + (3(2)-4) + (3(3)-4) + (3(4)-4) = -1+2+5+8 = 14$

23. $\displaystyle\sum_{k=10}^{10}36k = 36(10) = 360$

24. If the 5th term is $\frac{3}{2}$ and the 4th term is 3,
then the common ratio $r = \frac{3}{2} \div 3 = \frac{1}{2}$.
3rd term = 4th term $\div\, r = 3 \div \frac{1}{2} = 6$
2nd term = 3rd term $\div\, r = 6 \div \frac{1}{2} = 12$
1st term = 2nd term $\div\, r = 12 \div \frac{1}{2} = 24$
$24, 12, 6, 3, \frac{3}{2}$

25. nth term $= ar^{n-1}$
$= \dfrac{1}{8}(2)^{6-1}$
$= \dfrac{1}{8}(32) = 4$

26. 1st term $= -6$, 4th term $= 384$ nth term $= ar^{n-1}$
$$384 = -6r^{4-1}$$
$$-64 = r^3$$
$$-4 = r \Rightarrow -6, \boxed{24, -96}, 384$$

27. $a = 162, r = \frac{1}{3}, n = 7;\ S_n = \dfrac{a - ar^n}{1-r} = \dfrac{162 - 162\left(\frac{1}{3}\right)^7}{1 - \frac{1}{3}} = \dfrac{162 - 162\left(\frac{1}{2187}\right)}{-\frac{2}{3}} = \dfrac{\frac{4372}{27}}{-\frac{2}{3}} = \dfrac{2186}{9}$

28. $a = \frac{1}{8}, r = -2, n = 8;\ S_n = \dfrac{a - ar^n}{1-r} = \dfrac{\frac{1}{8} - \frac{1}{8}(-2)^8}{1 - (-2)} = \dfrac{\frac{1}{8} - \frac{1}{8}(256)}{3} = \dfrac{-\frac{255}{8}}{3} = -\dfrac{85}{8}$

29. $a = 25, r = \frac{4}{5};\ S = \dfrac{a}{1-r} = \dfrac{25}{1 - \frac{4}{5}} = \dfrac{25}{\frac{1}{5}} = 125$

30. $0.\overline{05} = \frac{5}{100} + \frac{5}{10,000} + \frac{5}{1,000,000} + \cdots \Rightarrow a = \frac{5}{100}, r = \frac{1}{100};\ S = \frac{a}{1-r} = \dfrac{\frac{5}{100}}{1 - \frac{1}{100}} = \dfrac{\frac{5}{100}}{\frac{99}{100}} = \dfrac{5}{99}$

31. $17 \cdot 8 = 136$

32. $P(7,7) = \dfrac{7!}{(7-7)!} = \dfrac{7!}{0!} = 7! = 5{,}040$

33. $P(7,0) = \dfrac{7!}{(7-0)!} = \dfrac{7!}{7!} = 1$

34. $P(8,6) = \dfrac{8!}{(8-6)!} = \dfrac{8!}{2!} = \dfrac{40{,}320}{2} = 20{,}160$

35. $\dfrac{P(9,6)}{P(10,7)} = \dfrac{\frac{9!}{(9-6)!}}{\frac{10!}{(10-7)!}} = \dfrac{\frac{9!}{3!}}{\frac{10!}{3!}} = \dfrac{9!}{3!} \cdot \dfrac{3!}{10!} = \dfrac{9!}{10!} = \dfrac{9!}{10 \cdot 9!} = \dfrac{1}{10}$

36. $C(7,7) = \dfrac{7!}{7!(7-7)!} = \dfrac{7!}{7!0!} = \dfrac{7!}{7!} = 1$ **37.** $C(7,0) = \dfrac{7!}{0!(7-0)!} = \dfrac{7!}{0!7!} = \dfrac{7!}{7!} = 1$

38. $\dbinom{8}{6} = \dfrac{8!}{6!(8-6)!} = \dfrac{8 \cdot 7 \cdot 6!}{6!2!} = \dfrac{56}{2} = 28$ **39.** $\dbinom{9}{6} = \dfrac{9!}{6!(9-6)!} = \dfrac{9 \cdot 8 \cdot 7 \cdot 6!}{6!3!} = \dfrac{504}{6} = 84$

40. $C(6,3) \cdot C(7,3) = \dfrac{6!}{3!(6-3)!} \cdot \dfrac{7!}{3!(7-3)!} = \dfrac{6!}{3!3!} \cdot \dfrac{7!}{3!4!} = 20 \cdot 35 = 700$

41. $\dfrac{C(7,3)}{C(6,3)} = \dfrac{\frac{7!}{3!(7-3)!}}{\frac{6!}{3!(6-3)!}} = \dfrac{\frac{7!}{3!4!}}{\frac{6!}{3!3!}} = \dfrac{35}{20} = \dfrac{7}{4}$

42. Sequence of amounts: $5000, 5000(0.80), 5000(0.80)^2, \dots$
$a = 5000, r = 0.80, n = 6 \Rightarrow n\text{th term} = ar^{n-1} = 5000(0.80)^5 \approx \$1{,}638.40$

43. Sequence of amounts: $25700, 25700(1.18), 25700(1.18)^2, \dots$
$a = 25700, r = 1.18, n = 11 \Rightarrow n\text{th term} = ar^{n-1} = 25700(1.18)^{10} \approx \$134{,}509.57$

44. $a = 300, d = 75;$ $n\text{th term} = a + (n-1)d$
$1200 = 300 + (n-1)(75)$
$900 = 75n - 75$
$975 = 75n \Rightarrow n = 13 \Rightarrow$ in 12 years

45. $16, d = 32.$
$n\text{th term} = a + (n-1)d = 16 + (10-1)(32) = 16 + 9(32) = 304$
$S_n = \dfrac{n(a+l)}{2} = \dfrac{10(16+304)}{2} = 1600 \text{ ft}$

46. $5! = 120$ **47.** $5! \cdot 3! = 720$

48. $\dbinom{10}{3} = \dfrac{10!}{3!7!} = 120$ **49.** $\dbinom{5}{2}\dbinom{6}{2} = 10 \cdot 15 = 150$

Chapter 11 Test (page 727)

1. $\dfrac{7!}{4!} = \dfrac{7 \cdot 6 \cdot 5 \cdot 4!}{4!} = 7 \cdot 6 \cdot 5 = 210$ **2.** $0! = 1$

3. In the 2nd term, the exponent on $-y$ is 1.

Variables: $x^4(-y)^1 = -x^4y$

Coef. $= \dfrac{n!}{r!(n-r)!} = \dfrac{5!}{1!4!} = 5$

Term $= 5(-x^4y) = -5x^4y$

4. In the 3rd term, the exponent on $2y$ is 2.

Variables: $(x)^2(2y)^2 = 4x^2y^2$

Coef. $= \dfrac{n!}{r!(n-r)!} = \dfrac{4!}{2!2!} = 6$

Term $= 6(4x^2y^2) = 24x^2y^2$

5. $a = 3, d = 7, n = 10$; nth term $= a + (n-1)d = 3 + (10-1)(7) = 3 + 9(7) = 66$

6. $a = -2, d = 5, n = 12$; nth term $= a + (n-1)d = -2 + (12-1)(5) = -2 + 11(5) = 53$

$S_n = \dfrac{n(a+l)}{2} = \dfrac{12(-2+53)}{2} = \dfrac{12(51)}{2} = 306$

7. Form an arithmetic sequence with a 1st term of 2 and a 4th term of 98:

nth term $= a + (n-1)d$

$98 = 2 + (4-1)d$

$96 = 3d$

$32 = d \Rightarrow 2, \boxed{34, 66}, 98$

8. $\displaystyle\sum_{k=1}^{3}(2k-3) = (2(1)-3) + (2(2)-3) + (2(3)-3) = -1 + 1 + 3 = 3$

9. $a = -\frac{1}{9}, r = 3, n = 7$; nth term $= ar^{n-1} = -\frac{1}{9}(3)^{7-1} = -\frac{1}{9}(3)^6 = -\frac{1}{9}(729) = -81$

10. $a = \frac{1}{27}, r = 3, n = 6$; $S_n = \dfrac{a - ar^n}{1-r} = \dfrac{\frac{1}{27} - \frac{1}{27}(3)^6}{1-3} = \dfrac{\frac{1}{27} - \frac{1}{27}(729)}{-2} = \dfrac{-\frac{728}{27}}{-2} = \dfrac{364}{27}$

11. 1st term $= 3$, 4th term $= 648$

nth term $= ar^{n-1}$

$648 = 3r^{4-1}$

$216 = r^3$

$6 = r \Rightarrow 3, \boxed{18, 108}, 648$

12. $a = 9, r = \dfrac{1}{3}$

$S = \dfrac{a}{1-r} = \dfrac{9}{1-\frac{1}{3}} = \dfrac{9}{\frac{2}{3}} = \dfrac{27}{2}$

13. $P(5,4) = \dfrac{5!}{(5-4)!} = \dfrac{5!}{1!} = 5! = 120$

14. $P(8,8) = \dfrac{8!}{(8-8)!} = \dfrac{8!}{0!} = 8! = 40{,}320$

15. $C(6,4) = \dfrac{6!}{4!2!} = \dfrac{6 \cdot 5 \cdot 4!}{4! \cdot 2 \cdot 1} = \dfrac{30}{2} = 15$

16. $C(8,3) = \dfrac{8!}{3!5!} = \dfrac{8 \cdot 7 \cdot 6 \cdot 5!}{3 \cdot 2 \cdot 1 \cdot 5!} = \dfrac{56 \cdot 6}{6} = 56$

17. $C(6,0) \cdot P(6,5) = \dfrac{6!}{0!6!} \cdot \dfrac{6!}{1!} = 1 \cdot 6!$

$= 720$

18. $P(8,7) \cdot C(8,7) = \dfrac{8!}{1!} \cdot \dfrac{8!}{7!1!} = 8! \cdot 8$

$= 322{,}560$

19. $\dfrac{P(6,4)}{C(6,4)} = \dfrac{\frac{6!}{2!}}{\frac{6!}{4!2!}} = \dfrac{6!}{2!} \cdot \dfrac{4!2!}{6!} = 4! = 24$

20. $\dfrac{C(9,6)}{P(6,4)} = \dfrac{\frac{9!}{6!3!}}{\frac{6!}{2!}} = \dfrac{84}{360} = \dfrac{7}{30}$

CHAPTER 11 TEST

21. $\dbinom{7}{3} = \dfrac{7!}{3!\,4!} = 35$

22. $\dbinom{5}{1}\dbinom{4}{2} = 5 \cdot 6 = 30$

Cumulative Review Exercises (page 728)

1. $\begin{cases} 2x + y = 5 \\ x - 2y = 0 \end{cases}$

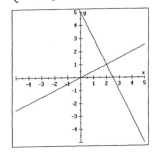

Solution: $(2, 1)$

2. $\begin{cases} (1) \quad 3x + y = 4 \\ (2) \quad 2x - 3y = -1 \end{cases}$

Substitute $y = 4 - 3x$ from (1) into (2):
$$2x - 3y = -1$$
$$2x - 3(4 - 3x) = -1$$
$$2x - 12 + 9x = -1$$
$$11x = 11$$
$$x = 1$$

Substitute this and solve for y:
$$y = 4 - 3x = 4 - 3(1) = 1$$
Solution: $(1, 1)$

3.
$$\begin{array}{l} x + 2y = -2 \\ 2x - y = 6 \Rightarrow \times (2) \end{array} \quad \begin{array}{r} x + 2y = -2 \\ 4x - 2y = 12 \\ \hline 5x = 10 \\ x = 2 \end{array} \quad \begin{array}{l} x + 2y = -2 \\ 2 + 2y = -2 \\ 2y = -4 \\ y = -2 \end{array}$$

Solution: $\boxed{(2, -2)}$

4.
$$\begin{array}{l} \dfrac{x}{10} + \dfrac{y}{5} = \dfrac{1}{2} \Rightarrow \times 10 \\ \dfrac{x}{2} - \dfrac{y}{5} = \dfrac{13}{10} \Rightarrow \times 10 \end{array} \quad \begin{array}{r} x + 2y = 5 \\ 5x - 2y = 13 \\ \hline 6x = 18 \\ x = 3 \end{array} \quad \begin{array}{l} x + 2y = 5 \\ 3 + 2y = 5 \\ 2y = 2 \\ y = 1 \Rightarrow \text{Solution: } \boxed{(3, 1)} \end{array}$$

5. $\begin{vmatrix} 3 & -2 \\ 1 & -1 \end{vmatrix} = 3(-1) - (-2)(1)$
$$= -3 + 2 = -1$$

6. $y = \dfrac{\begin{vmatrix} 4 & -1 \\ 3 & -7 \end{vmatrix}}{\begin{vmatrix} 4 & -3 \\ 3 & 4 \end{vmatrix}} = \dfrac{4(-7) - (-1)(3)}{4(4) - (-3)(3)}$
$$= \dfrac{-25}{25} = -1$$

7.

(1) $x + y + z = 1$	(1) $x + y + z = 1$	(2) $2x - y - z = -4$
(2) $2x - y - z = -4$	(2) $2x - y - z = -4$	(3) $x - 2y + z = 4$
(3) $x - 2y + z = 4$	(4) $3x \qquad = -3$	(5) $3x - 3y \qquad = 0$
	$x \qquad = -1$	

$$3x - 3y = 0 \qquad\qquad x + y + z = 1$$
$$3(-1) - 3y = 0 \qquad -1 + (-1) + z = 1$$
$$-3 - 3y = 0 \qquad\qquad -2 + z = 1$$
$$-3y = 3 \qquad\qquad\qquad z = 3 \quad \boxed{\text{The solution is } (-1, -1, 3).}$$
$$y = -1$$

8. $z = \dfrac{\begin{vmatrix} 1 & 2 & 6 \\ 3 & 2 & 6 \\ 2 & 3 & 6 \end{vmatrix}}{\begin{vmatrix} 1 & 2 & 3 \\ 3 & 2 & 1 \\ 2 & 3 & 1 \end{vmatrix}} = \dfrac{1\begin{vmatrix} 2 & 6 \\ 3 & 6 \end{vmatrix} - 2\begin{vmatrix} 3 & 6 \\ 2 & 6 \end{vmatrix} + 6\begin{vmatrix} 3 & 2 \\ 2 & 3 \end{vmatrix}}{1\begin{vmatrix} 2 & 1 \\ 3 & 1 \end{vmatrix} - 2\begin{vmatrix} 3 & 1 \\ 2 & 1 \end{vmatrix} + 3\begin{vmatrix} 3 & 2 \\ 2 & 3 \end{vmatrix}} = \dfrac{1(-6) - 2(6) + 6(5)}{1(-1) - 2(1) + 3(5)} = \dfrac{12}{12} = 1$

9. $\begin{cases} 3x - 2y < 6 \\ y < -x + 2 \end{cases}$

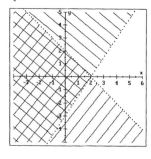

10. $\begin{cases} y < x + 2 \\ 3x + y \leq 6 \end{cases}$

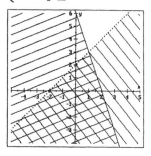

11. $y = \left(\dfrac{1}{2}\right)^x$

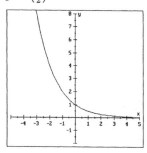

12. $y = \log_2 x \Rightarrow 2^y = x$

13. $\log_x 25 = 2 \Rightarrow x^2 = 25 \Rightarrow x = 5$

14. $\log_5 125 = x \Rightarrow 5^x = 125 \Rightarrow x = 3$

15. $\log_3 x = -3 \Rightarrow 3^{-3} = x \Rightarrow x = \frac{1}{27}$

16. $\log_5 x = 0 \Rightarrow 5^0 = x \Rightarrow x = 1$

17. $y = 2^x$

18. x

19. $\log 98 = \log(14 \cdot 7) = \log 14 + \log 7 = 1.1461 + 0.8451 = 1.9912$

20. $\log 2 = \log \frac{14}{7} = \log 14 - \log 7 = 1.1461 - 0.8451 = 0.3010$

21. $\log 49 = \log 7^2 = 2 \log 7 = 2(0.8451) = 1.6902$

22. $\log \frac{7}{5} = \log \frac{14}{10} = \log 14 - \log 10 = 1.1461 - 1 = 0.1461$

23.
$$2^{x+5} = 3^x$$
$$\log 2^{x+2} = \log 3^x$$
$$(x+5)\log 2 = x \log 3$$
$$x \log 2 + 5 \log 2 = x \log 3$$
$$5 \log 2 = x \log 3 - x \log 2$$
$$5 \log 2 = x(\log 3 - \log 2)$$
$$\frac{5 \log 2}{\log 3 - \log 2} = x$$

24.
$$\log 5 + \log x - \log 4 = 1$$
$$\log \frac{5x}{4} = 1$$
$$10^1 = \frac{5x}{4}$$
$$40 = 5x$$
$$8 = x$$

25. $A = P\left(1 + \frac{r}{k}\right)^{kt} = 9000\left(1 + \frac{-0.12}{1}\right)^{1(9)} \approx \$2,848.31$

26. $\log_6 8 = \frac{\log 8}{\log 6} \approx 1.16056$

27. $\frac{6!7!}{5!} = \frac{6 \cdot 5! \cdot 7!}{5!} = 6 \cdot 7! = 30,240$

28. $(3a - b)^4$
$$= (3a)^4 + \frac{4!}{1!(4-1)!}(3a)^3(-b) + \frac{4!}{2!(4-2)!}(3a)^2(-b)^2 + \frac{4!}{3!(4-3)!}(3a)(-b)^3 + (-b)^4$$
$$= 81a^4 + \frac{4!}{1!3!}(-27a^3b) + \frac{4!}{2!2!}(9a^2b^2) + \frac{4!}{3!1!}(-3ab^3) + b^4$$
$$= 81a^4 - \frac{4 \cdot 3!}{1!3!}(27a^3b) + \frac{4 \cdot 3 \cdot 2!}{2! \cdot 2 \cdot 1}(9a^2b^2) - \frac{4 \cdot 3!}{3!1!}(3ab^3) + b^4$$
$$= 81a^4 - \frac{4}{1}(27a^3b) + \frac{12}{2}(9a^2b^2) - \frac{4}{1}(3ab^3) + b^4$$
$$= 81a^4 - 108a^3b + 54a^2b^2 - 12ab^3 + b^4$$

29. In the 7th term, the exponent on $-y$ is 6.

Variables: $(2x)^2(-y)^6 = 4x^2y^6$

Coef. $= \frac{n!}{r!(n-r)!} = \frac{8!}{6!2!} = 28$

Term $= 28(4x^2y^6) = 112x^2y^6$

30. $a = -11, d = 6, n = 20$

nth term $= a + (n-1)d$
$$= -11 + (19)(6)$$
$$= -11 + 114 = 103$$

CUMULATIVE REVIEW EXERCISES

31. $a = 6, d = 3, n = 20;$ nth term $= a + (n-1)d = 6 + (20-1)(3) = 6 + 19(3) = 63$

$$S_n = \frac{n(a+l)}{2} = \frac{20(6+63)}{2} = \frac{20(69)}{2} = 690$$

32. 1st term $= -3$; 4th term $= 30$:

nth term $= a + (n-1)d$

$30 = -3 + (4-1)d$

$33 = 3d$

$11 = d \Rightarrow -3, \boxed{8, 19}, 30$

33. $\sum_{k=1}^{3} 3k^2 = 3(1)^2 + 3(2)^2 + 3(3)^2$

$$= 3 + 12 + 27 = 42$$

34. $\sum_{k=3}^{5} (2k+1) = (2(3)+1) + (2(4)+1) + (2(5)+1) = 7 + 9 + 11 = 27$

35. $a = \frac{1}{27}, r = 3, n = 7;$ nth term $= ar^{n-1} = \frac{1}{27}(3)^{7-1} = \frac{1}{27}(3)^6 = \frac{1}{27}(729) = 27$

36. $a = \frac{1}{64}, r = 2, n = 10;$ $S_n = \dfrac{a - ar^n}{1 - r} = \dfrac{\frac{1}{64} - \frac{1}{64}(2)^{10}}{1 - 2} = \dfrac{\frac{1}{64} - \frac{1}{64}(1024)}{-1} = \dfrac{-\frac{1023}{64}}{-1} = \dfrac{1023}{64}$

37. 1st term $= -3$, 4th term $= 192$

nth term $= ar^{n-1}$

$192 = -3r^{4-1}$

$-64 = r^3$

$-4 = r \Rightarrow -3, \boxed{12, -48}, 192$

38. $a = 9, r = \dfrac{1}{3}$

$$S = \frac{a}{1-r} = \frac{9}{1 - \frac{1}{3}} = \frac{9}{\frac{2}{3}} = \frac{27}{2}$$

39. $P(9, 3) = \dfrac{9!}{(9-3)!} = \dfrac{9!}{6!} = \dfrac{9 \cdot 8 \cdot 7 \cdot 6!}{6!} = 9 \cdot 8 \cdot 7 = 504$

40. $C(7, 4) = \dfrac{7!}{4!(7-4)!} = \dfrac{7!}{4!3!} = \dfrac{7 \cdot 6 \cdot 5 \cdot 4!}{4! \cdot 3 \cdot 2 \cdot 1} = \dfrac{210}{6} = 35$

41. $\dfrac{C(8, 4)C(8, 0)}{P(6, 2)} = \dfrac{\frac{8!}{4!4!} \cdot \frac{8!}{0!8!}}{\frac{6!}{4!}} = \dfrac{70 \cdot 1}{30} = \dfrac{7}{3}$

42. $C(n, n) = 1$ is smaller than $P(n, n) = n!$.

43. $7! = 5,040$

44. $\dbinom{9}{3} = \dfrac{9!}{3!6!} = 84$

Appendix 1 (page A-4)

1.

$$y = x^2 - 1$$

x-axis	y-axis	origin
$-y = x^2 - 1$	$y = (-x)^2 - 1$	$-y = (-x)^2 - 1$
not equivalent: no symmetry	$y = x^2 - 1$	$-y = x^2 - 1$
	equivalent: $\boxed{\text{symmetry}}$	not equivalent: no symmetry

3.

$$y = x^5$$

x-axis	y-axis	origin
$-y = x^5$	$y = (-x)^3$	$-y = (-x)^3$
not equivalent: no symmetry	$y = -x^5$	$-y = -x^5$
	not equivalent: no symmetry	$y = x^5$
		equivalent: $\boxed{\text{symmetry}}$

5.

$$y = -x^2 + 2$$

x-axis	y-axis	origin
$-y = -x^2 + 2$	$y = (-x)^2 + 2$	$-y = (-x)^2 + 2$
not equivalent: no symmetry	$y = -x^2 + 2$	$-y = x^2 + 2$
	equivalent: $\boxed{\text{symmetry}}$	not equivalent: no symmetry

7.

$$y = x^2 - x$$

x-axis	y-axis	origin
$-y = x^2 - x$	$y = (-x)^2 - (-x)$	$-y = (-x)^2 - (-x)$
not equivalent: no symmetry	$y = x^2 + x$	$-y = x^2 + x$
	not equivalent: no symmetry	not equivalent: no symmetry

9.

$$y = -|x + 2|$$

x-axis	y-axis	origin
$-y = -|x + 2|$	$y = -|-x + 2|$	$-y = -|-x + 2|$
not equivalent: no symmetry	not equivalent: no symmetry	not equivalent: no symmetry

11.

$$|y| = x$$

x-axis	y-axis	origin
$|-y| = x$	$|y| = -x$	$|-y| = -x$
$|-1||y| = x$	not equivalent: no symmetry	$|-1||y| = -x$
$|y| = x$		$|y| = -x$
equivalent: $\boxed{\text{symmetry}}$		not equivalent: no symmetry

13. $y = x^4 - 4$

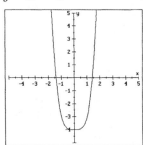

D $(-\infty, \infty)$; R $[-4, \infty)$

15. $y = -x^3$

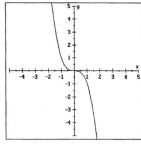

D $(-\infty, \infty)$; R $(-\infty, \infty)$

17. $y = x^4 + x^2$

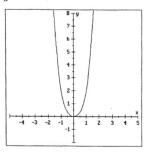

D $(-\infty, \infty)$; R $[0, \infty)$

19. $y = x^3 - x$

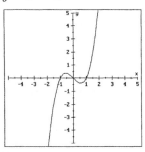

D $(-\infty, \infty)$; R $(-\infty, \infty)$

21. $y = \frac{1}{2}|x| - 1$

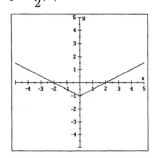

D $(-\infty, \infty)$; R $[-1, \infty)$

23. $y = -|x + 2|$

D $(-\infty, \infty)$; R $(-\infty, 0]$

Sample Final Examination (page A-6)

1. Prime numbers from 40 and 50: 41, 43, 47
    Answer: a

2. Commutative property of $\times$: $ab = ba$
    Answer: a

3. $\dfrac{c - ab}{bc} = \dfrac{6 - 3(-2)}{-2(6)} = \dfrac{12}{-12} = -1$
    Answer: c

4. $\dfrac{1}{2} + \dfrac{3}{4} \div \dfrac{5}{6} = \dfrac{1}{2} + \dfrac{3}{4} \cdot \dfrac{6}{5}$
$= \dfrac{1}{2} + \dfrac{9}{10}$
$= \dfrac{5}{10} + \dfrac{9}{10} = \dfrac{14}{10} = \dfrac{7}{5}$
    Answer: d

5. $\left(\dfrac{a^2}{a^5}\right)^{-5} = \left(\dfrac{a^5}{a^2}\right)^5 = (a^3)^5 = a^{15}$
    Answer: c

6. $0.0000234 = 2.34 \times 10^{-5}$
    Answer: a

7. $P(-1) = 2(-1)^2 - (-1) - 1 = 2(1) + 1 - 1 = 2 \Rightarrow$ **Answer: c**

8. $(3x + 2) - (2x - 1) + (x - 3) = 3x + 2 - 2x + 1 + x - 3 = 2x \Rightarrow$ **Answer: b**

9. $(3x - 2)(2x + 3) = 6x^2 + 9x - 4x - 6$

$$= 6x^2 + 5x - 6$$

Answer: a

10.

$$2x + 1 \overline{\smash{\big)}\ \begin{array}{r} x - 2 \\ 2x^2 - 3x - 2 \\ \end{array}}$$

$$\begin{array}{r} 2x^2 + x \\ \hline - 4x - 2 \\ - 4x - 2 \\ \hline 0 \end{array}$$

Answer: c

11. $5x - 3 = -2x + 10$

$$7x = 13$$

$$x = \frac{13}{7}$$

Answer: d

12. Let x and $x + 2$ represent the integers.

$$x + x + 2 = 44$$

$$2x = 42$$

$$x = 21$$

The integers are 21 and 23. Product $= 483$

Answer: c

13. $2ax - a = b + x$

$$2ax - x = a + b$$

$$x(2a - 1) = a + b$$

$$x = \frac{a + b}{2a - 1}$$

Answer: b

14.

$$|2x + 5| = 13$$

$$2x + 5 = 13 \quad \text{or} \quad 2x + 5 = -13$$

$$2x = 8 \qquad\qquad 2x = -18$$

$$x = 4 \qquad\qquad x = -9$$

Sum of solutions $= 4 + (-9) = -5$

Answer: d

15. $-2x + 5 > 9$

$$-2x > 4$$

$$\frac{-2x}{-2} < \frac{4}{-2}$$

$$x < -2$$

Answer: c

16. $|2x - 5| \leq 9$

$$-9 \leq 2x - 5 \leq 9$$

$$-9 + 5 \leq 2x - 5 + 5 \leq 9 + 5$$

$$-4 \leq 2x \leq 14$$

$$-2 \leq x \leq 7$$

Answer: d

17. $3ax^2 + 6a^2x = 3ax(x + 2a)$

Answer: d

18. $x^4 - 16 = \left(x^2 + 4\right)\left(x^2 - 4\right)$

$$= \left(x^2 + 4\right)(x + 2)(x - 2)$$

$$x^2 + 4 + x + 2 + x - 2 = x^2 + 2x + 4$$

Answer: b

19. $8x^2 - 2x - 3 = (4x - 3)(2x + 1)$

$$4x - 3 + 2x + 1 = 6x - 2$$

Answer: a

20. $27a^3 + 8 = (3a)^3 + 2^3$

$$= (3a + 2)\left(9a^2 - 6a + 4\right)$$

Answer: d

21. $6x^2 - 5x - 6 = 0$

$$(2x - 3)(3x + 2) = 0$$

$$2x - 3 = 0 \quad \text{or} \quad 3x + 2 = 0$$

$$x = \tfrac{3}{2} \qquad\qquad x = -\tfrac{2}{3}$$

Answer: c

22.

$$\frac{x^2 + 5x + 6}{x^2 - 9} = \frac{(x + 2)(x + 3)}{(x + 3)(x - 3)}$$

$$= \frac{x + 2}{x - 3}$$

Answer: b

23. $\dfrac{3x+6}{x+3} - \dfrac{x^2-4}{x^2+x-6} = \dfrac{3x+6}{x+3} - \dfrac{(x+2)(x-2)}{(x+3)(x-2)} = \dfrac{3x+6}{x+3} - \dfrac{x+2}{x+3}$

$$= \dfrac{3x+6-x-2}{x+3} = \dfrac{2x+4}{x+3}$$

Answer: d

24. $\dfrac{y}{x+y} + \dfrac{x}{x-y} = \dfrac{y(x-y)}{(x+y)(x-y)} + \dfrac{x(x+y)}{(x-y)(x+y)} = \dfrac{xy-y^2+x^2+xy}{(x+y)(x-y)} = \dfrac{x^2+2xy-y^2}{(x+y)(x-y)}$

Answer: a

25. $\dfrac{\frac{1}{x}+\frac{1}{y}}{\frac{1}{y}} = \dfrac{\left(\frac{1}{x}+\frac{1}{y}\right)xy}{\frac{1}{y}\cdot xy} = \dfrac{y+x}{x}$

Answer: c

26. $\dfrac{2}{y+1} = \dfrac{1}{y+1} - \dfrac{1}{3}$

$$3(y+1)\cdot\dfrac{2}{y+1} = 3(y+1)\left(\dfrac{1}{y+1} - \dfrac{1}{3}\right)$$
$$6 = 3 - (y+1)$$
$$6 = 3 - y - 1$$
$$y = -4$$

Answer: c

27.

$x=0$	$y=0$
$2x+3y=6$	$2x+3y=6$
$2(0)+3y=6$	$2x+3(0)=6$
$3y=6$	$2x=6$
$y=2$	$x=3$

Sum $= 2+3 = 5$

Answer: c

28. $m = \dfrac{y_2-y_1}{x_2-x_1} = \dfrac{-1-(-2)}{5-3} = \dfrac{1}{2}$

Answer: d

29.

$2x-3y=4$	$3x+2y=1$
$-3y=-2x+4$	$2y=-3x+1$
$y=\frac{2}{3}x-\frac{4}{3}$	$y=-\frac{3}{2}x+\frac{1}{2}$
$m=\frac{2}{3}$	$m=-\frac{3}{2}$

perpendicular

Answer: b

30. $m = \dfrac{y_2-y_1}{x_2-x_1} = \dfrac{7-5}{6-(-2)} = \dfrac{2}{8} = \dfrac{1}{4}$

$$y-y_1 = m(x-x_1)$$
$$y-7 = \dfrac{1}{4}(x-6)$$
$$y-7 = \dfrac{1}{4}x - \dfrac{3}{2}$$
$$y = \dfrac{1}{4}x + \dfrac{11}{2}$$

Answer: b

31. $g(t+1) = (t+1)^2 - 3$
$$= t^2+2t+1-3$$
$$= t^2+2t-2$$

Answer: d

32. $d = kt$
$$12 = k(3)$$
$$4 = k$$

Answer: b

33. $x^{a/2}x^{a/5} = x^{a/2+a/5} = x^{5a/10+2a/10} = x^{7a/10} \Rightarrow$ **Answer: b**

34. $\left(x^{1/2} + 2\right)\left(x^{-1/2} - 2\right) = x^{1/2}x^{-1/2} - 2x^{1/2} + 2x^{-1/2} - 4$

$$= x^0 - 2x^{1/2} + 2x^{-1/2} - 4$$

$$= 1 - 2x^{1/2} + 2x^{-1/2} - 4 = -3 - 2x^{1/2} + 2x^{-1/2}$$

Answer: c

35. $\sqrt{112a^3} = \sqrt{16a^2}\sqrt{7a} = 4a\sqrt{7a} \Rightarrow$ **Answer: d**

36. $\sqrt{50} - \sqrt{98} + \sqrt{128} = \sqrt{25}\sqrt{2} - \sqrt{49}\sqrt{2} + \sqrt{64}\sqrt{2} = 5\sqrt{2} - 7\sqrt{2} + 8\sqrt{2} = 6\sqrt{2}$

Answer: b

37. $\dfrac{3}{2 - \sqrt{3}} = \dfrac{3\left(2 + \sqrt{3}\right)}{\left(2 - \sqrt{3}\right)\left(2 + \sqrt{3}\right)} = \dfrac{3\left(2 + \sqrt{3}\right)}{4 + 2\sqrt{3} - 2\sqrt{3} - 3} = \dfrac{3\left(2 + \sqrt{3}\right)}{1} = 6 + 3\sqrt{3}$

Answer: b

38. $d = \sqrt{(x_2 - x_1)^2 + (y_2 - y_1)^2}$

$$= \sqrt{(6 - (-2))^2 + (-8 - 3)^2}$$

$$= \sqrt{8^2 + (-11)^2}$$

$$= \sqrt{64 + 121} = \sqrt{185}$$

Answer: a

39. $\sqrt{x + 7} - 2x = -1$

$$\sqrt{x + 7} = 2x - 1$$

$$\left(\sqrt{x + 7}\right)^2 = (2x - 1)^2$$

$$x + 7 = 4x^2 - 4x + 1$$

$$0 = 4x^2 - 5x - 6$$

$$0 = (x - 2)(4x + 3)$$

$$x - 2 = 0 \quad \textbf{or} \quad 4x + 3 = 0$$

$$x = 2 \qquad\qquad x = -\tfrac{3}{4}$$

doesn't check

Answer: a

40. $y > 3x + 2$

Answer: d

41. $x = \dfrac{-b \pm \sqrt{b^2 - 4ac}}{2a}$

Answer: d

42. $(2+3i)^2 = (2+3i)(2+3i)$
$\qquad = 4 + 6i + 6i + 9i^2$
$\qquad = 4 + 12i + 9(-1)$
$\qquad = 4 + 12i - 9$
$\qquad = -5 + 12i$
Answer: a

43. $\dfrac{i}{3+i} = \dfrac{i(3-i)}{(3+i)(3-i)}$
$\qquad = \dfrac{3i - i^2}{9 - 3i + 3i - i^2}$
$\qquad = \dfrac{3i - (-1)}{9 - (-1)}$
$\qquad = \dfrac{1 + 3i}{10} = \dfrac{1}{10} + \dfrac{3}{10}i$
Answer: b

44. $y = 2x^2 + 4x - 3$
$y + 3 = 2(x^2 + 2x)$
$y + 3 + 2 = 2(x^2 + 2x + 1)$
$\qquad y = 2(x+1)^2 - 5$
Vertex: $(-1, -5)$
Answer: c

45. $\dfrac{2}{x} < 3$
$\dfrac{2}{x} - 3 < 0$
$\dfrac{2}{x} - \dfrac{3x}{x} < 0$
$\dfrac{2 - 3x}{x} < 0$

$2 - 3x \quad +++++++++0\ ---$
$x \qquad\quad ---0+++++++++$

$\longleftarrow\!)\!\!-\!\!-\!\!-\!\!(\!\longrightarrow$
$\qquad 0 \qquad \frac{2}{3}$

solution set: $(-\infty, 0) \cup \left(\dfrac{2}{3}, \infty\right)$
Answer: c

46. $f(3) = 2(3)^2 + 1 = 2(9) + 1 = 19$
Answer: b

47. $y = 3x + 2$
$\quad x = 3y + 2$
$x - 2 = 3y$
$\dfrac{x-2}{3} = y$
Answer: c

48. $(x+2)^2 + (y-4)^2 = 16$
Answer: a

SAMPLE FINAL EXAMINATION

49. $\dfrac{4}{x}+\dfrac{2}{y}=2 \Rightarrow 4m+2n=2 \Rightarrow \times 3 \quad 12m+6n=6$

$\dfrac{2}{x}-\dfrac{3}{y}=-1 \Rightarrow 2m-3n=-1 \Rightarrow \times 2 \quad 4m-6n=-2$

$$16m=4$$
$$m=\tfrac{1}{4}$$

Solve for n: **Solve for x:**   **Solve for y:**   **Solution:** $\boxed{(4,2)}$ Sum $=4+2=6$

$4m+2n=2$

$4(\tfrac{1}{4})+2n=2 \qquad m=\dfrac{1}{x} \qquad n=\dfrac{1}{y}$

$1+2n=2 \qquad \dfrac{1}{4}=\dfrac{1}{x} \qquad \dfrac{1}{2}=\dfrac{1}{y}$

$2n=1 \qquad 4=x \qquad 2=y$

$n=\tfrac{1}{2}$

Answer: a

50. $y=\dfrac{\begin{vmatrix}4&5\\8&3\end{vmatrix}}{\begin{vmatrix}4&6\\8&-9\end{vmatrix}}=\dfrac{12-40}{-36-48}=\dfrac{-28}{-84}=\dfrac{1}{3}$

Answer: b

51. $\begin{vmatrix}2&-3\\4&4\end{vmatrix}=8-(-12)=20$

Answer: b

52. $z=\dfrac{\begin{vmatrix}1&1&4\\2&1&6\\3&1&8\end{vmatrix}}{\begin{vmatrix}1&1&1\\2&1&1\\3&1&2\end{vmatrix}}=\dfrac{1\begin{vmatrix}1&6\\1&8\end{vmatrix}-1\begin{vmatrix}2&6\\3&8\end{vmatrix}+4\begin{vmatrix}2&1\\3&1\end{vmatrix}}{1\begin{vmatrix}1&1\\1&2\end{vmatrix}-1\begin{vmatrix}2&1\\3&2\end{vmatrix}+1\begin{vmatrix}2&1\\3&1\end{vmatrix}}=\dfrac{1(2)-1(-2)+4(-1)}{1(1)-1(1)+1(-1)}=\dfrac{0}{-1}=0$

Answer: a

53. $\log_a N=x \Rightarrow a^x=N$
Answer: a

54. $\log_2\dfrac{1}{32}=x \Rightarrow 2^x=\dfrac{1}{32} \Rightarrow x=-5$
Answer: c

55. $\log 7+\log 5=\log(7\cdot5)=\log 35$
Answer: d

56. $b^{\log_b x}=x$
Answer: b

57. $\log y+\log(y+3)=1$
$\log y(y+3)=1$
$10^1=y^2+3y$
$0=y^2+3y-10$
$0=(y+5)(y-2)$
$y+5=0$ **or** $y-2=0$
$y=-5 \qquad\qquad y=2$
Answer: b

58. In the 3rd term, the exponent on b is 2.
Variables: a^4b^2
Coef. $=\dfrac{n!}{r!(n-r)!}=\dfrac{6!}{2!4!}=15$
Term $=15a^4b^2$
Answer: c

SAMPLE FINAL EXAMINATION

59. $P(7,3) = \dfrac{7!}{(7-3)!} = \dfrac{7!}{4!} = \dfrac{7 \cdot 6 \cdot 5 \cdot 4!}{4!}$

$\qquad\qquad\qquad = 7 \cdot 6 \cdot 5 = 210$

Answer: b

60. $C(7,3) = \dfrac{7!}{3!(7-3)!}$

$\qquad = \dfrac{7!}{3!4!}$

$\qquad = \dfrac{7 \cdot 6 \cdot 5 \cdot 4!}{3 \cdot 2 \cdot 1 \cdot 4!} = \dfrac{7 \cdot 6 \cdot 5}{6} = 35$

Answer: a

61. $a = 2, d = 3, n = 100$

nth term $= a + (n-1)d$

$\qquad = 2 + (100-1)(3)$

$\qquad = 2 + 99(3)$

$\qquad = 2 + 297 = 299$

Answer: b

62. $a = 1, r = \dfrac{1}{3}$

$S = \dfrac{a}{1-r} = \dfrac{1}{1-\frac{1}{3}} = \dfrac{1}{\frac{2}{3}} = \dfrac{3}{2}$

Answer: c